Lecture Notes in Artificial Intelligence 8537

Subseries of Lecture Notes in Computer Science

T0212188

Marzena Kryszkiewicz   Chris Cornelis
Davide Ciucci   Jesús Medina-Moreno
Hiroshi Motoda   Zbigniew W. Raś (Eds.)

Rough Sets and Intelligent Systems Paradigms

Second International Conference, RSEISP 2014
Held as Part of JRS 2014
Granada and Madrid, Spain, July 9-13, 2014
Proceedings

 Springer

Volume Editors

Marzena Kryszkiewicz
Warsaw University of Technology, Poland
E-mail: mkr@ii.pw.edu.pl

Chris Cornelis
University of Granada, Spain
E-mail: chris.cornelis@decsai.ugr.es

Davide Ciucci
University of Milano-Bicocca, Milano, Italy
E-mail: ciucci@disco.unimib.it

Jesús Medina-Moreno
University of Càdiz, Puerto Real, Spain
E-mail: jesus.medina@uca.es

Hiroshi Motoda
Osaka University, Japan
E-mail: motoda@ar.sanken.osaka-u.ac.jp

Zbigniew W. Raś
University of North Carolina, Charlotte, NC, USA
and Warsaw University of Technology, Poland
E-mail: ras@uncc.edu

ISSN 0302-9743   e-ISSN 1611-3349
ISBN 978-3-319-08728-3   e-ISBN 978-3-319-08729-0
DOI 10.1007/978-3-319-08729-0
Springer Cham Heidelberg New York Dordrecht London

Library of Congress Control Number: Applied for

LNCS Sublibrary: SL 7 – Artificial Intelligence

Typesetting: Camera-ready by author, data conversion by Scientific Publishing Services, Chennai, India

Printed on acid-free paper

Springer is part of Springer Science+Business Media (www.springer.com)

Preface

This volume contains the papers accepted for presentation at the Second International Conference on Rough Sets and Emerging Intelligent Systems Paradigms (RSEISP 2014), which, along with the Ninth International Conference on Rough Sets and Current Trends in Computing (RSCTC 2014), was held as a major part of the 2014 Joint Rough Set Symposium (JRS 2014), during July 9–13, 2014, in Granada and Madrid, Spain. In addition, JRS 2014 also hosted the workshop on Rough Sets: Theory and Applications (RST&A), held in Granada on July 9, 2014.

JRS was organized for the first time in 2007 and was re-established in 2012 as the major flagship IRSS-sponsored event gathering different rough-set-related conferences and workshops every year. This year, it provided a forum for researchers and practitioners interested in rough sets, fuzzy sets, intelligent systems, and complex data analysis.

RSEISP was organized for the first time in 2007. The conference was dedicated to the memory of Prof. Zdzisław Pawlak, who in 1981 had proposed rough set theory as a model of approximate reasoning. RSEISP 2007 was focused on various forms of soft and granular computing such as rough and fuzzy sets, knowledge technology and discovery, data processing and mining, as well as their applications in information systems. It provided researchers, practitioners, and students interested in emerging information technologies a forum to share innovative theories, methodologies, and applications in rough set theory and related areas.

JRS 2014 received 120 submissions that were carefully reviewed by three or more Program Committee members or external reviewers. Papers submitted to special sessions were subject to the same reviewing procedure as those submitted to regular sessions. After the rigorous reviewing process, 40 regular papers (acceptance rate 33.3%) and 37 short papers were accepted for presentation at the symposium and publication in two volumes of the JRS proceedings. This volume contains the papers accepted for the conference RSEISP 2014, as well as the invited papers by Jerzy Stefanowski (plenary speaker), Piero Pagliani, Jakub Wróblewski and Sebastian Stawicki (tutorial organizers).

It is truly a pleasure to thank all those people who helped this volume to come into being and to turn JRS 2014 into a successful and exciting event. In particular, we would like to express our appreciation for the work of the JRS 2014 Program Committee members who helped assure the high standards of accepted papers. Also, we are grateful to the organizers of special sessions of JRS 2014: Enrique Herrera-Viedma, Francisco Javier Cabrerizo, Ignacio Javier Pérez, Lluis Godo, Thomas Vetterlein, Manuel Ojeda-Aciego, Sergei O. Kuznetsov, Pablo Cordero, Isaac Triguero, Salvador García, Robert Bembenik, Dariusz Gotlib, Grzegorz Protaziuk, Bing Zhou, Hong Yu, and Huaxiong Li. Furthermore, we want to thank the RST&A 2014 workshop chairs (Nele Verbiest and Piotr Artiemjew), and we also gratefully acknowledge the generous help of the remaining JRS 2014 chairs - Dominik Ślęzak, Ernestina Menasalvas Ruiz, Rafael Bello, Lin Shang, Consuelo Gonzalo, and Salvador García, as well as of the Steering Committee members - Henryk Rybiński, Roman Słowiński, Guoyin Wang, and Yiyu Yao, for their hard work and valuable suggestions with respect to the preparation of the proceedings and conference organization. Also, we want to pay tribute to our honorary chairs, Lotfi Zadeh and Andrzej Skowron, whom we deeply respect for their countless contributions to the field. We also wish to express our thanks to Bernard De Baets, Francisco Herrera, and Jerzy Stefanowski for accepting to be plenary speakers of JRS 2014. Last but not least, we would like to thank all the authors of JRS 2014, without whose high-quality contributions it would not have been possible to organize the symposium.

We also want to take this opportunity to thank our sponsors, in particular, Infobright Inc. for being the industry sponsor of the entire event, and Springer for contributing the best paper award.

We are very thankful to Alfred Hofmann and the excellent LNCS team at Springer for their help and co-operation. We would also like to acknowledge the use of EasyChair, a great conference management system.

Finally, it is our sincere hope that the papers in the proceedings may be of interest to the readers and inspire them in their scientific activities.

July 2014

Marzena Kryszkiewicz
Chris Cornelis
Davide Ciucci
Jesús Medina-Moreno
Hiroshi Motoda
Zbigniew W. Raś

JRS 2014 Organization

General Chairs

Ernestina Menasalvas Ruiz   Polytechnic University of Madrid, Spain
Dominik Ślęzak   University of Warsaw, Poland & Infobright Inc., Poland

Steering Committee

Henryk Rybiński   Warsaw University of Technology, Poland
Roman Słowiński   Poznań University of Technology, Poland
Guoyin Wang   Chongqing University of Posts and Telecommunications, China
Yiyu Yao   University of Regina, Canada

Program Chairs

Chris Cornelis   University of Granada, Spain (RSCTC 2014)
Marzena Kryszkiewicz   Warsaw University of Technology, Poland (RSEISP 2014)

Program Co-chairs for RSCTC 2014

Rafael Bello   Central University of Las Villas, Cuba
Lin Shang   Nanjing University, China

Program Co-chairs for RSEISP 2014

Hiroshi Motoda   Osaka University, Japan
Zbigniew W. Raś   University of North Carolina at Charlotte, USA & Warsaw University of Technology, Poland

Special Session and Tutorial Chairs

Davide Ciucci   University of Milano-Bicocca, Italy
Jesús Medina-Moreno   University of Cádiz, Spain

RST&A Workshop Chairs

Piotr Artiemjew

University of Warmia and Masuria in Olsztyn,
Poland

Nele Verbiest

Ghent University, Belgium

Local Chairs

Salvador García

University of Jaén, Spain (Granada)

Consuelo Gonzalo

Polytechnic University of Madrid, Spain
(Madrid)

Program Committee

Arun Agarwal
Piotr Andruszkiewicz
Piotr Artiemjew
S. Asharaf
Sanghamitra Bandyopadhyay
Mohua Banerjee
Andrzej Bargiela
Luis Baumela Molina
Jan G. Bazan
Robert Bembenik
José Manuel Benítez
Jerzy Błaszczyński
Nizar Bouguila
Humberto Bustince
Francisco Javier Cabrerizo
Yongzhi Cao
Mihir K. Chakraborty
Chien-Chung Chan
Pablo Cordero
Oscar Cordón
Zoltán Csajbók
Bijan Davvaz
Martine De Cock
Antonio de La Torre
Dayong Deng
Lipika Dey
Fernando Diaz
Maria Do Carmo Nicoletti
Didier Dubois
Ivo Düntsch

Zied Elouedi
Javier Fernandez
Wojciech Froelich
Salvador García
Piotr Gawrysiak
Guenther Gediga
Lluis Godo
Joao Gomes
Anna Gomolińska
Santiago Gonzalez Tortosa
Consuelo Gonzalo Martín
Dariusz Gotlib
Salvatore Greco
Jerzy Grzymała-Busse
Jonatan Gómez
Jun He
Francisco Herrera
Enrique Herrera-Viedma
Pilar Herrero
Chris Hinde
Shoji Hirano
Władyslaw Homenda
Qinghua Hu
Masahiro Inuiguchi
Ryszard Janicki
Andrzej Janusz
Richard Jensen
Jouni Järvinen
Janusz Kacprzyk
Etienne Kerre

Beata Konikowska
Jacek Koronacki
Vladik Kreinovich
Yasuo Kudo
Sergei O. Kuznetsov
Huaxiong Li
Tianrui Li
Jiye Liang
Churn-Jung Liau
Antoni Ligęza
Mario Lillo Saavedra
Pawan Lingras
Dun Liu
Dickson Lukose
Neil Mac Parthalain
Luis Magdalena
Victor Marek
María José Martín-Bautista
Benedetto Matarazzo
Luis Mengual
Jusheng Mi
Duoqian Miao
Tamás Mihálydeák
Evangelos Milios
Fan Min
Sadaaki Miyamoto
Javier Montero
Jesús Montes
Mikhail Moshkov
Hiroshi Motoda
Santiago Muelas Pascual
Tetsuya Murai
Michinori Nakata
Amedeo Napoli
Hung Son Nguyen
Sinh Hoa Nguyen
Vilem Novak
Hannu Nurmi
Manuel Ojeda-Aciego
Jose Angel Olivas Varela
Piero Pagliani
Krzysztof Pancerz
Gabriella Pasi
Witold Pedrycz
Georg Peters

James F. Peters
Frederick Petry
Jose María Peña
Jonas Poelmans
Lech Polkowski
Henri Prade
Grzegorz Protaziuk
Ignacio Javier Pérez
Keyun Qin
Anna Maria Radzikowska
Vijay V. Raghavan
Sheela Ramanna
Francisco P. Romero
Dominik Ryżko
Wojciech Rząsa
Yvan Saeys
Hiroshi Sakai
Abdel-Badeeh Salem
Miguel Ángel Sanz-Bobi
Gerald Schaefer
Steven Schockaert
Jesús Serrano-Guerrero
B. Uma Shankar
Qiang Shen
Marek Sikora
Arul Siromoney
Myra Spiliopoulou
Jerzy Stefanowski
John G. Stell
Jarosław Stepaniuk
Zbigniew Suraj
Piotr Synak
Andrzej Szałas
Marcin Szczuka
Domenico Talia
Vicenç Torra
Isaac Triguero
Li-Shiang Tsay
Shusaku Tsumoto
Athina Vakali
Nam Van Huynh
Nele Verbiest
José Luis Verdegay
Thomas Vetterlein
Krzysztof Walczak

Xin Wang
Junzo Watada
Geert Wets
Alicja Wieczorkowska
Arkadiusz Wojna
Marcin Wolski
Michał Woźniak
Jakub Wróblewski
Wei-Zhi Wu

Yong Yang
Jingtao Yao
Hong Yu
Sławomir Zadrożny
Danuta Zakrzewska
Yan-Ping Zhang
Bing Zhou
William Zhu
Wojciech Ziarko

External Reviewers

Ola Amayri
Fernando Bobillo
María Eugenia Cornejo Piñero
Lynn D'eer
Subasish Das
Glad Deschrijver
Waldemar Koczkodaj
Adam Lenarcic
Tetsuya Murai
Petra Murinova
Julian Myrcha

Seyednaser Nourashrafeddin
Robert Olszewski
Stanisław Oszczak
Wojciech Pachelski
Emanuele Panzeri
Sandra Sandri
Andrzej Stateczny
Zbigniew Szymański
Peter Vojtáš
Xiangru Wang

Table of Contents

Applications of Rough Sets

Induction of Decision Rules - Theory and Practice

Information Extraction from Images

The Impact of Local Data Characteristics on Learning from Imbalanced Data

Jerzy Stefanowski

Institute of Computing Science, Poznań University of Technology,
60-965 Poznań, Poland

Abstract. Problems of learning classifiers from imbalanced data are discussed. First, we look at different data difficulty factors corresponding
to complex distributions of the minority class and show that they could
be approximated by analysing the neighbourhood of the learning examples from the minority class. We claim that the results of this analysis
could be a basis for developing new algorithms. In this paper we show
such possibilities by discussing modifications of informed pre-processing
method LN–SMOTE as well as by incorporating types of examples into
rule induction algorithm BRACID.

1 Introduction

Many difficult learning problems, from a wide variety of domains, involve learning from imbalanced data, where at least one of the target classes contains a
much smaller number of examples than the other classes. This class is usually
referred to as the *minority class*, while the remaining classes are denoted as *majority ones*. For instance, in medical problems the number of patients requiring
special attention is much smaller than the number of patients who do not need it.
Similar situations occur in such domains as: fraud detection, risk management,
technical diagnostics, image recognition, text categorization or information filtering. In all those problems, the correct recognition of the minority class is of
key importance. However, class imbalance constitutes a great difficulty for most
learning algorithms and often resulting classifiers are biased toward the majority
classes and fail to recognize examples from the minority class.

While the difficulty with learning classifiers from imbalanced data has been
known earlier from applications, this challenging problem has received a growing
research interest in the last decade and a number of specialized methods have
already been proposed, for their review see, e.g., [7,18,19,10].

In general, they may be categorized into *data level* and *algorithm level* ones.
The first group includes classifier-independent methods that are used in the
pre-processing step to modify the balance between classes, so that standard algorithms can be used to learn better classifiers. These methods are usually based
on either adding examples to the minority class (called *over-sampling*) or removing examples from the majority class (*under-sampling*). The other main category
of, so called, algorithmic methods involves modifications of either: learning phase

M. Kryszkiewicz et al. (Eds.): RSEISP 2014, LNAI 8537, pp. 1–13, 2014.

of the algorithm or classification strategy, construction of specialized ensembles, or adaptation of cost sensitive learning. For their reviews see [7,18,19,44].

Although several specialized methods already exist, the identification of conditions for their efficient use is still an open research problem. It is also related to more fundamental issues of better understanding the nature of the imbalance data and key properties of its underlying distribution.

Following related works [21,22,14,28] and earlier studies of J. Stefanowski and K.Napierala, J.Blaszczynski or Sz. Wilk [41,32,34,31] we claim that the high imbalance ratio between the minority and majority classes is not the only and not even the main reason of these difficulties. Other, as we call them, *data difficulty factors*, referring to characteristics of class distributions, are also influential. They include: decomposition of the minority class into many rare sub-concepts, the effect of too strong overlapping between the classes [36,14] or a presence of too many minority examples inside the majority class region. When these data difficulty factors occur *together* with class imbalance, they may seriously hinder the recognition of the minority class, see e.g. a study [28].

In our earlier paper [34] we propose to capture these data difficulty factors by considering the local characteristics of learning examples from the minority class. More precisely, it is achieved by analyzing the class distribution of examples from different classes inside a *local neighborhood* of the considered example. Finding how many examples from opposite classes are the neighbours of this example, the degree of its difficulty could be estimated.

We claim that the proper analyzing of this neighborhood of learning examples from the minority class could be the basis for developing new specialized algorithms for imbalanced data. In this paper we "implement" this postulate by considering representatives of two main categories of methods specialized for imbalanced data. Firstly, we will apply the analysis of neighbours into the new generalization of the most popular informed pre-processing method SMOTE. Secondly, we will show that data difficulty factors modeled by types of minority examples could be used inside the rule candidate generation phase of the rule induction algorithm BRACID. Finally, we will discuss other possible options of using the local information, in particular for ensembles and highlight other future research directions of studying imbalanced data.

2   Local Characteristics of Data Difficulty Factors and Identification of Example Types

Although many authors have experimentally shown that standard classifiers met difficulties while recognizing the minority class, it has also been observed that in some problems characterized by strong imbalance between classes standard classifiers are sufficiently accurate. Moreover, the discussion of data difficulty in imbalanced data still goes on, for its current review see, e.g., [10,19,29,34,40].

Some researchers have already noticed, that the *global class imbalance ratio* (i.e. the cardinality of the majority class referred to the total number of minority class examples) is not necessarily the only, or even the main, problem causing the decrease of classification performance and focusing only on this ratio may be insufficient for improving classification performance. Besides the imbalanced ratio other data difficulty factors may cause a severe deterioration of classifiers.

The experimental studies by Japkowicz *et al* with many artificial data sets have clearly demonstrated that the degradation of classification performance is also linked to the decomposition of the minority class into many sub-parts containing very few examples [21,22]. They have shown that the minority class does not form a homogeneous, compact distribution of the target concept but it is scattered into many smaller sub-clusters surrounded by majority examples. In other words, minority examples form, so called, *small disjuncts*, which are harder to learn and cause more classification errors than larger sub-concepts.

Other factors related to the class distribution are also linked to the effect of too strong *overlapping* between minority and majority class. Experiments with artificial data have shown that increasing overlapping has been more influential than changing the class imbalance ratio [36,14]. Yet another data factor, which influences degradation of classifiers performance on imbalanced data, is presence of noisy examples [2]. Experiments presented in [32] have also shown that single minority examples located inside the majority class regions cannot be treated as noise since their proper treatment by informed pre-processing may improve classifiers. Moreover, studies as [40] emphasize that several data factors usually occur together in real world imbalanced data sets.

These studies stress the role of the *local characteristics* of the class distribution. However it could be modeled in different ways. Here, we follow earlier works [24,25,32,31,29] and link data difficulty factors to *different types of examples* forming the minority class distribution. It leads us to a differentiation between safe and unsafe examples. *Safe examples* are ones located in the homogeneous regions populated by examples from one class only. Other examples are *unsafe* and more difficult for learning. Unsafe examples are categorized into *borderline* (placed close to the decision boundary between classes), *rare cases* (isolated groups of few examples located deeper inside the opposite class), or *outliers*. As the minority class can be highly under-represented in the data, we claim that the rare examples or outliers, could represent a very small but valid sub-concepts of which no other representatives could be collected for training. Therefore, they cannot be considered as noise examples which typically are then removed or re-labeled. Moreover, earlier works with graphical visualizations of real-world imbalanced data sets [34] have confirmed this categorization of example types.

The next question is how to automatically *identify these types of examples* in real world data sets (with unknown underlying class distributions). We keep the hypotheses [34] on role of the mutual positions of examples and the idea of assessing the type of example by analyzing class labels of the other examples in its *local neighbourhood*. This neighbourhood of the minority example could be modeled in different ways. In further considerations we will use an analysis of the class labels among *k-nearest neighbours* [34,31]. This approach requires choosing the value of k and the *distance function*. In our previous considerations we have followed results of analyzing different distance metrics [27] and chose the

HVDM metric (*Heterogeneous Value Difference Metric*) [45]. Its main advantage for mixed attributes is that it aggregates normalized distances for qualitative and quantitative attributes. In particular, comparing to other metrics it provides more appropriate handling of qualitative attributes as instead of simple value matching it calculates attribute value conditional probabilities by using a Stanfil and Valtz value difference metric [45]. Then, due to complexity of the distribution of the minority class, k should be rather a small value. Experiments from [34,31] over many UCI data sets have showed that $k = 5$ or 7 have led to good results.

Depending on the number of examples from the majority class in the neighbourhood of the minority example, we can evaluate whether this example could be safe or unsafe (difficult) to be learned. If all, or nearly all, its neighbours belong the minority class, this example is treated as the safe example, otherwise it is one of unsafe types. For instance, consider $k = 5$. In this case the type of example x is defined as: if 5 or 4 of its neighbours belong to the same class as x, it is treated as a safe example; if the numbers of neighbours from both classes are similar (proportions 3:2 or 2:3) – it is a borderline example; if it has only one neighbour with the same label (1:4) it is a rare case; finally if all neighbours come from the opposite class (0:5) – it is an outlier. Although this categorization is based on intuitive thresholding, its results are consistent with a probabilistic analysis of the neighbourhood, modeled with kernel functions, as shown in [31].

Our experiments with UCI imbalanced data sets [34,31] have also demonstrated that most of these real-world data do not include many safe minority examples. They rather contain all types of examples, but in different proportions. On the other hand, most of majority class examples have been identified as safe ones. Depending on the dominating type of identified minority examples, the considered datasets have been labeled as: safe, border, rare or outlier. As a large number of borderline examples often occurred in many data sets, some of these data sets could be assigned both to border and more difficult categories.

Moreover, the study [34] has shown that the classifier performance could be related to the category of data. First, for the safe data nearly compared single classifiers (SVM, RBF, k-NN, decision trees or rules) perform quite well with respect to sensitivity, F-measure or G-mean. The larger differentiation occurs for more unsafe data set. For instance, SVM and RBF work much better for safe category, while rare or outlier data strongly deteriorate their classification performance. On the other hand, unpruned decision trees and k-NN work quite well for these unsafe data sets. The similar analysis has been carried out for the most representative pre-processing approaches, showing that the competence area of each method depends on the data difficulty level, based on the types of minority class examples. Again in the case of safe data there are no significant differences between the compared methods - even random over-sampling works quite accurate. However, for borderline data sets Nearest Cleaning Rules (methods filtering difficult majority examples [25]) performs best. On the other hand, SMOTE [8] and SPIDER [41], which can add new examples to the data, have proved to be more suitable for rare and outlier data sets.

For more details on the competence of each studied single classifier and pre-processing methods see [31]. The similar analysis for different generalizations of bagging ensembles, included specialized solutions for class imbalances, have been carried out in the recent paper [5]. Finally, we will repeat our hypothesis that the appropriate treatment of these types of minority examples within new proposals of either pre-processing or classifiers should lead to improving classification performance. We will show it in the next sections.

3   Modifications of Informed Pre-processing Methods

The simplest data pre-processing techniques are random over-sampling, which replicates examples from the minority class, and random under-sampling, which randomly eliminates examples from the majority classes until a required degree of balance between classes is reached. However, random under-sampling may potentially remove some important examples and simple over-sampling may also lead to overfitting. Therefore, focused (also called *informed*) methods, which attempt to take into account internal characteristics of regions around minority class examples, were introduced, as e.g. SMOTE [8], one-side-sampling [24], NCR [25] or SPIDER[41] .

The most popular among the informed methods is SMOTE, which considers each example from the minority class and generates new synthetic examples along the lines between the selected example and some of its randomly selected k-nearest neighbors from the minority class. More precisely, let the training set S contain examples from the minority class P and other classes N. For each example $p_i \in P$ find its k nearest neighbours x from class P. Depending on the other parameter of this method – the amount of over-sampling – a given number of examples from these k nearest neighbours is randomly selected. Synthetic minority class examples are generated in the direction of each. For numerical attributes the new synthetic example is constructed as follows: compute the difference between attributes describing the example p_i and x – one of the selected k-nearest neighbours; multiply this feature vector difference by δ – a random number between 0 and 1; and add it to the attribute vector p_i creating a new vector $x_{new} = p_i + (x - p_i) \cdot \delta$. For qualitative attributes create a new example with the most common feature values among k nearest neighbours.

Although experiments have confirmed its usefulness (see e.g. [3,7]), some of the assumptions behind this technique could be still questioned. Two main shortcomings of SMOTE are: (1) treating all minority examples in the same way while they may not be equally important for learning classifiers (2) the possible over-generalization over the majority class regions as SMOTE blindly generalizes regions of the minority class without checking positions of the nearest examples from the majority classes. Some researchers solve these problem by integrating SMOTE with additional filtering steps (see e.g. [3,37]), while others modify SMOTE's internal strategies for selecting positions of synthetic examples.

In this paper we focus on the recent proposal called Local Neighbourhood extension of SMOTE (briefly LN-SMOTE) which is inspired by the analyzing local data characteristics and earlier modifications of SMOTE [4]. In this method the presence of the majority examples is taken into account before generating synthetic examples by calculating a special coefficient called a *safe level*. It is defined as the number of other minority class examples among its k nearest neighbours. The smaller its value, the more unsafe is this example. This level $sl(p)$ is calculated for the example p, which is a seed for oversampling and as $sl(x)$ for its randomly selected neighbour x. Unlike the standard SMOTE and its generalizations as [4,17], in LN-SMOTE the closest neighbours are calculated including also majority class examples. Having information about values of both safe levels $sl(p)$ and $sl(x)$, the range of positioning the synthetic example is modified. Only for equal both levels the examples will be generated along the whole line joining x and p in the same way as in the original SMOTE. If one safe level is greater than other the position new example will generated closer the safer example (more closer, the larger difference between these levels). Furthermore, in case of the neighbour from the majority class, the range of random overlapping is additionally limited not to come to close to the majority examples. Situations of outliers (safe level equal 0) are also distinguished by not putting the new examples for such a neighbour. Finally before starting over-sampling, all majority examples being outliers inside the minority class are identified by analysis content of k neighbourhood - they are removed from the learning set as they usually disturb the minority class distribution.

The LN-SMOTE was introduced and experimentally studied in [30]. Its comparative study against basic SMOTE and two other related generalizations called Borderline-SMOTE [17] and SL-SMOTE [4] applied with 3 different classifiers (J4.8, Naive Bayes and k-NN) showed that it was the best pre-processing method. For instance, Table 1 summarizes results of F-measure for these pre-processing methods applied to J4.8 decision tree, where Bord-SMOTE denotes Borderline SMOTE and SL-SM denotes Safe Level SMOTE.

Table 1. F-measure for the minority class for all compared methods used together with J48 classifier [%]

Data	None	SMOTE	Bord-SMOTE	SL-SM	LN-SMOTE
balance-scale	0.00	1.06	2.09	2.95	6.21
car	80.61	88.39	72.91	90.39	88.58
cleveland	19.29	21.35	22.89	21.33	26.70
cmc	40.81	41.46	42.05	41.66	44.75
ecoli	58.86	63.21	64.53	63.51	66.96
haberman	30.36	38.41	41.50	37.33	42.20
hepatitis	49.20	49.86	51.23	52.57	54.42
postoperative	5.84	11.90	15,06	12.08	16.18
solar-flare	28.79	27.13	28.32	29.62	31.6
transfusion	47.27	48.07	49.79	49.22	50.30
yeast	35.02	36.08	38.63	40.21	42.58

4 Incorporating Types of Examples in Rule Induction

Let us now consider using local characteristics of learning examples within approaches modifying the algorithms. Decision rules, being the most human readable knowledge representation, are particularly sensitive to class imbalance, see e.g. conclusions from [15,16,42]. Following some earlier methodological discussions [35,44] the standard algorithms for learning rule based classifiers share a number of their principles which are useful for classification with respect to the total accuracy but are limitations in case of class imbalances.

First, most algorithms induce rules using the top-down technique with *maximum generality bias*, which favors general rules however also hinders finding rules for smaller sets of learning examples, especially in the minority class. It is also connected with using *improper evaluation measures to guide the search* for best conditions. Typical measures, as presented, e.g., in [12,38], try to find a compromise between the accuracy and generality of the rule which achieve better values mainly for the majority class examples. Similar measures are also often used to *prune* induced rules, which are particularly inappropriate for small disjuncts or rare cases in the minority class where rules may be constructed as the conjunction of many elementary conditions. Thus, pruning of rules is guided mostly by measures referring more to the majority class examples, neglecting the minority class specific distributions [1].

Second, most algorithms use a *greedy sequential covering* approach [11], in which learning examples covered by the induced rule are removed from the current set of considered examples. This approach may increase the data fragmentation for the minority class and leads to the induction of *weaker* rules, i.e. supported by a smaller number of learning examples. The "weakness" of the minority rules could be also associated with a third factor: *classification strategies*, where minority rules have a smaller chance to contribute to the final classification decision, see discussions in [15,42,10].

The above limitations concerns difficulties inside the algorithm. Recall that one should also consider data difficulty factors referring to complex distributions of the minority class, as presented in section 2. Some researchers have already proposed the extensions of rule based approaches dedicated for class imbalance - for a comprehensive review see [35]. However, most of these proposals addresses only a single or at most a few of algorithmic or data factors.

Following these critical motivations K.Napierala and J.Stefanowski have introduced a new rule induction algorithm called BRACID (Bottom-up induction of Rules And Cases for Imbalanced Data) which is specialized for the classification of imbalanced data [35]. While constructing it we have addressed several limitations mentioned above.

First we have decided to induce rules by *bottom-up generalization* of the most specific rules representing single examples. Bearing in mind that local algorithms could better learn the difficult decision boundaries (which usually describe minority examples), this algorithm is a *hybrid* of rule-based and instance-based *knowledge representations*. Due to this representation it should also better deal with rare sub-concepts of the minority class. Overcoming the problem of data fragmentation is also connected with resigning from a greedy, sequential covering and top-down induction technique. Inside the crucial operation of the bottom-up generalization of the current rules the specific looking for the nearest example has been applied. The candidates for rules are temporarily added to the current classifier and evaluated with the F-measure in a leaving-one-out procedure. Such an approach allows us to evaluate and accept rules in a more appropriate way for recognizing the minority class. The final classifier uses a *nearest rule strategy* to classify new coming examples [39], which has proved to be more appropriate for recognizing minority classes than standard classification strategies too much oriented to majority examples.

An important component of BRACID is also using information about the nature of the neighbouring examples. Following the method presented in section 2, we identify types of learning examples in each class. Unsafe outlier examples from the majority classes are removed from the learning set as they may hinder fragmentation of the minority class and then the induction of more general minority rules. In case of an analogous situation for the minority class, this example is not removed but it is checked as a candidate for a rule generalization. Moreover for outliers, rare cases and borderline minority examples we allow to analyse k possible generalizations and to choose best ones according to the F-measure evaluation. It allows us to create more rules for the minority class in unsafe regions, as overlapping. It should diminish the possibility of overwhelming the minority class with the majority class rules in these difficult sub-regions.

Table 2. G-mean for BRACID and other rule induction algorithms [%]

Data	BRACID	CN2	RIPPER	MODLEM
abalone	65.0	39.6	42.1	48.4
balance-scale	56.7	2.9	1.9	0.0
cleveland	57.4	0.0	25.8	19.2
cmc	63.7	25.8	25.5	47.2
credit germ.	61.1	55.3	43.8	56.3
ecoli	83.1	28.4	58.7	56.8
haberman	57.6	34.5	35.6	40.1
ionosphere	91.2	87.0	87.4	89.2
vehicle	93.5	51.3	91.9	91.6
transfusion	63.9	34.2	26.6	52.9

The experimental studies with components of BRACID, in particular this way of incorporating types of examples have showed that it has improved the evaluation measures comparing to the plain option of treating all learning examples in the same way. For instance, for G-mean average improvements are 3.5 % [35]. Finally, BRACID has been compared to a number of state-of-the-art rule based classifiers (PART, RIPPER,C45rules,CN2, MODLEM, RISE), instance-based classifier (K-NN) and some approaches dedicated for class imbalances in the comprehensive experimental study over 22 imbalanced data sets. The results showed that BRACID can better recognize the minority classes than other compared algorithm - which is reflected by measures as F-measure and G-mean. The selected results of G-mean are presented in Table 2.

Furthermore, using the categorization of data sets obtained with analysis their local characteristics presented in section 2, we have found out that the best improvements of BRACID are observed for unsafe data sets containing many borderline examples from the minority class.

5   Other Perspectives of Applying Local Data Characteristics

The main message of this paper is to promote incorporating the information about the local neighbourhood of a chosen minority class example in the process of constructing and analyzing methods for learning classifiers from imbalanced data. Although we have shown two such possibilities, still other new directions are worth to be studied.

For instance, the proposed method for analysing k-neighbourhood of minority examples as well as practically all informed pre-processing methods have been considered for data with not so high number of attributes. However, some applications of data mining in bio-medicine, text or multimedia processing involve highly dimensional data sets. The use of typical dissimilarity measures and k-nearest neighbor classification on such data sets may suffer from the curse of dimensionality problem as it has been recently showed by Tomasev's research on, so called, *hubness*-aware shared neighbor distances for high-dimensional k-nearest neighbor classification [43]. Thus, studying generalizations of the presented methods for high dimensional data is still an open research challenge.

In case of pre-processing methods as SMOTE, there still remain some interesting questions on creation strategies for synthetic data, amount of new data to create, better identification of the most appropriate sub-regions of the minority class where to add new examples, avoidance of introducing noise instead of valuable instances of the under-represented class, distinguishing between noise and valuable outliers, and checking whether synthetic examples are equally important as the real ones by employing new evaluation measures while constructing and evaluating classifiers.

Another point of view on informed pre-processing methods and the role of generating large amounts of synthetic data could also come from specific applications. Such studies as [32,41] show that the degree of over-sampling is quite high for many data sets, even if it is tuned just to obtain a balance of cardinality of minority and majority classes. However, in medical problems physicians could be reluctant to analyse so a high number of artificial patients in their data set (i.e. their class of interest could include more non existing patient's descriptions than real clinical cases). Moreover, clinical experts often prefer to induce symbolic classifiers due to their potential interpretability. For instance, if rule induction algorithms are applied to data transformed by SMOTE, many rules in the final classifier could be supported just by synthetic learning examples. In spite of better classification performance such rules can be rejected by clinical experts due to their artificial supports and expert could still prefer to analyse rule referring to real facts in the original data.

These limitations open other perspectives on new informed preprocessing methods that do not introduce synthetic examples as well as on new specialized rule induction algorithms. Within the first perspective one can consider generalizing hybrid methods as SPIDER [41], which should better identify sub-regions where unsafe minority class examples should be amplified with different degrees as well as better filter the majority examples which too strongly influence these sub-regions. The other perspective is partly considered in such rule induction algorithms as EXPLORE [42], BRACID [35] or ABMODLEM [33] (which allows to directly incorporate expert explanations as to classifying difficult examples into the rule induction). However more extensive research and medical practical case studies are still needed.

Considering types of example could be also applied to new ensembles specialized for class imbalance. Most of current proposals are generalizations of known techniques as bagging, boosting or random forests; see their review in [13,26]. Their experimental results show that modifications of bagging often outperform boosting generalizations or more complex ensembles. While analyzing existing extensions of bagging one can also notice that most of them employ the simplest random re-sampling technique, as under-sampling or over-sampling, and, what is even more important, they just modify bootstraps to simply balance the cardinalities of minority and majority class. However, in all of these extensions (see, e.g., the Roughly Balanced Bagging [20]), all examples are treated as equally important while sampling them into bootstrap samples. We think that drawing of minority examples should not be done in a pure blind random way but it could be partly directed depending on the difficulty type of example.

In [5] we have already proposed to change probability of drawing different types of examples depending on the class distribution in the neighbourhood of the example. This has led us to the new type of bagging ensemble, called Nearest Neighborhood Bagging. The recent experiments [6] show that this new ensemble is significantly better than existing over-sampling bagging extensions and it is competitive to Roughly Balanced Bagging, which according to experiments [20,23] is the most accurate under-sampling extension of bagging. Nevertheless, several issues on: how much bootstrap samples should be modified, the influence of filtering majority class examples, diversity of bootstrap samples and the constructed classifiers, new techniques of their aggregation should be still studied. The research on this paper is partially supported by NCN grant.

References

1. An, A.: Learning classification rules from data. Computers and Mathematics with Applications 45, 737–748 (2003)
2. Anyfantis, D., Karagiannopoulos, M., Kotsiantis, S., Pintelas, P.: Robustness of learning techniques in handling class noise in imbalanced datasets. In: Boukis, C., Pnevmatikakis, A., Polymenakos, L. (eds.) AIAI 2007. IFIP, vol. 247, pp. 21–28. Springer, Boston (2007)

3. Batista, G., Prati, R., Monard, M.: A study of the behavior of several methods for balancing machine learning training data. ACM SIGKDD Explorations Newsletter 6(1), 20–29 (2004)
4. Bunkhumpornpat, C., Sinapiromsaran, K., Lursinsap, C.: Safe-Level-SMOTE: Safe-Level-Synthetic Minority Over-Sampling TEchnique for Handling the Class Imbalanced Problem. In: Theeramunkong, T., Kijsirikul, B., Cercone, N., Ho, T.-B. (eds.) PAKDD 2009. LNCS, vol. 5476, pp. 475–482. Springer, Heidelberg (2009)
5. Błaszczyński, J., Stefanowski, J., Idkowiak, L.: Extending bagging for imbalanced data. In: Burduk, R., Jackowski, K., Kurzynski, M., Wozniak, M., Zolnierek, A. (eds.) CORES 2013. AISC, vol. 226, pp. 273–282. Springer, Heidelberg (2013)
6. Błaszczyński, J., Stefanowski, J., Szajek, M.: Local Neighbourhood in Generalizing Bagging for Imbalanced Data. In: Proc. of COPEM 2013 - Solving Complex Machine Learning Problems with Ensemble Methods Workshop at ECML PKDD 2013, Praque, pp. 10–24 (2013)
7. Chawla, N.: Data mining for imbalanced datasets: An overview. In: Maimon, O., Rokach, L. (eds.) The Data Mining and Knowledge Discovery Handbook, pp. 853–867. Springer, Heidelberg (2005)
8. Chawla, N., Bowyer, K., Hall, L., Kegelmeyer, W.: SMOTE: Synthetic Minority Over-sampling Technique. J. of Artificial Intelligence Research 16, 341–378 (2002)
9. Cost, S., Salzberg, S.: A Weighted Nearest Neighbor Algorithm for Learning with Symbolic Features. Machine Learning Journal 10(1), 1213–1228 (1993)
10. Fernández, A., García, S., Herrera, F.: Addressing the Classification with Imbalanced Data: Open Problems and New Challenges on Class Distribution. In: Corchado, E., Kurzyński, M., Woźniak, M. (eds.) HAIS 2011, Part I. LNCS, vol. 6678, pp. 1–10. Springer, Heidelberg (2011)
11. Furnkranz, J.: Separate-and-conquer rule learning. Artificial Intelligence Review 13(1), 3–54 (1999)
12. Furnkranz, J., Gamberger, D., Lavrac, N.: Foundations of Rule Learning. Springer (2012)
13. Galar, M., Fernandez, A., Barrenechea, E., Bustince, H., Herrera, F.: A Review on Ensembles for the Class Imbalance Problem: Bagging-, Boosting-, and Hybrid-Based Approaches. IEEE Transactions on Systems, Man, and Cybernetics, Part C: Applications and Reviews 99, 1–22 (2011)
14. García, V., Sánchez, J., Mollineda, R.A.: An empirical study of the behavior of classifiers on imbalanced and overlapped data sets. In: Rueda, L., Mery, D., Kittler, J. (eds.) CIARP 2007. LNCS, vol. 4756, pp. 397–406. Springer, Heidelberg (2007)
15. Grzymala-Busse, J.W., Goodwin, L.K., Grzymala-Busse, W., Zheng, X.: An approach to imbalanced data sets based on changing rule strength. In: Proceedings of Learning from Imbalanced Data Sets, AAAI Workshop at the 17th Conference on AI, pp. 69–74 (2000)
16. Grzymala-Busse, J.W., Stefanowski, J., Wilk, S.: A comparison of two approaches to data mining from imbalanced data. Journal of Intelligent Manufacturing 16(6), 565–574 (2005)
17. Han, H., Wang, W., Mao, B.: Borderline-SMOTE: A New Over-Sampling Method in Imbalanced Data Sets Learning. In: Huang, D.-S., Zhang, X.-P., Huang, G.-B. (eds.) ICIC 2005. LNCS, vol. 3644, pp. 878–887. Springer, Heidelberg (2005)
18. He, H., Garcia, E.: Learning from imbalanced data. IEEE Transactions on Data and Knowledge Engineering 21(9), 1263–1284 (2009)
19. He, H., Yungian, M. (eds.): Imbalanced Learning. Foundations, Algorithms and Applications. IEEE - Wiley (2013)

20. Hido, S., Kashima, H.: Roughly balanced bagging for imbalance data. Statistical Analysis and Data Mining 2(5-6), 412–426 (2009)
21. Japkowicz, N.: Class imbalance: Are we focusing on the right issue? In: Proc. II Workshop on Learning from Imbalanced Data Sets, ICML Conf., pp. 17–23 (2003)
22. Jo, T., Japkowicz, N.: Class Imbalances versus small disjuncts. ACM SIGKDD Explorations Newsletter 6(1), 40–49 (2004)
23. Khoshgoftaar, T., Van Hulse, J., Napolitano, A.: Comparing boosting and bagging techniques with noisy and imbalanced data. IEEE Transactions on Systems, Man, and Cybernetics–Part A 41(3), 552–568 (2011)
24. Kubat, M., Matwin, S.: Addresing the curse of imbalanced training sets: one-side selection. In: Proc. of the 14th Int. Conf. on Machine Learning, ICML 1997, pp. 179–186 (1997)
25. Laurikkala, J.: Improving identification of difficult small classes by balancing class distribution. Tech. Report A-2001-2, University of Tampere (2001)
26. Liu, A., Zhu, Z.: Ensemble methods for class imbalance learning. In: He, H., Yungian, M. (eds.) Imbalanced Learning. Foundations, Algorithms and Apllications, pp. 61–82. Wiley (2013)
27. Lumijarvi, J., Laurikkala, J., Juhola, M.: A comparison of different heterogeneous proximity functions and Euclidean distance. Stud Health Technol. Inform. 107 (pt. 2), 1362–1366 (2004)
28. Lopez, V., Fernandez, A., Garcia, S., Palade, V., Herrera, F.: An Insight into Classification with Imbalanced Data: Empirical Results and Current Trends on Using Data Intrinsic Characteristics. Information Sciences 257, 113–141 (2014)
29. Lopez, V., Triguero, I., Garcia, S., Carmona, C., Herrera, F.: Addressing imbalanced classification with instance generation techniques: IPADE-ID. Neurocomputing 126, 15–28 (2014)
30. Maciejewski, T., Stefanowski, J.: Local neighbourhood extension of SMOTE for mining imbalanced data. In: Proc. IEEE Symp. on Computational Intelligence and Data Mining, pp. 104–111 (2011)
31. Napierala, K.: Improving rule classifiers for imbalanced data. Ph.D. Thesis. Poznan University of Technology (2013)
32. Napierała, K., Stefanowski, J., Wilk, S.: Learning from Imbalanced Data in Presence of Noisy and Borderline Examples. In: Szczuka, M., Kryszkiewicz, M., Ramanna, S., Jensen, R., Hu, Q. (eds.) RSCTC 2010. LNCS, vol. 6086, pp. 158–167. Springer, Heidelberg (2010)
33. Napierała, K., Stefanowski, J.: Argument Based Generalization of MODLEM Rule Induction Algorithm. In: Szczuka, M., Kryszkiewicz, M., Ramanna, S., Jensen, R., Hu, Q. (eds.) RSCTC 2010. LNCS (LNAI), vol. 6086, pp. 138–147. Springer, Heidelberg (2010)
34. Napierala, K., Stefanowski, J.: Identification of different types of minority class examples in imbalanced data. In: Corchado, E., Snášel, V., Abraham, A., Woźniak, M., Graña, M., Cho, S.-B. (eds.) HAIS 2012, Part II. LNCS (LNAI), vol. 7209, pp. 139–150. Springer, Heidelberg (2012)
35. Napierala, K., Stefanowski, J.: BRACID: a comprehensive approach to learning rules from imbalanced data. Journal of Intelligent Information Systems 39(2), 335–373 (2012)
36. Prati, R., Batista, G., Monard, M.: Class imbalance versus class overlapping: An analysis of a learning system behavior. In: Proc. 3rd Mexican Int. Conf. on Artificial Intelligence, pp. 312–321 (2004)

37. Ramentol, E., Caballero, Y., Bello, R., Herrera, F.: SMOTE-RSB *: A hybrid pre-processing approach based on oversampling and undersampling for high imbalanced data-sets using SMOTE and rough sets theory. Knowledge Inform. Systems 33(2), 245–265 (2012)

38. Sikora, M., Wrobel, L.: Data-driven adaptive selection of rule quality measures for improving rule induction and filtration algorithms. Int. J. General Systems 42(6), 594–613 (2013)

39. Stefanowski, J.: On combined classifiers, rule induction and rough sets. In: Peters, J.F., Skowron, A., Düntsch, I., Grzymała-Busse, J.W., Orłowska, E., Polkowski, L. (eds.) Transactions on Rough Sets VI. LNCS, vol. 4374, pp. 329–350. Springer, Heidelberg (2007)

40. Stefanowski, J.: Overlapping, rare examples and class decomposition in learning classifiers from imbalanced data. In: Ramanna, S., Jain, L.C., Howlett, R.J. (eds.) Emerging Paradigms in Machine Learning, pp. 277–306 (2013)

41. Stefanowski, J., Wilk, S.: Selective pre-processing of imbalanced data for improving classification performance. In: Song, I.-Y., Eder, J., Nguyen, T.M. (eds.) DaWaK 2008. LNCS, vol. 5182, pp. 283–292. Springer, Heidelberg (2008)

42. Stefanowski, J., Wilk, S.: Extending rule-based classifiers to improve recognition of imbalanced classes. In: Ras, Z.W., Dardzinska, A. (eds.) Advances in Data Management. SCI, vol. 223, pp. 131–154. Springer, Heidelberg (2009)

43. Tomasev, N., Mladenic, D.: Class imbalance and the curse of minority hubs. Knowledge-Based Systems 53, 157–172 (2013)

44. Weiss, G.M.: Mining with rarity: a unifying framework. ACM SIGKDD Explorations Newsletter 6(1), 7–19 (2004)

45. Wilson, D.R., Martinez, T.R.: Improved heterogeneous distance functions. Journal of Artifical Intelligence Research 6, 1–34 (1997)

The Relational Construction of Conceptual Patterns - Tools, Implementation and Theory

Piero Pagliani

Research Group on Knowledge and Communication Models
pier.pagliani@gmail.com

Abstract. Different conceptual ways to analyse information are here
defined by means of the fundamental notion of a relation. This approach
makes it possible to compare different mathematical notions and tools
used in qualitative data analysis. Moreover, since relations are repre-
sentable by Boolean matrices, computing the conceptual-oriented operators is straightforward. Finally, the relational-based approach makes it
possible to conceptually analyse not only sets but relations themselves.

1 Relations, Concepts and Information

The world is made of relations. In a sense, every entity or thing is nothing else
but a sheaf of relations which occur with other sheaves of relations. Without
offending Immanuel Kant, from this point of view the concept of a "monads"
seems to be an expedient to bypass some philosophical problem. On the contrary,
our point is inspired by another monumental assumption of Kant's philosophy:
relations hold between entities, but entities themselves are phenomenological
relations. That is, they are relations between *noumena* (not observed entities or
events) and their manifestation through observed properties, which transforms
noumena into *phenomena* (observable entities or events). In this way we arrive
at the pair *intension-extension* which is at the very heart of conceptual data
analysis.

Relations connect entities with their properties, in what we call an *obser-
vation system* or *property system* $\mathbf{P} = \langle U, M, R \rangle$, where U is the universe of
entities, M the set of properties, and $R \subseteq U \times M$ is the "manifestation" relation so that $\langle g, m \rangle \in R$ means that entity g fulfills property m. From now on,
instead of "entity" we shall use the term "object" in the sense of the German
term *Gegenstand* which means an object before interpretation. The symbol M is
after *Merkmal*, which means "property" or "characteristic feature". Relations in
property systems induce derived relations between objects themselves or between
properties. Indeed, sometimes phenomena can be perceived by directly observing relations occurring within objects or within properties. What is important
is the coherence of the entire framework. Thus, relations assemble *concepts*, for
instance by associating together those properties which are observed of a given
set of objects. Vice-versa, relations assemble *extensions of concepts* by grouping
together the objects which fulfill a given set of properties. In the former case an

M. Kryszkiewicz et al. (Eds.): RSEISP 2014, LNAI 8537, pp. 14–27, 2014.
© Springer International Publishing Switzerland 2014

intension is derived from an extension. In the latter an extension is derived from an intension. Therefore, we shall call *intensional* the operators which transform extensions into intensions, and, vice-versa, *extensional* if the construction operates in the opposite direction. The former kind of operators will be decorated by an "i" and the latter by an "e". Obviously, mutual constructions are in order and we shall explore the properties of intensions of extensions and extensions of intensions.

As much as Category Theory advocates that what is relevant are the *morphisms* between elements, and not the elements standing alone, one can maintain that in conceptual data analysis ontological commitments should be avoided because the fundamental ingredients of the analysis are relations.

However, some ontological feature comes into the picture if U or M are equipped with some relational structure (for instance a preference relation). Anyway, this structure is in principle given by some intensional or extensional operator derivable from other property systems.

The paper will discuss some fundamental topics related to the pair *intension-extension* as defined by relations:

- The logical schemata which define the basic extensional and intensional operators and, thus, their meanings.
- How to use these operators in conceptual data analysis (in particular, in approximation analysis).
- How to compute the operators.
- How to implement the above procedures by manipulating Boolean matrices.

The theses and/or the proofs of the proved results are new.

2   Relations, Closures, Interiors and Modalities

The first natural step is collecting together the properties fulfilled by a set of objects, and the objects which fulfill a given set of properties:

Definition 1. *Given a property system* $\mathbf{P} = \langle U, M, R \rangle$, $A \subseteq U, B \subseteq M$:

$$\langle i \rangle(A) = \{m : \exists g(\langle g, m \rangle \in R \wedge g \in A)\} \tag{1}$$
$$\langle e \rangle(B) = \{g : \exists m(\langle g, m \rangle \in R \wedge m \in B)\} \tag{2}$$

We call these operators *constructors*. Some observations are in order:

OBSERVATION 1. In Relation Algebra these two constructors are well known and are denoted by $R(A)$ and, respectively, $R^{\smile}(B)$. The first is called *the left Peirce product of R and A*, while the second is the *right Peirce product of R and B*, or the left Peirce product of $R^{\smile}$ and B, where $R^{\smile} = \{\langle m, g \rangle : \langle g, m \rangle \in R\}$ is the *inverse relation* of R. Indeed $\langle e \rangle$ is the same as $\langle i \rangle$ applied to the inverse relation: $\langle e \rangle(B) = \{g : \exists m(\langle m, g \rangle \in R^{\smile} \wedge m \in B)\}$. In this way the quantified variable takes the first place in the ordered pair of the definition of $\langle e \rangle$, as it happens in the definition of $\langle i \rangle$. The role of a quantified variable in a relation is a formality which will be useful to compare different definitions. For any singleton $\{x\}$, instead of $R(\{x\})$ we shall write $R(x)$.

OBSERVATION 2. The definition of the constructor $\langle e \rangle$ is the same as that of the operator $\Diamond$ (*possibility*) in Modal Logic. In fact, a *Kripke model* is a triple $\langle W, R, \models \rangle$, where W is a set of *possible worlds*, $R \subseteq W \times W$ is an *accessibility relation*, and for any formula α, $w \models \Diamond(\alpha)$ if and only if there exists a possible world w' which is accessible to w and such that $w' \models \alpha$, that is, $\exists w'(\langle w, w' \rangle \in R \wedge w' \models \alpha)$. If in Definition 1 we set $M = U$ and identify a subset A of U with the domain of validity of a formula α (i. e. $A = \{g \in U : g \models \alpha\}$), and if we denote the modal operator by $\langle R \rangle$, we obtain $\langle R \rangle(A) = \langle e \rangle(A) = R^{\smallsmile}(A)$ while $\langle i \rangle(A) = \langle R^{\smallsmile} \rangle(A) = R(A)$. Indeed, with respect to our constructors, one has the following modal reading: if $g \in A$ then it is *possible* that g fulfills properties in $\langle i \rangle(A)$, because if $m \in \langle i \rangle(A)$, then $R^{\smallsmile}(m)$ has non void intersection with A. Analogously, if $b \in B$ then it is *possible* that m is fulfilled by entities in $\langle e \rangle(B)$.

OBSERVATION 3. The logical structure of the definitions (1) and (2) is given by the combination $(\exists, \wedge)$. This is the logical core of a number of mathematical concepts. Apart from the above notion of "possibility" in Modal Logic, notably one finds it in the definition of a *closure* operator. Recalling that our framework is the Boolean lattice $\wp(U)$ or $\wp(M)$, we remind the following definitions:

Definition 2. *An operator ϕ on a lattice $\mathbf{L}$ is said to be a* closure *(resp. interior) operator if for any $x, y \in \mathbf{L}$ it is (i) increasing: $x \leq \phi(x)$ (resp. decreasing: $\phi(x) \leq x$), (ii) monotone: $x \leq y$ implies $\phi(x) \leq \phi(y)$, and (iii) idempotent: $\phi(\phi(x)) = \phi(x)$. Moreover, it is topological if it is (iv) additive: $\phi(x \vee y) = \phi(x) \vee \phi(y)$ (resp. multiplicative: $\phi(x \wedge y) = \phi(x) \wedge \phi(y)$) and (v) normal: $\phi(0) = 0$ (resp, conormal: $\phi(1) = 1$).*

We call a property system such that $U = M$ a *square relational system*, SRS. Intuitively, in a SRS an object $g \in U$ is closed to a subset A of U with respect to R, if in A there exists a g' such that $\langle g, g' \rangle \in R$, so that g is linked to A in this way. We then call $R(g)$ the *R-neighborhood* of g and if $g' \in R(g)$ then g' will be called an *R-neighbor* of g. Thus g is closed to a set A if $R(g) \cap A \neq \emptyset$. The closure of A is the operation of embedding all the entities which are closed to A.

Now, we illustrate how simple is the computation of a closure. Any finite property system can be represented by a Boolean matrix such that the entry (g, m) is 1 if $\langle g, m \rangle \in R$, 0 otherwise. To compute $\langle i \rangle(\{x_1, x_2, ..., x_n\})$ one has just to collect the elements of U that display 1 in the rows $x_1, x_2, ..., x_n$. Vice-versa, to compute $\langle e \rangle(\{x_1, x_2, ..., x_n\})$ one has to collect the elements of U that display 1 in the columns $x_1, x_2, ..., x_n$. We shall denote the Boolean matrix corresponding to a relation R by $\mathbf{R}$. If $X \subseteq U$, $\mathbf{R}(\mathbf{X})$ shall denote the Boolean array corresponding to $R(X)$ and $\mathbf{R}^{\smallsmile}(\mathbf{X})$ the array corresponding to $R^{\smallsmile}(X)$. $\mathbf{R} \restriction \mathbf{X}$ is the matrix representing the subrelation $\{\langle x, y \rangle : x \in X \wedge y \in R(x)\}$ The example runs in a SRS, but with generic property systems the story is the same

EXAMPLE 1. $U = \{a, b, c, d\}$. Let us manipulate the subset $\{a, c\}$.

R	a	b	c	d
a	1	0	0	1
b	0	1	1	1
c	0	1	0	0
d	0	1	0	1

$R \upharpoonright \{a,c\}$	a	b	c	d
a	1	0	0	1
b				
c	0	1	0	0
d				

$R^{\smile} \upharpoonright \{a,c\}$	a	b	c	d
a	1	0
b	0	1
c	0	0
d	0	0

$$\mathbf{R}(\mathbf{a,c}) = \overset{a\,b\,c\,d}{[1101]}$$
$$\langle i \rangle(\{a,c\}) = \{a,b,d\}$$

$$\mathbf{R^{\smile}}(\mathbf{a,c}) = \overset{a\,b\,c\,d}{[1100]}$$
$$\langle e \rangle(\{a,c\}) = \{a,b\}$$

Notice that $\mathbf{R}(\mathbf{X}) = \bigvee \mathbf{R}(\mathbf{x})_{x \in X}$, where $\vee$ is the element-wise Boolean sum of the arrays (for instance, $\mathbf{R}(\{\mathbf{a,c}\}) = \mathbf{R}(\mathbf{a}) \vee \mathbf{R}(\mathbf{c}) = [1 \vee 0, 0 \vee 1, 0 \vee 0, 1 \vee 0]$).

Since $\{g : g \in R^{\smile}(B)\} = \{g : R(g) \cap B \neq \emptyset\}$, $\langle e \rangle$ has a definition formally similar to that of a closure and, also, of an upper approximation operator. In fact, if R is an equivalence relation on U, $\langle e \rangle(X)$ is the upper approximation $(uR)(X)$ of Pawlak's Rough Set Theory. In turn, independently of the properties of R the *lower approximation* is defined as $(lR)(X) = \{x : R(x) \subseteq X\}$. That is, $(lR)(X) = \{x : \forall x'(\langle x,x' \rangle \in R \Longrightarrow x' \in X)\}$. Thus, we set in any property system $\mathbf{P} = \langle U, M, R \rangle, A \subseteq U, B \subseteq M$:

$$[i](A) = \{m : \forall g(\langle g,m \rangle \in R \Longrightarrow g \in A)\} \tag{3}$$
$$[e](B) = \{g : \forall m(\langle g,m \rangle \in R \Longrightarrow m \in B)\} \tag{4}$$

It is immediate to see that in a SRS, $(lR)(X) = [e](X)$.

OBSERVATION 4. As we shall see, if a set X is represented by a particular kind of relation (a *right cylinder*), then in Relation Algebra $[i](X)$ coincides with the *right residual* of R and X, while $[e](X)$ is the right residual of $R^{\smile}$ and X.

OBSERVATION 5. Again, one verifies a correspondence with Modal Logic. Given a Kripke model $\langle W, R, \models \rangle$ the forcing clause for a necessary formula $\square(\alpha)$ is:

$$w \models \square(\alpha) \;\; \textit{iff} \;\; \forall w'(\langle w,w' \rangle \in R \Longrightarrow w' \models \alpha) \tag{5}$$

Again, if $A = \{g : g \models \alpha\}$, then $[R](\alpha) = [e](A)$. Indeed, the modal reading of the above constructors is: in order to fulfill properties in $[i](A)$ it is necessary to be an object in A. Dually, in order to be fulfilled by objects in $[e](B)$, it is necessary to be a property of B.

OBSERVATION 6. The logical core of Definitions (3) and (4) is the combination $(\forall, \Longrightarrow)$. We remind that the logical core of the *possibility* operators is $(\exists, \wedge)$. To exploit these facts we need a strategic notion:

Definition 3. *Let* $\mathbf{O}$ *and* $\mathbf{O'}$ *be two preordered sets and* $\sigma : \mathbf{O} \longmapsto \mathbf{O'}$ *and* $\iota : \mathbf{O'} \longmapsto \mathbf{O}$ *be two maps such that for all* $p \in \mathbf{O}$ *and* $p' \in \mathbf{O'}$

$$\iota(p') \leq p \;\; \textit{iff} \;\; p' \leq' \sigma(p) \tag{6}$$

then σ *is called the* upper adjoint *of* ι *and* ι *is called the* lower adjoint *of* σ. *This fact is denoted by* $\mathbf{O'} \dashv^{\iota,\sigma} \mathbf{O}$.

Now, in a Heyting algebra $\mathbf{H}$, $\wedge$ is lower adjoint to $\Longrightarrow$ in the sense that for all elements $x, y, z \in \mathbf{H}$, $\wedge_x(y) \leq z$ iff $y \leq \Longrightarrow_x (z)$, where $\wedge_x(y)$ is a parameterized formulation of $x \wedge y$ and $\Longrightarrow_x (z)$ of $x \Longrightarrow z$. Moreover, $\exists$ and $\forall$ are, respectively, lower and upper adjoints to the pre-image $f^{-1} : \wp(Y) \longmapsto \wp(X)$ of a function $f : X \mapsto Y$: for all $A \subseteq X, B \subseteq Y$ one has $\exists_f(A) \subseteq B$ iff $A \subseteq f^{-1}(B)$ and $B \subseteq \forall_f(A)$ iff $f^{-1}(B) \subseteq A$, where $\exists_f(A) = \{b \in B : \exists a(f(a) = b \wedge a \in A\}$ and $\forall_f(A) = \{b \in B : \forall a(f(a) = b \Longrightarrow a \in A\}$. Therefore, since i and e constructors operate in opposite directions, it is not surprise if it can be proved that the following adjointness properties hold in any property system $\mathbf{P} = \langle U, M, R \rangle$, for $\mathbf{M} = \langle \wp(M), \subseteq \rangle$ and $\mathbf{U} = \langle \wp(U), \subseteq \rangle$ (see [17]):

$$(a)\mathbf{M} \dashv^{\langle e \rangle, [i]} \mathbf{U} \quad (b)\mathbf{U} \dashv^{\langle i \rangle, [e]} \mathbf{M} \tag{7}$$

From this we immediately obtain that $\langle \cdot \rangle$ constructors are additive (as like as any lower adjoint), while $[\cdot]$ constructors are multiplicative (as any upper adjoint)[1]. Moreover, $\langle \cdot \rangle(A \cap B) \subseteq \langle \cdot \rangle(A) \cap \langle \cdot \rangle(B)$ and $[\cdot](A \cup B) \supseteq [\cdot](A) \cup [\cdot](B)$. Again, this is a consequence of adjointness, but can be easily verified using the distributive properties of quantifiers[2].

To compute the necessity constructors, the following result is exploited, which can be proved by means of the equivalences $\neg \exists \equiv \forall \neg$ and $\neg(A \wedge \neg B) \equiv A \Longrightarrow B$:

$$\forall X \subseteq U, [\cdot](X) = -\langle \cdot \rangle(-X) \tag{8}$$

Let us continue the previous example and compute $[\cdot](\{a, b\})$: $-\{a, b\} = \{c, d\}$.

R	a b c d		R⌣ ↾ {c,d}	a b c d		R ↾ {c,d}	a b c d		
a	1 0 0 1		a	0 1		a			$-\mathbf{R}^{\smallsmile}(\mathbf{c,d}) = \overset{a\,b\,c\,d}{[0010]}$
b	0 1 1 1		b	1 1		b			$[e](\{a,b\}) = \{c\}$
c	0 1 0 0		c	0 0		c	0 1 0 0		
d	0 1 0 1		d	0 1		d	0 1 0 1		$-\mathbf{R}(\mathbf{c,d}) = \overset{a\,b\,c\,d}{[1010]}$
									$[i](\{a,b\}) = \{a,c\}$

$-\mathbf{R}$ is the element-wise Boolean complement of the matrix $\mathbf{R}$.

3   Pretopologies, Topologies and Coincidence of Operators

It may sound surprising, but substantially the above procedures are all the machinery we need in order to compute the operators required by relation-based conceptual data analysis. Notice that the procedures to compute $\langle \cdot \rangle$ and $[\cdot]$ are independent of the properties of the relation R. On the contrary, the properties of these constructors strictly depend on those of R. For a generic binary relation R, $\langle \cdot \rangle$ may fail to be increasing or idempotent: $\{a, c\} \not\subseteq \langle i \rangle(\{a, c\})$ and $\langle i \rangle(\langle i \rangle(\{a, c\})) \neq \langle i \rangle(\{a, c\})$. Anyway, additivity gives monotonicity. In turn, notwithstanding the formal analogy with the definition of an interior operator, $[\cdot]$ may be neither decreasing nor idempotent: $[e](\{a, b\}) \not\subseteq \{a, b\}$ and

---

[1] Often, a lower adjoint is called "left adjoint" and an upper adjoint is called "right adjoint". We avoid the terms "right" and "left" because they could make confusion with the position of the arguments of the operations on binary relations.

[2] For instance one has $\forall x A(x) \vee \forall x B(x) \Longrightarrow \forall x(A(x) \vee B(x))$, but not the opposite. This proves that $\forall$ cannot have an upper adjoint, otherwise it should be additive.

$[e]([e](\{a,b\})) \neq [e](\{a,b\})$, although multiplicativity guarantees monotonicity. We shall see that given a generic R, $\langle \cdot \rangle$ and $[\cdot]$ behave like pretopological closure and, respectively, interior operators induced by neighborhood families which are lattice filters (with respect to $\subseteq$ and $\cap$). Actually, in real-world situations R is derived from observations (for instance data collected by sensors) and one cannot expect R to enjoy "nice properties" necessarily. Unfortunately, in view of the failure of the increasing property, $\langle \cdot \rangle$ cannot in general be used to compute any sort of upper approximation and $[\cdot]$ cannot provide any lower approximation because of the failure of the decreasing property. Thus, we need more structured operators. One approach is equipping R with particular properties. Modal Logic, then, tells us the new behaviors of the possibility and necessity operators. But one obtains very interesting operators if adjoint constructors are combined. Let us then set, for all $A \subseteq U, B \subseteq M$:

$$(a) \;\; int(A) = \langle e \rangle([i](A)) \qquad\qquad (b) \;\; cl(A) = [e](\langle i \rangle(A)). \qquad (9)$$
$$(c) \;\; \mathcal{C}(B) = \langle i \rangle([e](B)) \qquad\qquad (d) \;\; \mathcal{A}(B) = [i](\langle e \rangle(B)). \qquad (10)$$

Notice that int and cl map $\wp(U)$ on $\wp(U)$, while $\mathcal{A}$ and $\mathcal{C}$ map $\wp(M)$ on $\wp(M)$.

OBSERVATION 7.(see [17]) Since these operators are combinations of adjoint functors, they fulfill a number of properties: (i) int and $\mathcal{C}$ are interior operators; (ii) cl and $\mathcal{A}$ are closure operators. This means that $int(A) \subseteq A \subseteq cl(A)$, any $A \subseteq U$ and $\mathcal{C}(B) \subseteq B \subseteq \mathcal{A}(B)$, any $B \subseteq M$. Thus, they are veritable approximations. However, they are not topological: int and $\mathcal{C}$ are not multiplicative, because the external constructor $\langle \cdot \rangle$ is not, and cl and $\mathcal{A}$ are not additive, because the external constructor $[\cdot]$ is not.

The interpretation of the above operators is intuitive in property systems. $\mathcal{C}(B)$ displays those properties which are fulfilled by objects which fulfills at most properties in B. That is, first we select from U the objects which fulfill at most properties in B, then we check the properties from B which are effectively fulfilled by the selected objects. Dually, $\mathcal{A}(B)$ displays all the properties which are fulfilled only by the elements which fulfill some property in B. So to say, one selects the objects which fulfill at least one property of B, and, after that, the properties which are exclusively fulfilled by the selected objects. For the opposite direction just substitute "objects" for "properties" and "fulfill" for "fulfilled".

On the contrary, the interpretation for SRSs is not that clear. For instance, $\mathcal{C}(B)$ displays all and only the elements which are R-related to some element whose R-neighborhood is included in B. And so on. We do not go further into this interpretation, but just list in parallel the set-theoretic shapes of pair-wise related operators:

Modal constructors	Pre-topological operator
$[e](X) = \{x : R(x) \subseteq X\}$	$\mathcal{C}(X) = \bigcup\{R(x) : R(x) \subseteq X\}$
$[i](X) = \{x : R^{\smile}(x) \subseteq X\}$	$int(X) = \bigcup\{R^{\smile}(x) : R^{\smile}(x) \subseteq X\}$
$\langle e \rangle(X) = \{x : x \in R^{\smile}(X)\}$	$\mathcal{A}(X) = \{x : R^{\smile}(x) \subseteq R^{\smile}(X)\}$
$\langle i \rangle(X) = \{x : x \in R(X)\}$	$cl(X) = \{x : R(x) \subseteq R(X)\}$

In our Example 1, $\mathcal{C}(\{a,b\}) = \langle i\rangle(\{c\}) = \{b\}$ and $\mathcal{A}(\{a,d\}) = [i](\{a,b,d\}) = \{a,c,d\}$. Notice that $\mathcal{C}$ and $\mathcal{A}$ are dual, that is, $\mathcal{C}(X) = -\mathcal{A}(-X)$. This is due to the fact that $\langle i\rangle$ and $[i]$ are dual. Moreover, $\mathcal{A}(X) = \{x : \forall y(x \in R(y) \Longrightarrow R(y) \cap X \neq \emptyset)\}$ and $cl(X) = \{x : \forall y(x \in R^{\smile}(y) \Longrightarrow R^{\smile}(y) \cap X \neq \emptyset)\}$.

Now we want to understand when the operators on the right column are equivalent to the corresponding constructors on the left column. We shall deal with int and $[i]$. But first an interesting brief excursus is in order.

EXCURSUS: COVERINGS. Generalisations of Pawlak's approximation operators have been introduced that are based on coverings instead of partitions. The relational machinery can simplify this approach. Given a set U, a *covering* is a family $C \subseteq \wp(U)$ such that $\bigcup C = U$. If one assigns all the elements of a component K of C to a unique element m from a set M, through a relation R, one obtains a property system $\langle U, M, R\rangle$ such that $\{R^{\smile}(m) : m \in M\} = C$. It is straightforward to prove that the operator $\underline{C_1}(X) = \bigcup\{K \in C : K \subseteq X\}$ introduced in [24], coincides with int. Therefore, its dual operator $\overline{C_1}$ coincides with cl and all the properties of $\underline{C_1}$ and $\overline{C_1}$ are provided for free by the adjunction properties. For further considerations see the final remarks.

Let us came back to our goal. We start with noticing that in view of Observation 7, $[i]$ must be an interior operator, in order to coincide with int. Moreover, it must be topological because of multiplicativity (similarly, if $\langle i\rangle$ is a closure operator, it is necessarily topological because of additivity). Indeed, we now prove that $[i] = int$ if and only if R is a preorder. It is well-known that in this case Kripkean necessity modalities are topological interior operators. However, if part of the result is well-known, the proof will be developed in a novel way which provides relevant information about the operators.

The proof is made of two parts. In the first part we prove that R must be a preorder to make int and $[i]$ coincide. After that, we complete the proof in a more specific manner: it will be proved that if R is a preorder, then $\mathcal{C}$ (thus $[e]$) is the interior operator of a particular topology induced by R.

Lemma 1. *Let $\langle U, U, R\rangle$ be a SRS. Then $\forall x \in U, x \in [i](R^{\smile}(x))$.*

Proof. Trivially, $x \in [i](R^{\smile}(x))$ iff $R^{\smile}(x) \subseteq R^{\smile}(x)$.

Theorem 1. *Let $\langle U, U, R\rangle$ be a SRS. Then for all $A \subseteq U, int(A) = [i](A)$ if and only if R is a preorder (i.e. R is reflexive and transitive).*

Proof. A) If $\exists A \subseteq U, int(A) \neq [i](A)$ then R is not a preorder (either reflexivity or transitivity fail). Proof. The antecedent holds in two cases: (i) $\exists x \in [i](A), x \notin int(A)$; (ii) $\exists x \in int(A), x \notin [i](A)$. In case (i) $\forall y \in [i](A), x \notin R^{\smile}(y)$. In particular, $x \notin R^{\smile}(x)$, so that reflexivity fails. In case (ii) $\exists y \in [i](A)$ such that $x \in R^{\smile}(y)$. Therefore, since $\langle x, y\rangle \in R$ and $y \in [i](A)$, x must belong to A. Moreover, it must exists $z \notin A, \langle z, x\rangle \in R$, otherwise $x \in [i](A)$. If R were transitive, $\langle z, y\rangle \in R$, so that $y \notin [i](A)$. Contradiction.

B) If R is not a preorder, then $\exists A \subseteq U, int(A) \neq [i](A)$. Proof. (i) Take $A = R^\smile(x)$. From Lemma 1, $x \in [i](R^\smile(x))$. Suppose R is not reflexive with $\langle x, x \rangle \notin R$. Thus $x \notin R^\smile(x)$. Hence, it cannot exists an y such that $x \in R^\smile(y)$ and $R^\smile(y) \subseteq R^\smile(x)$. So, $x \notin int(R^\smile(x))$. (ii) Suppose transitivity fails, with $\langle x, y \rangle, \langle y, z \rangle \in R, \langle x, z \rangle \notin R$. From Lemma 1, $z \in [i](R^\smile(z))$, but $y \notin [i](R^\smile(z))$, because $x \in R^\smile(y)$ while $x \notin R^\smile(z)$ so that $R^\smile(y) \not\subseteq R^\smile(z)$. On the contrary, $y \in R^\smile(z)$ and $R^\smile(z) \subseteq R^\smile(z)$. Therefore, $y \in int(R^\smile(z))$. We conclude that $int(R^\smile(z)) \neq [i](R^\smile(z))$.

Corollary 1. *In a SRS $\langle U, U, R \rangle$, the following are equivalent: (i) R is a preorder, (ii) $\mathcal{C} = [e]$, (iii) $int = [i]$, (iv) $int, [i], [e]$ and $\mathcal{C}$ are topological interior operators.*

We recall that a topological interior operator $\mathcal{I}$ on a set X induces a *specialisation preorder* defined as follows: $\forall x, y \in X, x \preceq y$ iff $\forall A \subseteq X, x \in \mathcal{I}(A)$ implies $y \in \mathcal{I}(A)$. However, in what follows we extend this definition to any monadic operator on sets. If $R \subseteq X \times X$ is a preorder, then the topology with bases the family $\{R(x) : x \in X\}$ is called the *Alexandrov topology induced by R*. The specialisation preorder induced by such a topology coincides with R itself.

Lemma 2. *If $R \subseteq X \times X$ is transitive, then $\forall x, y \in X, \langle x, y \rangle \in R$ implies $R(y) \subseteq R(x)$. If R is reflexive, then $R(y) \subseteq R(x)$ implies $\langle x, y \rangle \in R$.*

Proof. Suppose $\langle x, y \rangle \in R$ and $z \in R(y)$. Then $\langle y, z \rangle \in R$ and by transitivity $\langle x, z \rangle \in R$ so that $z \in R(x)$. Thus, $R(y) \subseteq R(x)$. Vice-versa, if $R(y) \subseteq R(x)$ then for all $z, \langle y, z \rangle \in R$ implies $\langle x, z \rangle \in R$. In particular $\langle y, y \rangle \in R$ by reflexivity. Hence $\langle x, y \rangle \in R$.

Theorem 2. *Let $\langle U, U, R \rangle$ be a SRS such that R is preorder. Then the specialization preorder induced by $[i]$ coincides with $R^\smile$ and that induced by $[e]$ coincides with R.*

Proof. If $x \preceq y$ then for all $A \subseteq X, x \in [i](A)$ implies $y \in [i](A)$. Therefore, $R^\smile(x) \subseteq A$ implies $R^\smile(y) \subseteq A$, all A. In particular, $R^\smile(x) \subseteq R^\smile(x)$ implies $R^\smile(y) \subseteq R^\smile(x)$. But the antecedent is true, so the consequence must be true, too, so that $R^\smile(y) \subseteq R^\smile(x)$. Since R is reflexive, so is $R^\smile$ and from Lemma 2, $\langle x, y \rangle \in R^\smile$. The opposite implication is proved analogously by transitivity. The thesis for $[e]$ and R is a trivial consequence.

Corollary 2. *Let $\mathcal{C}$ be a topological interior operator induced by a SRS $\langle U, U, R \rangle$. Then $\mathcal{C}$ is the interior operator of the Alexandrov topology induced by R.*

Proof. If $\mathcal{C}$ is a topological interior operator, then from Corollary 1, R is a preorder and $\mathcal{C} = [e]$. Therefore, from Theorem 2, the specialisation preorder induced by $\mathcal{C}$ coincides with R which, in turn, coincides with the specialisation preorder of the Alexandrov topology induced by R.

Obviously, if R is symmetric (as for equivalence relations, thus in Pawlak Rough Set Theory), then $R = R^\smile$, with all the simplifications due to this fact.

4 Approximation by Means of Neighborhoods

Let us now see the relationships between the operators so far discussed and those induced by neighborhoods. Consider a relational structure $\mathbf{N} = \langle U, \wp(U), R \rangle$, with $R \subseteq U \times \wp(U)$. We call it a *relational neighborhood structure*. If $u' \in N \in R(u)$, we say that u' is a *neighbor* and N a *neighborhood* of u. We set $\mathcal{N}_u = R(u)$ and call it the *neighborhood family* of u. The family $\mathcal{N}(U) = \{\mathcal{N}_u : u \in U\}$ is called a *neighborhood system*. Let us define on $\wp(U)$:

(a) $G(X) = \{u : X \in \mathcal{N}_u\}$; (b) $(X) = -G(-X) = \{u : -X \notin \mathcal{N}_u\}$.

Consider the following conditions on $\mathcal{N}(U)$, for any $x \in U$, $A, N, N' \subseteq U$:
1: $U \in \mathcal{N}_x$; **0**: $\emptyset \notin \mathcal{N}_x$; **Id**: if $x \in G(A)$ then $G(A) \in \mathcal{N}_x$;
N1: $x \in N$, for all $N \in \mathcal{N}_x$; **N2**: if $N \in \mathcal{N}_x$ and $N \subseteq N'$, then $N' \in \mathcal{N}_x$;
N3: if $N, N' \in \mathcal{N}_x$, then $N \cap N' \in \mathcal{N}_x$. **N4**: $\exists N, \mathcal{N}_x = \uparrow N = \{N' : N \subseteq N'\}$.
They induce the following properties of the operators G and F (see [14] or [17]):

Condition	Equivalent properties of G	Equivalent properties of F
1	$G(U) = U$	$F(\emptyset) = \emptyset$
0	$G(\emptyset) = \emptyset$	$F(U) = U$
Id	$G(X) \subseteq G(G(X))$	$F(F(X)) \subseteq F(X)$
N1	$G(X) \subseteq X$	$X \subseteq F(X)$
N2	$X \subseteq Y \Rightarrow G(X) \subseteq G(Y)$ $G(X \cap Y) \subseteq G(X) \cap G(Y)$	$X \subseteq Y \Rightarrow F(X) \subseteq F(Y)$ $F(X \cup Y) \supseteq F(X) \cup F(Y)$
N3	$G(X \cap Y) \supseteq G(X) \cap G(Y)$	$F(X \cup Y) \subseteq F(X) \cup F(Y)$

A neighborhood system can be defined by means of a property system $\mathbf{P} = \langle U, M, R \rangle$, in different ways. For instance by setting, for all $g \in U$, $\mathcal{N}_g = \{R^\smile(m) : m \in R(g)\}$. The properties of the operators G and F induced by such neighborhood systems will be studied in another paper. Here we just notice that the philosophy behind this choice is intuitive: we consider neighbors the elements which fulfill the same property. Hence, the extension of a property is a neighborhood, so that a neighborhood family $\mathcal{N}_g$ groups the neighborhoods determined by the properties fulfilled by g. However, in application contexts in which SRSs are involved, it is natural to consider $R(x)$ as the basic neighborhood of x. In [14] (see also [17]), families $\{R_i\}_{i \in I}$ of relations on the same domain are considered, so that one gathers neighborhood families by setting $\mathcal{N}_x = \{R_i(x)\}_{i \in I}$.

If one deals with just one SRS, an obvious way to obtain a neighborhood family is setting $\mathcal{N}_x^R = \uparrow R(x) = \{A : R(x) \subseteq A\}$. The family $\mathcal{N}_{F(R)}(U) = \{\mathcal{N}_x^R : x \in U\}$ will be called *principal neighborhood system generated by R*, by analogy with "principal filter", or *R-neighborhood system*, briefly. We now prove that the operator G induced by an R-neighborhood system coincides with the operator $[e]$ induced by R itself.

Theorem 3. *Let $\mathbf{P} = \langle U, U, R \rangle$ be a SRS and $\mathcal{N}_{F(R)}(U)$ its R-neighborhood system. Let G be the operator induced by $\mathcal{N}_{F(R)}(U)$ and $[e]$ the constructor defined by $\mathbf{P}$. Then for all $A \subseteq U, G(A) = [e](A)$.*

Proof. By definition, in $\mathcal{N}_{F(R)}(U)$ **N4** holds. In any neighborhood system with this property, if $\mathcal{N}_x = \uparrow Z_x$, then $x \in G(A)$ iff $Z_x \subseteq A$, because in this case $A \in \mathcal{N}_x$, too. But in $\mathcal{N}_{F(R)}(U)$, $Z_x = R(x)$. Hence, $G(A) = \{x : R(x) \subseteq A\} = [e](A)$.

An alternative proof runs as follows if R is a preorder:

Lemma 3. *Let* $\mathbf{P} = \langle U, U, R \rangle$ *be a SRS such that R is a preorder, and let $\mathcal{N}_{F(R)}(U)$ be its R-neighborhood system. Let $\preceq$ be the specialisation preorder induced by G. Then $\preceq$ coincides with R.*

Proof. If $x \preceq y$, then $\forall A \subseteq U, x \in G(A)$ implies $y \in G(A)$, so that $A \in \mathcal{N}_x^R$ implies $A \in \mathcal{N}_y^R$. In particular, $R(x) \in \mathcal{N}_x^R$. Hence, $R(x) \in \mathcal{N}_y^R$, so that $R(y) \subseteq R(x)$, because $\mathcal{N}_y^R = \uparrow R(y)$. Since R is a preorder, from Lemma 2 $\langle x, y \rangle \in R$. We omit the obvious reverse implication.

Then from Lemma 3 and Theorem 2 one obtains Theorem 3. Notice, however, that Theorem 3 holds independently of the properties of R. One can verify that in Example 1, $\{a, b\}$ belongs just to $\mathcal{N}_c^R = \uparrow \{b\}$. Hence, $G(\{a, b\}) = \{c\} = [e](\{a, b\})$. On the contrary, $\mathcal{C}(\{a, b\}) = \{b\}$. In fact, R is not a preorder ($\langle c, b \rangle, \langle b, c \rangle \in R$ but $\langle c, c \rangle \notin R$).

To understand the role of the properties of a neighborhood system, as to idempotence we notice that **Id** does not hold in the R-neighborhood system of Example 1: $c \in G(\{a, b\}) = \{c\}$, but $\{c\} \notin \mathcal{N}_c^R = \uparrow \{b\}$. As to deflation, notice that **N1** does not hold: $\{b\} \in \mathcal{N}_c^R$ but $c \notin \{b\}$.

Coming back to relational neighborhood systems one can notice that $\mathbf{N} = \langle U, \wp(U), R \rangle$ is a property system. So it is possible to define *int* and *cl*. What are the relations between the operator G defined on $\mathbf{N}$ *qua* relational neighborhood structure, and the operators *int* and $\mathcal{C}$ defined on $\mathbf{N}$ *qua* property system?

Lemma 4. *(see [17]) Let $\mathbf{N} = \langle U, \wp(U), R \rangle$ be a relational neighborhood structure. For all $X \subseteq U, G(X) = R^\smile(\{X\})$.*

Proof. $G(X) = \{x : X \in \mathcal{N}_x\}$. But $X \in \mathcal{N}_x$ iff $\langle x, X \rangle \in R$ iff $x \in R^\smile(\{X\})$. Hence, $G(X) = \{x : x \in R^\smile(\{X\})\} = R^\smile(\{X\})$.

Theorem 4. *Let $\mathbf{N} = \langle U, \wp(U), R \rangle$ induce a neighborhood system such that **Id**, **N1** and **N2** hold. Then for any $A \subseteq U, int(A) = G(A)$.*

Proof. If $x \in int(A), x \in \langle e \rangle(\{X : R^\smile(\{X\}) \subseteq A\})$. Thus $x \in \bigcup\{R^\smile(\{X\}) : R^\smile(\{X\}) \subseteq A\} = \bigcup\{G(X) : G(X) \subseteq A\}$. Now we prove that $\bigcup\{G(X) : G(X) \subseteq A\} = G(A)$, provided the three conditions of the hypothesis hold. Let $x \in \bigcup\{G(X) : G(X) \subseteq A\}$. Then $x \in G(N)$ for some $N \subseteq U$. Since **Id** holds, $G(N) \in \mathcal{N}_x$. From **N2**, $A \in \mathcal{N}_x$, too, so that $x \in G(A)$. Vice-versa, suppose $x \in G(A)$. But $G(A) \subseteq A$, because **N1** holds. We conclude that $x \in \bigcup\{G(X) : G(X) \subseteq A\}$.

Actually, this is a simplified proof of Lemma 15.14.4 of [17].

5   The Full Relational Environment

We have mentioned that given a SRS $\langle U, U, R \rangle$ if a subset $X \subseteq U$ is represented as a particular relation, then the entire computational machinery can be embedded in the Algebra of Relations.

Definition 4. *A full algebra of binary relations* over a set U, *is an algebra*

$$fullREL(U) = (\wp(U \times U), \cup, \cap, -, \mathbf{1}, \otimes, \smile, \mathbf{1}')$$

where $(\wp(U \times U), \cup, \cap, -, \mathbf{1})$ *is a Boolean algebra of sets,* $\otimes$ *is the relational composition,* $\smile$ *is the inverse and* $\mathbf{1}'$ *is the identity relation.*

Clearly, all elements of $\wp(U \times U)$ are binary relations. The unit $\mathbf{1}'$ is represented by the identity matrix, where the element at row i - column j is 1 if and only if $i = j$. Let $R, S \in \wp(U \times U)$, the composition is defined as follows:

$R \otimes S = \{\langle x, y \rangle \in U \times U : \exists z (\langle x, z \rangle \in R \text{ and } \langle z, y \rangle \in S)\}$.

Composition is simply the Boolean multiplication of matrices. Thus to obtain $\mathbf{R} \otimes \mathbf{S}$ we multiply pointwise row i with column j; if the pointwise Boolean multiplication gives 1 for at least one point, then element at row i and column j of $\mathbf{R} \otimes \mathbf{S}$ is 1. It is 0 otherwise.

EXAMPLE 2. Let $U = \{a, b, c, d\}$.

R	a b c d	**S**	a b c d	**R** $\otimes$ **S**	a b c d
a	1 1 1 1	a	1 1 1 0	a	1 1 1 1
b	0 1 1 0	b	0 0 0 1	b	0 1 0 1
c	0 0 1 0	c	0 1 0 0	c	0 1 0 0
d	0 0 0 1	d	1 0 0 1	d	1 0 0 1

To compute, for instance, the element at row c column b of $\mathbf{R} \otimes \mathbf{S}$, first we take row $\mathbf{R(c)}$, [0010], and column $\mathbf{S}^{\smile}(b)$, [1010]. Then we apply component-wise the logical multiplication to these two Boolean arrays obtaining [0010]. Finally we apply the logical summation to the resulting array and obtain 1.

We have enough instruments to introduce two fundamental operations. We define them on arbitrary binary relations. Assume $R : W \times W'$ and $S : U \times U'$.

$$R \longrightarrow S = -(R^{\smile} \otimes -S), \quad \textit{right residuation of } S \textit{ with respect to } R. \quad (11)$$
$$S \longleftarrow R = -(-S \otimes R^{\smile}), \quad \textit{left residuation of } S \textit{ with respect to } R. \quad (12)$$

The operation (11) is defined only if $|W| = |U|$; (12) is defined only if $|W'| = |U'|$. In particular, if R and S are binary relations on a set U, then (see [17]):

$$R \longrightarrow S = \{\langle a, b \rangle \in U \times U : \forall c \in U (\langle c, a \rangle \in R \Longrightarrow \langle c, b \rangle \in S)\} \quad (13)$$
$$S \longleftarrow R = \{\langle a, b \rangle \in U \times U : \forall c \in U (\langle b, c \rangle \in R \Longrightarrow \langle a, c \rangle \in S)\} \quad (14)$$

It can be shown that $R \longrightarrow S$, is the largest relation Z on U such that $R \otimes Z \subseteq S$, while $S \longleftarrow R$, is the largest relation Z such that $Z \otimes R \subseteq S$.

In [11] (see also [17]), it is possible to see how useful these operations are to compute and analyse, for instance, dependency relations between properties or choices based upon a set of properties. In this paper we want just to show how to use them to compute the operators $[\cdot]$, $\langle\cdot\rangle$.

First, we remind that a set $A \subseteq U$ may be represented by a *right cylinder* $A^c = A \times U = \{\langle a, x\rangle : a \in A, x \in U\}$. Its matrix $\mathbf{A^c}$ has dimensions $|U| \times |U|$ and $\mathbf{R}(a) = [1, 1, ..., 1]$ only if $a \in A$, otherwise $\mathbf{R}(a) = [0, 0, ..., 0]$. In this way sets turn into elements of full algebras of binary relations.

Let us now reconsider definitions (1), (2), (3) and (4). Since $-C$ and $R \otimes C$ output right cylinders whenever C is a right cylinder, we can turn any element x of the sets which appear in the definitions into a pair $\langle x, z\rangle$, where z is a dummy variable representing any element of U. For instance, $\{m : ...\}$, turns into $\{\langle m, z\rangle...\}$ and $g \in X$ turns into $\langle g, z\rangle \in X^C$. In this way one obtains:

$$\langle i\rangle(X^c) = \{\langle m, z\rangle : \exists g(\langle g, m\rangle \in R \wedge \langle g, z\rangle \in X^c)\} = R^\smile \otimes X^c \tag{15}$$

$$\langle e\rangle(X^c) = \{\langle g, z\rangle : \exists m(\langle g, m\rangle \in R \wedge \langle m, z\rangle \in X^c)\} = R \otimes X^c \tag{16}$$

$$[i](X^c) = \{\langle m, z\rangle : \forall g(\langle g, m\rangle \in R \Longrightarrow \langle g, z\rangle \in X^c)\} = R \longrightarrow X^c \tag{17}$$

$$[e](X^c) = \{\langle g, z\rangle : \forall m(\langle g, m\rangle \in R \Longrightarrow \langle m, z\rangle \in X^c)\} = R^\smile \longrightarrow X^c \tag{18}$$

After that, we can compute for instance $[e](\{a, b\})$ in Example 2 using (11): $[e](\{a, b\}^c) = -((\mathbf{R}^\smile)^\smile \otimes -\{a, b\}^c) = -(\mathbf{R} \otimes -\{a, b\}^c)$:

R	a b c d
a	1 0 0 1
b	0 1 1 1
c	0 1 0 0
d	0 1 0 1

{a,b}ᶜ	a b c d
a	1 1 1 1
b	1 1 1 1
c	0 0 0 0
d	0 0 0 0

−{a,b}ᶜ	a b c d
a	0 0 0 0
b	0 0 0 0
c	1 1 1 1
d	1 1 1 1

R ⊗ −{a,b}ᶜ	a b c d
a	1 1 1 1
b	1 1 1 1
c	0 0 0 0
d	1 1 1 1

−(R ⊗ −{a,b}ᶜ)	a b c d
a	0 0 0 0
b	0 0 0 0
c	1 1 1 1
d	0 0 0 0

$(-(\mathbf{R} \otimes -\{a, b\}^c))^\smile(U) = \{c\}$

To obtain $[e](\{a, b\})$, one applies the right Pierce product of the resulting relation to U (or any subset of U).

Relations and manipulations of relations provide all the ingredients for qualitative data analysis: the operations to perform the analysis and the object to be analysed. Central to this task are the implications (residuations) between relations and the composition of relations, which make it possible to define the basic analytic tools. Moreover, the objects to be analysed can be relations themselves. Indeed, right cylinders are just particular instances of relations. For some additional considerations see the next section.

6   Final Remarks and Bibliographic Notes

The bibliography on Modal Logic, Intuitionistic Logic, Adjointness, Kripke models, and Topology is huge. Thus we prefer to address the reader to the comprehensive bibliography and historical notes that can be found in [17].

The basic constructors have been introduced in different fields. We were inspired by the works on formal topology by G. Sambin (see for instance [20]). Together with the sufficiency constructors they have been analysed in [6] and in the context of property systems and neighborhood systems in ([13]). Moreover in [5] two of them where used to define "property oriented concepts", while in [22] the other two have been used to define "object oriented concepts". Eventually, they were fully used in approximation theory in [16]. In the present paper, however, we have not considered the *sufficiency constructors* which are used in R. Wille's Formal Concept Analysis. They are obtained by swapping the positions in the implicative parts of the [·] constructors. Since the [·] and ⟨·⟩ constructors form a square of duality, application direction, isomorphisms and adjointness (see for instance [17]), by adding the sufficiency operators one enters into the cube of oppositions discussed in [3].

A survey on covering-based approximation operators is [23]. In [19] these operators are studied from the point of view of duality and adjoint pairs. In [15] twenty one covering-based approximation operators are interpreted exclusively by means of the four basic constructors. In this way duality and adjointness properties are immediate consequences of the properties of the constructors. Moreover, in view of the clear logical and topological meaning of the four constructors, the meaning of these approximation operators is explained as well.

Finally, the operators introduced in Section 5 have been extended to deal with conceptual patterns within *multi-adjoint formal contexts* in [4] (see also [10] and [2]). This is a promising generalization which on one side is linked to the problem of multi property systems (see [14] and [9] for a first look at the topic), and on the other side to the problem of approximation of relations, which was introduced in [21] and solved for the case of two relations in [12], together with a comparison of rough sets and formal concepts developed within relation algebra (see also [17], Chap. 15.18). Our approach, however, was inspired by the notions of a *weakest pre-specification* and a *weakest post-specification* introduced in [7], by Lambek Calculus and Non Commutative Linear Logic (see [1]).

References

1. Abrusci, V.M.: Lambek Syntactic Calculus and Noncommutative Linear Logic. In: Atti del Convegno Nuovi Problemi della Logica e della Filosofia della Scienza, Viareggio, Italia, Bologna, Clueb, vol. II, pp. 251–258 (1991)
2. Antoni, L., Krajči, S., Krídlo, O., Macek, B., Pisková, L.: Relationship between two FCA approaches on heterogeneous formal contexts. In: Szathmary, L., Priss, U. (eds.) CLA 2012, Universidad de Malaga, pp. 93–102 (2012)
3. Ciucci, D., Dubois, D., Prade, H.: Oppositions in Rough Set Theory. In: Li, T., Nguyen, H.S., Wang, G., Grzymala-Busse, J., Janicki, R., Hassanien, A.E., Yu, H. (eds.) RSKT 2012. LNCS, vol. 7414, pp. 504–513. Springer, Heidelberg (2012)
4. Díaz, J.C., Medina, J.: Multi-adjoint relation equations. Definition, properties and solutions using concept lattices. Information Sciences 252, 100–109 (2013)
5. Düntsch, I., Gegida, G.: Modal-style operators in qualitative data analysis. In: Proc. of the 2002 IEEE Int.al Conf. on Data Mining, pp. 155–162 (2002)
6. Düntsch, I., Orłowska, E.: Mixing modal and sufficiency operators. Bulletin of the Section of Logic, Polish Academy of Sciences 28, 99–106 (1999)

7. Hoare C. A. R., Jifeng H.: The weakest prespecification. Parts 1 and 2. Fundamenta Informaticae 9, 51–84, 217–262 (1986).
8. Huang, A., Zhu, W.: Topological characterizations for three covering approximation operators. In: Ciucci, D., Inuiguchi, M., Yao, Y., Ślęzak, D., Wang, G. (eds.) RSFDGrC 2013. LNCS, vol. 8170, pp. 277–284. Springer, Heidelberg (2013)
9. Khan, M. A., Banerjee, M.: A study of multiple-source approximation systems. In: Peters, J.F., Skowron, A., Słowiński, R., Lingras, P., Miao, D., Tsumoto, S. (eds.) Transactions on Rough Sets XII. LNCS, vol. 6190, pp. 46–75. Springer, Heidelberg (2010)
10. Medina, J., Ojeda-Aciego, M., Ruiz-Calviño, J.: Formal concept analysis via multiadjoint concept lattices. Fuzzy Sets and Systems 160(2), 130–144 (2009)
11. Pagliani, P.: A practical introduction to the modal relational approach to Approximation Spaces. In: Skowron, A. (ed.) Rough Sets in Knowledge Discovery, pp. 209–232. Physica-Verlag (1998)
12. Pagliani, P.: Modalizing Relations by means of Relations: A general framework for two basic approaches to Knowledge Discovery in Database. In: Gevers, M. (ed.) Proc. of the 7th Int. Conf. on Information Processing and Management of Uncertainty in Knowledge-Based Systems, IPMU 1998, Paris, France, July 6-10, pp. 1175–1182. Editions E.D.K (1998)
13. Pagliani, P.: Concrete neighbourhood systems and formal pretopological spaces. In: Calcutta Logical Circle Conference on Logic and Artificial Intelligence, Calcutta, India, October 13-16 (2003); Now Chap. 15.14 of [17]
14. Pagliani, P.: Pretopology and Dynamic Spaces. Proc. of RSFSGRC 2003, Chongqing, R. P. China, 2003. Extended version in Fundamenta Informaticae 59(2-3), 221–239 (2004)
15. Pagliani, P.: Covering-based rough sets and formal topology. A uniform approach (Draft - available at Academia.edu) (2014)
16. Pagliani, P., Chakraborty, M.K.: Information Quanta and Approximation Spaces. I: Non-classical approximation operators. In: Proc. of the IEEE International Conference on Granular Computing, Beijing, R. P. China, July 25-27, vol. 2, pp. 605–610. IEEE, Los Alamitos (2005)
17. Pagliani, P., Chakraborty, M.K.: A geometry of Approximation. Trends in Logic, vol. 27. Springer (2008)
18. Qin, K., Gao, Y., Pei, Z.: On Covering Rough Sets. In: Yao, J., Lingras, P., Wu, W.-Z., Szczuka, M.S., Cercone, N.J., Ślęzak, D. (eds.) RSKT 2007. LNCS (LNAI), vol. 4481, pp. 34–41. Springer, Heidelberg (2007)
19. Restrepo, M., Cornelis, C., Gomez, J.: Duality, conjugacy and adjointness of approximation operators in covering-based rough sets. International Journal of Approximate Reasoning 55, 469–485 (2014)
20. Sambin, G., Gebellato, S.: A Preview of the Basic Picture: A New Perspective on Formal Topology. In: Altenkirch, T., Naraschewski, W., Reus, B. (eds.) TYPES 1998. LNCS, vol. 1657, pp. 194–207. Springer, Heidelberg (1999)
21. Skowron, A., Stepaniuk, J.: Approximation of Relations. In: Proc. of the Int. Workshop on Rough Sets and Knowledge Discovery, Banff, pp. 161–166. Springer (October 1993)
22. Yao, Y.Y., Chen, Y.H.: Rough set approximations in formal concept analysis. In: Proc. of 2004 Annual Meeting of the North American Fuzzy Inf. Proc. Society (NAFIPS 2004), IEEE Catalog Number: 04TH8736, pp. 73–78 (2004)
23. Yao, Y.Y., Yao, B.: Covering based rough sets approximations. Information Sciences 200, 91–107 (2012)
24. Zakowski, W.: Approximations in the space (U, Π). Demonstratio Mathematica 16, 761–769 (1983)

SQL-Based KDD with Infobright's RDBMS: Attributes, Reducts, Trees*

Jakub Wróblewski[1] and Sebastian Stawicki[2]

[1] Infobright Inc.
ul. Krzywickiego 34, lok. 219, 02-078 Warsaw, Poland
[2] Institute of Mathematics, University of Warsaw
ul. Banacha 2, 02-097 Warsaw, Poland
jakubw@infobright.com, stawicki@mimuw.edu.pl

Abstract. We present a framework for KDD process implemented using SQL procedures, consisting of constructing new attributes, finding rough set-based reducts and inducing decision trees. We focus particularly on attribute reduction, which is important especially for high-dimensional data sets. The main technical contribution of this paper is a complete framework for calculating short reducts using SQL queries on data stored in a relational form, without a need of any external tools generating or modifying their syntax. A case study of large real-world data is presented. The paper also recalls some other examples of SQL-based data mining implementations. The experimental results are based on the usage of Infobright's analytic RDBMS, whose performance characteristics perfectly fit the requirements of presented algorithms.

Keywords: KDD, Rough sets, Reducts, Decision trees, Feature extraction, SQL, High-dimensional data.

1 Introduction

The process of knowledge discovery in databases consists of many stages, such as data integration and cleaning, feature extraction, dimension reduction, model construction. With growing sizes of data, there are a number of approaches attempting to mine the available sources (often in a form of multitable relational databases) by means of database query languages rather than operating directly on files. Such approaches utilize additional tables containing some precalculated or partial results used in implemented algorithms' runs; the other trend is to split the main task into subtasks solved by relatively basic SQL queries. Our work concentrates on implementation of known KDD tools (feature extraction, decision reducts, decision trees) directly in SQL-based environment.

* The second author was partly supported by Polish National Science Centre (NCN) grants DEC-2011/01/B/ST6/03867 and DEC-2012/05/B/ST6/03215, and by National Centre for Research and Development (NCBiR) grant SP/I/1/77065/10 by the strategic scientific research and experimental development program: "Interdisciplinary System for Interactive Scientific and Scientific-Technical Information".

M. Kryszkiewicz et al. (Eds.): RSEISP 2014, LNAI 8537, pp. 28–41, 2014.

Most of data mining tools are based on a paradigm of one decision table. On the other hand, large data sets often have a form of relational, multi-table databases. In these cases a data mining process is preceded by preprocessing steps (based on domain-dependent knowledge), usually done interactively by a user. After these steps one joint data table is created. In [28] we presented an algorithm for automatic induction of such table, employing genetic algorithms and SQL-based attribute quality measures.

The next step in KDD process is an attribute subset selection [12,13]. It provides the basis for efficient classification, prediction and approximation models. It also provides the users with an insight into data dependencies. In this paper, we concentrate on attribute subset selection methods originating from the theory of rough sets [16]. There are numerous rough set-based algorithms aimed at searching for so called decision reducts – irreducible subsets of attributes, which (according to one of its most popular formulations) enable to discern sufficient amount of pairs of objects belonging to different decision classes. For a number of purposes, it is also useful to search for multiple decision reducts, containing possibly least overlapping sets of attributes [25]. This way, the users can observe multiple subsets of attributes providing roughly the same or complementary information. This may be important, e.g., for a design of ensembles of classifiers learnt over particular decision reducts, as combining classifiers is efficient especially if they are substantially different from each other [11].

There are numerous approaches to search for the shortest (or optimal with respect to various evaluation measures) reducts in large data [3]. One can combine basic top-down and bottom-up heuristics controlling a flow of adding and removing attributes with some elements of randomization, parallelism and evolution [17]. One can also use some modern computational paradigms, such as MapReduce [4], in order to scale computations with respect to the number of objects [30]. Finally, when analyzing texts [6], signals [24], images [26], microarrays [27] etc., some techniques aimed at attribute sampling and attribute clustering can help us to scale with respect to the number of dimensions [7].

In this paper, we follow another way of scaling decision reduct computations, namely, utilizing SQL-based scripts running against analytic databases. The idea is not new [5,10]. However, it seems that so far it did not get enough attention with respect to such aspects as: 1) easiness of operating with SQL while implementing initial designs of new attribute reduction approaches, 2) importance of choosing appropriate database systems and an efficient form of storing data information, and 3) performance benefits while searching for multiple reducts. Our methodology bases on EAV (Entity-Attribute-Value) data representation, which is suitable for storing sparse, multidimensional data in RDBMS.

The next step is an induction of predictive model in a form of decision rules or a decision tree. The idea of decision tree induction using SQL is not new [14]. In [10] the tree induction process is rewritten in a form of queries executed against EAV. In each induction step a new level of tree is added and the approach is based on a rule of choosing the most promising attribute from data. The authors introduce a measure which evaluates attributes and expresses their usefulness according to a simple heuristic approach. A measure value is computed using a single SQL statement that evaluates usefulness of attributes (all at once) based on analysis of diversity of their values' averages in particular decision classes. The major advantage of the heuristic is the speed of execution, especially for the RDBMS solutions designed for such types of aggregations. Then using the method a cut on its range is created in order to possibly maximally separate objects from different decision classes. One of the main assets of using EAV model is used here: it enables to significantly limit number of queries by assigning the cuts for all nodes at a given level with a single SQL query.

The paper is organized as follows: Section 2 describes Infobright's database system – the data storage engine of our choice. Section 3 outlines the task of feature extraction. Section 4 discusses an EAV format with a special case of so called discernibility tables. Section 5 presents queries realizing some atomic rough set operations for EAV representation. Section 6 introduces some SQL-based attribute reduction scripts. Section 7 refers to the task of decision tree induction. Section 8 outlines our experimental framework and results obtained for the task of attribute reduction. Section 9 concludes the paper.

2   RDBMS Engine

In our research, we used the Infobright's database system, whose algorithms accelerate performance of SQL statements by an internal usage of rough set principles of computing with granulated data [21,22]. It is also important to mention about Infobright's approximate query extensions [8,19]. Although those extensions were not yet utilized in our attribute reduction procedures, they can be useful for bigger data sets in future. As for our current framework, the following aspects of Infobright are the most significant:

- Data compression [1,9]: Storing and processing large amount of data is feasible. For example, for the AAIA'2014 data set (see section 8.1), a discernibility table in EAV format of raw size 40 GB was compressed to 1.7 GB after loading into the Infobright's data table.
- SQL performance: Infobright's layer of rough information about the contents of granulated data is automatically employed by internal mechanisms to minimize the data access intensity while executing queries. These mechanisms turn out to be especially efficient for complex analytic types of SQL statements, such as those reported in this paper.
- Ease of usage: Unlike some other SQL and NoSQL data processing platforms, Infobright does not require any kind of sophisticated tuning. In particular, all existing parallel and caching mechanisms are ready to be used automatically. It helped a lot in an iterative design of our SQL scripts.

3   Construction of New Attributes

In many real-life KDD tasks the input data is given as a multitable relational database rather than a single decision table. In general we can assume, that any descriptive or predictive problem in KDD [12] is concerned with a selected (or added) decision attribute in one of existing or newly created data tables (i.e. concerns one entity in relational database). If so, we have a distinguished table, which can be analyzed using standard data mining tools. On the other hand, only a part of information available in the database is collected in this table. Most of knowledge is distributed over relations and other tables.

In [28] we proposed a framework to mine dimension tables in a star (or snowflake) schema. The framework consists of a set of possible transitions (joins, aggregations), defined based on domain-dependent expert knowledge as well as on the interpretation of relations in database, and also quality measures to evaluate potential new attributes. Such attributes may be evaluated without actual materialization, basing on definitions expressed in SQL.

The main problem one will face using this scheme is a huge number of possible transitions (new attributes). However, since the process of attributes' quality calculation is faster than creation of classification algorithm, one can try to find good examples of attributes in reasonable time. An adaptive process of discovering new features based on genetic algorithms was proposed and implemented in [28]. To formalize this approach basing on rough set theory, a new notion of relational information system was introduced. An adaptive classification system based on predictive attributes' subsets quality measure was successfully used to analyze a real-life database. It is worth noting that analysis needs only a little preprocessing by a user; there is no need to create joint tables etc. since all operations are performed in SQL on original data. Only the final decision table consisting all derived attributes should be materialized in either classic or EAV form to perform further steps of data mining.

4   EAV and Discernibility Tables

In a case of Entity-Attribute-Value (EAV) form of data storage, a decision table is stored (e.g. in a relational database) as a collection of tuples (e, a, v), where e is an identifier of object (e.g. a row number in a classical decision table), a is an identifier of attribute, and v is an attribute value. EAV table is a convenient form for some cases of data mining/KDD tasks. In our algorithms, the main advantages of EAV are as follows:

- It is easily expandable, both in terms of entities (rows in a classical version of decision table) and attributes (columns). Adding new attributes does not require changing database schema nor reloading data. In particular, it may be useful to flexibly write down the results of the SQL-based feature extraction procedures referred in the previous section.
- Although it requires more space in general (three numbers for row/column, comparing to one in a classical case), but for sparse data containing a lot of NULL (unknown) values or for imbalanced attributes (when only positive information is actually stored), this form may occupy much less space. This makes various extensions of EAV format quite convenient in projects related to, e.g., scientific document indexing [20].

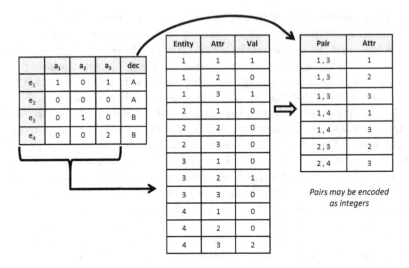

Fig. 1. Preparing a discernibility table (right) based on decision attribute of the original data table (left) and attribute values stored in EAV form (middle)

- Attributes are identified as numbers, and may be processed by SQL, allowing e.g. to operate on subsets of them, constructed dynamically or stored in tables. There is no need to construct a modified SQL query by external tools when another subset of attributes is analyzed. This makes EAV quite popular in various machine learning implementations, e.g., various extensions of association/decision rule mining procedures [18].

The above advantages has led also to EAV-based implementations of some rough set methods [10,23]. As for the task of calculation of decision reducts, we actually do not need to use an original decision table content. It is sufficient to store so called discernibility table in an EAV-like form, containing an identifier of an object pair and a number of the attribute:

```
CREATE TABLE discern (pair BIGINT, a INT);
```

For an input decision table of the form $\mathbf{A} = (U, A \cup \{dec\})$ [16], the content of discernibility EAV table is defined as follows:

$$\forall_{e_1,e_2 \in U; a \in A} \; a(e_1) \neq a(e_2) \wedge dec(e_1) \neq dec(e_2) \Leftrightarrow (id(e_1, e_2), a) \in discern \quad (1)$$

where dec is a distinguished decision attribute and $id(e_1, e_2) = e_1 + |U|e_2$ is a function encoding a pair of row numbers into one unique identifier. See Figure 1 for an outline of the whole process. Note that such created data table straightforwardly corresponds to the content of a discernibility matrix for $\mathbf{A}$ and, therefore, enables to implement analogous operations of attribute reduction [3].

In many cases the discernibility table is too large to be explicitly stored. As its size is $O(|U|^2|A|)$, data sets with too many objects must be analyzed by methods avoiding explicit generation of discernibility structures. On the other hand, even high-dimensional data (with large $|A|$) with moderate number of objects ($|U|$ up to an order of tens of thousands) is still easy to store in an ordinary RDBMS. Additionally, if attribute values are sparse (majority of 0 or NULL values) and the decision classes are imbalanced, then the actual number of pairs of objects to be discerned, i.e., the rows in the EAV form of the discernibility table may be far less than the upper bound.

5   Rough Set Notions in SQL

Many of rough set related tasks may be easily done by SQL queries operating on the EAV discernibility tables. Below we present some examples.

5.1   Superreducts

Let @*all_pairs* be a variable storing a number of all pairs to be discerned:

```
SET @all_pairs = SELECT COUNT(DISTINCT pair) FROM discern;
```

Then the following query will check whether a subset of attributes $a_1, a_2, ...a_k \in A$ is sufficient to discern all pairs of objects:

```
SELECT COUNT(DISTINCT pair) FROM discern WHERE a IN (a1,...,ak);
```

If the answer is equal to @*all_pairs*, then the subset is a superreduct.

As already mentioned, some SQL-based reduct calculation techniques are already known [5]. However, those methods usually require dynamic construction of queries by an external tool (e.g. a Java or C++ program). In our case, analogously to other EAV-based approaches [10,23], all queries are of the same form regardless of the considered subsets (or families of subsets) of attributes. A subset to be checked can be stored in a separate table and provided in form of a subquery or join. Below we denote such table as *candidates*:

```
SELECT COUNT(DISTINCT pair) FROM discern
   WHERE a IN (SELECT a FROM candidates);
```

This observation could make the process of data analysis much simpler for SQL-based users. The whole process of attribute reduction can be done automatically or semi-automatically by a set of queries and SQL procedures.

5.2   Decision Reducts and Approximate Reducts

To make sure a superreduct is actually a reduct, we should try to exclude all attributes one by one and make sure that the remaining subset is not sufficient. It can be done by single query, where $t2.a$ is an attribute to be excluded:

```
SELECT t2.a AS a_excl, COUNT(DISTINCT pair) AS cnt
  FROM discern AS t1 JOIN candidates AS t2
  WHERE t1.a IN (SELECT a FROM candidates) AND t1.a <> t2.a
GROUP BY a_excl HAVING cnt = @all_pairs;
```

If the answer is empty, then the *candidates* set is a reduct. Similar procedure may be applied to check whether a subset is an approximate (super)reduct [15]. It is sufficient to replace @*all_pairs* in the above queries with @*nearly_all_pairs* = @*all_pairs* $\cdot$ $(1 - \varepsilon)$ for a given error rate ε.

5.3 Pairwise Cores

Unlike classical cores which do not occur in data sets so often, pairwise cores are pairs of attributes such that each reduct needs to include at least one of them [29]. Such pairs can be easily calculated as follows:

```
SELECT t1.pair, MIN(a), MAX(a) FROM discern AS t1,
  (SELECT pair, COUNT(*) AS cnt FROM discern
   GROUP BY pair HAVING cnt=2) AS t2
WHERE t1.pair=t2.pair GROUP BY t1.pair;
```

5.4 Quality Measures

One of popular methods of evaluating usefulness of attributes is to calculate how many pairs an attribute can discern. When some attributes are already chosen, then we should count only pairs which are not discerned yet.

Let X be a set of attributes already selected. Generating a measure value for the rest of attributes can be done as follows: get all pairs which are discerned by X; find all remaining pairs (using WHERE NOT EXISTS or an outer join with IS NULL); find all attributes discerning these pairs, weighted by a number of pairs an attribute can discern. An outer join version of SQL query generating all candidate attributes is presented below (by ..X.. we denote an explicit list or a subquery generating attribute numbers from set X):

```
SELECT t1.a, COUNT(DISTINCT pair) AS weight
FROM discern AS t1 LEFT JOIN
(SELECT DISTINCT pair FROM discern WHERE a IN (..X..)) AS t2
ON t1.pair = t2.pair WHERE t2.pair IS NULL
GROUP BY t1.a ORDER BY weight DESC;
```

In [10] another attribute measure was proposed, calculated directly from the decision table stored as EAV and suitable especially for continuous values. Assume that t is a data table (e, a, v), and a decision value for every object e is stored in a separate table *decisions*:

```
SELECT a, STDDEV(avg_val) AS quality FROM
  ( SELECT t.a, t1.dec, avg(t.v) AS avg_val FROM t, decisions t1
    WHERE t.e = t1.e GROUP BY a, dec ) AS t2
GROUP BY a ORDER BY quality DESC;
```

Input: discernibility table $discern(pair, a)$
  initial set of attributes in table $candidates(set\_id, a)$
  where set_id is 0 for the initial set
Intermediate table: $intermediate(set\_id, a)$
Output: additional attributes in table $candidates$

```
CREATE PROCEDURE bottom_up()
BEGIN
  REPEAT
    DELETE FROM intermediate;
    // Find and add the best attribute
    INSERT INTO intermediate
      SELECT 0, a_to_add FROM
        (SELECT t1.a AS a_to_add, COUNT(DISTINCT t1.pair) AS c
        FROM discern AS t1 LEFT JOIN
          (SELECT DISTINCT pair FROM discern
          WHERE a IN (SELECT a FROM candidates)) AS t2
        ON t1.pair = t2.pair WHERE t2.pair IS NULL
        GROUP BY t1.a ORDER BY c DESC LIMIT 1
        ) t3;
    INSERT INTO candidates SELECT * FROM intermediate;
  UNTIL (SELECT COUNT(*) FROM intermediate) = 0 END REPEAT;
END
```

Fig. 2. SQL procedure for generating a good superreduct by a greedy algorithm

6 SQL-Based Attribute Reduction

6.1 Bottom-Up Building of a Candidate Set of Attributes

A known method of attribute selection in data mining [3] is to start with a set of attributes (which may consists of core attributes, other arbitrary set or even be empty) and then to add the best (with respect to some measure) attributes one by one, until the resulting set is sufficient to discern all pairs of objects in the discernibility table.

We adopted an attribute weight given by the query presented at the end of Section 5 as a measure for greedy bottom-up construction of a possibly small set of attributes. In each step we are adding an attribute which discerns the largest amount of pairs which are not discerned yet. We stop when all pairs are discerned, as shown in Figure 2.

The above procedure results with a superreduct, often containing redundant elements. Thus, the attribute set stored in *candidates* must be reduced.

6.2 Top-Down Reduction of a Candidate Set of Attributes

The procedure presented in Figure 4 is a complete, working algorithm generating all reducts – subsets of a given candidate set. See also Figure 3 for an outline of a single step. As the resulting set of reducts (as well as computation time) can be exponential, it is advisable to start with good candidate sets ("nearly reducts"), e.g. found by a bottom-up procedure described in Section 6.1. An outline of this algorithm is as follows:

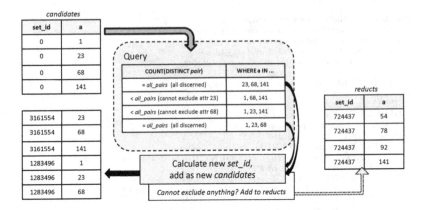

Fig. 3. A single step of the attribute reduction process

1. Select the first set of attributes from a queue (table *candidates*).
2. Use it to generate all superreducts shorter by a single attribute.
3. Insert these superreducts into *intermediate* table.
4. If the result of step 2 was empty, then the set chosen in step 1 is a reduct –
   insert it into an output *reduct* table (omit repetitions).
5. Add the new subsets from *intermediate* table into *candidates*.
6. Continue from step 1, until the *candidate* table is empty.

Steps 2 and 3 can be done by a single SQL query. Note that analyzing a whole family of subsets in a single query may provide performance advantages connected with internal caching mechanisms of a query engine.

The only nontrivial technique that was not described in earlier sections is a mechanism of ensuring uniqueness of subsets stored in *candidates* and *reducts*. In general, it is hard in SQL to operate on a set of rows and to compare it with another set. However, every subset of attributes in *candidates* and *reducts* can be labeled with an arbitrary integer identifier *set_id*. The initial candidate set may be labeled by 0, and any other set is numbered as *parent_id XOR CRC32(a)*, where a is an attribute just excluded from the parent set. As binary *XOR* is symmetrical, each subset has a unique identifier, no matter what was the order of attribute exclusion which led to it (assuming *CRC32* is strong enough as the hashing function). Figure 4 displays the complete algorithm.

7   Induction of Decision Trees

A set of good attributes (e.g. a reduct) may be used to generate a predictive data model in a form of decision rules or trees. An idea of using SQL-based solution in the process of a decision tree construction was presented in [14]. More specifically, the authors investigated the problem of minimizing the number of database SQL queries needed in order to find the binary partition of continuous attribute domain that is close to the optimal (with respect to a given quality measure) one.

Input: discernibility table $discern(pair, a)$
 initial set of attributes in table $candidates(set\_id, a)$
 where set_id is 0 for the initial set
 $@all\_pairs$ is equal to a number of discerned pairs of objects
Intermediate table: $intermediate(set\_id, a)$
Output: table $reducts(set\_id, a)$, initially empty

```
CREATE PROCEDURE reduce()
BEGIN
  WHILE((SELECT COUNT(*) FROM candidates) ≠ 0) DO
    SET @cur_id = (SELECT MIN(set_id) FROM candidates);
    // Exclude single attributes, if possible:
    INSERT INTO intermediate
      SELECT (@cur_id XOR CRC32(a_excl)), a FROM candidates JOIN
      (SELECT candidates.a AS a_excl, COUNT(DISTINCT pair) AS cnt
        FROM discern, candidates WHERE
        discern.a IN (SELECT a FROM candidates WHERE set_id = @cur_id)
        AND discern.a ≠ candidates.a AND set_id = @cur_id
        GROUP BY candidates.a HAVING cnt = @all_pairs
      ) t1
      WHERE set_id = @cur_id AND a ≠ a_excl;
    // Reduct, if nothing was excluded:
    INSERT INTO reducts SELECT * FROM candidates
      WHERE set_id = @cur_id
      AND (SELECT COUNT(*) FROM intermediate) = 0
      AND set_id NOT IN (SELECT DISTINCT set_id FROM reducts);
    // Cleanup:
    DELETE FROM candidates WHERE set_id = @cur_id;
    INSERT INTO candidates SELECT * FROM intermediate
      WHERE set_id NOT IN (SELECT DISTINCT set_id FROM candidates);
    DELETE FROM intermediate;
  END WHILE;
END
```

Fig. 4. SQL procedure for generating all reducts which are subsets of a given set

The approach presented is highly dependent on the number of attributes due to the fact that the computations are performed for each attribute independently. A different approach presenting a procedure of decision tree induction, based on SQL and EAV form, was proposed in [10]. It was based on a greedy rule of choosing the most promising attribute from data. Then we create a cut on its range in order to possibly maximally separate objects from different decision classes. One of the main assets of using EAV model is used here: it enables to significantly limit number of queries by assigning the cuts for all nodes at a given level with a single SQL query. The algorithm consists of:

 i Finding the best cut for the whole data set in the root;
 ii For a tree of depth $i - 1$, finding the best cuts producing nodes at depth i;
 iii Stopping if a predefined depth k of the tree is reached.

We consider binary decision class to avoid introducing potentially complicated rule of cut creation for the chosen attribute. For attribute which was chosen using the measure from Section 5.4 we calculate the cut as the arithmetic average of all averages from decision classes.

   The whole process consists of four database tables: the main table t in EAV format, the additional table *decisions* storing decision labels for objects, a temporary table *tmp* for statistics in subgroups, and *distr* table for storing distribution (assignment) of objects to the deepest level of the tree. For a detailed description and complete SQL process please refer to [10].

8   Experimental Framework for Reduct Finding

8.1   A Data Set

The AAIA'2014 data set used in our experiments consists of 50,000 rows corresponding to reports from actions carried out by the Polish State Fire Service.[1] It has 11,852 sparse attributes and an imbalanced decision attribute. The original table (CSV source) has 1.1 GB while its discernibility counterpart occupies 40 GB, which is still feasible. It makes it a suitable case study for the SQL-based algorithmic framework proposed in this paper.

Let us also add that the data mining competition based on the AAIA'2014 data set has been aimed at searching for ensembles of possibly small attribute subsets that are able to cooperate with each other by means of locally induced classifiers. This is quite a difference comparing to other high-dimensional data mining competitions related primarily to classification accuracy, with no explicit evaluation of the stage of attribute selection [6,27].

8.2   Discernibility Sampling

Pairs of objects (which correspond to a single cell in discernibility table) are more important if they are hard to be discerned, i.e. if there is few attributes discerning them. Thus, to obtain a sample of discernibility table, it is reasonable to select the most important pairs.

The example below generates a 1% sample (100,000 pairs) from an original discernibility table (nearly 10,000,000 pairs) of AAIA'2014 data:

```
INSERT INTO disc_sample
SELECT t1.pair, t1.a FROM discern AS t1,
  (SELECT pair, COUNT(*) cnt FROM discern
   GROUP BY pair ORDER BY cnt LIMIT 100000) AS t2
WHERE t1.pair=t2.pair ORDER BY 2,1;
```

Such table does not reflect to any sampling of original rows, but it gathers all the pairs of rows which are hard to be discerned. If a set $R \subseteq A$ is a reduct of such *disc_sample*, then it is usually a subreduct of the original *discern*, but usually close enough to be a good base for bottom-up building of a candidate. On the other hand, if such R happens to be a superreduct of *discern*, then it is for sure a proper reduct. Checking discernibility (see Section 5) is much cheaper than ensuring that R is a reduct.

---

[1] https://fedcsis.org/2014/dm_competition

8.3 Performance Results

The AAIA'2014 data set was loaded into the EAV table and then transformed into the discernibility table as described in Section 4. The procedure of calculating reducts consisted of generating good candidates on a sample and then extending them using the whole discernibility table. For a sampled case, the bottom-up procedure takes 18 seconds, the top-down reduction takes about 40 seconds, and it produces a few reducts of length 15.

The bottom-up greedy procedure on the whole table takes 45-60 minutes and produces sets of about 20 attributes. The top-down reduction started from a superreduct found by the greedy algorithm takes about 30 minutes and produces one or more reducts usually consisting of 18 attributes. For comparison, the procedure started from a subset chosen randomly from 30 attributes with the best scores of previously discussed quality measure lasts 2-3 times longer and usually produces bigger reducts (19-20 attributes).

9 Conclusions

The presented ideas may be treated as a step toward understanding to what extent SQL-level interfaces can be useful when designing, implementing and performing KDD operations over large data sets, particularly in relation to rough set-based methods of attribute reduction.

We noted that it is possible to construct fast-performing scripts based on automatically generated analytic queries, given a proper choice of an underlying RDBMS technology. The obtained performance results suggest that it may be useful for database systems to introduce approximate query extensions [8]. Moreover, using SQL-based algorithms as a starting point, one can develop their more optimal realizations using lower-level languages [30].

In some cases, specialized implementations or solutions that use large computational clusters may outperform the presented SQL-based algorithms both in terms of the achieved performance or the time complexity of a model induction. However, this will usually require much more effort to implement them than to use the embedded features of the data storage system. Therefore, these two approaches cannot be considered as competing with each other but rather as complementary solutions. The same complementarity between SQL-based and non-SQL-based methodologies can be seen also for other data mining tasks [2] and complex deployments of data processing environments [20].

In future, we will continue working on aggregate SQL statements evaluating, e.g., multiple candidate subsets of attributes. It may be also interesting to consider more dynamic scenarios, where new attributes are extracted and added while running the attribute selection procedures [26,28]. Another important future aspect refers to the already-mentioned usage of approximate extensions of standard SQL language in the KDD tasks. Last but not least, we are intending to develop SQL-based procedures for learning more complex decision models from data such as, e.g., ensembles of reducts or decision forests [7].

40   J. Wróblewski and S. Stawicki

References

1. Apanowicz, C., Eastwood, V., Ślęzak, D., Synak, P., Wojna, A., Wojnarski, M., Wróblewski, J.: Method and system for data compression in a relational database. US Patent 8,700,579 (2014)
2. Bae, S.-H., Choi, J.Y., Qiu, J., Fox, G.C.: High Performance Dimension Reduction and Visualization for Large High-dimensional Data Analysis. In: Proc. of HPDC, pp. 203–214 (2010)
3. Bazan, J.G., Nguyen, H.S., Nguyen, S.H., Synak, P., Wróblewski, J.: Rough Set Algorithms in Classification Problem. In: Rough Set Methods and Applications, pp. 49–88. Physica-Verlag (2000)
4. Dean, J., Ghemawat, S.: MapReduce: Simplified Data Processing on Large Clusters. In: OSDI, pp. 137–150 (2004)
5. Hu, X., Han, J., Lin, T.Y.: A New Rough Sets Model Based on Database Systems. Fundamenta Informaticae 59(2-3), 135–152 (2003)
6. Janusz, A., Nguyen, H.S., Ślęzak, D., Stawicki, S., Krasuski, A.: JRS'2012 Data Mining Competition: Topical Classification of Biomedical Research Papers. In: Yao, J., Yang, Y., Słowiński, R., Greco, S., Li, H., Mitra, S., Polkowski, L. (eds.) RSCTC 2012. LNCS, vol. 7413, pp. 422–431. Springer, Heidelberg (2012)
7. Janusz, A., Ślęzak, D.: Rough Set Methods for Attribute Clustering and Selection. Applied Artificial Intelligence 28(3), 220–242 (2014)
8. Kowalski, M., Ślęzak, D., Synak, P.: Approximate Assistance for Correlated Subqueries. In: Proc. of FedCSIS, pp. 1455–1462 (2013)
9. Kowalski, M., Ślęzak, D., Toppin, G., Wojna, A.: Injecting Domain Knowledge into RDBMS – Compression of Alphanumeric Data Attributes. In: Kryszkiewicz, M., Rybinski, H., Skowron, A., Raś, Z.W. (eds.) ISMIS 2011. LNCS, vol. 6804, pp. 386–395. Springer, Heidelberg (2011)
10. Kowalski, M., Stawicki, S.: SQL-Based Heuristics for Selected KDD Tasks over Large Data Sets. In: Proc. of FedCSIS, pp. 303–310 (2012)
11. Kuncheva, L.I.: Combining Pattern Classifiers: Methods and Algorithms. Wiley (2004)
12. Liu, H., Motoda, H. (eds.): Feature extraction, construction and selection – a data mining perspective. Kluwer Academic Publishers, Dordrecht (1998)
13. Liu, H., Motoda, H.: Computational Methods of Feature Selection. Chapman & Hall/CRC (2008)
14. Nguyen, H.S., Nguyen, S.H.: Fast split selection method and its application in decision tree construction from large databases. Int. J. Hybrid Intell. Syst. 2(2), 149–160 (2005)
15. Nguyen, H.S., Ślęzak, D.: Approximate reducts and association rules. In: Zhong, N., Skowron, A., Ohsuga, S. (eds.) RSFDGrC 1999. LNCS (LNAI), vol. 1711, pp. 137–145. Springer, Heidelberg (1999)
16. Pawlak, Z., Skowron, A.: Rudiments of Rough Sets. Information Sciences 177(1), 3–27 (2007)
17. Rahman, M.M., Ślęzak, D., Wróblewski, J.: Parallel Island Model for Attribute Reduction. In: Pal, S.K., Bandyopadhyay, S., Biswas, S. (eds.) PReMI 2005. LNCS, vol. 3776, pp. 714–719. Springer, Heidelberg (2005)
18. Sarawagi, S., Thomas, S., Agrawal, R.: Integrating Association Rule Mining with Relational Database Systems: Alternatives and Implications. Data Min. Knowl. Discov. 4(2/3), 89–125 (2000)

19. Ślęzak, D., Kowalski, M.: Towards approximate SQL – infobright's approach. In: Szczuka, M., Kryszkiewicz, M., Ramanna, S., Jensen, R., Hu, Q. (eds.) RSCTC 2010. LNCS, vol. 6086, pp. 630–639. Springer, Heidelberg (2010)
20. Ślęzak, D., Stencel, K., Nguyen, H.S.: (No)SQL Platform for Scalable Semantic Processing of Fast Growing Document Repositories. ERCIM News 2012(90) (2012)
21. Ślęzak, D., Synak, P., Wojna, A., Wróblewski, J.: Two Database Related Interpretations of Rough Approximations: Data Organization and Query Execution. Fundamenta Informaticae 127(1-4), 445–459 (2013)
22. Ślęzak, D., Wróblewski, J., Eastwood, V., Synak, P.: Brighthouse: An Analytic Data Warehouse for Ad-hoc Queries. PVLDB 1(2), 1337–1345 (2008)
23. Świeboda, W., Nguyen, H.S.: Rough Set Methods for Large and Spare Data in EAV Format. In: Proc. of RIVF, pp. 1–6 (2012)
24. Szczuka, M.S., Wojdyłło, P.: Neuro-wavelet classifiers for EEG signals based on rough set methods. Neurocomputing 36(1-4), 103–122 (2001)
25. Widz, S., Ślęzak, D.: Rough Set Based Decision Support – Models Easy to Interpret. In: Selected Methods and Applications of Rough Sets in Management and Engineering, pp. 95–112. Springer (2012)
26. Widz, S., Ślęzak, D.: Granular attribute selection: A case study of rough set approach to MRI segmentation. In: Maji, P., Ghosh, A., Murty, M.N., Ghosh, K., Pal, S.K. (eds.) PReMI 2013. LNCS, vol. 8251, pp. 47–52. Springer, Heidelberg (2013)
27. Wojnarski, M., et al.: RSCTC'2010 Discovery Challenge: Mining DNA Microarray Data for Medical Diagnosis and Treatment. In: Szczuka, M., Kryszkiewicz, M., Ramanna, S., Jensen, R., Hu, Q. (eds.) RSCTC 2010. LNCS, vol. 6086, pp. 4–19. Springer, Heidelberg (2010)
28. Wróblewski, J.: Analyzing relational databases using rough set based methods. In: Proc. of IPMU, vol. 1, pp. 256–262 (2000)
29. Wróblewski, J.: Pairwise Cores in Information Systems. In: Proc. of RSFDGrC, vol. 1, pp. 166–175 (2005)
30. Zhang, J., Li, T., Ruan, D., Gao, Z., Zhao, C.: A parallel method for computing rough set approximations. Information Sciences 194, 209–223 (2012)

From Vagueness to Rough Sets
in Partial Approximation Spaces

Zoltán Ernő Csajbók[1] and Tamás Mihálydeák[2]

[1] Department of Health Informatics, Faculty of Health, University of Debrecen,
Sóstói út 2-4, H-4400 Nyíregyháza, Hungary
csajbok.zoltan@foh.unideb.hu
[2] Department of Computer Science, Faculty of Informatics, University of Debrecen,
Kassai út 26, H-4028 Debrecen, Hungary
mihalydeak.tamas@inf.unideb.hu

Abstract. Vagueness has a central role in the motivation basis of rough set theory. Expressing vagueness, after Frege, Pawlak's information-based proposal was the boundary regions of sets. In rough set theory, Pawlak represented boundaries by the differences of upper and lower approximations and defined exactness and roughness of sets via these differences. However, defining exactness/roughness of sets have some possibilities in general. In this paper, categories of vagueness, i.e., different kinds of rough sets, are identified in partial approximation spaces. Their formal definitions and intuitive meanings are given under sensible restrictions.

Keywords: Vagueness, partial approximation spaces, categories of rough sets.

1 Introduction

Although two sets can be unequal in set theory, they can actually be equal *approximately* depending on what we know about them [12]. Pawlak has already constituted this idea in the notion of "rough equality of sets" in his seminal paper [11]. However, defining *exactness* or, from the other side, *roughness* of sets in general settings have many difficulties.

This paper follows a mixed attitude in the sense of [2], an "a priori" one with regard to definability and an "a posteriori" one with regard to exactness/roughness. Roughly speaking, the set of definable sets is given beforehand. Meanwhile, exact/rough sets are defined by means of properties of lower/upper approximations. Then, the question is what properties of the set of exact/rough sets have.

Pawlak's approach is basically set theoretical [11, 12]. On the other hand, algebraic aspects of rough sets have been studied intensively from the beginning as well. A survey of classical algebraic results can be found, e.g., in [7], whereas [1] contains a comprehensive extension of the initiative proposed in [8].

In Section 2, some reasonable generalizations of the boundary of sets are investigated in partial approximation spaces. In Section 3, first, the categories of rough sets are summarized from the point of view of their possible generalizations. Then, the classification of rough sets in partial approximation spaces are discussed under reasonable constraints.

M. Kryszkiewicz et al. (Eds.): RSEISP 2014, LNAI 8537, pp. 42–52, 2014.

2   General Approximation Framework

General approximation framework relies on a *ground set*, a set of object of interest. *Domain* is a set of subsets of the ground set whose members are approximated. There are some beforehand detached sets from the domain serving as the basis for the approximation process. They constitute a *base system*, its members are the *base sets*. *Definable sets* are made up of the base sets and they are possible approximations of the members of the domain (base sets are always definable). Last, an *approximation pair* determines the lower and upper approximations of the sets as definable sets.

The intuitive meaning behind this scheme is the following (for more details, see [4]). The background knowledge about of objects of interest is represented by the base system. Base sets can be considered as *primary tools* in the approximation process. Definable sets are derived from base sets in some ways. They can be seen as *derived tools* or simply *tools*. Tools constitute all available knowledge about the ground set. They are used to describe approximately any sets belonging to the domain. After all, lower/upper approximations must be definable sets so that we are able to make any decision relying on primary and derived tools.

2.1   Generalized Approximation Spaces

Let U be a nonempty set and 2^U denote its power set. In general we can suppose that the domain is equal to 2^U.

The 5-tuple $\mathsf{GAS}(U) = \langle U, \mathfrak{B}, \mathfrak{D}_\mathfrak{B}, \mathsf{l}, \mathsf{u} \rangle$ is a *general approximation space* over the *ground set* U with the *domain* 2^U if

1. $\mathfrak{B}(\neq \emptyset) \subseteq 2^U$ and if $B \in \mathfrak{B}$, $B \neq \emptyset$ (*base system*);
2. $\mathfrak{B} \subseteq \mathfrak{D}_\mathfrak{B} \subseteq 2^U$ and $\emptyset \in \mathfrak{D}_\mathfrak{B}$ (*definable sets*);
3. lower and upper approximation mappings $\mathsf{l}, \mathsf{u} : 2^U \to 2^U$ form an ordered pair $\langle \mathsf{l}, \mathsf{u} \rangle$, called a *weak approximation pair*, with the following minimum requirements:
   - $\mathsf{l}(2^U), \mathsf{u}(2^U) \subseteq \mathfrak{D}_\mathfrak{B}$ (l, u are *definable*);
   - if $S_1 \subseteq S_2$ ($S_1, S_2 \in 2^U$), $\mathsf{l}(S_1) \subseteq \mathsf{l}(S_2)$, $\mathsf{u}(S_1) \subseteq \mathsf{u}(S_2)$ (*monotonicity*);
   - $\mathsf{u}(\emptyset) = \emptyset$ (*normality* of u);
   - $\mathsf{l}(S) \subseteq \mathsf{u}(S)$ ($S \in 2^U$) (*weak approximation property*).

By the weak approximation property, $\mathsf{l}(\emptyset) \subseteq \mathsf{u}(\emptyset) = \emptyset$, thus $\mathsf{l}(\emptyset) = \emptyset$ also holds which property is called the *normality* of l.

$\mathsf{GAS}(U)$ is *total*, if the base system $\mathfrak{B}$ covers the universe, i.e., $\bigcup \mathfrak{B} = U$ and *partial* otherwise.

If $B \in \mathfrak{B}$ is a union of a family of sets $\mathfrak{B}' \subseteq \mathfrak{B} \setminus \{B\}$, B is called *reducible* in $\mathfrak{B}$, otherwise B is *irreducible* in $\mathfrak{B}$. A base system $\mathfrak{B}$ is *single-layered* if every base set is irreducible, and *one-layered* if the base sets are pairwise disjoint.

Many times the base sets can be exactly approximated from "lower side". In certain cases, it also holds for definable sets. Accordingly, a weak approximation pair $\langle \mathsf{l}, \mathsf{u} \rangle$ is *granular* if $\mathsf{l}(B) = B$ ($B \in \mathfrak{B}$) and *standard* if $\mathsf{l}(D) = D$ ($D \in \mathfrak{D}_\mathfrak{B}$).

It is an important question how lower and upper approximations relate to the approximated set itself. A weak approximation pair $\langle l, u \rangle$ is *lower semi-strong* if $l(S) \subseteq S$ ($S \in 2^U$) (i.e., l is *contractive*), *upper semi-strong* if $S \subseteq u(S)$ ($S \in 2^U$) (i.e., u is *extensive*), and *strong* if it is lower and upper semi-strong.

The general approximation space $\mathsf{GAS}(U)$ is a *weak/granular/standard/lower semi-strong/upper semi-strong/strong* approximation space if the approximation pair $\langle l, u \rangle$ is weak/granular/standard/lower semi-strong/upper semi-strong/ strong, respectively.

For the further details about partial approximation spaces, see [3, 5].

The nature of approximation spaces depends on

1. the base system,
2. how definable sets are made up of base sets, and
3. how lower and upper approximations are determined.

Let us note that these three conditions mutually interact.

2.2 Boundaries of Sets

The boundary of a set which represents vagueness comes from restricted nature of available knowledge. It is a crucial notion in rough set theory. However, its definition has many possibilities in different generalizations of the Pawlakian approximation space. It seems that without any condition for the base system, definable sets and lower/upper approximations, only an insufficient description of exactness/roughness can be reached [2].

Let $\mathsf{GAS}(U) = \langle U, \mathfrak{B}, \mathfrak{D}_\mathfrak{B}, l, u \rangle$ be a generalized approximation space. Since our experience about the objects of interest is bounded, available knowledge can be represented finitely, and so the base system $\mathfrak{B}$ is finite. In addition, $\mathfrak{D}_\mathfrak{B}$ is defined with the following inductive definition:

- $\emptyset \in \mathfrak{D}_\mathfrak{B}$, $\mathfrak{B} \subseteq \mathfrak{D}_\mathfrak{B}$;
- if $D_1, D_2 \in \mathfrak{D}_\mathfrak{B}$, then $D_1 \cup D_2 \in \mathfrak{D}_\mathfrak{B}$.

The approximation space of this type is denoted by $\mathsf{GAS}_\cup(U)$ and called *strictly union type*. In such spaces, the empty set and unions of base sets are definable and there is not any other definable set. It seems that this construction is *the weakest condition* for building up derived tools from primary ones.

It is easy to check that in strictly union type approximation spaces, each definable set is the union of all base sets which are included in it. In particular, the next statement holds.

Proposition 1. *Let $\mathsf{GAS}_\cup(U) = \langle U, \mathfrak{B}, \mathfrak{D}_\mathfrak{B}, l, u \rangle$ be a strictly union type approximation space with finite base system. Then,*

1. $l(S) = \bigcup \mathbb{L}(S)$, *where* $\mathbb{L}(S) = \{B \in \mathfrak{B} \mid B \subseteq l(S)\}$;
2. $u(S) = \bigcup \mathbb{U}(S)$, *where* $\mathbb{U}(S) = \{B \in \mathfrak{B} \mid B \subseteq u(S)\}$.

for all $S \in 2^U$.

In $\mathsf{GAS}_\cup(U)$, boundary for each $S \in 2^U$ can be defined in three natural ways:

1. $\mathsf{b}(S) = \mathsf{u}(S) \setminus \mathsf{l}(S)$;
2. $\mathsf{bn}(S) = \bigcup(\mathbb{U}(S) \setminus \mathbb{L}(S))$, where mappings $\mathbb{L}$ and $\mathbb{U}$ are defined as before;
3. $\mathsf{b_P}(S) = \bigcup \mathbb{B}_P(S)$, where $\mathbb{B}_P(S) = \{B \in \mathfrak{B} \mid B \cap S \neq \emptyset, B \not\subseteq S\}$.

In Pawlakian rough set theory, these definitions lead to the same boundary notion. $\mathsf{b}(S)$ is always definable in classical rough set theory, but it is not in general. However, it is commonly expected that boundaries have to be definable. Since $\mathfrak{D}_\mathfrak{B}$ is closed under unions, bn and $\mathsf{b_P}$ are definable in $\mathsf{GAS}_\cup(U)$, i.e., $\mathsf{bn}(2^U), \mathsf{b_P}(2^U) \subseteq \mathfrak{D}_\mathfrak{B}$. Thus, the other two boundary definitions may be used in general approximation spaces.

The next proposition deals with the relations between bn and b and it is examined when b become definable in $\mathsf{GAS}_\cup(U)$.

Proposition 2. *Let* $\mathsf{GAS}_\cup(U) = \langle U, \mathfrak{B}, \mathfrak{D}_\mathfrak{B}, \mathsf{l}, \mathsf{u} \rangle$ *be a strictly union type approximation space with finite base system. Then,*

1. $\mathsf{b}(S) \subseteq \mathsf{bn}(S) \subseteq \mathsf{u}(S)$ *($S \in 2^U$);*
2. $\mathsf{bn}(S) = \mathsf{b}(S)$ *if and only if* $\mathsf{bn}(S) \cap \mathsf{l}(S) = \emptyset$ *($S \in 2^U$);*
3. *if $\mathfrak{B}$ is one-layered,* $\mathsf{bn}(S) = \mathsf{b}(S)$ *($S \in 2^U$).*

Proof. 1. Provided that $\mathfrak{B}$ is not one-layered, if $u \in \mathsf{b}(S)$, there may exist a $B \in \mathfrak{B}$ such that $u \in B$, $B \in \mathbb{U}(S)$, $B \notin \mathbb{L}(S)$ but $u \in \bigcup \mathbb{L}(S)$. Hence,

$$u \in \mathsf{b}(S) = \mathsf{u}(S) \setminus \mathsf{l}(S) \Leftrightarrow u \in \bigcup \mathbb{U}(S) \wedge u \notin \bigcup \mathbb{L}(S)$$
$$\Rightarrow \exists B \in \mathfrak{B} \, (u \in B \wedge B \in \mathbb{U}(S) \wedge B \notin \mathbb{L}(S))$$
$$\Leftrightarrow \exists B \in \mathfrak{B} \, (u \in B \wedge B \in \mathbb{U}(S) \setminus \mathbb{L}(S))$$
$$\Leftrightarrow u \in \bigcup(\mathbb{U}(S) \setminus \mathbb{L}(S)) = \mathsf{bn}(S).$$

It is straightforward that $\mathsf{bn}(S) = \bigcup(\mathbb{U}(S) \setminus \mathbb{L}(S)) \subseteq \bigcup \mathbb{U}(S) = \mathsf{u}(S)$.

2. ($\Rightarrow$) $\mathsf{bn}(S) \cap \mathsf{l}(S) = \mathsf{b}(S) \cap \mathsf{l}(S) = (\mathsf{u}(S) \setminus \mathsf{l}(S)) \cap \mathsf{l}(S) = \emptyset$.
($\Leftarrow$) $\mathsf{bn}(S) = (\mathsf{bn}(S) \cap \mathsf{l}(S)) \cup (\mathsf{bn}(S)) \cap (\mathsf{l}(S))^c) = \mathsf{bn}(S) \cap (\mathsf{l}(S))^c$
$\subseteq \mathsf{u}(S) \cap (\mathsf{l}(S))^c = \mathsf{u}(S) \setminus \mathsf{l}(S) = \mathsf{b}(S)$.[1]
Compared it to (1), we get $\mathsf{b}(S) = \mathsf{bn}(S)$.

3. Provided that $\mathfrak{B}$ is one-layered, if $u \in \mathsf{b}(S)$, there is only one $B \in \mathfrak{B}$ in such a way that $u \in B$, $B \in \mathbb{U}(S)$ but $B \notin \mathbb{L}(S)$. Therefore, "$\Rightarrow$" can be changed for "$\Leftrightarrow$" in the formula of point (1) from which the statement follows.   □

Corollary 1. *Let* $\mathsf{GAS}_\cup(U) = \langle U, \mathfrak{B}, \mathfrak{D}_\mathfrak{B}, \mathsf{l}, \mathsf{u} \rangle$ *be a strictly union type approximation space with finite base system. If $\mathfrak{B}$ is one-layered,*

1. b *is definable, i.e.,* $\mathsf{b}(2^U) \subseteq \mathfrak{D}_\mathfrak{B}$;
2. $\mathsf{b}(S) \cap \mathsf{l}(S) = \emptyset$ *($S \in 2^U$).*

---

[1] If $S \in 2^U$, S^c denotes the complement of S in U, i.e., $S^c = U \setminus S$.

The third version of the boundary definition, i.e., the mapping b_P, can be used consistently only in lower semi-strong approximation spaces. To this end, e.g., let us define the lower approximation mapping l_P in $\mathsf{GAS}_\cup(U)$ as follows:

$$l_P(S) = \bigcup \mathbb{L}_P(S), \text{ where } \mathbb{L}_P(S) = \{B \in \mathfrak{B} \mid B \subseteq S\} \quad (S \in 2^U).$$

It is straightforward that l_P is lower semi-strong, i.e., $l_P(S) \subseteq S$ $(S \in 2^U)$.

After these, a natural definition of the upper approximation u_P immediately comes from the mappings l_P and b_P:

$$u_P(S) = \bigcup \mathbb{U}_P(S), \text{ where } \mathbb{U}_P(S) = \mathbb{L}_P(S) \cup \mathbb{B}_P(S) \quad (S \in 2^U).$$

It is easy to check that l_P and u_P are definable and monotone, $u_P(\emptyset) = \emptyset$, and $l_P(S) \subseteq u_P(S)$ $(S \in 2^U)$, that is $\langle l_P, u_P \rangle$ forms a weak approximation pair. It is called a *(generalized) Pawlakian approximation pair*. Let us denote the approximation space obtained in this way by $\mathsf{GAS}_P(U) = \langle U, \mathfrak{B}, \mathfrak{D}_\mathfrak{B}, l_P, u_P \rangle$. (Thus, $\mathsf{GAS}_P(U)$ is strictly union type with finite base system and Pawlakian approximation pair).

The next statement deals with the connections between the lower/upper approximation mappings and the different notions of boundaries.

Corollary 2. *1. In $\mathsf{GAS}_\cup(U)$, $u(S) = l(S) \cup b(S) = l(S) \cup bn(S)$ $(S \in 2^U)$.*
2. In $\mathsf{GAS}_P(U)$, $u_P(S) = l_P(S) \cup b_P(S)$ $(S \in 2^U)$.

In general, intersections of lower approximations and boundaries are not empty. However, if the base system one-layered, the next statement holds.

Corollary 3. *1. In $\mathsf{GAS}_\cup(U)$, $l(S) \cap b(S) = l(S) \cap bn(S) = \emptyset$ $(S \in 2^U)$ provided that $\mathfrak{B}$ is one-layered.*
2. In $\mathsf{GAS}_P(U)$, $l_P(S) \cap b_P(S) = \emptyset$ $(S \in 2^U)$ provided that $\mathfrak{B}$ is one-layered.

3   Rough Sets in Partial Approximation Spaces

3.1   Categories of Vagueness in Pawlakian Rough Set Theory

Let U be a finite set and R be an equivalence relation on U. Furthermore, let U/R denote the partition of U pertaining to R and $\mathfrak{D}_{U/R}$ the set of any unions of equivalence classes generated by R and $\emptyset \in \mathfrak{D}_{U/R}$. The members of U/R and $\mathfrak{D}_{U/R}$ are called *elementary* sets and *definable* sets, respectively.

Denoting Pawlakian lower and upper approximation mappings by $\underline{R}$ and $\overline{R}$, $S_1, S_2 \in 2^U$ is *roughly equal*, in notation $S_1 \approx S_2$, if $\underline{R}(S_1) = \underline{R}(S_2)$ and $\overline{R}(S_1) = \overline{R}(S_2)$. Rough equality is an equivalence relation on 2^U, and Pawlak originally defined rough sets as the equivalence classes in $2^U / \approx$. There are further equivalent definitions of classical rough sets, see, e.g., in [1, 7–10] and many additional references therein.

$\underline{R}(S)$ and $\overline{R}(S)^c$ $(S \in 2^U)$ are also called *positive* and *negative* regions of S and denoted them by $POS_R(S)$ and $NEG_R(S)$, respectively. Their intuitive meaning is the following: considering knowledge R, the members of $POS_R(S)$ are the objects of U which *certainly belong to* S, and the members of $NEG_R(S)$ are the objects of U which *certainly do not belong to* S.

Of course, $POS_R(S)$, $NEG_R(S)$ are definable, $POS_R(S) \cap NEG_R(S) = \emptyset$.

S is called *crisp* if $\underline{R}(S) = \overline{R}(S)$, or equivalently $POS_R(S) \cup NEG_R(S) = U$, and it is *rough* otherwise.

Within classical rough set theory, four well-known categories of vagueness, i.e., four well-known different kinds of rough sets, are identified [9, 13].

If $S \in 2^U$,

- S is *roughly definable* if $\underline{R}(S) \neq \emptyset$ and $\overline{R}(S) \neq U$,
  in other words, $POS_R(S) \neq \emptyset$ and $NEG_R(S) \neq \emptyset$.
- S is *externally undefinable* if $\underline{R}(S) \neq \emptyset$ and $\overline{R}(S) = U$,
  in other words, $POS_R(S) \neq \emptyset$ and $NEG_R(S) = \emptyset$.
- S is *internally undefinable* if $\underline{R}(S) = \emptyset$ and $\overline{R}(S) \neq U$,
  in other words, $POS_R(S) = \emptyset$ and $NEG_R(S) \neq \emptyset$.
- S is *totally undefinable* if $\underline{R}(S) = \emptyset$ and $\overline{R}(S) = U$,
  in other words, $POS_R(S) = \emptyset$ and $NEG_R(S) = \emptyset$.

3.2 Rough Sets in Partial Approximation Spaces: Some General Remarks

Let $\mathsf{GAS}(U)$ be a general approximation space and let $S \in 2^U$.

Like Pawlakian rough set theory,

- both $\mathsf{l}(S) = \emptyset$ and $\mathsf{l}(S) \neq \emptyset$ are possible if $S \neq \emptyset$, and
- if $\mathsf{l}(S) \neq \emptyset$, $\mathsf{u}(S) \neq \emptyset$ also holds by the weak approximation property.

Unlike Pawlakian rough set theory, $\mathsf{u}(S) = \emptyset$ may occur even if $S \neq \emptyset$. Then, of course, $\mathsf{l}(S) = \emptyset$ by the weak approximation property. This special case will be a new category of rough sets in partial approximation space context.

In $\mathsf{GAS}(U)$, the base system does not cover the ground set necessarily, i.e., $\bigcup \mathfrak{B} \subseteq U$. Thus, upper approximations have to be related to $\bigcup \mathfrak{B}$.

The positive region of S can immediately be defined as $POS_{\langle \mathsf{l}, \mathsf{u} \rangle}(S) = \mathsf{l}(S)$. However, $\mathsf{u}(S)^c$ cannot form negative region because it is not definable in general. $NEG_{\langle \mathsf{l}, \mathsf{u} \rangle}(S) = \mathsf{l}(\mathsf{u}(S)^c)$ may be a natural negative region in general approximation spaces.

Proposition 3. *If* $\mathsf{GAS}(U)$ *is lower semi-strong,* $POS_{\langle \mathsf{l}, \mathsf{u} \rangle}(S) \cap NEG_{\langle \mathsf{l}, \mathsf{u} \rangle}(S) = \emptyset$ $(S \in 2^U)$.

Proof. The weak approximation property $\mathsf{l}(S) \subseteq \mathsf{u}(S)$ implies $\mathsf{l}(S) \cap \mathsf{u}(S)^c = \emptyset$. Since l is lower semi-strong, $\mathsf{l}(\mathsf{u}(S)^c) \subseteq \mathsf{u}(S)^c$, and so $\mathsf{l}(S) \cap \mathsf{l}(\mathsf{u}(S)^c) = \emptyset$. $\square$

Corollary 4. *In* $\mathsf{GAS_P}(U)$, $POS_{\langle \mathsf{l_P}, \mathsf{u_P} \rangle}(S) \cap NEG_{\langle \mathsf{l_P}, \mathsf{u_P} \rangle}(S) = \emptyset$ $(S \in 2^U)$.

3.3 Classifying Rough Sets in Partial Approximation Spaces with Pawlakian Approximation Pair

Let us consider the approximation space $\mathsf{GAS_P}(U) = \langle U, \mathfrak{B}, \mathfrak{D}_{\mathfrak{B}}, \mathsf{lp}, \mathsf{up} \rangle$. Lower and upper approximations, positive and negative regions in $\mathsf{GAS_P}(U)$ are illustrated in Fig. 1.

By Corollary 4, $POS_{\langle \mathsf{lp},\mathsf{up}\rangle}(S) \cap NEG_{\langle \mathsf{lp},\mathsf{up}\rangle}(S) = \emptyset$. The construction of $\langle \mathsf{lp}, \mathsf{up} \rangle$ also implies that $S \cap NEG_{\langle \mathsf{lp},\mathsf{up}\rangle}(S) = \emptyset$.

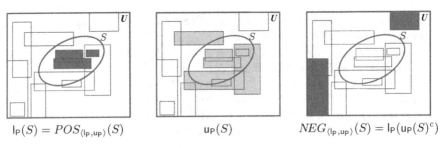

$\mathsf{lp}(S) = POS_{\langle \mathsf{lp},\mathsf{up}\rangle}(S)$ \qquad $\mathsf{up}(S)$ \qquad $NEG_{\langle \mathsf{lp},\mathsf{up}\rangle}(S) = \mathsf{lp}(\mathsf{up}(S)^c)$

Fig. 1. Generalized Pawlakian approximation pair $\langle \mathsf{lp}, \mathsf{up} \rangle$

Positive and negative regions are determined by different conditions together. For adequate interpretation of the classification of rough sets, all these conditions have to be taken into account. However, for the sake of simplicity, they are not mentioned explicitly in the following reformulation and reinterpretation of rough set categories.

In $\mathsf{GAS_P}(U)$, the formal classification of vagueness and its intuitive meaning, considering the previous remark, are the following:

1. S is *roughly definable* if $POS_{\langle \mathsf{lp},\mathsf{up}\rangle}(S) \neq \emptyset$ and $NEG_{\langle \mathsf{lp},\mathsf{up}\rangle}(S) \neq \emptyset$.
   It means that
   - there are objects in $\bigcup \mathfrak{B}$ and so in U which certainly belong to S in such a way that there is at least one base set containing them and completely including in S, and
   - there are other objects in $\bigcup \mathfrak{B}$ and so in U which certainly do not belong to S in such a way that there is at least one base set containing them and completely including in $\mathsf{up}(S)^c$.
2. S is *externally undefinable* if $POS_{\langle \mathsf{lp},\mathsf{up}\rangle}(S) \neq \emptyset$ and $NEG_{\langle \mathsf{lp},\mathsf{up}\rangle}(S) = \emptyset$.
   It means that
   - there are objects in $\bigcup \mathfrak{B}$ and so in U which certainly belong to S in such a way that there is at least one base set containing them and completely including S, and
   - there is not any object in $\bigcup \mathfrak{B}$ and so in U which certainly does not belong to S in such a way that there is at least one base set containing it and completely including in $\mathsf{up}(S)^c$.

3. S is *internally undefinable* if $POS_{\langle lp,up \rangle}(S) = \emptyset$ and $NEG_{\langle lp,up \rangle}(S) \neq \emptyset$, $NEG_{\langle lp,up \rangle}(S) \neq \bigcup \mathfrak{B}$.
   It means that
   - there is not any object in $\bigcup \mathfrak{B}$ and so in U which certainly belongs to S in such a way that there is at least one base set containing it and completely including in S, and
   - there are objects in $\bigcup \mathfrak{B}$ and so in U which certainly do not belong to S in such a way that there is at least one base set containing them and completely including in $up(S)^c$.
4. S is *totally undefinable* if $POS_{\langle lp,up \rangle}(S) = \emptyset$ and $NEG_{\langle lp,up \rangle}(S) = \emptyset$.
   It means that
   - there is not any object in $\bigcup \mathfrak{B}$ and so in U which certainly belongs to S in such a way that there is at least one base set containing it and completely including in S, and
   - there is not any object in $\bigcup \mathfrak{B}$ and so in U which certainly does not belong to S in such a way that there is at least one base set containing it and completely including in $up(S)^c$,
5. S is *negatively definable* if $POS_{\langle lp,up \rangle}(S) = \emptyset$ and $NEG_{\langle lp,up \rangle}(S) = \bigcup \mathfrak{B}$.
   This is a new category of rough sets.
   It means that
   - there is not any object in $\bigcup \mathfrak{B}$ and so in U which certainly belongs to S, in such a way that there is at least one base set containing it and completely including in S, and
   - any members of U of which something are known certainly do not belong to S.

3.4 Classifying Rough Sets in Partial Approximation Spaces with Pessimistic Approximation Pair

In [6], two not Pawlakian approximation pairs were constructed. One of them is called pessimistic which is used here to build a general approximation space.

Let us suppose that U is finite, $\mathfrak{D}_\mathfrak{B}$ is strictly union type and closed under set differences.

For any $u \in U$, let $\mathcal{N}_\mathfrak{B}(u) = \{B \in \mathfrak{B} \mid u \in B\}$ be the *(reflexive) neighborhood system* of u with respect to $\mathfrak{B}$ [13]. Its members are called *neighborhoods* of u. First, let us define a *partial pessimistic membership functions*:

$$\mu_S^p : U \to [0,1], \quad \mu_S^p(u) = \begin{cases} \min\left\{ \frac{|B \cap S|}{|B|} \mid B \in \mathcal{N}_\mathfrak{B}(u) \right\}, & \text{if } u \in \bigcup \mathfrak{B}, \\ \text{undefined}, & \text{otherwise}, \end{cases}$$

where $|\cdot|$ denotes the cardinality of a set.

Proposition 4 ([6], Proposition 4). *For any $u \in U$,*

- $\mu_S^p(u) = 1$ *if and only if* $\forall B \in \mathcal{N}_\mathfrak{B}(u) \, (B \subseteq S)$;
- $\mu_S^p(u) > 0$ *if and only if* $\forall B \in \mathcal{N}_\mathfrak{B}(u) \, (B \cap S \neq \emptyset)$;
- $\mu_S^p(u) = 0$ *if and only if* $\exists B \in \mathcal{N}_\mathfrak{B}(u) \, (B \cap S = \emptyset)$.

Last, the so-called *pessimistic approximation pair* is defined:

$$\mathsf{l}^p(S) = \{u \in U \mid \mu_S^p(u) = 1\} = \{u \in U \mid \forall B \in \mathcal{N}_\mathfrak{B}(u)\,(B \subseteq S)\}$$
$$\mathsf{u}^p(S) = \{u \in U \mid \mu_S^p(u) > 0\} = \{u \in U \mid \forall B \in \mathcal{N}_\mathfrak{B}(u)\,(B \cap S \neq \emptyset)\}$$

Proposition 5 ([6], Proposition 6). $\langle \mathsf{l}^p, \mathsf{u}^p \rangle$ *is a weak approximation pair.*

The construction also implies immediately that $\langle \mathsf{l}^p, \mathsf{u}^p \rangle$ is lower semi-strong, i.e., $\mathsf{l}^p(S) \subseteq S$ $(S \in 2^U)$.

Let us denote the approximation space obtained in this way by $\mathsf{GAS}^p(U) = \langle U, \mathfrak{B}, \mathfrak{D}_\mathfrak{B}, \mathsf{l}^p, \mathsf{u}^p \rangle$. (Thus, $\mathsf{GAS}^p(U)$ is strictly union type, lower semi-strong, and $\mathfrak{D}_\mathfrak{B}$ is closed under set differences.)

$\mathsf{GAS_P}(U)$ can also be formed on U with the same $\mathfrak{B}$ and $\mathfrak{D}_\mathfrak{B}$. It is clear that $\mathsf{l}^p(S) \subseteq \mathsf{l_P}(S)$ and $\mathsf{u}^p(S) \subseteq \mathsf{u_P}(S)$ $(S \in 2^U)$, This property may indicate the adjective "pessimistic" in the name of the approximation pair $\langle \mathsf{l}^p, \mathsf{u}^p \rangle$.

In $\mathsf{GAS}^p(U)$, positive and negative regions can be defined in the usual way: $POS_{\langle \mathsf{l}^p, \mathsf{u}^p \rangle}(S) = \mathsf{l}^p(S)$ and $NEG_{\langle \mathsf{l}^p, \mathsf{u}^p \rangle}(S) = \mathsf{l}^p(\mathsf{u}^p(S)^c)$ $(S \in 2^U)$.

By Proposition 3, $POS_{\langle \mathsf{l}^p, \mathsf{u}^p \rangle}(S) \cap NEG_{\langle \mathsf{l}^p, \mathsf{u}^p \rangle}(S) = \emptyset$ $(S \in 2^U)$. The construction of $\langle \mathsf{l}^p, \mathsf{u}^p \rangle$ also implies that $S \cap NEG_{\langle \mathsf{l}^p, \mathsf{u}^p \rangle}(S) = \emptyset$ $(S \in 2^U)$.

Lower and upper approximations, positive and negative regions are illustrated in Fig. 2.

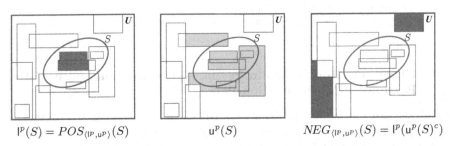

$\mathsf{l}^p(S) = POS_{\langle \mathsf{l}^p, \mathsf{u}^p \rangle}(S)$ $\qquad$ $\mathsf{u}^p(S)$ $\qquad$ $NEG_{\langle \mathsf{l}^p, \mathsf{u}^p \rangle}(S) = \mathsf{l}^p(\mathsf{u}^p(S)^c)$

Fig. 2. Pessimistic approximation pair $\langle \mathsf{l}^p, \mathsf{u}^p \rangle$

Let $S \in 2^U$. Then, the formal definition and their intuitive meaning of rough set classes in $\mathsf{GAS}^p(U)$ are the following:

1. S is *roughly definable* if $POS_{\langle \mathsf{l}^p, \mathsf{u}^p \rangle}(S) \neq \emptyset$ and $NEG_{\langle \mathsf{l}^p, \mathsf{u}^p \rangle}(S) \neq \emptyset$.
   It means that
   – there are objects in $\bigcup \mathfrak{B}$ and so in U which certainly belong to S in such a way that all base sets containing them are included in S, and
   – there are other objects in $\bigcup \mathfrak{B}$ and so in U which certainly do not belong to S in such a way that all base sets containing them are included in $\mathsf{u}^p(S)^c$.

2. S is *externally undefinable* if $POS_{\langle l^p, u^p \rangle}(S) \neq \emptyset$ and $NEG_{\langle l^p, u^p \rangle}(S) = \emptyset$.
   It means that
   - there are objects in $\bigcup \mathfrak{B}$ and so in U which certainly belong to S in such a way that all base sets containing them are included in S, and
   - there is not any object in $\bigcup \mathfrak{B}$ and so in U which certainly does not belong to S in such a way that all base sets containing it are included in $u^p(S)^c$.
3. S is *internally undefinable* if $POS_{\langle l^p, u^p \rangle}(S) = \emptyset$ and $NEG_{\langle l^p, u^p \rangle}(S) \neq \emptyset$, $NEG_{\langle l^p, u^p \rangle}(S) \neq \bigcup \mathfrak{B}$.
   It means that
   - there is not any object in $\bigcup \mathfrak{B}$ and so in U which certainly belongs to S in such a way that all base sets containing it is included in S, and
   - there are objects in $\bigcup \mathfrak{B}$ and so in U which certainly do not belong to S in such a way that all base sets containing them are included in $u^p(S)^c$.
4. S is *totally undefinable* if $POS_{\langle l^p, u^p \rangle}(S) = \emptyset$ and $NEG_{\langle l^p, u^p \rangle}(S) = \emptyset$.
   It means that
   - there is not any object in $\bigcup \mathfrak{B}$ and so in U which certainly belongs to S in such a way that all base sets containing it is included in S, and
   - there is not any object in $\bigcup \mathfrak{B}$ and so in U which certainly does not belong to S in such a way that all base sets containing it are included in $u^p(S)^c$.
5. S is *negatively definable* if $POS_{\langle l^p, u^p \rangle}(S) = \emptyset$ and $NEG_{\langle l^p, u^p \rangle}(S) = \bigcup \mathfrak{B}$.
   This is a new category of rough sets.
   It means that
   - there is not any object in $\bigcup \mathfrak{B}$ and so in U which certainly belongs to S in such a way that all base sets containing it is included in S, and
   - any members of U of which something are known certainly does not belong to S.

4 Conclusion and Future Work

In this paper, categories of vagueness, i.e., different kinds of rough sets in partial approximation spaces have been reformulated. Their formal definitions have been given under reasonable constraints and their intuitive meanings have been reinterpreted.

The investigations presented in this paper can be extended in many directions. First of all, categories of vagueness for not necessarily Pawlakian type approximations pairs should be worth studying in order to gather more and more facts about the nature of rough sets in general setting.

Acknowledgements. The publication was supported by the TÁMOP–4.2.2.C–11/1/KONV–2012–0001 project. The project has been supported by the European Union, co-financed by the European Social Fund.

The authors are thankful to Davide Ciucci for his constructive remarks, and would like to thank the anonymous referees for their useful comments.

References

1. Banerjee, M., Chakraborty, M.: Algebras from rough sets. In: Pal, S., Polkowski, L., Skowron, A. (eds.) Rough-Neuro Computing: Techniques for Computing with Words, pp. 157–184. Springer, Berlin (2004)
2. Ciucci, D.: Approximation algebra and framework. Fundamenta Informaticae 94, 147–161 (2009)
3. Csajbók, Z., Mihálydeák, T.: Partial approximative set theory: A generalization of the rough set theory. International Journal of Computer Information Systems and Industrial Management Applications 4, 437–444 (2012)
4. Csajbók, Z., Mihálydeák, T.: A general set theoretic approximation framework. In: Greco, S., Bouchon-Meunier, B., Coletti, G., Fedrizzi, M., Matarazzo, B., Yager, R.R. (eds.) IPMU 2012, Part I. CCIS, vol. 297, pp. 604–612. Springer, Heidelberg (2012)
5. Csajbók, Z.E.: Approximation of sets based on partial covering. In: Peters, J.F., Skowron, A., Ramanna, S., Suraj, Z., Wang, X. (eds.) Transactions on Rough Sets XVI. LNCS, vol. 7736, pp. 144–220. Springer, Heidelberg (2013)
6. Csajbók, Z.E., Mihálydeák, T.: Fuzziness in partial approximation framework. In: Ganzha, M., Maciaszek, L.A., Paprzycki, M. (eds.) Proceedings of the 2013 Federated Conference on Computer Science and Information Systems, Kraków, Poland (FedCSIS 2013), September 8-11. Annals of Computer Science and Information Systems, Polish Information Processing Society, vol. 1, pp. 35–41. Polskie Towarzystwo Informatyczne (PTI), IEEE Computer Society Press, Warsaw, Poland (2013)
7. Järvinen, J.: Lattice theory for rough sets. In: Peters, J.F., Skowron, A., Düntsch, I., Grzymała-Busse, J.W., Orłowska, E., Polkowski, L. (eds.) Transactions on Rough Sets VI. LNCS, vol. 4374, pp. 400–498. Springer, Heidelberg (2007)
8. Komorowski, J., Pawlak, Z., Polkowski, L., Skowron, A.: Rough sets: A tutorial. In: Pal, S., Skowron, A. (eds.) Rough Fuzzy Hybridization. A New Trend in Decision-Making, pp. 3–98. Springer, Singapore (1999)
9. Pagliani, P.: A pure logic-algebraic analysis of rough top and rough bottom equalities. In: Ziarko, W. (ed.) Rough Sets, Fuzzy Sets and Knowledge Discovery, Proceedings of the International Workshop on Rough Sets and Knowledge Discovery (RSKD 1993), Banff, Alberta, Canada, October 12-15. Workshops in Computing, pp. 225–236. Springer (1993)
10. Pagliani, P., Chakraborty, M.: A Geometry of Approximation: Rough Set Theory Logic, Algebra and Topology of Conceptual Patterns (Trends in Logic). Springer Publishing Company, Incorporated (2008)
11. Pawlak, Z.: Rough sets. International Journal of Computer and Information Sciences 11(5), 341–356 (1982)
12. Pawlak, Z.: Rough Sets: Theoretical Aspects of Reasoning about Data. Kluwer Academic Publishers, Dordrecht (1991)
13. Yao, Y.Y.: Granular computing using neighborhood systems. In: Roy, R., Furuhashi, T., Chawdhry, P.K. (eds.) Advances in Soft Computing: Engineering Design and Manufacturing. The 3rd On-line World Conference on Soft Computing (WSC3), pp. 539–553. Springer, London (1999)

Random Probes in Computation
and Assessment of Approximate Reducts*

Andrzej Janusz[1] and Dominik Ślęzak[1,2]

[1] Institute of Mathematics, University of Warsaw,
ul. Banacha 2, 02-097, Warsaw, Poland
[2] Infobright Inc., Poland
ul. Krzywickiego 34 pok. 219, 02-078 Warsaw, Poland
{janusza,slezak}@mimuw.edu.pl

Abstract. We discuss applications of random probes in a process of computation and assessment of approximate reducts. By random probes we mean artificial attributes, generated independently from a decision vector but having similar value distributions to the attributes in the original data. We introduce a concept of a randomized reduct which is a reduct constructed solely from random probes and we show how to use it for unsupervised evaluation of attribute sets. We also propose a modification of the greedy heuristic for a computation of approximate reducts, which reduces a chance of including irrelevant attributes into a reduct. To support our claims we present results of experiments on high dimensional data. Analysis of obtained results confirms usefulness of random probes in a search for informative attribute sets.

Keywords: attribute selection, attribute reduction, high dimensional data, approximate reducts.

1 Introduction

Since the foundation of the rough set theory, the problem of finding informative attribute subsets has been in a scope of interest of numerous researchers from this field [1]. Zdzisław Pawlak, the founder of rough sets, wrote in [2] that discovering redundancy and dependencies between attributes is one of the most fundamental and challenging problems to the rough set philosophy. Although many attribute subset selection techniques were proposed in the related literature, this topic is still one of the most viable research directions [3, 4].

In the rough set approach, this task is commonly tackled using the notion of information or decision reducts. A reduct can be defined as a minimal set of attributes that sufficiently preserve information allowing to discern objects with

* Partly supported by Polish National Science Centre (NCN) grants DEC-2011/01/B/-ST6/03867 and DEC-2012/05/B/ST6/03215, and by National Centre for Research and Development (NCBiR) grant SP/I/1/77065/10 by the strategic scientific research and experimental development program: "Interdisciplinary System for Interactive Scientific and Scientific-Technical Information".

M. Kryszkiewicz et al. (Eds.): RSEISP 2014, LNAI 8537, pp. 53–64, 2014.

different properties, e.g. belonging to different decision classes. Techniques for computation of decision reducts have been widely discussed in literature related to data analysis and knowledge discovery [5, 6].

Due to their appealing properties, reducts can only consist of attributes that are relevant in a sense discussed in [7]. In practice, however, this property often fails to prevent an inclusion of unnecessary attributes into reducts. This happens mostly due to a fact that in a finite set of objects described by a large number of features there are often some dependencies between conditional and decision attributes, which are valid only for available training data but do not hold for new cases. Such dependencies are sometimes called *random* or *illusionary*. Even though extensions of the reduct concept, such as approximate reducts [8], were introduced to make it less sensitive to random disturbances in data, there is still a need for methods that can deal with random dependencies.

A similar problem was also recognized by other researchers working on development of feature selection methods [9, 10]. For instance, a question was risen how to determine the best number of attributes to be returned by an attribute filtering algorithm for a given data set [10, 11]. The filtering attribute selection methods create rankings of individual features or feature subsets based on some predefined scoring function. After a ranking of attributes is created, it is necessary to decide on a score threshold value that distinguishes attributes to be discarded from those which are to be selected. In other words, the threshold should indicate a value of the score, such that attributes which receive higher scores with a high probability do not represent a random dependency in data. One common technique to set this value uses, so called, *random probes*– artificial attributes whose values were generated independently from the classes of objects in data. By computing a distribution of scores obtained by random probes it is possible to choose the threshold, so that a probability of achieving a higher score by an attribute which is not truly relevant is extremely low [10].

In this paper, we propose a random probes-based technique aiming at detecting random dependencies between conditional and decision attributes during a computation of reducts. Since random probes exemplify such dependencies in data, we may utilize them in order to avoid selecting real attributes that seem to have analogous properties. We show how to apply the proposed method within a frame of the greedy heuristic [12]. As a result we come up with a new stopping criterion for the greedy computation of approximate reducts, which allows a dynamic adjustment of the approximation threshold for a specific data set. Moreover, we propose criteria for deciding whether reducts represent random dependencies, which do not require using additional validation data.

The remainder of the paper is organized as follows: In Section 2, we recall standard notions of decision reducts and approximate decision reducts, followed by introduction of random probes and randomized decision reducts. In Section 3, we compare a well-known greedy heuristic approach to computation of approximate decision reducts with a new algorithm, where the standard approximation threshold-based criterion for preserving decision information is replaced by a more dynamically adjusted method of comparing attributes with their random probe counterparts. In Section 4, we utilize the framework of random probes to establish assessment tools for comparison of reducts produced using different algorithms and parameter settings. In Section 5, we report implementation details and experimental results. In Section 6, we conclude the paper with some directions for further research.

2 Preliminary Notions and Notations

Let us first recall some of the basic notions from the rough set theory. We will assume that our data set is represented by a decision system understood as a tuple $\mathbb{S}_d = (U, A \cup \{d\})$, where U is a set of objects, A is a set of attributes of objects from U and d, $d \notin A$, is a distinguished decision attribute. The task in the attribute selection is to find an informative subset of attributes from A, which compactly represents relevant information about objects from U and allows an accurate prediction of decisions for new objects.

Typically, in the rough set theory selecting compact yet informative sets of attributes is conducted using the notion of discernibility, by computing so called reducts [13]. If for a decision system $\mathbb{S}_d = (U, A \cup \{d\})$ we consider a subset of attributes $B \subseteq A \cup \{d\}$, we will say that $u_1, u_2 \in U$ are satisfying an indiscernibility relation IND_B with regard to B, if and only if they have equal attribute values for every $a \in B$: $(u_1, u_2) \in IND_B \Leftrightarrow \forall_{a \in B}\, a(u_1) = a(u_2)$. Otherwise u_1 and u_2 will be regarded discernible. An indiscernibility class of an object u with regard to a subset B will be denoted by $[u]_B$. We can define a decision reduct DR of a decision system $\mathbb{S}_d$ as follows:

Definition 1 (Decision reduct)
Let $\mathbb{S}_d = (U, A \cup \{d\})$ be a decision system. A subset of attributes $DR \subseteq A$ will be called a decision reduct, if and only if the following conditions are met:

1. *For any $u \in U$ if the indiscernibility class of u relative to A is a subset of some decision class, its indiscernibility class relative to DR should also be a subset of that decision class, i.e. $[u]_A \subseteq [u]_d \Rightarrow [u]_{DR} \subseteq [u]_d$.*
2. *There is no proper subset $DR' \subsetneq DR$ for which the first condition holds.*

A decision system usually has many reducts which may have different properties. In particular, some of the reducts may contain attributes which are not really related to decision values, even though they discern some of the objects from U that are indiscernible using the other attributes from the reduct. This situation happens simply by a chance. It can be explained either by random disturbances in data or by the fact that even a randomly generated binary attribute would discern approximately a number of pairs of objects. Although such attributes may seem to be valuable, the dependency which they indicate is invalid for objects outside U and thus, they are useless from a point of view of the predictive data analysis. It is not easy to distinguish between truly relevant and irrelevant attributes. This is difficult especially in a case when some of the attributes are *partially* correlated with the decision, but this dependency is lost when they are considered in a context of some other attributes. These problems were addressed by authors of several extensions of the decision reduct concept [14, 15]. One of those is the following notion of approximate reduct [16]:

Definition 2 (Approximate decision reduct)

Let $\mathbb{S}_d = (U, A \cup \{d\})$ be a decision system and let $\phi_d : 2^A \to \mathbb{R}$ be an attribute subset quality measure which is nondecreasing with regard to the inclusion over its arguments. A subset $AR \subseteq A$ is called a (ϕ_d, ε)-approximate decision reduct for an approximation threshold $\varepsilon \in [0, 1)$, if and only if:

1. $\phi_d(AR) \geq (1 - \varepsilon)\phi_d(A)$.
2. There is no proper subset $AR' \subsetneq AR$ for which the first condition holds.

By softening the requirement for preservation of all dependencies between conditional and decision attributes, approximate reducts can deal with random disturbances in data and to some extent avoid selecting attributes that do not influence the decision. However, since the value of ε is chosen arbitrary, there is no guarantee that in a particular case it corresponds to the actual level of uncertainty in data. Typically, multiple tests are conducted to select an appropriate ε value. In those tests, performance of models constructed using approximate reducts with different approximation thresholds is evaluated on validation data. In some areas of applications, it is possible to anticipate optimal levels of ε basing on the observed quality of input data sets [17]. Nevertheless, even if a range of ε is partially narrowed down, the choice of its setting requires the availability of additional data and suffers from high computational burden.

In this paper we propose two approaches that can be used to overcome this problem. Both of them relay on the notion of random probes. Firstly, it is worth noticing that each attribute from a decision system $\mathbb{S}_d$ may be associated with a random variable that could have generated the values of this attribute for specific objects. Of course, in practice we do not have an access to distributions of the random variables that correspond to the attributes. In some real-life cases, it may be actually problematic to justify their presence. However, from a technical point of view, we can still proceed as if they existed, even though we can reason about them only through their valuation in the available data.

Definition 3 (Random probe)

Let $\mathbb{S}_d = (U, A \cup \{d\})$ be a decision system. A random probe for an attribute $a \in A$ is an artificial attribute $\hat{a}$, such that the corresponding random variable has an identical distribution as values of a but is independent from the random variables that correspond to the attribute a and the decision d.

A random probe for a given attribute can be easily obtained by randomly permuting its values for objects described in data. By its definition, a random probe has no influence on values of the decision attribute and as such, can be used to exemplify random dependencies that may occur in data. In particular, having a sufficient number of random probes for attributes from $\mathbb{S}_d$, we can select their subset such that it meets the conditions of a decision reduct.

Definition 4 (Randomized decision reduct)
Let $\mathbb{S}_d = (U, A \cup \{d\})$ *be a decision system and let* $\hat{A}$ *be a set of random probes for attributes from* A. $RR \subseteq \hat{A}$ *which is a decision reduct of* $\hat{\mathbb{S}}_d = (U, \hat{A} \cup \{d\})$ *will be called a randomized decision reduct of* $\mathbb{S}_d$.

Technically, a set of random probes corresponding to attributes from $\mathbb{S}_d$ does not need to be computed by permuting all of them. The same effect can be achieved by leaving the original attributes unchanged and permuting decision values [10]. Another advantage of this method is that it does not disturb internal dependencies between the conditional attributes in data. Analogical definitions can also be made for other types of reducts. In fact, some of attribute subset quality measures $\phi_d(AR)$ studied in [16] attempted to express robustness of decision determination while interchanging probability distributions of decision values induced by particular indiscernibility classes of IND_A within each of larger indiscernibility classes of IND_{AR}.

3   A Stopping Strategy for the Greedy Heuristic

Let us first recall a popular heuristic for computation of approximate decision reducts, namely the greedy approach. It is explained by Algorithm 1. The meaning of function $\phi_d : 2^A \rightarrow \mathbb{R}$ is here the same as in Definition 2. However, it is also worth noting that the function used in lines 4-11 can be different than the function used in the rest of the algorithm [5, 11].

Algorithm 1 needs a predefined value of the parameter ε. Its appropriate setting is crucial for the quality of the resulting reduct. If it is too low, the reduct is likely to contain unnecessary attributes which introduce a random dependency with the decision and can mislead a classification model. This can also be interpreted as so called over-fitting. On the other hand, if ε is too high, the approximations of decision classes defined by the resulting indiscernibility relation are too coarse. Then, as a consequence, classification algorithms are unable to predict decision classes with a sufficient accuracy.

To overcome this issue we propose an algorithm that tries to automatically select a good approximation threshold for the greedy computation of approximate decision reducts. In this algorithm, random probes are used to compute a conditional probability that, given a set of already selected attributes, the addition of a new attribute increases a value of the quality measure by more than it would when adding a random probe. An attribute is added only in a case when the computed probability is sufficiently high. In this way we can stop adding new attributes without a need to refer to the artificial value of ε. This approach is explained in more details by Algorithm 2.

In this algorithm $\hat{a}_{best}$ denotes a random probe corresponding to the attribute a_{best}. The probability $P\left(\phi_d\left(AR \cup \{a_{best}\}\right) > \phi_d\left(AR \cup \{\hat{a}_{best}\}\right)\right)$ from the condition in line 12 tells how probable is that the inclusion of the selected attribute did not increase the value of ϕ_d to its current level by a chance. Thus, it may be interpreted as a chance that the attribute is truly relevant in the context of

---

Algorithm 1. A greedy calculation of an approximate decision reduct

---

Input: a decision system $\mathbb{S}_d = (U, A \cup \{d\})$;   a quality measure $\phi_d : 2^A \to \mathbb{R}$;
an approximation threshold $\varepsilon \in [0, 1)$;
Output: an approximate decision reduct AR of $\mathbb{S}_d$

```
1  begin
2      AR := ∅; φ_max := −∞;
3      while φ_max < (1 − ε)φ_d(A) do
4          foreach a ∈ A \ AR do
5              AR′ := AR ∪ {a};
6              if φ_d(AR′) > φ_max then
7                  φ_max := φ_d(AR′);
8                  a_best := a;
9              end
10         end
11         AR := AR ∪ {a_best};
12     end
13     foreach a ∈ AR do
14         AR′ := AR \ {a};
15         if φ_d(AR′) ≥ (1 − ε)φ_d(A) then
16             AR := AR′;
17         end
18     end
19     return AR;
20 end
```

---

previously selected features. The probability value can be easily estimated using Algorithm 3. A quality of estimation depends on a number of random probes generated in the process, which can be treated as additional parameter of the algorithm. The higher number of the utilized random probes, the more accurate probability estimation can be obtained by the algorithm. However, a better estimation requires more computations and, in practice, some compromise between the quality and the performance is needed.

The computational cost of Algorithms 1 and 2 can be conveniently expressed in terms of a number of required evaluations of ϕ_d. In a worst-case scenario this complexity is $O(|A|^2)$ for the classical approximation reduct algorithm and $O\left(\left(|A| + nProbes\right)|A|\right)$ for the random probe-based method. However, the number of necessary evaluations of the function ϕ_d is usually much lower, since even in a case when all the attributes represent random dependencies in data, on average only about $log_2|U|$ binary features are needed to discern all objects from U. Accordingly, an anticipated number of required evaluations of the function ϕ_d for Algorithms 1 and 2 may be given as $O\left(|A| \min\left(log_2|U|, |A|\right)\right)$, and $O\left(\left(|A| + nProbes\right)\min\left(log_2|U|, |A|\right)\right)$, respectively.

The default values of the parameters p_{rel} and $nProbes$ can be set to $1/|A|$ and $|A|$, respectively. This setting seems secure even in a situation when only a

---

Algorithm 2. A dynamically adjusted approximate reduct calculation

---

Input: a decision system $\mathbb{S}_d = (U, A \cup \{d\})$;   a quality measure $\phi_d : 2^A \to \mathbb{R}$;
an acceptable probability of adding an irrelevant attribute $p_{rel} \in [0, 1)$;
Output: an approximate decision reduct AR of $\mathbb{S}_d$

```
1  begin
2  │  AR := ∅; φ_max := −∞;
3  │  stopFlag := FALSE;
4  │  while stopFlag = FALSE do
5  │  │  foreach a ∈ A \ AR do
6  │  │  │  AR' := AR ∪ {a};
7  │  │  │  if φ_d(AR') > φ_max then
8  │  │  │  │  φ_max := φ_d(AR');
9  │  │  │  │  a_best := a;
10 │  │  │  end
11 │  │  end
12 │  │  if P(φ_d(AR ∪ {a_best}) > φ_d(AR ∪ {â_best})) ≥ p_rel then
13 │  │  │  AR := AR ∪ {a_best};
14 │  │  end
15 │  │  else
16 │  │  │  stopFlag := TRUE;
17 │  │  end
18 │  end
19 │  foreach a ∈ AR do
20 │  │  AR' := AR \ {a};
21 │  │  if φ_d(AR') ≥ φ_d(AR) then
22 │  │  │  AR := AR';
23 │  │  end
24 │  end
25 │  return AR;
26 end
```

---

small fraction of attributes from A is truly relevant with regard to the decision. In particular, if we assume that all attributes have similar distributions and that after $k << |A|$ iterations of the algorithm all the remaining attributes become irrelevant (thus they begin to behave similarly to random probes), then the process is likely to stop before an inclusion of any redundant feature.

The proposed stopping condition can be modified to consider the probability that *all* attributes added to the reduct are relevant. In order to express it, the probabilities computed in each iteration need to be remembered. Let us denote the probability from the i-th iteration by P_i. The probability that we do not select any irrelevant attribute in k iterations equals $1 - \prod_{i=1}^{k} (1 - P_i)$.

Additionally, Algorithm 2 can be used to explicitly compute a value of ε that can be regarded as an appropriate approximation threshold for a given data set, i.e. a value for which the probability of the inclusion of an irrelevant attribute is

---

Algorithm 3. Estimation of $P\Big(\phi_d\big(B \cup \{a\}\big) > \phi_d\big(B \cup \{\hat{a}\}\big)\Big)$

Input: a decision system $\mathbb{S}_d = \big(U, A \cup \{d\}\big)$; a quality measure $\phi_d : 2^A \to \mathbb{R}$; an attribute subset $B \subseteq A$; an attribute $a \in A$; a positive integer $nProbes$;

Output: an estimation of $P\Big(\phi_d\big(B \cup \{a\}\big) > \phi_d\big(B \cup \{\hat{a}\}\big)\Big)$

```
 1  begin
 2  |   higherScoreCount = 0;
 3  |   foreach i ∈ 1 : nProbes do
 4  |   |   generate a random probe âᵢ;
 5  |   |   if φd(B ∪ {a}) > φd(B ∪ {âᵢ}) then
 6  |   |   |   higherScoreCount = higherScoreCount + 1;
 7  |   |   end
 8  |   end
 9  |   return higherScoreCount / nProbes;
10  end
```

---

desirably low. This value can be estimated by $\varepsilon_{p_{rel}} = 1 - \phi_d(AR)/\phi_d(A)$, where AR is an approximate reduct computed using the parameter p_{rel}. However, one needs to realize that it is just the first step toward grasping a relationship between the approximation thresholds and the random probe settings, if there exists any clear relationship at all.

4   Assessment of a Reduct Quality

The idea explained in the previous section may also be utilized for assessing a quality of reducts. In practice, this assessment is usually performed with a use of additional validation data and it requires computation of a classification model [5,11]. Then, the quality of a reduct is identified with the predictive performance of a classifier constructed using that reduct. Although such a method can accurately estimate usefulness of a reduct for a classification task, its downside is a considerable computational cost and the requirement for additional data. Moreover, this approach does not make sense when applied to reducts computed for different tasks than the supervised classification.

Two reducts can also be compared with regard to their basic characteristics, such as the cardinality or a number of generated indiscernibility classes. Reducts that minimize those indicators represent more general dependencies in data and are usually preferable. However, those simple statistics are quite coarse – reducts with the same cardinality may have completely different properties. Additionally, those characteristics do not allow to meaningfully compare different types of reducts, e.g. it is quite obvious that an approximate reduct computed for $\varepsilon = 0.1$ will usually be shorter and will generate less indiscernibility classes than a reduct generated for $\varepsilon = 0.01$, but we are unable to say, based only on those statistics, which of them is more reliable.

In order to address those issues, we introduce an evaluation method measuring the odds that a reduct does not involve random dependencies. We call the proposed measure the *relevanceLift*. Consider a decision system $\mathbb{S}_d = (U, A \cup \{d\})$ and let AR be an approximate decision reduct of $\mathbb{S}_d$, constructed using the attribute subset quality measure ϕ_d. Let us denote by a_{AR} a new attribute constructed by enumerating equivalence classes of the indiscernibility relation IND_{AR} and assigning the corresponding number to every object from U. For the decision system $\mathbb{S}_d$ we may compute a number of randomized reducts RR_i, $i = 1, \ldots, n$, using the same algorithm and parameter settings as were applied to compute AR. The *relevanceLift* of AR can be given as follows:

$$ relevanceLift(AR) = \frac{P\big(\phi_d(\{a_{AR}\}) > \phi_d(\{\hat{a}_{AR}\})\big)}{\frac{1}{n}\sum_{i=1}^{n} P\big(\phi_d(\{a_{RR_i}\}) > \phi_d(\{\hat{a}_{RR_i}\})\big)} $$

In simple words, the above formula expresses a ratio between the probability that a given reduct corresponds to a relevant attribute and the probability that a relevant attribute can also be generated by a chance from a randomized reduct. The latter of those probabilities is estimated by the average from probabilities obtained for different randomized reducts. Importantly, the same algorithm and parameter settings as were used in the computation of AR need to be used for the derivation of the randomized reducts RR_i. This can be regarded as a normalization of the relevance probability with respect to the computation method. In this way it is possible to compare reducts computed using different algorithms or parameter settings, as long as some kind of attribute subset quality measure is used in the process.

5   Experimental Evaluation of the Proposed Algorithm

To verify usefulness of the methods proposed in this paper, we conducted a series of preliminary experiments on synthetic data sets. We decided to utilize only the artificial data because we wanted to have an explicit control on dependencies between conditional and decision attributes. This is a popular approach to testing attribute selection methods [10, 11].

In the experiments we generated 20 data sets, each composed of 10000 objects described by 1000 attributes. The attributes differed in a number of distinct values they could take (from 2 to 10 for individual attributes) and in a distribution of their values in data. For each set, the decision attribute was deterministically generated based on values of 10 first conditional attributes. The remaining attributes did not have any influence on decision classes. Next, some decisions were randomly changed in order to model disturbances in data. A fraction of changed decision values was a parameter of the data generation process. It was 10% for the first 10 data sets and to 25% for the remaining sets.

The experiments were conducted in R System [18]. For each data set, we followed Algorithm 1 to compute approximate decision reducts with different ε values. For this purpose we utilized the implementation from the *RoughSets*

Table 1. Average quality of approximate reducts computed using different methods

Criterion:	disturbance : 10%			disturbance : 25%		
	F_1-score	Precision	Recall	F_1-score	Precision	Recall
$\varepsilon = 0.01$	0.39	0.45	0.35	0.35	0.40	0.31
$\varepsilon = 0.10$	0.46	0.61	0.35	0.38	0.53	0.30
$\varepsilon = 0.25$	0.44	0.66	0.34	0.38	0.56	0.29
ProbeBased	0.50	0.96	0.34	0.46	1.00	0.30

package.[1] In particular, we used the function *FS.reduct.computation* with the parameter *method* set to *greedy.heuristic*. The ε values were set in the consecutive experiments to 0.01, 0.10 and 0.25. All other parameters of the algorithm were left to their defaults. We also computed approximate reducts using Algorithm 2. Its parameters were set to the defaults discussed in Section 3. Both reduct computation methods used the same attribute subset quality measure, i.e. the *Gini Gain* [19]. This measure can be understood as an expected decrease in the impurity of decisions after a division of data into indiscernibility classes defined by the attribute subset. It was also used in [16] as one of possible functions ϕ_d in the formulation of (ϕ_d, ε)-approximate decision reducts.

We assessed the quality of obtained reducts using three well-known measures from the information retrieval domain: F_1-score, *Precision* and *Recall*. Their explanation can be found, e.g., in [20]. Average results of those tests for all data sets are shown in Table 1. They indicate that the proposed stopping criterion for the greedy heuristic is more reliable than the original approach. For nearly all data sets, random probes helped to avoid including irrelevant attributes into a reduct. The obtained average precisions for data sets with 10% and 25% of artificial noise were equal to 0.96 and 1.00, respectively. As a consequence, the average F_1-scores for the reducts computed using random probes were also much higher than for other reducts. In the experiments, the random probe-based heuristic achieved the best F_1-score on 19 out of 20 data sets.

We also measured a correlation between F_1-scores and *relevanceLifts* computed for the approximate reducts with different ε values. It turned out to be high, i.e. 0.69, which indicates a strong linear dependency between those two measures. Since normally, for real-life data, we would not be able to obtain the F_1-score values, the proposed reduct quality assessment method may be very useful in practical applications. Unfortunately, one downside of the *relevanceLift* measure is its relatively high computational cost. Further studies are needed to formulate a more computationally efficient version of *relevanceLift*.

6   Conclusions

Results of the preliminary experiments presented in Section 5 show a big potential hidden in the methods that use random probes in the computation of

---

[1] The package *RoughSets* is available through R-CRAN under the following link:
http://cran.r-project.org/web/packages/RoughSets/index.html

reducts. We plan to continue our research on this subject by performing more thorough tests on a larger number of synthetic and real-life data sets. We also intend to include the implementation of proposed reduct computation and assessment methods in a new version of the *RoughSets* package.

There are a number of research directions worth further investigation. First of all, Algorithm 2 may require more careful tuning of the threshold p_{rel}. Although its default settings discussed in Section 3 seem to work efficiently, some other thresholds could bring better outcomes for particular decision systems. For example, appropriate tuning may limit a risk of "false irrelevants", i.e., meaningful attributes that were not allowed to join a reduct because of a too restrictive comparison with their random probe counterparts.

It is also important to identify data sets, for which random probes are especially useful. One of hypotheses refers to the existence of partial dependencies between the attributes in data. Imagine two attributes a and a', which have the same values except a few fluctuations on objects in U. Then, for a given $AR \subseteq A$, it may happen that $\phi_d(AR \cup \{a, a'\})$ is slightly higher than $\phi_d(AR \cup \{a\})$ and $\phi_d(AR \cup \{a'\})$. If $\phi_d(AR \cup \{a, a'\})$ is less than $(1-\varepsilon)\phi_d(A)$, then both a and a' are added to the resulting reduct, although one of them is practically meaningless given an inclusion of the another one. In general, we expect that dynamically adjusted approximate reduct calculations may be helpful for data sets with many non-trivial approximate dependencies rather than for those with a few very informative attributes, surrounded by many other that provide little additional information about the decisions [20].

Let us also comment on utilizing random probes in other methods of deriving useful decision reducts from data. Indeed, in this paper, we consider only the basic example of a greedy approach. On the other hand, there are other methods based on randomly generated attribute subsets and permutations, optionally preceded by partitioning A into clusters of attributes providing interchangeable information about d [5]. In our opinion, the random probe-based attribute significance test applied in Algorithm 2 can be reconsidered for all those methods. In particular, the idea of attribute clustering may work well with the above-mentioned idea of avoiding attributes bringing negligible increase of the quantity of ϕ_d. Moreover, we believe that the methodology of reduct quality assessment developed in Section 4 can be successfully embedded into the algorithms searching for bigger ensembles of decision reducts [15, 17].

References

1. Modrzejewski, M.: Feature Selection Using Rough Sets Theory. In: Brazdil, P.B. (ed.) ECML 1993. LNCS, vol. 667, pp. 213–226. Springer, Heidelberg (1993)
2. Pawlak, Z.: Rough Sets: Present State and the Future. Foundations of Computing and Decision Sciences 18(3-4), 157–166 (1993)
3. Błaszczyński, J., Słowiński, R., Susmaga, R.: Rule-Based Estimation of Attribute Relevance. In: Yao, J., Ramanna, S., Wang, G., Suraj, Z. (eds.) RSKT 2011. LNCS, vol. 6954, pp. 36–44. Springer, Heidelberg (2011)

4. Jensen, R., Shen, Q.: New Approaches to Fuzzy-Rough Feature Selection. IEEE Transactions on Fuzzy Systems 17(4), 824–838 (2009)
5. Janusz, A., Ślęzak, D.: Rough Set Methods for Attribute Clustering and Selection. Applied Artificial Intelligence 28(3), 220–242 (2014)
6. Świniarski, R.W., Skowron, A.: Rough Set Methods in Feature Selection and Recognition. Pattern Recognition Letters 24(6), 833–849 (2003)
7. Kohavi, R., John, G.H.: Wrappers for Feature Subset Selection. Artificial Intelligence 97, 273–324 (1997)
8. Ślęzak, D.: Approximate Reducts in Decision Tables. In: Proceedings of IPMU 1996, vol. 3, pp. 1159–1164 (1996)
9. Abeel, T., Helleputte, T., de Peer, Y.V., Dupont, P., Saeys, Y.: Robust Biomarker Identification for Cancer Diagnosis with Ensemble Feature Selection Methods. Bioinformatics 26(3), 392–398 (2010)
10. Guyon, I., Gunn, S., Nikravesh, M., Zadeh, L. (eds.): Feature Extraction, Foundations and Applications. STUDFUZZ, vol. 207. Physica-Verlag, Springer (2006)
11. Janusz, A., Stawicki, S.: Applications of Approximate Reducts to the Feature Selection Problem. In: Yao, J., Ramanna, S., Wang, G., Suraj, Z. (eds.) RSKT 2011. LNCS, vol. 6954, pp. 45–50. Springer, Heidelberg (2011)
12. Nguyen, H.S.: Approximate Boolean Reasoning: Foundations and Applications in Data Mining. In: Peters, J.F., Skowron, A. (eds.) Transactions on Rough Sets V. LNCS, vol. 4100, pp. 334–506. Springer, Heidelberg (2006)
13. Skowron, A., Rauszer, C.: The Discernibility Matrices and Functions in Information Systems. In: Słowiński, R. (ed.) Intelligent Decision Support. Theory and Decision Library, vol. 11, pp. 331–362. Kluwer (1992)
14. Bazan, J.G., Skowron, A., Synak, P.: Dynamic Reducts as a Tool for Extracting Laws from Decisions Tables. In: Raś, Z.W., Zemankova, M. (eds.) ISMIS 1994. LNCS, vol. 869, pp. 346–355. Springer, Heidelberg (1994)
15. Ślęzak, D., Janusz, A.: Ensembles of Bireducts: Towards Robust Classification and Simple Representation. In: Kim, T.-h., Adeli, H., Slezak, D., Sandnes, F.E., Song, X., Chung, K.-I., Arnett, K.P. (eds.) FGIT 2011. LNCS, vol. 7105, pp. 64–77. Springer, Heidelberg (2011)
16. Ślęzak, D.: Various Approaches to Reasoning with Frequency-Based Decision Reducts: A Survey. In: Polkowski, L., Lin, T., Tsumoto, S. (eds.) Rough Sets in Soft Computing and Knowledge Discovery: New Developments, pp. 235–285. Physica-Verlag (2000)
17. Widz, S., Ślęzak, D.: Approximation Degrees in Decision Reduct-Based MRI Segmentation. In: Proceedings of FBIT 2007, pp. 431–436 (2007)
18. R Development Core Team: R: A Language and Environment for Statistical Computing. R Foundation for Statistical Computing, Vienna, Austria (2008)
19. Soman, K.P., Diwakar, S., Ajay, V.: Insight into Data Mining: Theory and Practice. Prentice-Hall (2006)
20. Janusz, A., Nguyen, H.S., Ślęzak, D., Stawicki, S., Krasuski, A.: JRS'2012 Data Mining Competition: Topical Classification of Biomedical Research Papers. In: Yao, J., Yang, Y., Słowiński, R., Greco, S., Li, H., Mitra, S., Polkowski, L. (eds.) RSCTC 2012. LNCS (LNAI), vol. 7413, pp. 422–431. Springer, Heidelberg (2012)

Applications of Boolean Kernels in Rough Sets[*]

Sinh Hoa Nguyen[1,2] and Hung Son Nguyen[1]

[1] The University of Warsaw, Banacha 2, 02-097, Warsaw, Poland
[2] Polish-Japanese Institute of Inf. Technology, Koszykowa 86, 02008, Warsaw, Poland

Abstract. Rough Sets (RS) and Support Vector Machine (SVM) are the two big and independent research areas in AI. Originally, rough set theory is dealing with the concept approximation problem under uncertainty. The basic idea of RS is related to lower and upper approximations, and it can be applied in classification problem. At the first sight RS and SVM offer different approaches to classification problem. Most RS methods are based on minimal decision rules, while SVM converts the linear classifiers into instance based classifiers. This paper presents a comparison analysis between these areas and shows that, despite differences, there are quite many analogies in the two approaches. We will show that some rough set classifiers are in fact the SVM with Boolean kernel and propose some hybrid methods that combine the advantages of those two great machine learning approaches.

Keywords: Rough sets, SVM, Boolean Kernel, Hybrid Systems.

1 Introduction

Searching for approximation of a concept is a fundamental problem in machine learning and data mining. Classification, clustering, association analysis, and many other tasks in data mining can be formulated as concept approximation problems. Let $\mathcal{X}$ be a given universe of objects, and let $\mathcal{L}$ be a predefined descriptive language consisting of such formulas that are interpretable as subsets of $\mathcal{X}$. Concept approximation problem can be understood as a problem of searching for a description ψ of a given concept $C \subset \mathcal{X}$ such that (i) ψ expressible in $\mathcal{L}$ and (ii) the interpretation of ψ should be as close to the original concept as possible. Usually, the concept to be approximated is given on a *finite set of examples* $U \subset \mathcal{X}$, called the training set, only.

From the view point of classification problem, both Rough set and Support Vector Machine approaches are managing the concept of "boundary region" or "margin". However these approaches differ one from the other by the way of controlling the boundary regions and the form of final classifiers. Rough set

[*] Research supported by Polish National Science Centre (NCN) grants DEC-2011-/01/B/ST6/03867 and DEC-2012/05/B/ST6/03215, and the grant SYNAT - SP/I/1/77065/10 in frame of strategic scientific research and experimental development program: "Interdisciplinary System for Interactive Scientific and Scientific-Technical Information" founded by Polish National Centre for Research and Development (NCBiR).

M. Kryszkiewicz et al. (Eds.): RSEISP 2014, LNAI 8537, pp. 65–76, 2014.

approach to classification is based on Minimal Description Length (MDL) principle and hierarchical learning method. Usually, it is formulated in a form of such problems like reduct, decision reduct or minimal decision rule calculation. One of the common approaches in rough set theory is the Boolean Reasoning methodology to problem solving.

Support Vector Machine proposes the maximal margin principle. There are two main interesting idea in SVM approach to classification. The first is to transform the problem of searching for the optimal linear classifier with maximal margin into a dual Quadratic Optimization problem by using Lagrange multipliers. The second important idea is related to the Kernel tricks that realize an efficient mapping the input data set into a very large feature space.

Most existing attempts to combine rough sets with SVM are related to (i) using rough set approach to data reduction as a preprocessing step for SVM [3], [4], [7], and (ii) using two SVM models to approximate the lower and upper approximations of a concept [1], [8], [9], [10], [18].

In the previous paper we presented the applications of Boolean reasoning approach to Rough set and data mining [11]. The main idea is to transform the rough set based methods including reducts, decision rules, discretization and many other problems into the problem of searching for minimal prime implicants of a Boolean function. In this paper we discuss a not very popular class of kernel functions called the Boolean kernels. We show the general claim, that many rough set methods are equivalent to SVM with a corresponding Boolean kernel.

The paper is organized as follows. In Section 2 we describe the family of Boolean kernels and the Support Vector Machine using such kernels. In Sections 3 and 4 we present the rough set methods based on SVM with Boolean kernels. Section 5 presents some properties of the proposed methods and an illustrative example. The conclusions and future plan are presented in Section 6.

2   Boolean Reasoning Approach to Rough Set Methods

Feature selection has been an active research area in pattern recognition, statistics, and data mining communities. The main idea of feature selection is to select a subset of most relevant attributes for classification task, or to eliminate features with little or no predictive information. Feature selection can significantly improve the comprehensibility of the resulting classifier models and often build a model that generalizes better to unseen objects. Further, it is often the case that finding the correct subset of predictive features is an important problem in its own right. In rough set theory, the feature selection problem is defined in terms of reducts [13]. We will generalize this notion and show an application of the ABR approach to this problem.

For any subset of attributes $B \subset A$ of a given information system $\mathcal{S} = (U, A)$, the B-indiscernibility relation (denoted by $IND_{\mathcal{S}}(B)$) is defined by

$$IND_{\mathcal{S}}(B) = \{(x, y) \in U \times U : inf_B(x) = inf_B(y)\}.$$

Relation $IND_{\mathcal{S}}(B)$ is an equivalence relation. Its equivalence classes can be used to define the lower and upper approximations of concepts in rough set theory

[12], [13]. The complement of indiscernibility relation is called B-*discernibility relation* and is denoted by $DISC_\mathcal{S}(B)$. Hence,

$$DISC_\mathcal{S}(B) = \{(x, y) \in U \times U : inf_B(x) \neq inf_B(y)\}$$

Intuitively, reduct (in rough set theory) is a minimal subset of attributes that preserves the discernibility between information vectors of objects. The following notions of reducts are often used in rough set theory.

Definition 1 (information reducts). *Any minimal subset B of A such that $DISC_\mathcal{S}(A) = DISC_\mathcal{S}(B)$ is called the* information reduct *(or reduct) of $\mathcal{S}$. The set of all reducts of a given information system $\mathcal{S}$ is denoted by $RED(\mathcal{S})$*

In the case of decision tables, we are interested in the ability of describing decision classes by using subsets of condition attributes. This ability can be expressed in terms of *generalized decision function* $\partial_B : U \to \mathcal{P}(V_{dec})$, where

$$\partial_B(x) = \{i : \exists_{x' \in U} [(x'IND(B)x) \wedge (d(x') = i)]\}$$

Definition 2 (decision reducts). *A set of attributes $B \subseteq A$ is called a decision oriented reduct (or a relative reduct) of decision table $\mathcal{S}$ if and only if $\partial_B(x) = \partial_A(x)$ for all object $x \in U$ and any proper subset of B does not satisfy the previous condition, i.e. B is a minimal subset (with respect to the inclusion relation $\subseteq$) of the attribute set satisfying the property $\forall_{x \in U} \partial_B(x) = \partial_A(x)$.*

A set $C \subset A$ of attributes is called *super-reduct* if there exists a reduct B such that $B \subset C$.

2.1 Boolean Reasoning Approach to Reduct Problem

There are two problems related to the notion of "reduct", which have been intensively explored in rough set theory by many researchers (see, e.g., [2], [15], [6], [16], [11]. It has been shown that the problem of searching for shortest reduct is NP-hard (see [15]) and the problem of searching for all reducts is at least NP-hard. Some heuristics have been proposed for those problems. Here we present the approach based on Boolean reasoning as proposed in [15] [11] (see Fig. 1).

Given a decision table $\mathcal{S} = (U, A \cup \{dec\})$, where $U = \{u_1, u_2, \ldots, u_n\}$ and $A = \{a_1, \ldots, a_k\}$. By *discernibility matrix* of the decision table $\mathcal{S}$ we mean the $(n \times n)$ matrix $\mathbf{M}(\mathcal{S}) = [M_{i,j}]_{ij=1}^{n}$, where $M_{i,j} \subset A$ is the set of attributes discerning u_i and u_j, i.e., $M_{i,j} = \{a_m \in A : a_m(u_i) \neq a_m(u_j)\}$. Let us denote by $VAR_\mathcal{S} = \{x_1, \ldots, x_k\}$ a set of boolean variables corresponding to attributes $a_1, \ldots, a_k$. For any subset of attributes $B \subset A$, we denote by $X(B)$ the set of boolean variables corresponding to attributes from B. We will encode reduct problem as a problem of searching for the corresponding set of variables.

For any two objects $u_i, u_j \in U$ such that $M_{i,j} \neq \emptyset$, the boolean clause χ_{u_i, u_j}, called *discernibility clause*, is defined as follows:

$$\chi_{u_i, u_j}(x_1, \ldots, x_k) = \sum_{a_m \in M_{i,j}} x_m \tag{1}$$

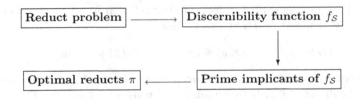

Fig. 1. The Boolean reasoning scheme for solving reduct problem

The objective is to create a boolean function of which any prime implicant corresponds to a reduct of $\mathcal{S}$. In case of the relative reduct problem, it is called the *decision oriented discernibility function* and denoted by $f_{\mathcal{S}}^{dec}$:

$$f_{\mathcal{S}}^{dec}(x_1, \ldots, x_k) = \prod_{dec(u_i) \neq dec(u_j)} \left(\chi_{u_i, u_j}(x_1, \ldots, x_k) \right)$$

The following properties were proved in [15]:

1. A subset of attributes B is a reduct of $\mathcal{S}$ if and only if the term $T_{X(B)} = \prod_{a_i \in B} x_i$ is a prime implicant of the discernibility function $f_{\mathcal{S}}$.
2. A subset of attributes B is a relative reduct of decision table $\mathcal{S}$ if and only if $T_{X(B)}$ is the prime implicant of the relative discernibility function $f_{\mathcal{S}}^{dec}$.

2.2   Decision Rule Induction

Decision rules are important components of many rough classifiers, see [2]. Therefore, searching for a good collection of rules is the fundamental problem in rough sets. In this section we investigate the application of the boolean reasoning method to the rule induction problem.

Decision rules are logical formulas that indicate the relationship between conditional and decision attributes. In this paper we consider only those decision rules **r** whose the premise is a boolean monomial of descriptors, i.e.,

$$\mathbf{r} \equiv (a_{i_1} = v_1) \wedge \ldots \wedge (a_{i_m} = v_m) \Rightarrow (dec = k) \tag{2}$$

Every decision rule **r** of the form (2) can be characterized by:

$$length(\mathbf{r}) = \text{the number of descriptors in the premise of } \mathbf{r}$$
$$support(\mathbf{r}) = \text{the number of objects satisfying the premise of } \mathbf{r}$$
$$conf(\mathbf{r}) = \text{the confidence of } \mathbf{r}\text{: } confidence(\mathbf{r}) = \frac{|[\mathbf{r}] \cap DEC_k|}{|[\mathbf{r}]|}$$

The decision rule **r** is called *consistent* with $\mathbb{A}$ if $confidence(\mathbf{r}) = 1$.

In rough set approach to concept approximation, decision rules are also used to define rough approximations. Each decision rule is supported by some objects and, inversely, the information vector of each object can be reduced to obtain a minimal consistent decision rule.

The boolean reasoning approach to decision rule construction from a given decision table $S = (U, A \cup \{dec\})$ is very similar to the minimal reduct problem. The only difference occurs in the encoding step, i.e. for any object $u \in U$ in , we define a function $f_u(a_1^*, ..., a_m^*)$, called *discernibility function for u* by

$$f_u(a_1^*, ..., a_m^*) = \prod_{u_j : dec(u_j) \neq dec(u)} \chi_{u,u_j}(x_1, ..., x_k) \qquad (3)$$

where χ_{u,u_j} is the discernibility clause for objects u and u_j. (see Eq. (1)).

3   Support Vector Machine Using Boolean Kernels

In this section we recall the main idea of support vector machine approach to classification problem. Given a decision table which can be also as a training set of object-decision pairs $\mathbb{D} = \{(\mathbf{x}_1, y_1), (\mathbf{x}_2, y_2), ..., (\mathbf{x}_m, y_m)\}$, where $\mathbf{x}_i \in R^n$ and $y_i \in \{-1, 1\}$ for $1 \leq i \leq m$, the support vector machine (SVM) [17] requires the solution of the following optimization problem:

$$\min_{\mathbf{w},\xi,b} \frac{||\mathbf{w}||^2}{2} + C \sum \xi_i$$

$$\text{subject to } y_i(\mathbf{w} \cdot \Psi(\mathbf{x}_i) + b) \geq 1 - \xi_i; \text{ for } i = 1, ..., n \qquad (4)$$

Here Ψ is a function that maps the objects $\mathbf{x}_i$ into a higher (maybe infinite) dimensional space. SVM is searching for a linear separating hyperplanes with the maximal margin in this higher dimensional space. Furthermore $K(\mathbf{x}_i, \mathbf{x}_j) =_{def} \langle \Psi(\mathbf{x}_i), \Psi(\mathbf{x}_i) \rangle$ is called the kernel function.

The first mathematical tricks that make SVM one of the most efficient classifiers in recent years is related to the use of Lagrange functions to transform the SVM problem into the following quadratic optimization problem:

$$\max_{\alpha} \sum_{i=1}^{n} \alpha_i - \frac{1}{2} \sum_{ij=1}^{n} \alpha_i \alpha_j y_i y_j K(\mathbf{x}_i, \mathbf{x}_j)$$

$$\text{subject to } C \geq \alpha_i \geq 0, i = 1, .., n \quad and \quad \sum_{i=1}^{n} \alpha_i y_i = 0 \qquad (5)$$

This new problem is also called the dual representation SVM problem. The values new positive parameters $(\alpha_1, ... \alpha_m)$ can be interpreted as the significance weights of the objects. In particular, if $(\alpha_1^o, ... \alpha_m^o)$ are the solutions of the quadratic optimization problem (5), then the object $\mathbf{x}_i$ is called *the support vector* if $\alpha_i^o > 0$, otherwise the object is called *redundant* and can be removed from the training decision table.

The support vectors can be used to build the SVM decision function that classifies any new object $\mathbf{x}$ as follows:

$$d_{SVM}(\mathbf{x}) = \text{sgn} \left(\sum_{\text{sup. vectors}} y_i \alpha_i^o K(\mathbf{x}_i, \mathbf{x}) + b_0 \right) \qquad (6)$$

3.1   Boolean Kernels

There exist many efficient kernel functions, however all of these kernels are suitable for real value data only. In this paper we recall the learning algorithm for boolean functions in Disjunctive Normal Form (DNF) via Linear Threshold Elements (LTE) proposed in [5], [14].

Let $\mathbb{D} = \{(\mathbf{x}_1, y_1), (\mathbf{x}_2, y_2), ..., (\mathbf{x}_m, y_m)\}$ be the training decision table, where $\mathbf{x}_i \in \{0,1\}^n$ and $y_i \in \{-1, 1\}$ for $1 \le i \le m$. Usually the objects from $\mathbb{D}$ are not linearly separable and the straightforward application of LTE to solve the classification problem over $\mathbb{D}$ is not possible.

The idea of the solution to this problem is to define a mapping $\Psi : \{0,1\}^n \longrightarrow \{0,1\}^N$ from the original space of Boolean vectors $\{0,1\}^n$ into an expanded (or enriched) feature-space $\{0,1\}^N$, in which the objects become linearly separable. Similarly to rule-based classifiers, the expanded feature-space is determined by a set $\mathbb{M} = \{m_1, ..., m_N\}$ of Boolean monomials. Each Boolean vector $\mathbf{x} \in \{0,1\}^n$ is transformed to the extended vector $\langle y_1, ..., y_N \rangle \in \{0,1\}^N$ using the encoding function $y_i = \psi_{m_i}(\mathbf{x})$, which is inductively defined as follows:

$$\psi_{x_i}(\mathbf{x}) = x_i; \quad \psi_{\overline{x_i}}(\mathbf{x}) = 1 - x_i; \quad \psi_{f_1 \wedge f_2}(\mathbf{x}) = \psi_{f_1}(\mathbf{x}) \cdot \psi_{f_2}(\mathbf{x}).$$

For example, the monomial $m_i = x_1 \overline{x_3} x_5$ is encoded by the function $\psi_{m_i} = x_1(1 - x_3)x_5$. Therefore, for any family $\mathbb{M} = \{m_1, ..., m_N\}$ of Boolean monomials, the transformation Ψ is a vector of N single-valued Boolean functions:

$$\Psi(\mathbf{x}) = \langle \psi_{m_1}(\mathbf{x}), \dots, \psi_{m_N}(\mathbf{x}) \rangle$$

It has been shown (see [14]) that

Theorem 1 (Sadohara). *For any boolean function $g(\mathbf{x})$ there exists a hyperplane $L(\mathbf{z}) = \sum w_i z_i = 0$ such that*

$$g(\mathbf{x}) = 1 \Leftrightarrow L(\Psi(\mathbf{x})) = 1 \text{ and } g(\mathbf{x}) = 0 \Leftrightarrow L(\Psi(\mathbf{x})) = -1$$

for any $\mathbf{x} \in \{0,1\}^n$.

After the transformation of training examples into an expanded feature space, we can apply the LTE learning algorithm to find the optimal hyperplanes for the extended feature vectors of the objects, i.e., for the new training data set:

$$\{(\Psi(\mathbf{x}_1), y_1), (\Psi(\mathbf{x}_2), y_2), ..., (\Psi(\mathbf{x}_m), y_m)\}.$$

The described above representation of hyperplanes may have $3^n - 1$ terms for n-variable Boolean functions, it is difficult to deal with the representation directly from a computational viewpoint. The following kernel functions have been proposed [5], [14] in order to learn the different forms of DNF expressions:

$$K_1^{k,\varepsilon}(\mathbf{u}, \mathbf{v}) = \sum_{i=1}^{k} \binom{\langle \mathbf{u}, \mathbf{v} \rangle}{i} \varepsilon^i$$

$$K_2^{\varepsilon}(\mathbf{u}, \mathbf{v}) = (1 + \varepsilon)^{\langle \mathbf{u}, \mathbf{v} \rangle} - 1$$

$$K_3^{\varepsilon}(\mathbf{u}, \mathbf{v}) = (1 + \varepsilon)^{\langle \mathbf{u}, \mathbf{v} \rangle + \langle \overline{\mathbf{u}}, \overline{\mathbf{v}} \rangle} - 1$$

where $\overline{\mathbf{u}}$ and $\overline{\mathbf{v}}$ are the complements of binary vectors $\mathbf{u}, \mathbf{u} \in \{0,1\}^n$.

4   Construction of Rough Classifier by SVM

In this Section we prove a kind of correspondence result claiming that the Support Vector Machine with Boolean kernels is equivalent to some setting of rule-based classifiers in rough set theory.

First of all, let us recall the process of construction rule-based classifier in rough set theory. Usually, this process consists of the following steps:

P1: *Learning phase*: generates a set of decision rules $\mathcal{R}(\mathbb{S})$ (satisfying some predefined conditions) from a given decision table $\mathbb{S}$.

P2: *Rule selection phase*: selects from $\mathcal{R}(\mathbb{S})$ the set of such rules that can be supported by $\mathbf{x}$. We denote this set by $MatchRules(\mathbb{S}, \mathbf{x})$.

P3: *Post-processing phase*: makes a decision for $\mathbf{x}$ using some voting algorithm for decision rules from $MatchRules(\mathbb{A}, \mathbf{x})$

The set of decision rules $\mathcal{R}(\mathbb{S})$ can be generated in Step P1 by different heuristics including the methods based on Boolean reasoning presented in Section 2.2. Let $\mathcal{R}(\mathbb{S}) = \{R_1, \ldots, R_p\}$ and let w_i be the strength of rule R_i and d_i the decision class indicated by the decision rule R_i for $i = 1, ..., p$.

In rule-based classifiers we should implement the matching function, denoted by $Match(\mathbf{x}, R_i)$, which returns the matching degree of the object $\mathbf{x}$ to the decision rule R_i, where $\mathbf{x}$ is the object to be classified.

In the voting step, each decision rule R_i can be treated as a voter who give the vote for the decision class d_i. The power of his vote is proportional to his strength w_i and the matching degree $Match(\mathbf{x}, R_i)$. The decision class with the higher total voting power will be selected as the final decision made for the object $\mathbf{x}$. In other words, the final decision made by rule-based classifier in rough set theory can be rewritten as follows:

$$Dec_{\mathcal{R}}(\mathbf{x}) = S \left(\sum_{i=1}^{p} dec(R_i) \cdot w_i \cdot Match(\mathbf{x}, R_i) \right) \tag{7}$$

where S is an activate function that converts the total voting power into a decision class. One can see many analogies in Equations (6) and (7).

Secondly, let us consider the following interpretation of Boolean kernels:

Definition 3. *For any $\varepsilon > 0$ and a set $\mathbb{M} = \{m_1, ..., m_N\}$ of Boolean monomials, we can define a real ε-transformation $\Phi_{\mathbb{M}}^{\varepsilon} : \{0,1\}^n \longrightarrow \mathbb{R}^N$ such that $\Phi_{\mathbb{M}}^{\varepsilon}(\mathbf{x}) = \langle z_1, ..., z_N \rangle \in \mathbb{R}^N$, where the values z_i are defined by $z_i = \varepsilon^{\frac{|m_i|}{2}} \phi_{m_i}(\mathbf{x})$.*

The smaller is the value of ε, the lower is the importance of long monomials in the model. For this real ε-transformation, the following facts hold:

Theorem 2. *For the real ε-transformation $\Psi_{\mathbb{M}}^{\varepsilon}$ defined in Definition 3, if $k \geq \max\{|m_i| : i = 1...N\}$ we have the following identity:*

$$K_1^{k,\varepsilon}(\mathbf{u}, \mathbf{v}) = \langle \Phi_{\mathbb{M}}^{\varepsilon}(\mathbf{u}), \Phi_{\mathbb{M}}^{\varepsilon}(\mathbf{v}) \rangle$$

Moreover if $k \geq \langle \mathbf{u}, \mathbf{v} \rangle$ then $K_1^{k,\varepsilon}(\mathbf{u}, \mathbf{v}) = K_2^{\varepsilon}(\mathbf{u}, \mathbf{v})$.

The proof of this Theorem is quite straitforward, therefore it is omitted due to the lack of space. We will prove the following crucial result of the paper.

Theorem 3. *Let $S = (U, A \cup \{d\})$ be a symbolic value decision table with two decision classes(without lost of generality, we assume that $V_d = \{1, -1\}$). One can learn from S a support vector machine with Boolean kernels that can simulate a rule-based classifier for S.*

Proof. Due to the limited space, we present the main idea of the proof only.

Let $Desc(S)$ be the set of all possible descriptors of the form $(a = v)$, where $a \in A$ and $v \in V_a$, derived from S. Assume that $Desc(S) = \{D_1, \ldots, D_M\}$, then if $U = \{u_1, \ldots u_m\}$, then we define a new boolean training decision table

$$\mathbb{D}(S) = \{(\mathbf{x}_1, y_1), (\mathbf{x}_2, y_2), \ldots, (\mathbf{x}_m, y_m)\}, \tag{8}$$

where $\mathbf{x}_i = (x_1^{(i)}, x_2^{(i)}) \ldots x_M^{(i)}) \in \{0, 1\}^M$, where $x_j^{(i)} = 1$ if and only if u_i satisfies the descriptor D_j and $y_i = d(u_i) \in \{-1, 1\}$ for $i = 1, \ldots, m$.

Let $(\alpha_1^o, \ldots \alpha_m^o)$ be the optimal solution for the SVM quadratic optimization problem with Boolean kernel function K_2^{ε} for the training table $\mathbb{D}(S)$ described in (5), and let $SV = \{(\mathbf{x}_{i_1}, y_{i_1}, \alpha_{i_1}^o), (\mathbf{x}_{i_2}, y_{i_2}, \alpha_{i_2}^o), \ldots, (\mathbf{x}_{i_s}, y_{i_s}, \alpha_{i_s}^o)\}$ be the set of support vectors (i.e. corresponding to non-zero parameters α_i^o) for this SVM machine. We will show that there exists a set of decision rules $\mathcal{R}$ for which the rough classifier defined by Equation (7) is equivalent to the SVM classifier based on support vectors SV.

According to Definition 3 and Theorem 2, we can assume that for any support vector $(\mathbf{x}_{i_1}, y_{i_1}, \alpha_{i_1}^o) \in SV$ the ε-transformation Φ_M^{ε} maps the vector $\mathbf{x}$ into a vector $\langle z_1, \ldots, z_N \rangle \in \mathbb{R}^N$, where $z_j = \varepsilon^{\frac{|m_j|}{2}} \phi_{m_j}(\mathbf{x}_i)$, and m_j is a DNF formula defined over $Desc(S)$. Therefore, we can define by $\mathcal{R}(\mathbf{x}_i)$ the set of all decision rules of the form $R_{i,j} = (m_j \implies y_j)$, for which $z_j \neq 0$, i.e. $\phi_{m_j}(\mathbf{x}_i) \neq 0$. The strength of $R_{i,j}$ is defined by z_j. Now we can set $\mathcal{R}(SV) = \mathcal{R}(\mathbf{x}_i) \cup \ldots \mathcal{R}(\mathbf{x}_i)$ and one can see that the rule based classifier using $\mathcal{R}(SV)$ is equivalent to the SVM based on the support vectors from SV.

The previous Theorem is showing the correspondence result between Rough set and SVM approaches to classification. As the consequence we have the following practical observations:

1. *The kernel functions K_3^{ε} can be directly calculated from the discernibility matrix $\mathbf{M}(S)$:* Indeed, one can see that if for any $\mathbf{u}, \mathbf{v} \in \{0, 1\}^m$, we define the counting matrix $[n_{ij}(\mathbf{u}, \mathbf{v})]_{i,j \in \{0,1\}}$ by

$$n_{ij}(\mathbf{u}, \mathbf{v}) = card\left(\{k : \mathbf{u}[k] = i; \mathbf{v}[k] = j\}\right).$$

    It is obvious that $n_{00}(\mathbf{u}, \mathbf{v}) + n_{01}(\mathbf{u}, \mathbf{v}) + n_{10}(\mathbf{u}, \mathbf{v}) + n_{11}(\mathbf{u}, \mathbf{v}) = m$. Moreover

$$\langle \mathbf{u}, \mathbf{v} \rangle = n_{11}(\mathbf{u}, \mathbf{v}) \text{ and } \langle \overline{\mathbf{u}}, \overline{\mathbf{v}} \rangle = n_{00}(\mathbf{u}, \mathbf{v})$$

    According to this observation, we can see that if $\mathbf{x}_i$ and $\mathbf{x}_j$ are the boolean representation of $u_i, u_j \in U$ then

$$K_3^{\varepsilon}(\mathbf{x}_i, \mathbf{x}_j) = (1 + \varepsilon)^{|A| - |M_{i,j}|} - 1$$

2. *In case of real value decision table, there exists an SVM machine based on Boolean kernel that can simulate the Rough classifier using Discretization and Decision Rules*
3. *Although the SVM based on Boolean kernel does not output the decision rules, it can be applied as a preprocessing step to select the support vectors as the most important objects in rough set based algorithms*

5   The Illustrative Example

Let us consider the decision table "weather" presented in Fig. 2 (left). This table consists of 4 attributes: a_1, a_2, a_3, a_4, hence the set of corresponding boolean variables consists of $VAR_S = \{x_1, x_2, x_3, x_4\}$. The discernibility matrix can be treated as a $n \times n$ board. Noteworthy is the fact that discernibility matrix is symmetrical with respect to the main diagonal, and that rearranging the objects according to their decisions one can shift off all empty boxes nearby to the main diagonal. In case of decision table with two decision classes, the discernibility matrix can be presented in a more compact form as shown in Fig. 2 (right).

The discernibility function is constructed from discernibility matrix by taking conjunction of all discernibility clauses. After reducing repeated clauses we have:

$$\begin{aligned} f(x_1, x_2, x_3, x_4) =& (x_1)(x_1 + x_4)(x_1 + x_2)(x_1 + x_2 + x_3 + x_4)(x_1 + x_2 + x_4) \\ & (x_2 + x_3 + x_4)(x_1 + x_2 + x_3)(x_4)(x_2 + x_3)(x_2 + x_4) \\ =& (x_1 + x_3)(x_3 + x_4)(x_1 + x_2 + x_4). \end{aligned}$$

One can find relative reducts of the decision table by searching for its prime implicants. The straightforward method calculates all prime implicants by translation to DNF. After removing those clauses that are absorbed by shorter clauses (using absorption rule: $p(p + q) \equiv p$) we have $f = (x_1)(x_4)(x_2 + x_3)$. Now we can translate f from CNF into DNF: $f = x_1 x_4 x_2 + x_1 x_4 x_3$. Every monomial corresponds to a reduct. Thus we have 2 reducts: $R_1 = \{a_1, a_2, a_4\}$ and $R_2 = \{a_1, a_3, a_4\}$.

To demonstrate the rule induction step, let us consider the object number 1. The discernibility function is determined as follows:

$$\begin{aligned} f_1(x_1, x_2, x_3, x_4) =& x_1(x_1 + x_2)(x_1 + x_2 + x_3)(x_1 + x_2 + x_3 + x_4) \\ & (x_2 + x_3)(x_1 + x_2 + x_3)(x_2 + x_3 + x_4)(x_1 + x_2 + x_4) \end{aligned}$$

After transformation into DNF we have

$$f_1(x_1, x_2, x_3, x_4) = x_1(x_2 + x_3) = x_1 x_2 + x_1 x_3$$

Hence, there are two object oriented reducts, i.e., $\{a_1, a_2\}$ and $\{a_1, a_3\}$. The corresponding decision rules are $(a_1 = \text{sunny}) \wedge (a_2 = \text{hot}) \Rightarrow (dec = \text{no})$ and $(a_1 = \text{sunny}) \wedge (a_3 = \text{high}) \Rightarrow (dec = \text{no})$.

Let us notice that all rules have the same decision class, precisely the class of the considered object. If we wish to obtain minimal consistent rules for the other decision classes, we should repeat the algorithm for another object.

S	outlook	temp.	humidity	windy	play
ID	a_1	a_2	a_3	a_4	dec
1	sunny	hot	high	FALSE	no
2	sunny	hot	high	TRUE	no
3	overcast	hot	high	FALSE	yes
4	rainy	mild	high	FALSE	yes
5	rainy	cool	normal	FALSE	yes
6	rainy	cool	normal	TRUE	no
7	overcast	cool	normal	TRUE	yes
8	sunny	mild	high	FALSE	no
9	sunny	cool	normal	FALSE	yes
10	rainy	mild	normal	FALSE	yes
11	sunny	mild	normal	TRUE	yes
12	overcast	mild	high	TRUE	yes

M	1	2	6	8
3	a_1	a_1, a_4	A	a_1, a_2
4	a_1, a_2	a_1, a_2, a_4	a_2, a_3, a_4	a_1
5	a_1, a_2, a_3	A	a_4	a_1, a_2, a_3
7	A	a_1, a_2, a_3	a_1	A
9	a_2, a_3	a_2, a_3, a_4	a_1, a_4	a_2, a_3
10	a_1, a_2, a_3	A	a_2, a_4	a_1, a_3
11	a_2, a_3, a_4	a_2, a_3	a_1, a_2	a_3, a_4
12	a_1, a_2, a_4	a_1, a_2	a_1, a_2, a_3	a_1, a_4

Fig. 2. An example of decision table and a compact form of the discernibility matrix

One can see in Table 1 the set of all possible decision rules, and the process of classification for the object $x = \langle$ sunny, mild, high, TRUE $\rangle$.

The set $MatchRules(\mathbb{A}, x)$ of exactly matched rules consists of two rules (nr 3 and nr 13). However, if we can accept a partial matching, few more decision rules can be considered in the voting process. In any case the object x is classified into the class "no".

The illustration of our main idea, using kernel $K_2^\varepsilon(.,.)$, is shown in Table 2. One can see that there are 10 descriptors and are 8 support vectors (four objects for each decision class). We can see that in this case, the testing object is also classified into the decision class "no".

If we restrict the original decision table into those 8 objects, the corresponding discernibility matrix is shown in Table 3. It is easy to notice the fact that all reducts of the original decision table, i.e. $R_1 = \{a_1, a_2, a_4\}$ and $R_2 = \{a_1, a_3, a_4\}$, are still presented in this restricted decision table.

Table 1. The set of all minimal decision rules generated by RSES and the classification process for the object $x = \langle$ sunny, mild, high, TRUE $\rangle$

RId	Condition	⇒ Decision	supp.	match
1	outlook(overcast) ⇒	yes	4	0
2	humidity(normal) AND windy(FALSE) ⇒	yes	4	0
3	outlook(sunny) AND humidity(high) ⇒	no	-3	1
4	outlook(rainy) AND windy(FALSE) ⇒	yes	3	0
5	outlook(sunny) AND temp.(hot) ⇒	no	-2	1/2
6	outlook(rainy) AND windy(TRUE) ⇒	no	-2	1/2
7	outlook(sunny) AND humidity(normal) ⇒	yes	2	1/2
8	temp.(cool) AND windy(FALSE) ⇒	yes	2	0
9	temp.(mild) AND humidity(normal) ⇒	yes	2	1/2
10	temp.(hot) AND windy(TRUE) ⇒	no	-1	1/2
11	outlook(sunny) AND temp.(mild) AND windy(FALSE) ⇒	no	-1	2/3
12	outlook(sunny) AND temp.(cool) ⇒	yes	1	1/2
13	outlook(sunny) AND temp.(mild) AND windy(TRUE) ⇒	yes	1	1
14	temp.(hot) AND humidity(normal) ⇒	yes	1	0

Table 2. The boolean representation of the weather decision table and the support vectors (nr. 1, 4, 6, 8, 9, 11, 12, 14) corresponding to the optimal parameters α_i

ID	outlook	temp	humidity	windy	play	sunny	overcast	rainy	hot	mild	cool	high	normal	FALSE	TRUE	dec	alpha=0.2	sum
1	sunny	hot	high	FALSE	no	1	0	0	1	0	0	1	0	1	0	-1	0,22	-0,64
2	sunny	hot	high	TRUE	no	1	0	0	1	0	0	1	0	0	1	-1	0,00	0,00
3	overcast	hot	high	FALSE	yes	0	1	0	1	0	0	1	0	1	0	1	0,00	0,00
4	rainy	mild	high	FALSE	yes	0	0	1	0	1	0	1	0	1	0	1	6,20	17,87
5	rainy	cool	normal	FALSE	yes	0	0	1	0	0	1	0	1	1	0	1	0,00	0,00
6	rainy	cool	normal	TRUE	no	0	0	1	0	0	1	0	1	0	1	-1	2,60	-3,12
7	overcast	cool	normal	TRUE	yes	0	1	0	0	0	1	0	1	0	1	1	0,00	0,00
8	sunny	mild	high	FALSE	no	1	0	0	0	1	0	1	0	1	0	-1	6,12	-31,73
9	sunny	cool	normal	FALSE	yes	1	0	0	0	0	1	0	1	1	0	1	1,69	2,02
10	rainy	mild	normal	FALSE	yes	0	0	1	0	1	0	0	1	1	0	1	0,00	0,00
11	sunny	mild	normal	TRUE	yes	1	0	0	0	1	0	0	1	0	1	1	2,86	14,83
12	overcast	mild	high	TRUE	yes	0	1	0	0	1	0	1	0	0	1	1	2,79	14,45
13	overcast	hot	normal	FALSE	yes	0	1	0	1	0	0	0	1	1	0	1	0,00	0,00
14	rainy	mild	high	TRUE	no	0	0	1	0	1	0	1	0	0	1	-1	4,60	-23,84
15	sunny	mild	high	TRUE	?	1	0	0	0	1	0	1	0	0	1			-10,16

Table 3. The discernibility matrix for the support vectors (nr. 1, 4, 6, 8, 9, 11, 12, 14)

M	1	6	8	14
4	a_1, a_2	a_2, a_3, a_4	a_1	a_4
9	a_2, a_3	a_1, a_4	a_2, a_3	A
11	a_2, a_3, a_4	a_1, a_2	a_3, a_4	a_1, a_3
12	a_1, a_2, a_4	a_1, a_2, a_3	a_1, a_4	a_1

Moreover if we are generating decision rules for the support vectors (objects nr 1, 4, 6, 8, 9, 11, 12, 14) only, one can obtain 12 among 14 decision rules from Table 1. We lost only two rules (nr 10 and 14), which have the lowest supports. These facts confirm the our hypothesis that SVM can be applied as an object reduction preprocessing step for rough set methods.

6 Conclusions

We have presented some properties of Boolean kernel functions and proved the correspondence theorem between SVM and Boolean reasoning approach to rough set theory. These theoretical results are showing, that despite differences at the first sight, there are quite many analogies in the two approaches. We also proposed an idea how to combine these two great approaches to data analysis. In the future we will want to present the experiment results performed on benchmark data sets to evaluate the efficiency of the proposed idea.

References

1. Asharaf, S., Shevade, S.K., Murty, N.M.: Rough support vector clustering. Pattern Recognition 38(10), 1779–1783 (2005)
2. Bazan, J.: A Comparison of Dynamic and non-Dynamic Rough Set Methods for Extracting Laws from Decision Tables. In: Polkowski, L., Skowron, A. (eds.) Rough Sets in Knowledge Discovery 1: Methodology and Applications. STUD-FUZZ, vol. 18, pp. 321–365. Springer, Heidelberg (1998)
3. Chen, H.-L., Yang, B., Liu, J., Liu, D.-Y.: A support vector machine classifier with rough set-based feature selection for breast cancer diagnosis. Expert Systems with Applications 38(7), 9014–9022 (2011)
4. Honghai, F., Baoyan, L., Cheng, Y., Ping, L., Bingru, Y., Yumei, C.: Using Rough Set to Reduce SVM Classifier Complexity and Its Use in SARS Data Set. In: Khosla, R., Howlett, R.J., Jain, L.C. (eds.) KES 2005. LNCS (LNAI), vol. 3683, pp. 575–580. Springer, Heidelberg (2005)
5. Khardon, R., Servedio, R.A.: Maximum margin algorithms with boolean kernels. J. of Machine Learning Res. 6, 1405–1429 (2005)
6. Kryszkiewicz, M.: Maintenance of reducts in the variable precision rough set model. In: Lin, T.Y., Cercone, N. (eds.) Rough Sets and Data Mining – Analysis of Imperfect Data, pp. 355–372. Kluwer Academic Publishers, Boston (1997)
7. Li, Y., Cai, Y., Li, Y., Xu, X.: Rough sets method for SVM data preprocessing. In: Proceedings of the 2004 IEEE Conference on Cybernetics and Intelligent Systems, pp. 1039–1042 (2004)
8. Lingras, P., Butz, C.J.: Interval set classifiers using support vector machines. In: Proc. the North American Fuzzy Inform. Processing Society Conference, pp. 707–710 (2004)
9. Lingras, P., Butz, C.J.: Rough set based 1-v-1 and 1-v-r approaches to support vector machine multi-classification. Information Sciences 177, 3782–3798 (2007)
10. Lingras, P., Butz, C.J.: Rough support vector regression. European Journal of Operational Research 206, 445–455 (2010)
11. Nguyen, H.S.: Approximate boolean reasoning: foundations and applications in data mining. In: Peters, J.F., Skowron, A. (eds.) Transactions on Rough Sets V. LNCS, vol. 4100, pp. 334–506. Springer, Heidelberg (2006)
12. Pawlak, Z.: Rough sets. International Journal of Computer and Information Sciences 11, 341–356 (1982)
13. Pawlak, Z.: Rough Sets: Theoretical Aspects of Reasoning about Data. System Theory, Knowledge Engineering and Problem Solving, vol. 9. Kluwer Academic Publishers, Dordrecht (1991)
14. Sadohara, K.: Learning of boolean functions using support vector machines. In: Abe, N., Khardon, R., Zeugmann, T. (eds.) ALT 2001. LNCS (LNAI), vol. 2225, pp. 106–118. Springer, Heidelberg (2001)
15. Skowron, A., Rauszer, C.: The Discernibility Matrices and Functions in Information Systems. In: Słowiński, R. (ed.) Intelligent Decision Support – Handbook of Applications and Advances of the Rough Sets Theory, ch. 3, pp. 331–362. Kluwer Academic Publishers, Dordrecht (1992)
16. Ślęzak, D.: Approximate Entropy Reducts. Fundamenta Informaticae 53, 365–387 (2002)
17. Vapnik, V.: The Nature of Statistical Learning Theory. Springer, NY (1995)
18. Zhang, J., Wang, Y.: A rough margin based support vector machine. Inf. Sci. 178(9), 2204–2214 (2008)

Robust Ordinal Regression for Dominance-Based Rough Set Approach under Uncertainty

Roman Słowiński[1,2], Miłosz Kadziński[1], and Salvatore Greco[3,4]

[1] Institute of Computing Science, Poznań University of Technology, Poznań, Poland
[2] Systems Research Institute, Polish Academy of Sciences, Warsaw, Poland
[3] Department of Business and Economics, University of Catania, Catania, Italy
[4] Portsmouth Business School, University of Portsmouth, United Kingdom

Abstract. We consider decision under uncertainty where preference information provided by a Decision Maker (DM) is a classification of some reference acts, relatively well-known to the DM, described by outcomes to be gained with given probabilities. We structure the classification data using a variant of the Dominance-based Rough Set Approach. Then, we induce from this data all possible minimal-cover sets of rules which correspond to all instances of the preference model compatible with the input preference information. We apply these instances on a set of unseen acts, and draw robust conclusions about their quality using the Robust Ordinal Regression paradigm. Specifically, for each act we derive the necessary and possible assignments specifying the range of classes to which the act is assigned by all or at least one compatible set of rules, respectively, as well as class acceptability indices. The whole approach is illustrated by a didactic example.

1 Introduction

Decision under uncertainty is a classical topic of decision theory (see [3] for a review). The main approaches to modeling this decision are based on expected utility theory. Since many experiments showed systematic violation of expected utility hypotheses, alternative models weakening some original axioms have been proposed. In particular, Greco, Matarazzo and Słowiński [5] used an approach based on stochastic dominance, which is the weakest assumption possible. They applied for this Dominance-based Rough Set Approach (DRSA) [4], which adapts the rough set concept introduced by Pawlak [8] to handle ambiguity with respect to dominance.

The main difficulty in recommending a decision under uncertainty for some Decision Maker (DM) relies on such an aggregation of outcomes predicted for the considered acts with given probabilities, that an aggregated model, called preference model, indicates the act that best respects preferences of this DM. The preference model used in [5] has the form of a set of "if..., then..." decision rules. The input preference information to this model is a set of classification examples concerning some acts relatively well known to the DM, called reference acts. The decision rules are induced from these classification examples, however,

M. Kryszkiewicz et al. (Eds.): RSEISP 2014, LNAI 8537, pp. 77–87, 2014.

for the reason of possible ambiguity in the set of examples caused by DM's violation of the stochastic dominance principle, the classification data are structured
first using DRSA. DRSA permits distinguishing non-ambiguous from ambiguous classification examples, which corresponds to the lower approximation and
to the boundary of the classification, respectively. In consequence, decision rules
induced from data structured in this way are certain or ambiguous, respectively.

The most popular rule induction strategy consists in looking for a minimal set
of rules covering the classification examples [1]. This strategy is called minimal-
cover (MC) strategy. In practice, it is performed by a greedy heuristic of sequential covering type, giving an approximately-MC set of rules (later, for simplicity,
we drop the prefix "approximately" from MC). However, representation of the
DM's preferences with a MC set of rules is not unique, because there are in
general many MC sets of rules for a given rough approximation of classification
data, that reproduce equally well the provided preference information. Choosing
among them is either arbitrary or requires involvement of the DM, which is not
easy for most of them. This inconvenience has been already noticed in Multiple Criteria Decision Aiding (MCDA) and resulted in proposing Robust Ordinal
Regression (ROR) ([6]; for a survey see [2]; for application of ROR to DRSA see
[7]), which takes into account all preference model instances compatible with the
DM's preference information.

The aim of this paper is to adapt ROR, involving MC sets of rules as compatible instances of the preference model, to decision under uncertainty formulated
as a multi-attribute classification problem. In order to generate all MC sets of
rules, first an exhaustive set of rules is induced, and then the MC sets are constructed by solving a series of Integer Linear Programming (ILP) problems.

Note that although compatible MC sets of rules reproduce the classification
examples provided by the DM, the assignment of non-reference acts that result
from application of any of these MC sets of rules can vary significantly. We
investigate the diversity of the recommendations suggested by different MC sets
of rules by producing two types of assignment, necessary and possible, for each
act from the considered set of acts. Since all compatible MC sets of rules are
known, we are also able to compute class acceptability indices defined as the
share of compatible MC sets of rules assigning an act to a single class or a set
of contiguous classes.

The paper is organized as follows. Section 2 recalls basics of DRSA for decision
under uncertainty. Then, we present algorithms for generating all compatible
rules and all compatible MC sets of rules in Sections 3 and 4, respectively.
In Section 5, we show how to get the possible and the necessary assignments.
To illustrate the whole approach, throughout the paper we refer to a didactic
example of decision under uncertainty. Section 6 contains conclusions.

2   DRSA for Decision under Uncertainty

Following [5], we formulate the decision under uncertainty as a multi-attribute
classification problem. For this, we consider the following basic elements:

- a finite set $S=\{s_1, s_2, \ldots, s_u\}$ of states of the world, or simply *states*, which are mutually exclusive and exhaustive,
- an *a priori probability distribution* P over S: more precisely; the probabilities of states $s_1, s_2, \ldots, s_u$ are given by $p_1, p_2, \ldots, p_u$, respectively, $p_1 + p_2 + \ldots + p_u = 1$, $p_i \geq 0$, $i = 1, \ldots, u$,
- a set $A=\{a_1, a_2, \ldots, a_m\}$ of all considered *acts*, and a set $A^R \subset A$ of *reference acts* for which the DM expressed her/his preferences,
- a set $X=\{x_1, x_2, \ldots, x_r\}$ of *outcomes* that, for the sake of simplicity, we consider expressed in monetary terms ($X \subseteq \Re$),
- a function $g : A \times S \rightarrow X$ assigning to each pair act-state $(a_i, s_j) \in A \times S$ an outcome $x_k \in X$,
- a set of *quality classes* $Cl=\{Cl_1, Cl_2, \ldots, Cl_n\}$, such that $Cl_1 \cup Cl_2 \cup \ldots \cup Cl_n = A$, $Cl_r \cap Cl_q = \emptyset$ for each $r,q \in\{1,\ldots,n\}$ with $r \neq q$; the classes from $\boldsymbol{Cl}$ are preference-ordered according to the increasing order of their indices,
- a function $e\colon A \rightarrow \boldsymbol{Cl}$ assigning each act $a_i \in A$ to a quality class $Cl_j \in \boldsymbol{Cl}$.

On the basis of P, we can assign to each subset of states of the world $W \subseteq S$ the probability $P(W)$ that one of the states in W is verified, i.e., $P(W) = \sum\limits_{i:s_i \in W} p_i$, and then we can build up the set $\varPi$ of all possible values $P(W)$, i.e.,

$$\varPi = \{\pi \in [0,1]\colon \pi = P(W),\ W \subseteq S\}.$$

Let us define the following functions $z\colon A \times S \rightarrow \varPi$ and $z'\colon A \times S \rightarrow \varPi$ assigning to each act-state pair $(a_i, s_j) \in A \times S$ a probability $\pi \in \varPi$, as follows:

$$z(a_i, s_j) = \sum_{r:g(a_i,s_r) \geq g(a_i,s_j)} p_r \quad \text{and} \quad z'(a_i, s_j) = \sum_{r:g(a_i,s_r) \leq g(a_i,s_j)} p_r.$$

Therefore, $z(a_i,s_j)$ ($z'(a_i,s_j)$) represents the probability of obtaining an outcome whose value is *at least* (*at most*) $g(a_i,s_j)$ by act a_i.

On the basis of functions $z(a_i,s_j)$ and $z'(a_i,s_j)$, we can define functions, respectively, $\rho\colon A \times \varPi \rightarrow X$ and $\rho'\colon A \times \varPi \rightarrow X$ as follows:

$$\rho(a_i, \pi) = \max_{j:z(a_i,s_j) \geq \pi} \{g(a_i, s_j)\} \quad \text{and} \quad \rho'(a_i, \pi) = \min_{j:z'(a_i,s_j) \geq \pi} \{g(a_i, s_j)\}.$$

Thus, $\rho(a_i,\pi) = x$ ($\rho'(a_i,\pi) = x$) means that the outcome got by act a_i is greater (smaller) than or equal to x with a probability *at least* π. As observed in [5], information given by $\rho(a_i,\pi)$ and $\rho'(a_i,\pi)$ is related such that for all $a_i \in A$ and $\pi_{(j-1)}, \pi_{(j)} \in \varPi$:

$$\rho(a_i, \pi_{(j)}) = \rho'(a_i, 1 - \pi_{(j-1)}). \tag{1}$$

The above considerations lead us to formulation of the decision under uncertainty in terms of a multi-attribute classification problem, where the set of classified objects is the set of acts A, the set of condition attributes describing the acts is the set $\varPi$, cl denotes the decision attribute assigning the acts from A to classes from $\boldsymbol{Cl}$, the set X is a value set of condition attributes, the set $\boldsymbol{Cl}$ is the value set of the decision attribute, and $\boldsymbol{f}$ is an information function, such that $f(a_i, \pi) = \rho(a_i, \pi)$ and $f(a_i, cl)=e(a_i)$. Let us observe that due to the above stated equivalence, one can consider alternatively information function $f'(a_i, \pi) = \rho'(a_i, \pi)$.

The classification examples concerning the reference acts A^R constitute the DM's preference information considered in the context of decision under uncertainty. Formally, they are presented as an *information table* whose rows correspond to reference acts belonging to set A^R, and columns correspond to condition attributes from set Π and to the decision attribute cl. The entries of the information table are values $\rho(a_i, \pi)$ of the information function f, as well as class assignments of the acts.

These classification data are structured using DRSA. In DRSA, we are approximating the upward $Cl_t^{\geq} = \bigcup_{s \geq t} Cl_s$ and downward $Cl_t^{\leq} = \bigcup_{s \leq t} Cl_s$, unions of classes, $t = 1, \ldots, n$, using *dominance cones* defined in the condition attribute space for any subset of condition attributes $\Theta \subseteq \Pi$. The fact that act a_p stochastically dominates act a_q with respect to $\Theta \subseteq \Pi$ (i.e., $\rho(a_p, \pi) \geq \rho(a_q, \pi)$ for each $\pi \in \Theta$) is denoted by $xD_\Theta y$. Given $\Theta \subseteq \Pi$ and $a_i \in A^R$, the cones of dominance are:

- a set of acts dominating a_i: $D_\Theta^+(a_i) = \{a_j \in A^R : a_j D_\Theta a_i\}$,
- a set of objects dominated by a_i: $D_\Theta^-(a_i) = \{a_j \in A^R : a_i D_\Theta a_j\}$.

With respect to $\Theta \subseteq \Pi$, the set of all acts belonging to $Cl_t^{\geq}$ ($Cl_t^{\leq}$) without any ambiguity constitutes the Θ-*lower approximation* of $Cl_t^{\geq}$ ($Cl_t^{\leq}$), denoted by $\underline{\Theta}(Cl_t^{\geq})$ ($\underline{\Theta}(Cl_t^{\leq})$), and the set of all acts that could belong to $Cl_t^{\geq}$ ($Cl_t^{\leq}$) constitutes the Θ-*upper approximation* of $Cl_t^{\geq}$ ($Cl_t^{\leq}$), denoted by $\overline{\Theta}(Cl_t^{\geq})$ ($\overline{\Theta}(Cl_t^{\leq})$), i.e., for $t = 1, \ldots, n$:

$$\underline{\Theta}(Cl_t^{\geq}) = \{a \in A^R : D_\Theta^+(a) \subseteq Cl_t^{\geq}\} \text{ and } \underline{\Theta}(Cl_t^{\leq}) = \{a \in A^R : D_\Theta^-(a) \subseteq Cl_t^{\leq}\},$$
$$\overline{\Theta}(Cl_t^{\geq}) = \{a \in A^R : D_\Theta^-(a) \cap Cl_t^{\geq} \neq \emptyset\} \text{ and } \overline{\Theta}(Cl_t^{\leq}) = \{a \in A^R : D_\Theta^+(a) \cap Cl_t^{\leq} \neq \emptyset\}.$$

The Θ-*boundaries* (Θ-doubtful regions) of $Cl_t^{\geq}$ and $Cl_t^{\leq}$ are defined as:

$$Bn_\Theta(Cl_t^{\geq}) = \overline{\Theta}(Cl_t^{\geq}) - \underline{\Theta}(Cl_t^{\geq}) \text{ and } Bn_\Theta(Cl_t^{\leq}) = \overline{\Theta}(Cl_t^{\leq}) - \underline{\Theta}(Cl_t^{\leq}).$$

For every $t \in \{1, \ldots, n\}$ and for every $\Theta \subseteq \Pi$, we define the *quality of approximation of classification* Cl by set of attributes Θ, or in short, *quality of classification*:

$$\gamma_\Theta(Cl) = \frac{card\left(A - \bigcup_{t=1}^{n-1} Bn_\Theta(Cl_t^{\leq})\right)}{card(A)} = \frac{card\left(A - \bigcup_{t=2}^{n} Bn_\Theta(Cl_t^{\geq})\right)}{card(A)}.$$

Each minimal (with respect to inclusion) subset $\Theta \subseteq \Pi$ for which $\gamma_\Theta(Cl) = \gamma_\Pi(Cl)$, is called a *reduct* of Cl and denoted by RED_{Cl}. The intersection of all reducts is called the *core* and denoted by $CORE_{Cl}$.

Illustrative Study: Part 1. The following example illustrates the approach. Let us consider:

- a set $S = \{s_1, s_2, s_3\}$ of states of the world,
- an *a priori* probability distribution P over S defined as follows: $p_1 = 0.20$, $p_2 = 0.35$, $p_3 = 0.45$,
- a set $A = \{a_1, \ldots, a_{12}\}$ of acts, and a set of reference acts $A^R = \{a_1, \ldots, a_6\}$;
- a set $X = \{0, 10, 15, 20, 25, 30\}$ of possible consequences,
- a set of classes $\boldsymbol{Cl} = \{Cl_1, Cl_2, Cl_3\}$, where Cl_1 is the set of bad acts, Cl_2 is the set of medium acts, and Cl_3 is the set of good acts,
- a function $g : A \times S \to X$ assigning to each act-state pair $(a_i, s_j) \in A \times S$ a consequence $x_h \in X$ and a function $e : A^R \to \boldsymbol{Cl}$ assigning each reference act $a_i \in A^R$ to a class $Cl_j \in \boldsymbol{Cl}$, presented in Table 1a).

Table 1. a) Acts, consequences, and assignment to class from $\boldsymbol{Cl}$ by the DM. b) Acts, values of function $\rho(a_i, \pi)$ and assignment to class from $\boldsymbol{Cl}$.

		Part a)				Part b)							
	s_1	s_2	s_3	cl									
p_j	0.25	0.35	0.40		$\rho(a_i, \pi)$	0.20	0.35	0.45	0.55	0.65	0.80	1.00	cl
a_1	30	15	10	medium (2)	a_1	30	15	15	15	10	10	10	2
a_2	10	20	30	good (3)	a_2	30	30	30	20	20	20	10	3
a_3	15	0	20	bad (1)	a_3	20	20	20	15	15	0	0	1
a_4	0	15	20	bad (1)	a_4	20	20	20	15	15	15	0	1
a_5	20	10	30	medium (2)	a_5	30	30	30	20	20	10	10	3
a_6	0	10	30	good (3)	a_6	30	30	30	10	10	10	0	2
a_7	30	10	10	-	a_7	30	10	10	10	10	10	10	-
a_8	10	10	20	-	a_8	20	20	20	10	10	10	10	-
a_9	30	10	30	-	a_9	30	30	30	30	30	10	10	-
a_{10}	0	15	25	-	a_{10}	25	25	25	15	15	15	0	-
a_{11}	30	10	25	-	a_{11}	30	25	25	25	10	10	10	-
a_{12}	0	20	15	-	a_{12}	20	20	15	15	15	15	0	-

Table 1b) shows values of function $\rho(a_i, \pi)$. This information table is consistent. Indeed, there is no single reference act that would stochastically dominate some other reference act assigned to a better class. In consequence, lower and upper approximations of upward and downward unions of classes are equal to:

$$\underline{\Pi}(Cl_3^{\geq}) = \overline{\Pi}(Cl_3^{\geq}) = \{a_2, a_5\}, \quad \underline{\Pi}(Cl_2^{\geq}) = \overline{\Pi}(Cl_2^{\geq}) = \{a_1, a_6, a_2, a_5\},$$
$$\underline{\Pi}(Cl_1^{\leq}) = \overline{\Pi}(Cl_1^{\leq}) = \{a_3, a_4\}, \quad \underline{\Pi}(Cl_2^{\leq}) = \overline{\Pi}(Cl_2^{\leq}) = \{a_1, a_6, a_3, a_4\}.$$

The quality of approximation $\gamma(\boldsymbol{Cl}) = 1$. Moreover, there are four reducts of condition attributes (criteria) ensuring the same quality of sorting as the whole set Π of probabilities: $RED_{\boldsymbol{Cl}}^1 = \{0.45, 1.00\}$, $RED_{\boldsymbol{Cl}}^2 = \{0.35, 1.00\}$, $RED_{\boldsymbol{Cl}}^3 = \{0.20, 0.65\}$, and $RED_{\boldsymbol{Cl}}^4 = \{0.20, 0.55\}$. This means that we can explain the preferences of the DM using the probabilities from each $RED_{\boldsymbol{Cl}}$ only. The core is empty, i.e., each probability value can be removed individually from Π without deteriorating the quality of classification.

3 Generating All Compatible Minimal Rules

The dominance-based rough approximations of upward and downward unions of classes serve to induce a generalized description of acts contained in the information table in terms of "*if* ..., *then* ..." decision rules. In what follows, we focus on certain decision rules. For a given upward or downward union of classes, $Cl_t^{\geq}$ or $Cl_s^{\leq}$, the decision rules induced under a hypothesis that reference acts belonging to $\underline{\Pi}(Cl_t^{\geq})$ or $\underline{\Pi}(Cl_s^{\leq})$ are *positive* and all the other reference acts *negative*, suggest a *certain* assignment to "at least class Cl_t" or to "at most class Cl_s", respectively. The syntax of decision rules obtained from DRSA is the following:

- $D_{\geq}$-*decision rules*:
  if $\rho(a,\pi_{h1}) \geq x_{h1}$ *and*, ..., *and* $\rho(a,\pi_{hz}) \geq x_{hz}$, *then* $a \in Cl_r^{\geq}$
  (i.e. "if by act a the outcome is at least x_{h1} with probability at least π_{h1}, and, ..., and the outcome is at least x_{hz} with probability at least π_{hz}, then $a \in Cl_r^{\geq}$"), where $\pi_{h1}, \ldots, \pi_{hz} \in \Pi$, $x_{h1}, \ldots, x_{hz} \in X$, and $r \in \{2, \ldots, n\}$;
- $D_{\leq}$-*decision rules*:
  if $\rho'(a,p_{h1}) \leq x_{h1}$ *and*, ..., *and* $\rho'(a,p_{hz}) \leq x_{hz}$, *then* $a \in Cl_r^{\leq}$
  (i.e. "if by act a the outcome is at most x_{h1} with probability at least π_{h1}, and, ..., and the outcome is at most x_{hz} with probability at least π_{hz}, then $a \in Cl_r^{\leq}$"), where $\pi_{h1}, \ldots, \pi_{hz} \in \Pi$, $x_{h1}, \ldots, x_{hz} \in X$, and $r \in \{1, \ldots, n-1\}$.

In the following, we discuss an algorithm which generates all certain decision rules for $\underline{\Pi}(Cl_t^{\geq})$, $t = 2, \ldots, n$. The algorithm for $\underline{\Pi}(Cl_t^{\leq})$, $t = 1, \ldots, n-1$, can be formulated analogously.

In the first phase, we generate a set of elementary conditions C_1 to be used in the construction of decision rules. If there exists an act $a_i \in \underline{\Pi}(Cl_t^{\geq})$, such that $\rho(a_i, \pi_{h1}) = x_{h1}$ and $\rho(a_i, \pi_{h2}) = x_{h2}$ and $\ldots \rho(a_i, \pi_{hz}) = x_{hz}$, then a_i is called basis of the rule. Each $D_{\geq}$-decision rule having a basis is called robust because it is "built" on an existing reference act. Although algorithms which construct robust rules require less computational effort than algorithms which construct non-robust rules, they usually generate a greater number of less general rules. Thus, in this paper, we rather focus on "mix of conditions" rules, which are possibly founded by multiple reference acts. For this purpose, C_1 needs to be composed of conditions in form $\rho(a,\pi_h) \geq x_h$, such that there exists $a_i \in \underline{\Pi}(Cl_t^{\geq}) : \rho(a_i, \pi_h) = x_h$.

In the second phase, we generate a set of conjunctions of elementary conditions which cover at least one reference act in $\underline{\Pi}(Cl_t^{\geq})$. It is an iterative process in which conjunctions of size $k + 1$ are constructed from conjunctions of size k (i.e., first, a set C_2 of conjunctions of size 2 is constructed by combining elementary conditions from C_1; then, a set C_3 of conjunctions of size 3 is built from conjunctions of size 2, etc.). This procedure is repeated as long as it is possible to obtain conjunctions of a particular size. Precisely, each conjunction of size $k+1$ is obtained by merging a pair of conjunctions of size k which contain the same $k-1$ conditions, thus, differing by just a single elementary condition.

These differentiating conditions need to concern different criteria. At each stage, we neglect conjunctions of size k with negative support equal to 0, since they already contain all conditions necessary to discriminate reference acts in $\underline{\Pi}(Cl_t^{\geq})$ and $A^R \setminus \underline{\Pi}(Cl_t^{\geq})$. Moreover, the set C_{k+1} of conjunctions of size $k+1$ contains only these conjunctions whose positive support is greater than 0.

After generating all possible conjunctions of elementary conditions covering at least one reference act in $\underline{\Pi}(Cl_t^{\geq})$, we eliminate conjunctions covering any negative example in $A^R \setminus \underline{\Pi}(Cl_t^{\geq})$. Subsequently, we remove conjunctions of conditions which are not minimal, i.e. such that there exists some other conjunction:

- using a subset of elementary conditions or/and weaker elementary conditions,
- requiring in all elementary conditions the same cumulated outcome with less probability; for example, when considering two rules, r1 $\equiv$ if $\rho(a_i, 0.55) \geq 20$ then $a \in Cl_3^{\geq}$, and r2 $\equiv$ if $\rho(a_i, 0.8) \geq 20$ then $a \in Cl_3^{\geq}$, r1 is minimal among them, because it requires a cumulated outcome to be at least 20, but with less probability, 0.55 against 0.8. This allows removing the rules which are not minimal in the specific context of the DRSA analysis using stochastic dominance.

Thus filtered, the remained conjunctions are used to construct the rules with a decision part: $a \in Cl_t^{\geq}$. Let us denote by $\mathcal{R}_{all}^{\underline{\Pi}(Cl_t^{\geq})}$ $(\mathcal{R}_{all}^{\underline{\Pi}(Cl_t^{\leq})})$ the set of all compatible minimal rules induced from $\underline{\Pi}(Cl_t^{\geq})$ $(\underline{\Pi}(Cl_t^{\leq}))$.

Table 2. All compatible certain minimal rules

Symbol	Rule	Support
$r_{\geq 2}^1$	if $\rho(a_i, 1.00) \geq 10$ then $a \in Cl_2^{\geq}$	$\{a_1, a_2, a_5\}$
$r_{\geq 2}^2$	if $\rho(a_i, 0.55) \geq 20$ then $a \in Cl_2^{\geq}$	$\{a_2, a_5\}$
$r_{\geq 2}^3$	if $\rho(a_i, 0.20) \geq 30$ then $a \in Cl_2^{\geq}$	$\{a_1.a_2, a_5, a_6\}$
$r_{\geq 3}^1$	if $\rho(a_i, 0.55) \geq 20$ then $a \in Cl_3^{\geq}$	$\{a_2, a_5\}$
$r_{\geq 3}^2$	if $\rho(a_i, 0.35) \geq 30$ and $\rho(a_i, 1.00) \geq 10$ then $a \in Cl_3^{\geq}$	$\{a_2, a_5\}$
$r_{\leq 1}^1$	if $\rho(a_i, 0.80) \leq 0$ ($\rho'(a_i, 0.35) \leq 0$) then $a \in Cl_1^{\leq}$	$\{a_3\}$
$r_{\leq 1}^2$	if $\rho(a_i, 0.20) \leq 20$ ($\rho'(a_i, 1.0) \leq 20$) then $a \in Cl_1^{\leq}$	$\{a_3, a_4\}$
$r_{\leq 1}^3$	if $\rho(a_i, 0.45) \leq 20$ ($\rho'(a_i, 0.65) \leq 20$) and $\rho(a_i, 1.0) \leq 0$ ($\rho'(a_i, 0.20) \leq 0$) then $a \in Cl_1^{\leq}$	$\{a_3, a_4\}$
$r_{\leq 2}^1$	if $\rho(a_i, 1.00) \leq 0$ ($\rho'(a_i, 0.20) \leq 0$) then $a \in Cl_2^{\leq}$	$\{a_3, a_4, a_6\}$
$r_{\leq 2}^2$	if $\rho(a_i, 0.65) \leq 15$ ($\rho'(a_i, 0.45) \leq 15$) then $a \in Cl_2^{\leq}$	$\{a_1, a_3, a_4, a_6\}$
$r_{\leq 2}^3$	if $\rho(a_i, 0.45) \leq 20$ ($\rho'(a_i, 0.65) \leq 20$) then $a \in Cl_2^{\leq}$	$\{a_1, a_3, a_4\}$

Illustrative Study: Part 2. A set of all minimal decision rules describing the DM's preferences is provided in Table 2. Minimal decision rules constitute the most concise and non-redundant representation of knowledge contained in Tables 1a) and b). There are 11 certain rules overall (5 and 6 rules for the upward and downward class unions, respectively). When it comes to the number of elementary conditions, there are 9 rules with just a single condition and 2 rules with two ones.

4   Generating All Compatible Minimal-cover Sets of Rules

A set of certain decision rules is minimal-cover if and only if it is complete, i.e., it is able to cover all reference acts and non-redundant, i.e., exclusion of any rule from this set makes it non-complete. Finding a minimal set of rules covering the reference acts in $\underline{\Pi}(Cl_t^{\geq})$ $(\underline{\Pi}(Cl_t^{\leq}))$ is analogous to solving the minimum set cover problem. This classical problem in combinatorics and computer science can be solved as an Integer Linear Programming (ILP) problem (without loss of generality, we focus on $\underline{\Pi}(Cl_t^{\geq})$):

$$Minimize: f_k = \sum_{r_k \in \mathcal{R}_{all}^{\underline{\Pi}(Cl_t^{\geq})}} v_k, \tag{2}$$

s.t.
$$\left. \begin{array}{l} \sum_{r_k \in \mathcal{R}_{all}^{\underline{\Pi}(Cl_t^{\geq})} \text{ covering } a_i} v_k \geq 1, \text{ for } a_i \in A^R, \\[2ex] v_k \in \{0,1\}, \text{ for } r_k \in \mathcal{R}_{all}^{\underline{\Pi}(Cl_t^{\geq})}. \end{array} \right\}$$

Thus, v_k is a binary variable associated with each rule $r_k \in \mathcal{R}_{all}^{\underline{\Pi}(Cl_t^{\geq})}$. If $v_k = 1$, r_k is used in the set of rules covering all reference acts in $\underline{\Pi}(Cl_t^{\geq})$. The optimal solution of the above ILP indicates one of the MC sets of rules:

$$\mathcal{R}_k^{\underline{\Pi}(Cl_t^{\geq})} = \{r_k \in \mathcal{R}_{all}^{\underline{\Pi}(Cl_t^{\geq})}, \text{ such that } v_k^* = 1\},$$

where f_k^* is the optimal value of f_k and v_k^* are the values of the binary variables at the corresponding optimum found. Other sets can be identified by adding constraints that forbid finding again the solutions which have been already identified in the previously conducted optimizations:

$$\sum_{r_k \in \mathcal{R}_k^{\underline{\Pi}(Cl_t^{\geq})}} v_k \leq f_k^* - 1.$$

Let us denote by $\mathcal{R}_{mrc}^{\underline{\Pi}(Cl_t^{\geq})}$ $(\mathcal{R}_{mrc}^{\underline{\Pi}(Cl_t^{\leq})})$ all MC sets of rules for $\underline{\Pi}(Cl_t^{\geq})$ $(\underline{\Pi}(Cl_t^{\leq}))$. All compatible minimal sets of rules $\mathcal{R}^{A^R}$ are formed by the following product:

$$\mathcal{R}^{A^R} = \mathcal{R}_{mrc}^{\underline{\Pi}(Cl_2^{\geq})} \times \ldots \times \mathcal{R}_{mrc}^{\underline{\Pi}(Cl_n^{\geq})} \times \mathcal{R}_{mrc}^{\underline{\Pi}(Cl_1^{\leq})} \times \ldots \times \mathcal{R}_{mrc}^{\underline{\Pi}(Cl_{n-1}^{\leq})}. \tag{3}$$

When computing each MC rule set in $\mathcal{R}^{A^R}$ according to (3), we should eliminate decision rules from $\mathcal{R}_{mrc}^{\underline{\Pi}(Cl_t^{\geq})}$ or $\mathcal{R}_{mrc}^{\underline{\Pi}(Cl_t^{\leq})}$ with a consequent having at least the same strength (i.e., rules assigning objects to the same union or sub-union of classes) as some other rules from, respectively $\mathcal{R}_{mrc}^{\underline{\Pi}(Cl_h^{\geq})}$, $h > t$, or $\mathcal{R}_{mrc}^{\underline{\Pi}(Cl_h^{\leq})}$, $h < t$.

Illustrative Study: Part 3. All minimal-cover sets of rules for the lower approximation of each class union are presented in Table 3. In particular, there are two MC sets of rules for reference acts in $\underline{\Pi}(Cl_1^{\leq})$, $\underline{\Pi}(Cl_2^{\leq})$, and $\underline{\Pi}(Cl_3^{\geq})$, and a unique way of covering all reference acts in $\underline{\Pi}(Cl_2^{\geq})$. Combination of these minimal rule covers leads to 8 minimal sets of minimal rules $\mathcal{R}^{A^R}$ which reproduce the preference information provided by the DM.

Table 3. All minimal-cover sets of rules for the lower approximations of class unions

	Minimal-cover sets of rules		Minimal-cover sets of rules
$\mathcal{R}_{mrc}^{\underline{C}(Cl_2^{\geq})}$	$\{r_{\geq 2}^3\}$	$\mathcal{R}_{mrc}^{\underline{C}(Cl_1^{\leq})}$	$\{r_{\leq 1}^2\}, \{r_{\leq 1}^3\}$
$\mathcal{R}_{mrc}^{\underline{C}(Cl_3^{\geq})}$	$\{r_{\geq 3}^1\}, \{r_{\geq 3}^2\}$	$\mathcal{R}_{mrc}^{\underline{C}(Cl_2^{\leq})}$	$\{r_{\leq 2}^2\}, \{r_{\leq 2}^1, r_{\leq 2}^3\}$

5   Class Acceptability Indices

Each set of rules covering classification examples constitutes a preference model of the DM and can be used to classify new (non-reference) acts $A \setminus A^R$. We apply the following sorting method. Let us denote by $l^{\mathcal{R}}$ ($u^{\mathcal{R}}$) the lowest (highest) class of the intersection of suggested unions of all $D_{\geq}$- ($D_{\leq}$-) decision rules in $\mathcal{R}$ covering a. If $l^{\mathcal{R}}$ and/or $u^{\mathcal{R}}$ are undefined or $l^{\mathcal{R}} \leq u^{\mathcal{R}}$, then a sorting procedure driven by a compatible set of rules $\mathcal{R}$ assigns an act $a \in A$ to an interval of classes $[Cl_{L^{\mathcal{R}}(a)}, Cl_{R^{\mathcal{R}}(a)}]$ such that

- $L^{\mathcal{R}}(a) = l^{\mathcal{R}}$, if $l^{\mathcal{R}}$ is defined, and $L^{\mathcal{R}}(a) = 1$, otherwise,
- $R^{\mathcal{R}}(a) = u^{\mathcal{R}}$, if $u^{\mathcal{R}}$ is defined, and $R^{\mathcal{R}}(a) = n$, otherwise.

In case of inconsistency (i.e., if $u^{\mathcal{R}} < l^{\mathcal{R}}$), a is left without recommendation (i.e., the procedure indicates an empty set $\emptyset$ of classes).

Since all compatible sets of rules are known, for each range of contiguous classes $[Cl_{h_L}, Cl_{h_L+1}, \ldots, Cl_{h_R}]$, with $1 \leq h_L \leq h_R \leq n$, we can define *class range acceptability index* $CAI(a, [h_L, h_R])$ as the share of compatible sets of rules $\mathcal{R} \in \mathcal{R}^{A^R}$ that assign alternative a precisely to the range of classes $[Cl_{h_L}, Cl_{h_L+1}, \ldots, Cl_{h_R}]$. We can also compute the share of $\mathcal{R} \in \mathcal{R}^{A^R}$ for which Cl_h is within $[Cl_{L^{\mathcal{R}}(a)}, \ldots, Cl_{R^{\mathcal{R}}(a)}]$, i.e. the share of sets of rules that either precisely or imprecisely assign a to Cl_h. Let us call such a share the *cumulative class acceptability index* $CuCAI(a, h)$.

Note that if $CuCAI(a, h) > 0$, a is *possibly* assigned to Cl_h (let us denote it by $h \in C_P(a)$), because there exists at least one compatible set of rules assigning a to Cl_h. If $CuCAI(a, h) = 1$, a is *necessarily* assigned to Cl_h, because all compatible sets of rules assign a to Cl_h (then $h \in C_N(a)$).

Illustrative Study: Part 4. Among the non-reference acts $A^R = \{a_7, \ldots, a_{12}\}$, there are three ones $\{a_7, a_9, a_{12}\}$ which are possibly and necessarily assigned to just a single class (see Table 4). The remaining three acts are possibly assigned to two consecutive classes (i.e., $Cl_1 - Cl_2$ or $Cl_2 - Cl_3$). Note, however, that a_8 and a_{10} are necessarily assigned to, respectively, Cl_1 and a class interval Cl_1-Cl_2. For a_{11} there is no agreement with respect to recommendation between all compatible sets of rules; moreover, some models indicate $\emptyset$.

When it comes to class acceptability indices, the 5 acts that are necessarily assigned to some class or class interval have the respective $CuCAI$ 100%. In general, for acts that are possibly assigned to at least two consecutive classes, we can analyze $CAIs$ and $CuCAIs$ to indicate a recommendation suggested by most of compatible MC sets of rules. While for a_{11} the shares of compatible sets of rules indicating Cl_2 or Cl_3 (either precisely or imprecisely) are the same, for a_8 all compatible sets of rules suggest Cl_1, even though half of them indicate Cl_2 as well.

Table 4. Class acceptability indices (CAIs), cumulative class acceptability indices (CUCAIs), and possible C_P and necessary C_N assignments

Act	CAIs						CUCAIs				Assignments	
	$1-1$	$1-2$	$2-2$	$2-3$	$3-3$	$\emptyset$	1	2	3	$\emptyset$	C_P	C_N
a_7	–	–	100.0	–	–	–	–	100.0	–	–	2	2
a_8	50.0	50.0	–	–	–	–	100.0	50.0	–	–	$1-2$	1
a_9	–	–	–	–	100.0	–	–	–	100.0	–	3	3
a_{10}	–	100.0	–	–	–	–	100.0	100.0	–	–	$1-2$	$1-2$
a_{11}	–	–	25.0	25.0	25.0	25.0	–	50.0	50.0	25.0	$2-3$	
a_{12}	100.0	–	–	–	–	–	100.0	–	–	–	1	1

6   Conclusions

We integrated Robust Ordinal Regression into Dominance-based Rough Set Approach to modeling the decision under uncertainty. The whole approach is addressing decision situations where preference information provided by a Decision Maker (DM) is a set of classification examples concerning some reference acts relatively well-known to the DM. All considered acts are described by outcomes to be gained with given probabilities. The method proceeds as follows. First, we structure the classification data using the concept of dominance-based rough sets, discerning consistent from inconsistent classification examples. Then, we induce from this data all possible minimal-cover sets of rules which correspond to all instances of the preference model compatible with the input preference information. Precisely, the rules are induced from lower approximations of the unions of ordered quality classes of the reference acts, i.e., from consistent classification examples. A minimal-cover set of rules is covering all consistent classification examples using a minimal number of "if..., then..." decision rules, chosen from among all possible rules. Finally, we apply these multiple instances of the compatible preference model on a set of unseen acts, and we draw robust conclusions about their quality using the Robust Ordinal Regression paradigm. Specifically, for each act we derive the necessary and possible assignments, specifying the range of classes to which the act is assigned by all or at least one minimal-cover set of rules, respectively. We also provide class acceptability indices informing for each particular act about the distribution of assignment decisions by all minimal-cover sets of rules.

The whole approach can be extended in several ways, in particular, by:

– considering non-additive probability instead of additive one,
– accounting for consequences distributed over time,
– selecting a single representative minimal-cover set of rules among all compatible ones.

Acknowledgment. The first two authors wish to acknowledge financial support from the Polish National Science Center (PNSC), grant no. DEC-2011/01/B/ST6/07318.

References

1. Błaszczyński, J., Słowiński, R., Szelag, M.: Sequential covering rule induction algorithm for variable consistency rough set approaches. Information Sciences 181(5), 987–1002 (2011)
2. Corrente, S., Greco, S., Kadziński, M., Słowiński, R.: Robust ordinal regression in preference learning and ranking. Machine Learning 93(2-3), 381–422 (2013)
3. Fishburn, P.C.: Nonlinear Preferences and Utility Theory. The John Hopkins University Press (1988)
4. Greco, S., Matarazzo, B., Słowiński, R.: Rough Sets Theory for Multicriteria Decision Analysis. European Journal of Operational Research 129, 1–47 (2001)
5. Greco, S., Matarazzo, B., Słowiński, R.: Dominance-based Rough Set Approach to decision under uncertainty and time preference. Annals of Operations Research 176, 41–75 (2010)
6. Greco, S., Mousseau, V., Słowiński, R.: Ordinal regression revisited: multiple criteria ranking using a set of additive value functions. European Journal of Operational Research 191(2), 415–435 (2008)
7. Kadziński, M., Greco, S., Słowiński, R.: Robust Ordinal Regression for Dominance-based Rough Set Approach to Multiple Criteria Sorting. Information Sciences (Submitted for Publication, 2014)
8. Pawlak, Z.: Rough Sets. Theoretical Aspects of Reasoning about Data. Kluwer Academic Publishers (1991)

Monotonic Uncertainty Measures in Probabilistic Rough Set Model

Guoyin Wang[1], Xi'ao Ma[1,2,*], and Hong Yu[1]

[1] Chongqing Key Laboratory of Computational Intelligence,
Chongqing University of Posts and Telecommunications, Chongqing, 400065, China
[2] School of Information Science and Technology,
Southwest Jiaotong University, Chengdu, 610031, China
wanggy@ieee.org, maxiao73559@163.com, yuhong@cqupt.edu.cn

Abstract. Uncertainty measure is one of the key research issues in the rough set theory. In the Pawlak rough set model, the accuracy measure, the roughness measure and the approximation accuracy measure are used as uncertainty measures. Monotonicity is a basic property of these measures. However, the monotonicity of these measures does not hold in the probabilistic rough set model, which makes them not so reasonable to evaluate the uncertainty. The main objective of this paper is to address the uncertainty measure problem in the probabilistic rough set model. We propose three monotonic uncertainty measures which are called the probabilistic accuracy measure, the probabilistic roughness measure and the probabilistic approximation accuracy measure respectively. The monotonicity of the proposed uncertainty measures is proved to be held. Finally, an example is used to verify the validity of the proposed uncertainty measures.

Keywords: Uncertainty measures, approximation accuracy, Pawlak rough set model, probabilistic rough set model.

1 Introduction

The rough set theory, introduced by Pawlak [5], is a valid mathematical theory that deals well with imprecise, vague and uncertain information, and it has been applied in many research fields successfully, such as machine learning, data mining, knowledge discovery, intelligent data analysis [8].

Uncertainty measure is an important issue in rough set theory and has been receiving increasing attention from researchers [1,3,7,9]. Generally speaking, the methodologies dealing with uncertainty measure problem can be classified into two categories: the pure rough set approach and the information theory approach [2]. In this paper, we only discuss the pure rough set approach. For pure rough set approach, the accuracy measure, the roughness measure and the approximation accuracy measure are used to evaluate the uncertainty.

---

* Corresponding author.

M. Kryszkiewicz et al. (Eds.): RSEISP 2014, LNAI 8537, pp. 88–98, 2014.

Probabilistic rough set model is an important generalization of Pawlak rough set model. To date, few studies have been done on uncertainty measures in the probabilistic rough set model. In this paper, we focus on the uncertainty measure problem in the probabilistic rough set model. In the Pawlak rough set model, the lower and upper approximations are defined by requiring the set inclusion must be fully correct or certain, namely, the definitions of the lower and upper approximations do not allow any tolerance of errors. Hence, the accuracy measure, the roughness measure and the approximation accuracy measure based on the lower and upper approximations are monotonous with respect to the granularity of partitions. In other words, the finer the partition is, the more the available knowledge is, and correspondingly the smaller the uncertainty is. This process is much similar to reasoning of human's mind. Thus, these measures are reasonable to be used as uncertainty measures in the Pawlak rough set model [2].

Unfortunately, since the lower and upper approximations are defined by introducing the probabilistic threshold values in the probabilistic rough set model, we will find the accuracy measure, roughness measure and approximation accuracy measure by the parallel extension are not monotonic with respect to the granularity of partitions, which makes them not so reasonable to be used as uncertainty measures in the probabilistic rough set model. Moreover, uncertainty measures can be applied to rank attributes and evaluate the significance of attributes in attribute reduction [1]. However, they can not evaluate attribute subset effectively if the used measures do not satisfy the monotonicity. Thus, further studies on monotonic uncertainty measures in the probabilistic rough set model are necessary.

The rest of the paper is organized as follows. Some basic concepts regarding uncertainty measures in the Pawlak rough set model and the probabilistic rough set model are recalled in Section 2. In Section 3, the existing problems of uncertainty measures are analyzed in probabilistic rough set model. Three monotonic uncertainty measures are proposed and their monotonicity are also proved in Section 4. In Section 5, an illustration is used to show the effectiveness of the proposed uncertainty measures. The conclusions and future research topics are discussed in Section 6.

2 Preliminaries

In this section, we recall the basic notions related to uncertainty measures in the Pawlak rough set model, and several basic concepts are also reviewed in the probabilistic rough set model.

2.1 Uncertainty Measures in Pawlak Rough Set Model

An information system is a pair $IS = (U, A)$, where U is a finite nonempty set of objects called universe, A is a nonempty finite set of attributes.

Each subset of attributes $R \subseteq A$ determines a binary indistinguishable relationship $IND(R)$ as follows:

$$IND(R) = \{(x, y) \in U \times U | \forall a \in R, f(x, a) = f(y, a)\}.$$

It can be easily shown that $IND(R)$ is an equivalence relationship on U. For $R \subseteq A$, the equivalence relation $IND(R)$ partitions U into some equivalence classes, which is denoted by $U/IND(R) = \{[x]_R | u \in U\}$, for simplicity, $U/IND(R)$ will be replaced by U/R, where $[x]_R$ is an equivalence class determined by x with respect to R, i.e. $[x]_R = \{y \in U | (x, y) \in IND(R)\}$. $U/IND(R)$ denotes the classification or the knowledge induced by R. One can therefore use the terms "partition" and "knowledge" interchangeably.

For any given information system $IS = (U, A)$, $R \subseteq A$ and $X \subseteq U$, the lower and upper approximations of X with respect to R are defined as follows:

$$\underline{apr}_R(X) = \{x \in U | [x]_R \subseteq X\},$$
$$\overline{apr}_R(X) = \{x \in U | [x]_R \cap X \neq \emptyset\}.$$

Given an information system $IS = (U, A)$, $P, Q \subseteq A$, one can define a partial relation $\preceq$ on 2^A: $P \preceq Q \Leftrightarrow \forall x \in U, [x]_P \subseteq [x]_Q$ [2]. If $P \preceq Q$, Q is said to be coarser than P (or P is finer than Q). If $P \preceq Q$ and $P \neq Q$, Q is said to be strictly coarser than P or P is strictly finer than Q, denoted by $P \prec Q$. In fact, $P \prec Q \Leftrightarrow \forall x \in U$, we have that $[x]_P \subseteq [x]_Q$ and there exists $y \in U$ such that $[y]_P \subset [y]_Q$.

A decision system is a pair $DS = (U, C \cup D)$, where C is a set of condition attributes, D is decision attribute set, and $C \cap D = \emptyset$.

Let $DS = (U, C \cup D)$ be a decision system, $R \subseteq C$ be an attribute set and $U/D = \{Y_1, Y_2, \ldots, Y_m\}$ be a classification of the universe U. Then, the lower and upper approximations of U/D with respect to R are defined as follows:

$$\underline{apr}_R(U/D) = \underline{apr}_R(Y_1) \cup \underline{apr}_R(Y_2) \cup \cdots \cup \underline{apr}_R(Y_m),$$
$$\overline{apr}_R(U/D) = \overline{apr}_R(Y_1) \cup \overline{apr}_R(Y_2) \cup \cdots \cup \overline{apr}_R(Y_m).$$

Two numerical measures can be used to evaluate the uncertainty of a rough set: accuracy measure and roughness measure. Furthermore, approximation accuracy characterized the uncertainty of a rough classification. The definitions of the uncertainty measures are shown as follows:

Definition 1. [6] *Given an information system $IS = (U, A)$, $R \subseteq A$, $X \subseteq U$. The accuracy of X with respect to R is defined as follows:*

$$\alpha_R(X) = \frac{|\underline{apr}_R(X)|}{|\overline{apr}_R(X)|}.$$

The roughness of X with respect to R is defined as follows:

$$\rho_R(X) = 1 - \alpha_R(X) = 1 - \frac{|\underline{apr}_R(X)|}{|\overline{apr}_R(X)|}.$$

Definition 2. [6] *Given a decision system* $DS = (U, C \cup D)$, $R \subseteq C$ *and* $U/D = \{Y_1, Y_2, \ldots, Y_m\}$ *is a classification of the universe* U. *The approximation accuracy of* U/D *with respect to* R *is defined as follows:*

$$\alpha_R(U/D) = \frac{|\underline{apr}_R(U/D)|}{|\overline{apr}_R(U/D)|} = \frac{\sum_{Y_i \in U/D} |\underline{apr}_R(Y_i)|}{\sum_{Y_i \in U/D} |\overline{apr}_R(Y_i)|}.$$

The accuracy measure, the roughness measure and the approximation accuracy measure have the following two theorems.

Theorem 1. [6] *Given an information system* $IS = (U, A)$, $P, Q \subseteq A$, $X \subseteq U$. *The following properties hold:*

(1) $P \preceq Q \Rightarrow \alpha_P(X) \geq \alpha_Q(X)$,
(2) $P \preceq Q \Rightarrow \rho_P(X) \leq \rho_Q(X)$.

Theorem 2. [6] *Given a decision system* $DS = (U, C \cup D)$, $P, Q \subseteq C$ *and* $U/D = \{Y_1, Y_2, \ldots, Y_m\}$ *is a classification of the universe* U. *The following property holds:*

$$P \preceq Q \Rightarrow \alpha_P(U/D) \geq \alpha_Q(U/D).$$

Theorem 1 and Theorem 2 state that the accuracy of X with respect to R increases, the roughness of X with respect to R decreases and the approximation accuracy of U/D with respect to R increases with R becoming finer.

2.2 Probabilistic Rough Set Model

In practical applications, Pawlak rough set model can not deal with data sets which have some noisy data effectively because the definitions of lower and upper approximations do not allow any tolerance of errors. Lots of information in the boundary region will be abandoned, which may provide latent useful knowledge. To address this problem, probabilistic rough set model is presented [4,10].

Definition 3. *Given an information system* $IS = (U, A)$, *for any* $0 \leq \beta < \alpha \leq 1$, $R \subseteq A$ *and* $X \subseteq U$. *The probabilistic lower approximation and probabilistic upper approximation of* X *with respect to* R *are defined as follows:*

$$\underline{apr}_R^{(\alpha, \beta)}(X) = \{x \in U | p(X|[x]_R) \geq \alpha\},$$
$$\overline{apr}_R^{(\alpha, \beta)}(X) = \{x \in U | p(X|[x]_R) > \beta\}.$$

Definition 4. *Given a decision system* $DS = (U, C \cup D)$, $R \subseteq C$ *and* $U/D = \{Y_1, Y_2, \ldots, Y_m\}$ *is a classification of the universe* U. *The probabilistic lower approximation and probabilistic upper approximation of* U/D *with respect to* R *are defined as follows:*

$$\underline{apr}_R^{(\alpha, \beta)}(U/D) = \underline{apr}_R^{(\alpha, \beta)}(Y_1) \cup \underline{apr}_R^{(\alpha, \beta)}(Y_2) \cup \cdots \cup \underline{apr}_R^{(\alpha, \beta)}(Y_m),$$
$$\overline{apr}_R^{(\alpha, \beta)}(U/D) = \overline{apr}_R^{(\alpha, \beta)}(Y_1) \cup \overline{apr}_R^{(\alpha, \beta)}(Y_2) \cup \cdots \cup \overline{apr}_R^{(\alpha, \beta)}(Y_m).$$

3 Limitations of Uncertainty Measures in Probabilistic Rough Set Model

In this section, we will show that the accuracy measure, the roughness measure and the approximation accuracy measure are not appropriate to be used as uncertainty measures in the probabilistic rough set model.

Let us first give the parallel extension definitions of the accuracy measure, the roughness measure and the approximation accuracy measure in the probabilistic rough set model.

Definition 5. *Given an information system* $IS=(U, A)$, *for any* $0 \leq \beta < \alpha \leq 1$, $R \subseteq A$ *and* $X \subseteq U$. *The accuracy of* X *with respect to* R *is defined as follows:*

$$\alpha_R^{(\alpha,\beta)}(X) = \frac{|\underline{apr}_R^{(\alpha,\beta)}(X)|}{|\overline{apr}_R^{(\alpha,\beta)}(X)|}.$$

The roughness of X *with respect to* R *is defined as follows:*

$$\rho_R^{(\alpha,\beta)}(X) = 1 - \alpha_R^{(\alpha,\beta)}(X) = 1 - \frac{|\underline{apr}_R^{(\alpha,\beta)}(X)|}{|\overline{apr}_R^{(\alpha,\beta)}(X)|}.$$

If $\underline{apr}_R^{(\alpha,\beta)}(X) = \overline{apr}_R^{(\alpha,\beta)}(X) = \emptyset$, *then define* $\frac{0}{0} = 0$.

Definition 6. *Given a decision system* $DS = (U, C \cup D)$, *for any* $0 \leq \beta < \alpha \leq 1$, $R \subseteq C$ *and* $U/D = \{Y_1, Y_2, \ldots, Y_m\}$ *is a classification of the universe* U. *The approximation accuracy of* U/D *with respect to* R *is defined as follows:*

$$\alpha_R^{(\alpha,\beta)}(U/D) = \frac{\sum_{Y_i \in U/D} |\underline{apr}_R^{(\alpha,\beta)}(Y_i)|}{\sum_{Y_i \in U/D} |\overline{apr}_R^{(\alpha,\beta)}(Y_i)|}.$$

In the Pawlak rough set model, the accuracy measure, the roughness measure and the approximation accuracy measure are reasonable to be used as uncertainty measures, because they have the monotonicity with respect to the granularity of partitions [2]. In the probabilistic rough set model, we should have the same conclusion. Moreover, a major application of the uncertainty measures is evaluating the significance of attributes in attribute reduction. However, they can not evaluate attribute subset effectively if the used measures do not satisfy the monotonicity.

It can be easily seen that the accuracy measure, the roughness measure and the approximation accuracy measure in the Pawlak rough set model are defined by using the lower and upper approximations. We can obtain the monotonicity property of Pawlak lower and upper approximations with respect to the granularity of partitions. However, since the probabilistic lower and upper approximations do not satisfy the monotonicity with respect to the granularity of partitions, the monotonicity of the accuracy measure, the roughness measure and the approximation accuracy measure in the probabilistic rough set model do not hold.

A counter-example is showed as follows.

Example 1. Given a decision system $DS = (U, C \cup D)$ shown in Table 1, where $U = \{x_1, x_2, \cdots, x_{10}\}$ and $C = \{a_1, a_2, a_3, a_4, a_5, a_6\}$. Suppose that $\alpha = 0.75$ and $\beta = 0.60$, $P, Q \subseteq C$, where $P = \{a_1, a_2, a_3, a_4\}$ and $Q = \{a_1, a_2, a_3\}$. As we can see, $P \supseteq Q$, which means $P \preceq Q$.

Table 1. A decision system

Car	a_1	a_2	a_3	a_4	a_5	a_6	d
x_1	0	1	0	1	0	0	0
x_2	0	1	0	1	0	0	0
x_3	0	1	0	1	0	0	1
x_4	0	1	0	0	1	1	0
x_5	1	0	1	0	1	1	1
x_6	0	0	1	1	0	0	1
x_7	1	0	1	0	1	0	0
x_8	0	0	1	0	1	1	1
x_9	0	0	1	0	1	1	0
x_{10}	0	0	1	0	0	0	0

Let $X = \{x_1, x_2, x_4, x_7, x_9, x_{10}\}$, the accuracies of X with respect to P and Q are given by

$$\alpha_P^{(0.75,0.60)}(X) = \frac{|\underline{apr}_P^{(0.75,0.60)}(X)|}{|\overline{apr}_P^{(0.75,0.60)}(X)|} = \frac{1}{7},$$

$$\alpha_Q^{(0.75,0.60)}(X) = \frac{|\underline{apr}_Q^{(0.75,0.60)}(X)|}{|\overline{apr}_Q^{(0.75,0.60)}(X)|} = 1.$$

Note that $\alpha_P^{(0.75,0.60)}(X) \geq \alpha_Q^{(0.75,0.60)}(X)$ does not hold though $P \preceq Q$. The roughnesses of X with respect to P and Q are given by

$$\rho_P^{(0.75,0.60)}(X) = 1 - \alpha_P^{(0.75,0.60)}(X) = \frac{6}{7},$$

$$\rho_Q^{(0.75,0.60)}(X) = 1 - \alpha_Q^{(0.75,0.60)}(X) = 0.$$

Thus, $\rho_P^{(0.75,0.60)}(X) \leq \rho_Q^{(0.75,0.60)}(X)$ does not hold though $P \preceq Q$. By computing, we have $U/D = \{\{x_1, x_2, x_4, x_7, x_9, x_{10}\}, \{x_3, x_5, x_6, x_8\}\}$. The approximation accuracies of U/D with respect to P and Q are given by

$$\alpha_P^{(0.75,0.60)}(U/D) = \frac{\sum_{Y_i \in U/D} |\underline{apr}_P^{(0.75,0.60)}(Y_i)|}{\sum_{Y_i \in U/D} |\overline{apr}_P^{(0.75,0.60)}(Y_i)|} = \frac{1}{4},$$

$$\alpha_Q^{(0.75,0.60)}(U/D) = \frac{\sum_{Y_i \in U/D} |\underline{apr}_Q^{(0.75,0.60)}(Y_i)|}{\sum_{Y_i \in U/D} |\overline{apr}_Q^{(0.75,0.60)}(Y_i)|} = 1.$$

Thus, $\alpha_P^{(0.75,0.6)}(U/D) \geq \alpha_Q^{(0.75,0.6)}(U/D)$ does not hold though $P \preceq Q$.

Since the accuracy measure, the roughness measure and the approximation accuracy measure in Definition 7 and Definition 8 do not have the monotonicity with respect to the granularity of partitions, we can not use $\alpha_R^{(\alpha,\beta)}(X)$, $\rho_R^{(\alpha,\beta)}(X)$ and $\alpha_R^{(\alpha,\beta)}(U/D)$ to evaluate the uncertainty in the probabilistic rough set model. Thus, new measures need to be developed to evaluate the uncertainty in the probabilistic rough set model by using monotonously measure functions.

4 Monotonic Uncertainty Measures in Probabilistic Rough Set Model

In this subsection, the accuracy measure, the roughness measure and the approximation accuracy measure in the probabilistic rough set model are reformulated in order to satisfy the monotonicity with respect to the granularity of partitions, which are called the probabilistic accuracy measure, the probabilistic roughness measure and the probabilistic approximation accuracy measure respectively, and the monotonicity of these measures are discussed.

Definition 7. *Given an information system* $IS = (U, A)$, *for any* $0 \leq \beta < \alpha \leq 1$, $R \subseteq A$ *and* $X \subseteq U$. *The probabilistic accuracy of* X *with respect to* R *is defined as follows:*

$$\alpha_R^{(\alpha,\beta)}(X) = \frac{|\underline{apr}_R(\underline{apr}_A^{(\alpha,\beta)}(X))|}{|\overline{apr}_R(\overline{apr}_A^{(\alpha,\beta)}(X))|}.$$

The probabilistic roughness of X *with respect to* R *is defined as follows:*

$$\rho_R^{(\alpha,\beta)}(X) = 1 - \alpha_R^{(\alpha,\beta)}(X) = 1 - \frac{|\underline{apr}_R(\underline{apr}_A^{(\alpha,\beta)}(X))|}{|\overline{apr}_R(\overline{apr}_A^{(\alpha,\beta)}(X))|}.$$

Definition 8. *Given a decision system* $DS = (U, C \cup D)$, *for any* $0 \leq \beta < \alpha \leq 1$, $R \subseteq C$ *and* $U/D = \{Y_1, Y_2, \ldots, Y_m\}$ *is a classification of the universe* U. *The probabilistic approximation accuracy of* U/D *with respect to* R *is defined as follows:*

$$\alpha_R^{(\alpha,\beta)}(U/D) = \frac{\sum_{Y_i \in U/D} |\underline{apr}_R(\underline{apr}_C^{(\alpha,\beta)}(Y_i))|}{\sum_{Y_i \in U/D} |\overline{apr}_R(\overline{apr}_C^{(\alpha,\beta)}(Y_i))|}.$$

In the following, we show that the probabilistic accuracy measure, the probabilistic roughness measure and the probabilistic approximation accuracy measure have the monotonicity.

Theorem 3. *Given an information system* $IS = (U, A)$, *for any* $0 \leq \beta < \alpha \leq 1$, $P, Q \subseteq A$ *and* $X \subseteq U$. *The following properties hold:*

(1) $P \preceq Q \Rightarrow \alpha_P^{(\alpha,\beta)}(X) \geq \alpha_Q^{(\alpha,\beta)}(X)$,
(2) $P \preceq Q \Rightarrow \rho_P^{(\alpha,\beta)}(X) \leq \rho_Q^{(\alpha,\beta)}(X)$.

Proof. (1) Suppose $P \preceq Q$. In terms of the definition of $\preceq$, we have $[x]_P \subseteq [x]_Q$.

On the one hand, for $\forall x \in \underline{apr}_Q(\underline{apr}_A^{(\alpha,\beta)}(X))$, one can obtain $[x]_Q \subseteq \underline{apr}_A^{(\alpha,\beta)}(X)$, hence $[x]_P \subseteq \underline{apr}_A^{(\alpha,\beta)}(X)$, then we obtain that $x \in \underline{apr}_P(\underline{apr}_A^{(\alpha,\beta)}(X))$.

Thus, $\underline{apr}_P(\underline{apr}_A^{(\alpha,\beta)}(X)) \supseteq \underline{apr}_Q(\underline{apr}_A^{(\alpha,\beta)}(X))$.

In the other hand, for $\forall x \in \overline{apr}_P(\overline{apr}_A^{(\alpha,\beta)}(X))$, we have $[x]_P \cap \overline{apr}_A^{(\alpha,\beta)}(X) \neq \emptyset$, since $[x]_P \subseteq [x]_Q$, $[x]_Q \cap \overline{apr}_A^{(\alpha,\beta)}(X) \neq \emptyset$, then $x \in \overline{apr}_Q(\overline{apr}_A^{(\alpha,\beta)}(X))$. It follows that $\overline{apr}_P(\overline{apr}_A^{(\alpha,\beta)}(X)) \subseteq \overline{apr}_Q(\overline{apr}_A^{(\alpha,\beta)}(X))$.

As a result, we have that

$$\alpha_P^{(\alpha,\beta)}(X) = \frac{|\underline{apr}_P(\underline{apr}_A^{(\alpha,\beta)}(X))|}{|\overline{apr}_P(\overline{apr}_A^{(\alpha,\beta)}(X))|} \geq \frac{|\underline{apr}_Q(\underline{apr}_A^{(\alpha,\beta)}(X))|}{|\overline{apr}_Q(\overline{apr}_A^{(\alpha,\beta)}(X))|} = \alpha_Q^{(\alpha,\beta)}(X).$$

(2) The proof is similar to that of (1).

Theorem 4. *Given a decision system $DS = (U, C \cup D)$, for any $0 \leq \beta < \alpha \leq 1$, $P, Q \subseteq C$ and $U/D = \{Y_1, Y_2, \ldots, Y_m\}$ is a classification of the universe U. The following property holds:*

$$P \preceq Q \Rightarrow \alpha_P^{(\alpha,\beta)}(U/D) \geq \alpha_Q^{(\alpha,\beta)}(U/D).$$

Proof. According to the proof of (1) in Theorem 3, if $P \preceq Q$, then for $\forall Y_i \in U/D$, we have

$$|\underline{apr}_P(\underline{apr}_C^{(\alpha,\beta)}(Y_i))| \geq |\underline{apr}_Q(\underline{apr}_C^{(\alpha,\beta)}(Y_i))|,$$

$$|\overline{apr}_P(\overline{apr}_C^{(\alpha,\beta)}(Y_i))| \leq |\overline{apr}_Q(\overline{apr}_C^{(\alpha,\beta)}(Y_i))|.$$

Thus,

$$\sum_{Y_i \in U/D} |\underline{apr}_P(\underline{apr}_C^{(\alpha,\beta)}(Y_i))| \geq \sum_{Y_i \in U/D} |\underline{apr}_Q(\underline{apr}_C^{(\alpha,\beta)}(Y_i))|,$$

$$\sum_{Y_i \in U/D} |\overline{apr}_P(\overline{apr}_C^{(\alpha,\beta)}(Y_i))| \leq \sum_{Y_i \in U/D} |\overline{apr}_Q(\overline{apr}_C^{(\alpha,\beta)}(Y_i))|.$$

According to Definition 8, we have

$$\alpha_P^{(\alpha,\beta)}(U/D) \geq \alpha_Q^{(\alpha,\beta)}(U/D).$$

This completes the proof.

Theorem 3 and Theorem 4 state that the probabilistic accuracy of X with respect to R increases, the probabilistic roughness of X with respect to R decreases and the probabilistic approximation accuracy of U/D with respect to R increases with R becoming finer.

5   Case Study

Example 2. Continued from Example 1.
By calculating, we obtain

$$U/C = \{\{x_1, x_2, x_3\}, \{x_4\}, \{x_5\}, \{x_6\}, \{x_7\}, \{x_8, x_9\}, \{x_{10}\}\}.$$

Let $X = \{x_1, x_2, x_4, x_7, x_9, x_{10}\}$, then

$$\underline{apr}_C^{(0.75, 0.60)}(X) = \{x_4, x_7, x_{10}\},$$
$$\overline{apr}_C^{(0.75, 0.60)}(X) = \{x_1, x_2, x_3, x_4, x_7, x_{10}\}.$$

Hence, we have

$$\underline{apr}_P(\underline{apr}_C^{(0.75, 0.60)}(X)) = \{x_4\},$$
$$\overline{apr}_P(\overline{apr}_C^{(0.75, 0.60)}(X)) = \{x_1, x_2, x_3, x_4, x_5, x_7, x_8, x_9, x_{10}\},$$
$$\underline{apr}_Q(\underline{apr}_C^{(0.75, 0.60)}(X)) = \emptyset,$$
$$\overline{apr}_Q(\overline{apr}_C^{(0.75, 0.60)}(X)) = \{x_1, x_2, x_3, x_4, x_5, x_6, x_7, x_8, x_9, x_{10}\}.$$

According to Definition 7, the probabilistic accuracies of X with respect to P and Q are given by

$$\alpha_P^{(0.75, 0.60)}(X) = \frac{|\underline{apr}_P(\underline{apr}_C^{(0.75, 0.60)}(X))|}{|\overline{apr}_P(\overline{apr}_C^{(0.75, 0.60)}(X))|} = \frac{1}{9},$$
$$\alpha_Q^{(0.75, 0.60)}(X) = \frac{|\underline{apr}_Q(\underline{apr}_C^{(0.75, 0.60)}(X))|}{|\overline{apr}_Q(\overline{apr}_C^{(0.75, 0.60)}(X))|} = 0.$$

Thus, $P \preceq Q \Rightarrow \alpha_P^{(0.75, 0.60)}(X) \geq \alpha_Q^{(0.75, 0.60)}(X).$
The probabilistic roughnesses of X with respect to P and Q are given by

$$\rho_P^{(0.75, 0.60)}(X) = 1 - \alpha_P^{(0.75, 0.60)}(X) = 1 - \frac{|\underline{apr}_P(\underline{apr}_C^{(0.75, 0.60)}(X))|}{|\overline{apr}_P(\overline{apr}_C^{(0.75, 0.60)}(X))|} = \frac{8}{9},$$
$$\rho_Q^{(0.75, 0.60)}(X) = 1 - \alpha_Q^{(0.75, 0.60)}(X) = 1 - \frac{|\underline{apr}_Q(\underline{apr}_C^{(0.75, 0.60)}(X))|}{|\overline{apr}_Q(\overline{apr}_C^{(0.75, 0.60)}(X))|} = 1.$$

Therefore, $P \preceq Q \Rightarrow \rho_P^{(0.75, 0.60)}(X) \leq \rho_Q^{(0.75, 0.60)}(X).$
By computing, we have $U/D = \{\{x_1, x_2, x_4, x_7, x_9, x_{10}\}, \{x_3, x_5, x_6, x_8\}\}.$
According to Definition 8, the probabilistic approximation accuracies of X with respect to P and Q are given by

$$\alpha_P^{(0.75,0.60)}(U/D) = \frac{\sum_{Y_i \in U/D} |\underline{apr}_P(\underline{apr}_C^{(0.75,0.60)}(Y_i))|}{\sum_{Y_i \in U/D} |\overline{apr}_P(\overline{apr}_C^{(0..75,0.60)}(Y_i))|} = \frac{1}{6}$$

$$\alpha_Q^{(0.75,0.60)}(U/D) = \frac{\sum_{Y_i \in U/D} |\underline{apr}_Q(\underline{apr}_C^{(0.75,0.60)}(Y_i))|}{\sum_{Y_i \in U/D} |\overline{apr}_Q(\overline{apr}_C^{(0..75,0.60)}(Y_i))|} = 0$$

Hence, $P \preceq Q \Rightarrow \alpha_P^{(0.75,0.60)}(X) \geq \alpha_Q^{(0.75,0.60)}(X)$.

6 Conclusion

Uncertainty measure is an important problem in rough set theory. Probabilistic rough set model is generalization of Pawlak rough set model by introducing probabilistic threshold values. In this paper, the existing problem of uncertainty measures in the probabilistic rough set model is analyzed. We find that parallel extensions of the accuracy measure, the roughness measure and the approximation accuracy measure do not satisfy the monotonicity. Furthermore, a major application of the uncertainty measures is ranking attributes and evaluating the significance of attributes in attribute reduction. However, the monotonicity of uncertainty measures is very important for constructing reduct algorithms. Hence, we proposed three uncertainty measures which satisfy the monotonicity. In the future, we will design monotonic uncertainty measures based attribute reduction algorithms in the probabilistic rough set model.

Acknowledgments. This work is supported by the Natural Science Foundation of China under Grant (Nos. 61272060, 61379114), the Key Natural Science Foundation of Chongqing of China under Grant (Nos. CSTC2013jjB40003).

References

1. Dai, J.H., Wang, W.T., Xu, Q.: An uncertainty measure for incomplete decision tables and its applications. IEEE Transactions on Cybernetics 43, 1277–1289 (2012)
2. Dai, J.H., Xu, Q.: Approximations and uncertainty measures in incomplete information systems. Information Sciences 198, 62–80 (2012)
3. Hu, Q.H., Zhang, L., Chen, D.G., Pedrycz, W., Yu, D.R.: Gaussian kernel based fuzzy rough sets: Model, uncertainty measures and applications. International Journal of Approximate Reasoning 51(4), 453–471 (2010)
4. Katzberg, J.D., Ziarko, W.: Variable precision rough sets with asymmetric bounds. In: Ziarko, W. (ed.) Rough Sets, Fuzzy Sets and Knowledge Discovery, pp. 167–177. Springer, London (1994)
5. Pawlak, Z.: Rough sets. International Journal of Computer and Information Science 11, 341–356 (1982)
6. Pawlak, Z.: Rough sets: theoretical aspects of reasoning about data. Kluwer Academic Publishers, Boston (1991)

7. Qian, Y.H., Liang, J.Y., Wang, F.: A new method for measuring the uncertainty in incomplete information systems. International Journal of Uncertainty, Fuzziness and Knowledge-Based Systems 17(06), 855–880 (2009)
8. Shen, Q., Jensen, R.: Rough sets, their extensions and applications. International Journal of Automation and Computing 4, 217–228 (2007)
9. Xu, W.H., Zhang, X.Y., Zhang, W.X.: Knowledge granulation, knowledge entropy and knowledge uncertainty measure in ordered information systems. Applied Soft Computing 9(4), 1244–1251 (2009)
10. Ziarko, W.: Variable precision rough set model. Journal of Computer and System Sciences 46, 39–59 (1993)

Attribute Subset Quality Functions over a Universe of Weighted Objects*

Sebastian Widz[1] and Dominik Ślęzak[2,3]

[1] Systems Research Institute, Polish Academy of Sciences
ul. Newelska 6, 01-447 Warsaw, Poland
[2] Institute of Mathematics, University of Warsaw
ul. Banacha 2, 02-097 Warsaw, Poland
[3] Infobright Inc., Poland
ul. Krzywickiego 34 pok. 219, 02-078 Warsaw, Poland
sebastian.widz@ibspan.waw.pl, slezak@mimuw.edu.pl

Abstract. We consider a rough set inspired approach to deriving meaningful attribute subsets from data organized in a form of a decision system. We focus on quality functions measuring degrees in which particular attribute subsets determine the values of a decision attribute. We follow a well known idea of assigning weights to the training objects in order to reflect their importance in the attribute subset selection and new case classification processes. We discuss an example of an object weighting strategy related to probabilities of decision classes in the training data. We show that two attribute subset quality functions used in our earlier research are the same function computed using two different weighting techniques. We also investigate whether it is worth using the same weights during the processes of attribute selection and new case classification.

Keywords: Approximate decision reducts, Attribute subset quality functions, Strategies of weighting objects, Voting among decision rules.

1 Introduction

Attribute selection plays an important role in knowledge discovery. It establishes the basis for more efficient classification, prediction and approximation models. Attribute selection methods originating from the theory of rough sets put a special attention on providing users with a better insight into data dependencies [1]. There are numerous rough set based algorithms aimed at searching for so called decision reducts – irreducible subsets of attributes that satisfy predefined criteria for keeping *enough* information about decision classes [2]. Those criteria reflect more or less explicitly a chance of misclassification of objects by if-then decision rules with their antecedents referring to the values of attributes and their consequents referring to decisions.

Original definition of a reduct is quite restrictive, requiring that it should determine decisions or, if data inconsistencies do not allow for full determinism, provide the same level of information about decisions as the complete set of attributes. However, there are

---

* This research was partly supported by Polish National Science Centre (NCN) grants DEC-2011/01/B/ST6/03867 and DEC-2012/05/B/ST6/03215.

M. Kryszkiewicz et al. (Eds.): RSEISP 2014, LNAI 8537, pp. 99–110, 2014.

a number of formulations of approximate reducts, which *almost* preserve original decision information [3]. Such approximate criteria usually rely on functions measuring degrees of decision information induced by subsets of attributes and the corresponding thresholds specifying which subsets of attributes are *good enough*. Those degrees are usually computed from the training data, which makes the whole rough set based framework an example of a filter attribute selection approach [4].

Approximate decision reducts usually include less attributes than classical reducts. On the other hand, they may generate if-then rules that make mistakes even within the training samples. For noisy data sets it is to some extent desirable. Nevertheless, some methods for controlling those mistakes should be considered. For example, if the goal is to construct a classification model based on several approximate decision reducts, then – by following ideas taken from machine learning [5] – one may wish to assure that if-then rules generated by different reducts do not repeat the same mistakes on the training data. For this purpose, we can consider a mechanism aiming at diversification of importance of particular objects while searching for different approximate reducts. Objects' importance can be also controlled by a domain expert or in a more automatic way, according to some criteria related to the problem specification.

In this paper, we discuss the most straightforward approach to achieving the above functionality, that is, labeling objects with non-negative weights. We show that attribute subset quality functions used during the process of approximate reduct derivation can be employed in the same way for weighted and non-weighted objects. As an example, we investigate a weighting strategy designed for imbalanced data sets, where standard attribute selection criteria are usually unable to describe less frequent decision classes. We simply propose that the training objects supporting less frequent decisions could be given higher weights in order to increase their importance. Using this example, we show that quite a variety of attribute subset quality functions could be modeled within a more uniform framework, by considering specific object weights.

The paper is organized as follows: In Section 2, we outline foundations of approximate decision reducts and attribute quality functions. In Section 3, we extend rough set based attribute reduction and classifier construction framework onto decision systems with weighted objects. In Section 4, we compare two examples of object weighting mechanisms from the perspectives of their mathematical foundations and accuracy of the corresponding rule based classification models. In Section 5, we discuss other possible weighting strategies and outline our future research directions.

2 Approximate Reducts and Quality Functions

Let us introduce basic framework for representing qualitative data sets. By a decision system we mean a tuple $\mathbb{A} = (U, A \cup \{d\})$, where U is a finite set of objects, A is a finite set of attributes and $d \notin A$ is a distinguished decision attribute. We refer to elements of U using their ordinal numbers $i = 1, ..., |U|$. We treat attributes $a \in A$ as functions $a : U \to V_a$, where V_a denotes the a's domain. Values $v_d \in V_d$ correspond to decision classes that we want to describe using the values of attributes in A.

Each subset of attributes $B \subseteq A$ yields a set of decision rules based on all combinations of its values in U. Rules' left sides correspond to equivalence classes $E \in$

$U/IND(B)$ where $IND(B) = \{(i,j) \in U \times U : \forall_{a \in B}\, a(i) = a(j)\}$ is so called indiscernibility relation. For simplicity, we will write U/B instead of $U/IND(B)$. Rules' right sides can be defined using various methods, e.g., as the following decision classes $X_E \in U/\{d\}$ occurring most often within particular blocks $E \in U/B$:

$$X_E = \underset{X \in U/\{d\}}{\text{argmax}}\, |X \cap E| \tag{1}$$

The elements of $U/\{d\}$ correspond to the values of d. The ratio of objects in U, which are correctly classified by if-then rules produced by B, takes the following form:

$$M(B) = \frac{1}{|U|} \sum_{E \in U/B} |X_E \cap E| \tag{2}$$

We say that $B \subseteq A$ is a decision reduct for $\mathbb{A}$, iff it is an irreducible subset of attributes such that each pair $i, j \in U$ satisfying inequality $d(i) \neq d(j)$ is discerned by B, i.e., i and j belong to different elements of U/B. In inconsistent decision systems, where even the whole A is not enough to determine decisions, the constraint for B can be modified, e.g., subject to only those pairs $i, j \in U$, $d(i) \neq d(j)$, that can be discerned by A. Furthermore, one can consider irreducible subsets that discern *almost all* pairs of objects from different decision classes. In general, there is a great variety of criteria that can be followed while searching for meaningful subsets of attributes.

For a classical decision reduct $B \subseteq A$ in a consistent decision system, there is always equality $M(B) = 1$. However, it is not the case once we allow rules generated by B to be not necessarily deterministic. In [6], it was proposed to use M explicitly for the formulation of the conditions for approximate attribute reduction:

Definition 1. *[6] Let $\varepsilon \in [0,1)$ and $\mathbb{A} = (U, A \cup \{d\})$ be given. We say that $B \subseteq A$ is an (M, ε)-approximate decision reduct, iff it is an irreducible subset of attributes satisfying the following condition:*

$$M(B) \geq (1 - \varepsilon)M(A) \tag{3}$$

The problem of finding the smallest (M, ε)-approximate decision reducts is NP-hard, for each $\varepsilon \in [0,1)$ treated as a constant in the problem's specification [6]. On the other hand, M is monotonic, i.e. $M(C) \leq M(B)$ for any $C \subseteq B$, which makes it possible to adapt some classical reduct search heuristics discussed, e.g., in [2]. Surely, M is not the only possible attribute subset quality function that could be employed in the above inequality. Nevertheless, in the rest of this paper, we concentrate on function M as one of the most natural and generic choices, showing that it can be actually utilized to model quite many strategies of attribute selection and classifier construction.

3   Foundations of Object Weighted Reducts

Attribute subset quality functions provide good basis for evaluating degrees of determining decision classes. However, the original approximate reduct criteria do not allow for controlling which parts of data are problematic for particular reducts. They do not provide the means for expressing importance of objects in a data set either. Let us propose a general mechanism for expressing objects' importance, based on an arbitrary weight function $\omega : U \to [0, +\infty)$. Let us reformulate the notion of cardinality of a subset of objects $Y \subseteq U$ according to the following simple definition:

$$|Y|_\omega = \sum_{u \in Y} \omega(u) \qquad (4)$$

We rephrase the formula for function M from the previous section as follows, with implicit assumption that there is at least one $u \in U$ such that $\omega(u) > 0$:

$$M_\omega(B) = \frac{1}{|U|_\omega} \sum_{E \in U/B} |X_E^\omega \cap E|_\omega \qquad (5)$$

where

$$X_E^\omega = \underset{X \in U/\{d\}}{\operatorname{argmax}} |X \cap E|_\omega \qquad (6)$$

For a trivial constant function denoted as $1 : U \to \{1\}$ we have $M_1(B) = M(B)$. Also, for a classical decision reduct $B \subseteq A$ in a consistent decision system there is always $M_\omega(B) = 1$. More specific characteristics can be formulated for the cases of $\omega : U \to [0, +\infty)$ and $\omega : U \to (0, +\infty)$. Letting $\omega(u) = 0$ for some objects $u \in U$ may be actually compared to operations on bireducts considered in [7].

Let us formulate several basic properties of M_ω. We start by noting that, for a given $E \in U/B$, there may be several decision classes satisfying (6). Thus, we propose to consider function $\partial_\omega : U/B \to 2^{U/\{d\}}$ defined as follows, for $E \in U/B$:

$$\partial_\omega(E) = \left\{ X \in U/\{d\} : \forall_{X' \in U/\{d\}} |X \cap E|_\omega \geq |X' \cap E|_\omega \right\} \qquad (7)$$

Let us note that ∂_ω is a simple modification of so called generalized decision function introduced within the classical rough set framework [8]. For a given $B \subseteq A$ and $E \in U/B$, the choice of specific $X_E^\omega \in \partial_\omega(E)$ does not influence the quantity of $M_\omega(B)$. However, operating with the values of ∂_ω analogously to the original rough set discernibility criteria becomes important in the following characteristics:

Proposition 1. *Let* $\mathbb{A} = (U, A \cup \{d\})$ *and* $\omega : U \to [0, +\infty)$ *be given. Consider subsets of attributes* $B, C \subseteq A$ *such that* $C \subseteq B$. *There is the following inequality:*

$$M_\omega(B) \geq M_\omega(C) \qquad (8)$$

Moreover, the equality $M_\omega(B) = M_\omega(C)$ *takes place, iff for each* $E \in U/C$ *the intersection of all sets* $\partial_\omega(E')$ *such that* $E' \in U/B$ *and* $E' \subseteq E$ *is not empty.*

Proof. The proof is the same as in the case of function M in [6], now reformulated for non-constant weights of objects. Consider indiscernibility classes $E1, E2 \in U/B$. Assume that those classes would be merged for a smaller subset $C \subseteq B$, i.e. there is $E1 \cup E2 \in U/C$. The whole point is to notice the following:

$$\max_{X \in U/\{d\}} |X \cap (E1 \cup E2)|_\omega \leq \max_{X \in U/\{d\}} |X \cap E1|_\omega + \max_{X \in U/\{d\}} |X \cap E2|_\omega \qquad (9)$$

Figure 1 supports this inequality. It also shows how to deal with $M_\omega(B) = M_\omega(C)$. □

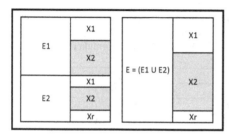 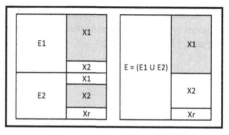

Fig. 1. Left (Right): Indiscernibility classes with the same (different) most significant decisions. The sets of the form $X_E^\omega \cap E$ considered in the equation (5) are marked with a gray color.

Proposition 2. *Let* $\mathbb{A} = (U, A \cup \{d\})$ *and* $\omega : U \to [0, +\infty)$ *be given. For any* $B \subseteq A$, *we have:*

$$\frac{1}{|V_d|} \leq M_\omega(B) \leq 1 \tag{10}$$

Proof. The first inequality can be derived from Proposition 1 by noticing that there is $M_\omega(\emptyset) = \max_{X \in U/\{d\}} |X|_\omega / |U|_\omega \geq 1/|V_d|$. The second inequality is obvious. $\qquad \square$

Definition 2. *Let* $\varepsilon \in [0, 1)$, $\mathbb{A} = (U, A \cup \{d\})$ *and* $\omega : U \to [0, +\infty)$ *be given. We say that* $B \subseteq A$ *is an* (ω, ε)*-approximate decision reduct, iff it is an irreducible subset of attributes satisfying the following condition:*

$$M_\omega(B) \geq (1 - \varepsilon)M_\omega(A) \tag{11}$$

Proposition 3. *Let* $\varepsilon \in [0, 1)$ *be given. The problem of finding an* (ω, ε)*-approximate decision reduct with minimum number of attributes for an input decision system* $\mathbb{A} = (U, A \cup \{d\})$ *and* $\omega : U \to [0, +\infty)$ *is NP-hard.*

Proof. Trivial reduction of the problem of finding minimal (M, ε)-approximate decision reducts [6], by setting ω constantly equal to 1. $\qquad \square$

There can be an exponential number of (ω, ε)-reducts for a given $\mathbb{A}$ and ω. Moreover, different object weights can yield different (ω, ε)-reducts for the same data. Modern rough set approaches to knowledge representation and classifier construction are usually based on ensembles of heuristically found reducts and the corresponding if-then rules with coefficients calculated from the training data. Those coefficients are then used while voting about new objects. – A new object is assigned to a decision class with the highest sum of coefficients produced by rules which match its values.

Table 1 illustrates six examples of coefficients that can be assigned by an if-then rule with its left side supported by indiscernibility class $E \in U/B$ to a decision class $X_E^\omega \in U/\{d\}$ identified by formula (6). They are analogous to some of other voting strategies already discussed in the rough set literature [2]. Precisely, we consider three possibilities to assign a voting degree to a given decision class: plain, ω-confidence and ω-coverage. We can additionally multiply the rule's vote by its normalized left side's support $|E|_\omega / |U|_\omega$ (antecedent voting type ω-support) or not (single).

Table 1. Six options of weighting decisions by if-then rules, corresponding to the consequent coefficient types plain, ω-confidence and ω-coverage, and antecedent coefficient types single and ω-support. $|E|_\omega$ denotes the support of a rule's left side. X_E^ω is defined by formula (6).

	single	ω-support
plain	1	$|E|_\omega/|U|_\omega$
ω-confidence	$|X_E^\omega \cap E|_\omega/|E|_\omega$	$|X_E^\omega \cap E|_\omega/|U|_\omega$
ω-coverage	$(|X_E^\omega \cap E|_\omega/|X_E^\omega|_\omega)/(|E|_\omega/|U|_\omega)$	$|X_E^\omega \cap E|_\omega/|X_E^\omega|_\omega$

Below we outline a simple example of the framework aimed at comparison of different approaches at the stages of attribute selection and new object classification. At the first stage, we can use a number of heuristic methods adopted, e.g., from [3] thanks to analogies between the properties of (M, ε)-reducts and (ω, ε)-reducts.

1. For each considered ω and ε, heuristically generate up to m (ω, ε)-reducts.
2. Choose k (ω, ε)-reducts including the smallest amounts of attributes.[1]
3. Use formula (6) to define the right sides of the if-then decision rules.[2]
4. Apply rules to vote about new objects using some of coefficients in Table 1.

Intuitively, object weights used to compute the voting coefficients should be the same as those used while searching for (ω, ε)-reducts. As in the case of weighting methods in machine learning [9], it seems to be better to keep more or less explicit link between those two general stages. This is because the mechanism of extracting (ω, ε)-reducts is based on removing attributes that – on top of the others – do not contribute significantly to ω-weighted confidence of the corresponding rules. On the other hand, as shown in the next section, sometimes different weights may be worth considering.

4   Comparison of Two Examples of Weighting Schemes

According to the results presented in Section 3, rough set based attribute selection computations can be conducted over data sets with weighted objects in the same way as in the standard non-weighted case. It greatly simplifies the algorithmic framework for searching for ensembles of possibly minimal subsets of attributes which generate pairwise complementary if-then rules. In this section, we investigate a very specific example of the weight function $\omega : U \to [0, +\infty)$ which turns out to correspond to an attribute subset quality measure known in the literature for a longer time.

Proposition 4. *Let* $\mathbb{A} = (U, A \cup \{d\})$ *be given. Consider function* $r : U \to [0, +\infty)$ *defined as follows:*

$$r(u) = \frac{1}{|\{x \in U : d(x) = d(u)\}|} \tag{12}$$

Then we obtain

$$M_r(B) = R(B) \tag{13}$$

---

[1] In our experiments presented in the next section we use $m = 100$ and $k = 10$.
[2] If there are multiple choices due to formula (7), choose one of them randomly.

Table 2. Symbolic benchmark data sets used in our experiments

data set	$|A|$	all objects	train objects	$|V_d|$	decision distribution(s) train/test
chess	36	3196	(CV5)	2	1527:1669
dna	20	3186	2000	3	464:485:1051/303:280:603
sem	256	1593	(CV5)	10	161:162:159:159:161:159:161:158:155:158
spect	22	267	80	2	54:26/103:84
zoo	16	101	(CV5)	7	41:20:5:13:4:8:10

where

$$R(B) = \frac{1}{|V_d|} \sum_{E \in U/B} \max_{X \in U/\{d\}} \frac{|X \cap E|}{|X|} \tag{14}$$

Proof.

$$
\begin{aligned}
M_r(B) &= \tfrac{1}{|U|_r} \sum_{E \in U/B} |X_E^r \cap E|_r \\
&= \tfrac{1}{|V_d|} \sum_{E \in U/B} |X_E^r \cap E|_r \\
&= \tfrac{1}{|V_d|} \sum_{E \in U/B} \max_{X \in U/\{d\}} |X \cap E|_r \\
&= \tfrac{1}{|V_d|} \sum_{E \in U/B} \max_{X \in U/\{d\}} \sum_{u \in X \cap E} \tfrac{1}{|\{x \in U : d(x) = d(u)\}|} \\
&= \tfrac{1}{|V_d|} \sum_{E \in U/B} \max_{X \in U/\{d\}} \tfrac{|X \cap E|}{|X|}
\end{aligned}
\tag{15}
$$

Function R was introduced in [10] for better adjusting classical rough set approximations to the nature of imbalanced decision classes. The idea behind R is to identify the consequents of if-then rules as decision values that occur within particular indiscernibility classes relatively most frequently comparing to their occurrence in the whole U. In other words, the underlying decision process attempts to increase importance of decision values that are less represented in data in order to equalize their chances for valid recognition. An interesting point is that so far we treated approximate attribute reduction constraints defined by M and R as two fully separate cases. Now it turns out that they correspond to two different object weighting strategies within the same methodology. One can even consider a wider class of measures with characteristics balancing between M and R, modeled by weights spanned between functions 1 and r.

Let us take a look at 1-weights and r-weights from a more experimental perspective, in order to verify whether it is worth following the same weighting strategy while searching for approximate reducts and constructing classification models derived from those reducts. As a reference, we recall our earlier experimental results, which correspond to utilization of r-weights during attribute reduction and 1-weights during classifier construction. We analyze five qualitative data sets available from UCI Machine Learning Repository [11]: *chess*, *dna*, *semeion*, *spect* and *zoo*. Two of them are divided into train/test subsets. For the rest of them we use standard CV5 technique. Table 2 summarizes characteristics of investigated data sets. We would like to pay a special attention on *dna* and *zoo*, since they are neither typical balanced sets nor typical imbalanced sets – they represent a kind of mixture of more or less frequent decision classes.

Let us start by illustrating that 1-weights and r-weights can indeed lead toward different results of attribute selection. Figure 2 presents the average lengths of $(1, \varepsilon)$-reducts and (r, ε)-reducts (Y-axis) in ensembles obtained for *dna* and *zoo*, for particular

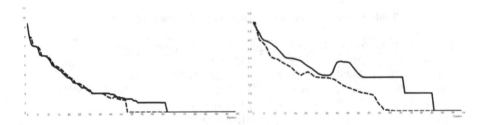

Fig. 2. Average (ω, ε)-reduct cardinalities for two data sets (left: *dna*; right: *zoo*). Dashed line: 1-weights. Solid line: r-weights.

Table 3. Classification accuracy percentages obtained for the ensembles of $(1, \varepsilon)$-reducts and (r, ε)-reducts. Average and best scores for the range of considered values of ε are reported for several data sets from [11], for different settings of the voting coefficients. The first two parts of results (for M_1 and M_r) reflect the framework proposed in this paper. The third part (for M_r but with 1-weights used during voting) corresponds to our earlier experimental results.

quality function	consequent coefficient	antecedent coefficient	chess set avg / best	dna set avg / best	sem set avg / best	spect set avg / best	zoo set avg / best
M_1 (M)	confidence$_1$	single	82,6/83,0	79,0/94,1	61,7/63,2	65,3/76,5	74,7/76,4
	confidence$_1$	support$_1$	76,2/76,4	79,2/90,2	54,6/56,5	62,9/74,3	66,5/67,8
	coverage$_1$	single	82,6/82,9	80,4/92,7	62,2/63,6	65,7/74,3	79,6/81,4
	coverage$_1$	support$_1$	76,7/77,1	72,6/83,7	55,2/56,9	64,9/76,5	71,0/72,2
	plain	single	81,0/81,3	81,8/94,9	49,9/51,4	65,4/75,4	72,1/73,1
	plain	support$_1$	74,8/75,1	76,5/86,3	51,8/53,5	62,9/74,3	66,8/68,2
M_r (R)	confidence$_r$	single	84,9/85,2	80,2/93,3	63,1/64,2	63,6/73,8	86,1/87,1
	confidence$_r$	support$_r$	78,4/78,5	62,7/79,9	56,4/57,8	64,7/73,3	75,6/78,1
	coverage$_r$	single	84,9/85,2	80,4/93,3	63,0/64,1	63,6/73,8	86,1/87,1
	coverage$_r$	support$_r$	78,4/78,5	62,7/79,9	56,4/57,8	64,7/73,3	76,0/78,5
	plain	single	82,9/82,9	79,3/93,9	51,5/52,5	64,3/72,2	80,0/82,2
	plain	support$_r$	76,7/76,8	60,4/79,2	53,6/54,8	63,8/73,8	66,5/69,7
M_r (R)	confidence$_1$	single	84,1/84,4	82,3/93,7	62,9/64,2	65,2/73,8	84,3/85,9
	confidence$_1$	support$_1$	77,9/78,0	74,3/86,1	55,9/57,6	63,6/74,9	75,9/79,0
	coverage$_1$	single	84,8/85,1	74,7/88,9	63,3/64,3	62,8/70,6	79,6/80,6
	coverage$_1$	support$_1$	78,4/78,5	62,7/79,9	56,4/57,8	64,7/73,3	76,0/78,5
	plain	single	82,9/82,9	79,3/93,9	51,5/52,5	64,3/72,2	80,0/82,2
	plain	support$_1$	76,6/76,7	65,5/84,6	53,3/54,7	63,8/74,3	73,1/75,4

settings of ε (X-axis). Going further, Table 3 reports results obtained for all considered data sets, for different voting mechanisms in conjunction with attribute subset quality functions used during approximate attribute reduction. Each result is described by average and maximum classification accuracies computed over the range $[0, 1)$ of approximation degrees. Quite often, the best classifiers apply coefficients calculated using 1-weights while voting between rules derived from $(1, \varepsilon)$-reducts, or coefficients based on r-weights together with (r, ε)-reducts. However, there are some counterexamples, such as combining (r, ε)-reducts with *confidence$_1$-single* for *dna* data set.

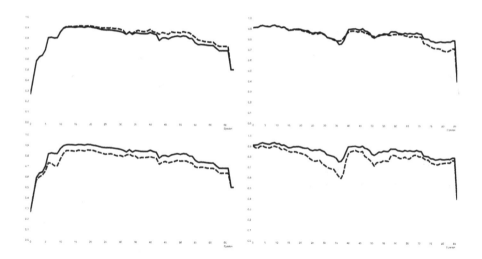

Fig. 3. Classification accuracies for two data sets (left: *dna*; right: *zoo*). Results for (r, ε)-reducts. Solid line (top): *confidence$_r$-single* voting. Dashed line (top): *confidence$_1$-single*. Solid line (bottom): *coverage$_r$-single*. Dashed line (bottom): *coverage$_1$-single*.

Certainly, average and maximum scores do not provide complete information about the dynamics of analyzed rule based classifiers. Figure 3 compares voting strategies *confidence$_r$-single* and *confidence$_1$-single*, as well as *coverage$_r$-single* and *coverage$_1$-single*, for two selected data sets. Here, X-axis and Y-axis correspond to approximation degrees and obtained accuracies, respectively. We can see that for *dna* the ensembles of (r, ε)-reducts should be used together with *confidence$_1$-single* voting indeed. In other cases, as illustrated in detail by Figure 3 and generally by Table 3, *confidence$_r$-single* and *coverage$_r$-single* options look like better (or at least not worse) choices.

The content of Table 4 is analogous to that of Table 3, but now representing so called balanced classification accuracy, that is, the mean of percentages of correctly classified objects within each of decision classes [12]. Balanced accuracy gives bigger weights to objects from minority classes in order to emphasize their importance in the classification process. It corresponds to introducing r-weights while computing classification accuracy over the test data. Thus, one might expect that (r, ε)-reducts and r-weighted voting coefficients should be the best combination in this case. In our study, it is true for four out of five sets. Surprisingly, *dna* becomes a counterexample again.

Figure 4 provides better insight into differences between two considered types of accuracies. Relative differences between r-weighted voting strategies remain roughly the same in both cases, except low values of ε for *dna* data set. Figure 4 brings also two other observations. In case of *dna*, differences in accuracy seem to depend on the choice between *single* and *r-support* antecedent coefficients. However, this is not true for *zoo*, especially for higher values of ε. The second observation, supported also by Figure 3,

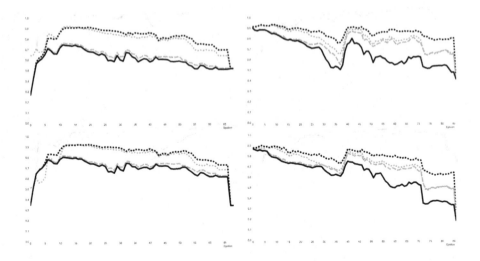

Fig. 4. Two types of classification accuracies (top: *standard*; bottom: *balanced*) for two data sets (left: *dna*; right: *zoo*). Results for (r, ε)-reducts. Black dotted line: $confidence_r$-*single* voting. Gray dotted: *plain-single*. Gray dashed: $coverage_r$-$support_r$. Solid black: $plain$-$support_r$.

Table 4. Balanced classification accuracy percentages obtained for the same reduct ensembles and the same voting settings as it was reported for standard classification accuracy in Table 3

quality function	consequent coefficient	antecedent coefficient	chess set avg / best	dna set avg / best	sem set avg / best	spect set avg / best	zoo set avg / best
M_1 (M)	$confidence_1$	single	82,3/82,7	76,9/93,7	62,2/63,2	63,5/75,2	60,1/64,0
	$confidence_1$	$support_1$	75,4/75,7	81,8/90,2	55,2/56,5	60,3/72,6	48,8/51,0
	$coverage_1$	single	82,3/82,7	82,2/93,3	62,5/63,6	65,2/74,2	69,2/71,4
	$coverage_1$	$support_1$	76,0/76,4	78,2/86,4	55,6/56,9	63,2/75,5	55,5/58,3
	plain	single	80,5/80,8	76,9/94,4	50,4/51,4	62,5/74,3	56,1/58,4
	plain	$support_1$	74,1/74,3	79,9/88,5	52,3/53,5	60,3/72,6	48,9/51,1
M_r (R)	$confidence_r$	single	84,6/85,0	67,4/94,2	62,9/64,2	63,4/73,7	78,8/83,2
	$confidence_r$	$support_r$	77,8/78,0	72,6/84,4	56,1/57,8	63,6/72,7	69,2/74,6
	$coverage_r$	single	84,6/85,0	75,0/94,2	62,9/64,1	63,3/73,7	78,8/83,2
	$coverage_r$	$support_r$	77,8/78,0	75,4/84,4	56,1/57,8	63,6/72,7	69,6/74,8
	plain	single	82,5/82,6	67,8/94,6	51,3/52,5	63,0/72,0	72,4/77,7
	plain	$support_r$	76,1/76,1	71,5/83,6	53,3/54,8	62,5/73,1	61,9/66,6
M_r (R)	$confidence_1$	single	83,8/84,1	82,0/94,4	62,9/64,2	64,1/73,3	74,9/78,4
	$confidence_1$	$support_1$	77,3/77,4	78,5/88,0	55,9/57,6	61,5/74,2	64,3/67,5
	$coverage_1$	single	84,6/84,9	79,7/90,9	63,0/64,3	63,3/71,0	75,9/79,8
	$coverage_1$	$support_1$	77,8/78,0	71,3/84,4	56,1/57,8	63,6/72,7	69,6/74,8
	plain	single	82,5/82,6	80,0/94,6	51,3/52,5	63,0/72,0	72,4/77,7
	plain	$support_1$	76,0/76,0	72,7/87,2	53,2/54,7	62,2/73,0	64,5/69,3

refers to dependency between accuracy and average cardinality of approximate reducts controlled by ε. In case of *zoo*, it looks like we should reduce attributes very carefully. On the other hand, *dna* may require relatively higher approximation degrees.

5 Conclusions and Future Work

We discussed a new approach to defining attribute subset quality functions based on assigning non-negative weights to the training objects. We showed that two different functions that we used so far in our research follow the same generic definition, for appropriate settings of weights. We adopted an experimental classifier construction framework comprising of the stages of searching for approximate decision reducts, selecting reducts with the smallest amounts of attributes, defining the right sides of if-then rules induced by selected reducts, and applying those rules to vote about new objects. For two considered examples of quality functions, we investigated whether the weights applied to calculate voting coefficients should (or should not) correspond to the weights applied during the process of extracting approximate reducts. We also showed how the proposed methodology refers to classical notions of the theory of rough sets [8].

The idea of weighting objects is certainly not new. Nevertheless, it opens some opportunities for rough set inspired approaches to modeling and utilizing approximate attribute dependencies. If appropriately designed at the level of graphical interfaces, it can let users interact with classification and data based knowledge representation methods, by iterative reevaluations of object weights during the process of attribute selection. For example, we could consider a system reporting some clusters of properly and wrongly classified data, where it is possible to increase weights of objects located inside and/or at the edges of those clusters. We can also consider some additional domain specific tools for describing important objects. For example, in applications related to functional magnetic resonance imaging [5], users might mark the whole areas of interesting voxels, triggering recalculations of weights of the corresponding objects.

Besides influencing the steps of attribute selection and new case classification, object weights can be also utilized to express user expectations with regard to performance of decision models. In Section 4, we observed it for the notion of balanced accuracy, which corresponds to evaluating classifiers with respect to less frequent but potentially more important decision classes [12]. Indeed, for a weighting strategy used during training, a question arises how to measure the results of classification. If we have any hints related to the weights of particular test objects, we may compute a kind of generalized balanced accuracy as the sum of weights of properly classified objects divided by the sum of weights of all observed objects. Going further, we can say that the process of extracting optimal rule based classifiers from data should also include learning an object weighting strategy which characterizes practical goals of classification.

In future, besides the above aspects, we will continue our studies on correspondence between weights applied at different stages of our rough set inspired classification framework. We will attempt to find more useful dependencies between characteristics of data sets and optimal weighting strategies as well. Obviously, we are also going to keep seeking for analogies between the problem of searching for ensembles of (ω, ε)-reducts and basic techniques of bagging and boosting developed for ensembles of weak classifiers [9]. Such analogies have been already investigated for ensembles of decision reducts with respect to minimizing occurrences of common attributes [4], as well as for ensembles of decision bireducts which are defined by both subsets of attributes and subsets of training objects that can be correctly classified by those attributes [7]. Operating with numerical weights of objects can further strengthen those analogies, with no harm for simplicity and clarity of rough set approach to attribute selection.

References

1. Świniarski, R.W., Skowron, A.: Rough Set Methods in Feature Selection and Recognition. Pattern Recognition Letters 24(6), 833–849 (2003)
2. Bazan, J., Szczuka, M.S.: The Rough Set Exploration System. In: Peters, J.F., Skowron, A. (eds.) Transactions on Rough Sets III. LNCS, vol. 3400, pp. 37–56. Springer, Heidelberg (2005)
3. Ślęzak, D.: Rough Sets and Functional Dependencies in Data: Foundations of Association Reducts. In: Gavrilova, M.L., Tan, C.J.K., Wang, Y., Chan, K.C.C. (eds.) Transactions on Computational Science V. LNCS, vol. 5540, pp. 182–205. Springer, Heidelberg (2009)
4. Widz, S., Ślęzak, D.: Rough Set Based Decision Support – Models Easy to Interpret. In: Peters, G., Lingras, P., Ślęzak, D., Yao, Y. (eds.) Rough Sets: Selected Methods and Applications in Management & Engineering. Advanced Information and Knowledge Processing, pp. 95–112. Springer (2012)
5. Kuncheva, L.I., Diez, J.J.R., Plumpton, C.O., Linden, D.E.J., Johnston, S.J.: Random Subspace Ensembles for fMRI Classification. IEEE Transactions on Medical Imaging 29(2), 531–542 (2010)
6. Ślęzak, D.: Normalized Decision Functions and Measures for Inconsistent Decision Tables Analysis. Fundamenta Informaticae 44(3), 291–319 (2000)
7. Stawicki, S., Widz, S.: Decision Bireducts and Approximate Decision Reducts: Comparison of Two Approaches to Attribute Subset Ensemble Construction. In: Ganzha, M., Maciaszek, L.A., Paprzycki, M. (eds.) Federated Conference on Computer Science and Information Systems – FedCSIS 2012, Wrocław, Poland, September 9-12, pp. 331–338. IEEE (2012)
8. Pawlak, Z., Skowron, A.: Rudiments of Rough Sets. Information Sciences 177(1), 3–27 (2007)
9. Skurichina, M., Duin, R.P.W.: Bagging, Boosting and the Random Subspace Method for Linear Classifiers. Pattern Analysis and Applications 5(2), 121–135 (2002)
10. Ślęzak, D., Ziarko, W.: The Investigation of the Bayesian Rough Set Model. International Journal of Approximate Reasoning 40(1-2), 81–91 (2005)
11. Frank, A., Asuncion, A.: UCI Machine Learning Repository (2010), http://archive.ics.uci.edu/ml
12. Brodersen, K.H., Ong, C.S., Stephan, K.E., Buhmann, J.M.: The Balanced Accuracy and Its Posterior Distribution. In: 20th International Conference on Pattern Recognition, ICPR 2010, Istanbul, Turkey, August 23-26, pp. 3121–3124. IEEE (2010)

A Definition of Structured Rough Set Approximations

Yiyu Yao* and Mengjun Hu

Department of Computer Science, University of Regina
Regina, Saskatchewan, Canada S4S 0A2
{yyao,hu258}@cs.uregina.ca

Abstract. Pawlak lower and upper approximations are unions of equivalence classes. By explicitly expressing individual equivalence classes in the approximations, Bryniarski uses a pair of families of equivalence classes as rough set approximations. Although the latter takes into consideration of structural information of the approximations, it has not received its due attention. The main objective of this paper is to further explore the Bryniarski definition and propose a generalized definition of structured rough set approximations by using a family of conjunctively definable sets. The connections to covering-based rough sets and Grzymala-Busse's LERS systems are investigated.

1 Introduction

A fundamental construct of rough set theory is the approximation of an undefinable set by a pair of definable sets called the lower and upper approximations [2, 8, 10, 13]. At least three senses of the definability of sets appeared in rough set literature. In one sense, a set is said to be definable, originally called describable by Marek and Pawlak [7], if the set is exactly the set of objects satisfying a formula in a description language. That is, we can use a logic formula to describe the set. In another sense, a set is said to be definable if it is the union of a family of equivalence classes induced by an equivalence relation [9], where the family is a subset of the partition of the equivalence relation. Finally, a set is said to be definable if its lower and upper approximations are the same as the set itself. Although the three definitions of definability are mathematically equivalent, they are very different semantically. The first two definitions treat definability as a primitive notion that motivates the introduction of rough set approximations. In other words, approximations are needed in order to make inference about undefinable sets [8]. In comparison, the description-language based definition is advantageous over equivalence-relation based definition, because the latter does not provide a satisfactory answer to the question: why is the union of a family of equivalence classes a definable set? The third definition treats approximations as primitive notions and definability as a derived notion. This

---

* This work is partially supported by a Discovery Grant from NSERC, Canada. The authors would like to thank reviewers for their constructive comments.

M. Kryszkiewicz et al. (Eds.): RSEISP 2014, LNAI 8537, pp. 111–122, 2014.

definition fails to explain the reason for introducing rough set approximations. By taking advantages of the semantically superior description-language based definition, we propose the notion of structured rough set approximations.

Let $E \subseteq U \times U$ denote an equivalence relation on a nonempty and finite set of objects U. The equivalence class containing x is given by $[x] = \{y \in U \mid xEy\}$. The partition U/E induced by the equivalence relation E contains all equivalence classes which are the building blocks to construct rough set approximations. For a subset of objects $X \subseteq U$, Pawlak lower and upper approximations [8] are defined through equivalence classes as follows:

$$\underline{apr}(X) = \bigcup \{[x] \in U/E \mid [x] \subseteq X\},$$
$$\overline{apr}(X) = \bigcup \{[x] \in U/E \mid [x] \cap X \neq \emptyset\}. \tag{1}$$

Both lower and upper approximations are subsets of U. Bryniarski [1] proposed an alternative definition by removing the union in the Pawlak approximations:

$$\underline{bapr}(X) = \{[x] \in U/E \mid [x] \subseteq X\},$$
$$\overline{bapr}(X) = \{[x] \in U/E \mid [x] \cap X \neq \emptyset\}. \tag{2}$$

These lower and upper approximations are no longer subsets of objects, but are families of equivalence classes, namely, $\underline{bapr}(X) \subseteq U/E$ and $\overline{bapr}(X) \subseteq U/E$. While the Pawlak definition is widely used in the main stream research, only a few studies [3] consider the Bryniarski definition.

The two definitions differ slightly in form. They define each other as follows:

$$\underline{apr}(X) = \bigcup \underline{bapr}(X),$$
$$\overline{apr}(X) = \bigcup \overline{bapr}(X), \tag{3}$$

and

$$\underline{bapr}(X) = \{[x] \in U/E \mid [x] \subseteq \underline{apr}(X)\},$$
$$\overline{bapr}(X) = \{[x] \in U/E \mid [x] \subseteq \overline{apr}(X)\}. \tag{4}$$

However, they are very different semantically. By using the equivalence classes in the lower and upper approximations, $\underline{bapr}(X)$ and $\overline{bapr}(X)$ preserve the structural information about rough set approximations, which was missing from the Pawlak definition. In this paper, we call $(\underline{apr}(X), \overline{apr}(X))$ unstructured Pawlak approximations or Pawlak rough set induced by X, and $(\underline{bapr}(X), \overline{bapr}(X))$ Bryniarski approximations or structured Pawlak approximations.

There are several reasons to retain structural information. When we construct rules from an approximation, we use each equivalence class to form one classification rule. An explicit representation of the structural information makes it much easier to explain and understand such a learning task. Each equivalence class may be viewed as a granule and a structured approximation explicitly shows the composition of a family of granules in forming a rough set approximation. This interpretation connects rough set theory and granular computing [14].

The main objective of this paper is to propose a generalized definition of structured rough set approximations by considering the family of all conjunctively definable sets, of which U/E is a subset. In the Pawlak framework, equivalence classes are pair-wise disjoint and there do not exist redundant equivalence classes in a Bryniarski approximation. On the other hand, in the new framework, there may exist redundant conjunctively definable sets in a structured approximation. It is therefore necessary to remove redundancy. The notion of ⊔-reduct is adopted and two methods for deriving reduced structured approximations are introduced. One method directly finds an ⊔-reduct of a structured approximation. The other method first computes the family of maximal conjunctively definable sets in a structured approximation and then finds an ⊔-reduct, which enables us to derive more general classification rules.

2   Structured Rough Set Approximations

For Pawlak rough sets, the upper approximation of a set is the complement of the lower approximation of the complement of the set, that is, $\overline{apr}(X) = (\underline{apr}(X^c))^c$. In this paper, we demand such a duality and only consider the lower approximation.

2.1   Definable and Conjunctively Definable Sets

As a theory for data analysis, rough sets deal with data in a tabular form, in which each row represents an object, each column represents an attribute, and a cell represents the value of an object on an attribute. Pawlak [9] originally called such a table an information system. We call it an information table to avoid confusions with other commonly associated meanings of an information system.

Definition 1. *An information table is defined by a tuple:*

$$T = (U, AT, \{V_a \mid a \in AT\}, \{I_a : U \to V_a \mid a \in AT\}), \tag{5}$$

where U is a finite nonempty set of objects called the universe, AT is a finite nonempty set of attributes, V_a is the domain of an attribute a and I_a is a description function that assigns a value from V_a to each object.

Based on an information table, we use a sublanguage of the description language proposed by Marek and Pawlak [7] by considering only logic conjunction and disjunction. This sublanguage is sufficient for the present study.

Definition 2. *Formulas of a description language (DL) are defined as follows:*

(1) *$(a = v) \in DL$, where $a \in AT$ and $v \in V_a$. Such formulas are called atomic formulas.*

(2) *If $p, q \in DL$, then $p \wedge q, p \vee q \in DL$.*

Intuitively, an atomic formula $a = v$ corresponds to the condition that the value of an object on the attribute $a \in AT$ is v. A formula is constructed from these conditions by using logic conjunction and disjunction. The meaning of a formula is interpreted in terms of a subset of objects that satisfy the formula.

Definition 3. *For a formula $p \in DL$, let $x \models p$ denote that an object x satisfies p. The satisfiability of formulas by an object is defined by:*

$$
\begin{aligned}
&(1) \quad x \models (a = v), \text{ if } I_a(x) = v, \\
&(2) \quad x \models p \wedge q, \text{ if } x \models p \text{ and } x \models q, \\
&(3) \quad x \models p \vee q, \text{ if } x \models p \text{ or } x \models q.
\end{aligned}
\tag{6}
$$

According to the satisfiability of objects, the following set of objects:

$$
m(p) = \{x \in U \mid x \models p\},
\tag{7}
$$

is called the meaning set of a formula p in DL.

As shown by the following theorem [8], one can interpret logic conjunction and disjunction using set intersection and union through the meaning sets of formulas.

Theorem 1. *The meaning sets of formulas in DL can be computed by:*

$$
\begin{aligned}
&(1) \quad m(a = v) = \{x \in U \mid I_a(x) = v\}, \\
&(2) \quad m(p \wedge q) = m(p) \cap m(q), \\
&(3) \quad m(p \vee q) = m(p) \cup m(q).
\end{aligned}
\tag{8}
$$

An arbitrary subset of objects may fail to be the meaning set of a formula. Those subsets that are meaning sets of formulas are useful in constructing rough set approximations.

Definition 4. *A subset of objects $X \subseteq U$ is called a definable set with respect to a description language DL, if there exists a formula $p \in DL$ such that*

$$
X = m(p).
\tag{9}
$$

That is, a definable set X is the meaning set of a formula p and the formula p is a description of objects in X. The set X is called a conjunctively definable set if the formula p only contains the logic conjunction $\wedge$ and the formula is called a conjunctive formula.

In the classical view of concepts [12, 15], a concept is jointly characterized by a pair of an intension (i.e., properties of instances of the concept) and an extension (i.e., instances of the concept). The description language DL provides a formal way to represent a concept. A definable set X may be viewed as the extension of a concept and a formula p with $m(p) = X$ may be viewed as an intension of the concept. There may exist more than one formula that defines the same definable set.

Let CDEF(U) denote the family of all conjunctively definable sets, which is closed under set intersection. A reviewer of this paper pointed out that CDEF(U) is in fact "the $\cap$-semilattice of 2^U generated by the classes of the equivalence relations associated with the single attributes. Using the lattice point of view, the description language seems unnecessary overhead, and the results are imminent." In response to the comments, it is crucial to establish the necessity of using a description language. The reason for adopting a description language DL is to demonstrate explicitly that each set in CDEF(U) is a conjunctively definable set, corresponding to a conjunctive concept in concept formulation and learning [6]. The use of equivalence classes does not explicitly provide such a semantically sound interpretation. To a large extent, the proposed notion of structured approximations is motivated from a semantical consideration. As will be shown later, a structured approximation covers the same set of objects as Pawlak approximation. The clarification of semantics offered by a description language is the major contribution of this paper.

Now we show that equivalence classes are conjunctively definable sets. Given a subset of attributes $A \subseteq AT$, we define an equivalence relation:

$$E_A = \{(x, y) \in U \times U \mid \forall a \in A, I_a(x) = I_a(y)\}. \tag{10}$$

The equivalence class containing object x is given by:

$$[x] = [x]_{E_A} = \{y \in U \mid y E_A x\}. \tag{11}$$

All equivalence classes form a partition of U:

$$U/E_A = \{[x] \mid x \in U\}. \tag{12}$$

The equivalence class $[x]$ is in fact the meaning set of the formula:

$$\bigwedge_{a \in A} (a = I_a(x)), \tag{13}$$

where $I_a(x) = v \in V_a$ denotes the value of x on attribute $a \in AT$ and $a = I_a(x)$ is the atomic formula $a = v$. Thus, $[x]$ is a conjunctively definable set. It follows that the union of a family of equivalence classes is a definable set. Furthermore, Pawlak lower approximation is the greatest definable set contained in X and Pawlak upper approximation is the least definable set containing X. This offers a semantically sound interpretation of rough set approximations [10, 13].

2.2   A Definition of Structured Lower Approximations

The family of nonempty sets in CDEF(U) is a covering of the universe U and, furthermore, $U/E_A \subseteq$ CDEF(U). By replacing U/E with CDEF(U) in Equation (2), we propose a generalized definition of structured lower approximations.

Definition 5. *For a subset of objects $X \subseteq U$, the structured lower approximation of X is defined by:*

$$\underline{sapr}(X) = \{G \in \text{CDEF}(U) \mid G \neq \emptyset, G \subseteq X\}. \tag{14}$$

That is, $\underline{sapr}(X)$ consists of all nonempty conjunctively definable subsets of X.

By definition and the fact that equivalence classes are conjunctively definable sets, one can establish connections between structured lower approximations, Pawlak approximations and Bryniarski approximations.

Theorem 2. *The following properties hold:*

$$\text{(i)} \quad \bigcup \underline{sapr}(X) = \underline{apr}(X),$$

$$\text{(ii)} \quad \underline{bapr}(X) \subseteq \underline{sapr}(X), \tag{15}$$

where $\underline{apr}(X)$ is Pawlak lower approximation and $\underline{bapr}(X)$ is Bryniarski lower approximation or structured Pawlak lower approximation.

Property (i) shows that our structured lower approximation is consistent with Pawlak lower approximation in the sense that both of them contain the same set of objects if the structural information in $\underline{sapr}(X)$ is ignored. Property (ii) states that $\underline{sapr}(X)$ contains the Bryniarski lower approximation.

The structured lower approximations are related to covering-based rough sets [11, 17–19]. In many studies on covering-based rough sets, it is assumed that a covering of the universe is given and various rough set approximations are introduced. The meaning of subsets in a covering and the covering is not explicitly stated. In our formulation, each nonempty set in CDEF(U) is a conjunctively definable set and has a well-defined semantics interpretation. This is consistent with Pawlak formulation in which each equivalence class is a conjunctively definable set. Consequently, our formulation provides a semantically sound approach to covering-based rough sets.

2.3   Reduced Structured Lower Approximations

In Bryniarski lower approximation $\underline{bapr}(X)$, for any proper subset $\mathbb{B} \subsetneq \underline{bapr}(X)$, we have

$$\bigcup \mathbb{B} \neq \bigcup \underline{bapr}(X) = \underline{apr}(X). \tag{16}$$

That is, to cover all objects in $\underline{apr}(X)$, we cannot remove any equivalence class in $\underline{bapr}(X)$. For the structured lower approximation $\underline{sapr}(X)$, it is possible to remove some conjunctively definable sets in $\underline{sapr}(X)$ and, at the same time, to cover $\underline{apr}(X)$. We may find a proper subset $\mathbb{R} \subsetneq \underline{sapr}(X)$ such that

$$\bigcup \mathbb{R} = \bigcup \underline{sapr}(X) = \underline{apr}(X). \tag{17}$$

This suggests that $\underline{sapr}(X)$ contains redundant conjunctively definable sets. By removing those redundant sets, we may obtain a reduced structured lower approximation. For this purpose, let us first recall the notion of an ∪-reduct proposed by Pawlak [8, 16].

Definition 6. *Suppose $\mathbb{S}$ is a family of subsets of U, a subset $\mathbb{R} \subseteq \mathbb{S}$ is called an union-reduct or ∪-reduct of $\mathbb{S}$ if it satisfies the following two conditions:*

$$\text{(1)} \quad \bigcup \mathbb{R} = \bigcup \mathbb{S},$$

$$\text{(2)} \quad \forall S \in \mathbb{R}, \bigcup (\mathbb{R} - \{S\}) \neq \bigcup \mathbb{S}. \tag{18}$$

Condition (1) indicates that $\mathbb{R}$ is sufficient for covering all objects covered by $\mathbb{S}$, and condition (2) states that each set in $\mathbb{R}$ is necessary. That is, an $\cup$-reduct is a minimal family of sets in $\mathbb{S}$ that covers the same objects as $\mathbb{S}$. There may exist more than one $\cup$-reduct of $\mathbb{S}$.

The notion of $\cup$-reduct can be applied to a covering. However, an $\cup$-reduct of a covering is different from the set of all $\cup$-irreducible elements of a covering. The latter is unfortunately called an $\cup$-reduct of a covering by some authors [17, 19]. We adopt Pawlak's definition of $\cup$-reduct for constructing reduced structured lower approximations.

Definition 7. *An $\cup$-reduct of $\underline{sapr}(X)$ is called a reduced structured lower approximation.*

Let $R_\cup(\underline{sapr}(X))$ denote the family of all reduced structured lower approximations of a subset $X \subseteq U$. The following easy-to-prove theorem shows the properties of $R_\cup(\underline{sapr}(X))$.

Theorem 3. *The following properties hold for $R_\cup(\underline{sapr}(X))$:*

(i)   For any $\mathbb{R} \in R_\cup(\underline{sapr}(X)), \bigcup \mathbb{R} = \bigcup \underline{sapr}(X) = \underline{apr}(X)$,

(ii) $\underline{bapr}(X) \in R_\cup(\underline{sapr}(X))$.   (19)

Property (i) states that all $\cup$-reducts of $\underline{sapr}(X)$ are equivalent in the sense that each of them covers the same set of objects in $\underline{apr}(X)$. Property (ii) states that the Bryniarski lower approximation $\underline{bapr}(X)$ is in fact a reduced structured lower approximation.

When constructing classification rules, one prefers to use more general rules induced by larger conjunctively definable sets. An equivalence class induced by the entire set of attributes is a minimal nonempty conjunctively definable set, that is, no proper nonempty subset of an equivalence class is conjunctively definable. An $\cup$-reduct of $\underline{sapr}(X)$ may not necessarily use larger conjunctively definable subsets of X. That is, an $\cup$-reduct of $\underline{sapr}(X)$ may contain a family of smaller conjunctively definable sets whose union is a larger conjunctively definable set. One can solve this problem by introducing an extra step before constructing an $\cup$-reduct.

2.4   Reduced Structured Lower Approximations with Maximal Conjunctively Definable Sets

Let $\mathbb{M}(\mathbb{S})$ denote the set of all maximal elements of a family $\mathbb{S}$ of subsets of U. The family $\mathbb{M}(\mathbb{S})$ can be easily obtained by removing all subsets $S \in \mathbb{S}$ such that there exists another set $S' \in \mathbb{S}$ with $S \subsetneq S'$. The family $\mathbb{M}(\mathbb{S})$ covers the same set of objects as $\mathbb{S}$, that is,

$$\bigcup \mathbb{M}(\mathbb{S}) = \bigcup \mathbb{S}.   (20)$$

An $\cup$-reduct of $\mathbb{M}(\underline{sapr}(X))$ leads to a reduced structured lower approximation with maximal conjunctively definable sets.

Definition 8. *An* ∪-*reduct of* $\mathbb{M}(\underline{sapr}(X))$ *is called a reduced structured lower approximation with maximal conjunctively definable sets.*

An advantage of this method is that we not only reduce the redundancy, but also use maximal conjunctively definable sets. By definition, we have the following theorem.

Theorem 4. *Let* $R_\cup(\mathbb{S})$ *denote the family of all* ∪-*reducts of a family* $\mathbb{S}$ *of subsets of* U. *The following property holds:*

$$R_\cup(\mathbb{M}(\underline{sapr}(X))) \subseteq R_\cup(\underline{sapr}(X)). \tag{21}$$

Bryniarski lower approximation, although a member of $R_\cup(\underline{sapr}(X))$, may not be a member of $R_\cup(\mathbb{M}(\underline{sapr}(X)))$. This stems from the fact that some equivalence classes in $\underline{sapr}(X)$ may not be maximal elements of $\underline{sapr}(X)$.

2.5 Relationships to LERS

With reference to the classical view of concepts, the formulation of Grzymala-Busse [4, 5] in the LERS (Learning from Examples based on Rough Sets) systems and our formulation are complementary to each other, focusing on the intension and extension of a concept, respectively. In LERS, Grzymala-Busse considers the set-inclusion relation ⊆ between sets of atomic formulas, reflecting the complexity or generality of the intensions of conjunctive concepts, but he does not consider the set-inclusion relationship between conjunctively definable sets, namely, extensions of conjunctive concepts, defined by sets of atomic formulas. In our maximal conjunctively definable sets based formulation, we consider the set-inclusion relation ⊆ between conjunctively definable sets, but we do not consider the complexity or generality of conjunctive formulas used in defining conjunctively definable sets.

Table 1. Correspondences between our and Grzymala-Busse formulations

Our formulation	Grzymala-Busse formulation
atomic formula $a = v$	attribute-value pair $t = (a, v)$
meaning set $m(a = v)$	block $[(a, v)]$
conjunctive formula p	set T of attribute-value pairs (corresponding to atomic formulas in p)
maximal conjunctively definable subset of X	minimal complex of X
∪-reduct $\mathbb{R}$ of $\mathbb{M}(\underline{sapr}(X))$	local covering $\mathbb{T}$ of $\underline{apr}(X)$

Table 1 shows some relevant correspondences between notions of the two formulations. An attribute-value pair (a, v) corresponds to an atomic formula $a = v$. The block of an attribute-value pair $t = (a, v)$, denoted $[t]$, is defined as the set of all objects that take value v on attribute a which is the meaning set of atomic formula $a = v$. That is, $[(a, v)] = \{x \in U \mid I_a(x) = v\} = m(a = v)$. A set T of attribute-value pairs corresponds to the set of all atomic formulas in a formula of DL, and the block of T, denoted $[T]$, is the meaning set of the corresponding formula. In other words, T is another representation of the intension of a conjunctively definable set. A set T of attribute-value pairs is a minimal complex of $\underline{apr}(X)$ if it satisfies the conditions:

$$(1) \quad \emptyset \neq [T] = \bigcap_{t \in T} [t] \subseteq \underline{apr}(X),$$

$$(2) \quad \forall T' \subsetneq T, [T'] = \bigcap_{t \in T'} [t] \not\subseteq \underline{apr}(X). \tag{22}$$

That is, Grzymala-Busse considers the intensions of concepts by using minimal sets of attribute-value pairs; we consider the extensions of concepts by using maximal conjunctively definable sets. A family $\mathbb{T}$ of sets of attribute-value pairs is called a local covering of $\underline{apr}(X)$ if and only if it satisfies the conditions:

(1) each member T of $\mathbb{T}$ is a minimal complex of $\underline{apr}(X)$,

(2) $\bigcup_{T \in \mathbb{T}} [T] = \underline{apr}(X)$,

(3) $\mathbb{T}$ is minimal, i.e., $\mathbb{T}$ has the smallest possible number of members.

A local covering is a family of sets of attribute-value pairs and an $\cup$-reduct of $M(\underline{sapr}(X))$ is a family of maximal conjunctively definable sets. That is, from the viewpoint of intension, Grzymala-Busse uses a family of minimal complexes whose blocks are subsets of X; from the viewpoint of extension, we use a family of maximal conjunctively definable subsets of X.

3   An Example

We use an example to demonstrate the basic ideas of structured lower approximations. Consider an information table given by Table 2. The family of all conjunctively definable sets, except the empty set $\emptyset$, is given by Table 3. For each set, we only give one most general formula that has a minimal number of atomic formulas.

The partition induced by the entire set of attributes is given by $U/E = \{\{o_1\}, \{o_2\}, \{o_3\}, \{o_4\}, \{o_5, o_6\}\}$. According to Equations (1) and (14), we obtain the Pawlak and structured lower approximations of $X = \{o_1, o_2, o_3, o_4, o_5\}$, respectively, as:

$$\underline{apr}(X) = \bigcup\{[x] \in U/E \mid [x] \subseteq X\}$$
$$= \{o_1, o_2, o_3, o_4\}; \tag{23}$$

Table 2. An information table

Object	Height	Hair	Eyes
o_1	short	blond	blue
o_2	short	blond	brown
o_3	short	red	blue
o_4	tall	blond	blue
o_5	tall	red	brown
o_6	tall	red	brown

Table 3. The family of conjunctively definable sets CDEF(U)

formulas	conjunctively definable sets
Height=short	$\{o_1, o_2, o_3\}$
Height=tall	$\{o_4, o_5, o_6\}$
Hair=blond	$\{o_1, o_2, o_4\}$
Hair=red	$\{o_3, o_5, o_6\}$
Eyes=blue	$\{o_1, o_3, o_4\}$
Eyes=brown	$\{o_2, o_5, o_6\}$
Height=short $\wedge$ Hair=blond	$\{o_1, o_2\}$
Height=short $\wedge$ Hair=red	$\{o_3\}$
Height=tall $\wedge$ Hair=blond	$\{o_4\}$
Height=tall $\wedge$ Hair=red	$\{o_5, o_6\}$
Height=short $\wedge$ Eyes=blue	$\{o_1, o_3\}$
Height=short $\wedge$ Eyes=brown	$\{o_2\}$
Hair=blond $\wedge$ Eyes=blue	$\{o_1, o_4\}$
Height=short $\wedge$ Hair=blond $\wedge$ Eyes=blue	$\{o_1\}$

$$\underline{sapr}(X) = \{G \in \text{CDEF}(U) \mid G \neq \emptyset, G \subseteq X\}$$
$$= \{\{o_1, o_2, o_3\}, \{o_1, o_2, o_4\}, \{o_1, o_3, o_4\}, \{o_1, o_2\}, \{o_3\}, \{o_4\},$$
$$\{o_1, o_3\}, \{o_2\}, \{o_1, o_4\}, \{o_1\}\}. \tag{24}$$

One can see that $\bigcup \underline{sapr}(X) = \underline{apr}(X) = \{o_1, o_2, o_3, o_4\}$.

Consider now reduced structured lower approximations. For the first method, we compute a reduced structured lower approximation by finding an $\cup$-reduct of $\underline{sapr}(X)$. Two examples of $\cup$-reducts of $\underline{sapr}(X)$ are:

$$\mathbb{R}_1 = \{\{o_1, o_2\}, \{o_1, o_3\}, \{o_4\}\};$$
$$\mathbb{R}_2 = \{\{o_1\}, \{o_2\}, \{o_3\}, \{o_4\}\}. \tag{25}$$

The second $\cup$-reduct $\mathbb{R}_2$ is in fact the Bryniarski lower approximation of X.

For the second method, we first find the set of all the maximal conjunctively definable sets in $\underline{sapr}(X)$:

$$\mathbb{M}(\underline{sapr}(X)) = \{\{o_1, o_2, o_3\}, \{o_1, o_2, o_4\}, \{o_1, o_3, o_4\}\}. \tag{26}$$

Then, we find an $\cup$-reduct of $\mathbb{M}(\underline{sapr}(X))$, which is a reduced structured lower approximation with maximal conjunctively definable sets. In this case, we have three $\cup$-reducts of $\mathbb{M}(\underline{sapr}(X))$:

$$\begin{aligned}
\mathbb{R}_3 &= \{\{o_1, o_2, o_3\}, \{o_1, o_2, o_4\}\}, \\
\mathbb{R}_4 &= \{\{o_1, o_2, o_3\}, \{o_1, o_3, o_4\}\}, \\
\mathbb{R}_5 &= \{\{o_1, o_2, o_4\}, \{o_1, o_3, o_4\}\}.
\end{aligned} \tag{27}$$

Compared with $\mathbb{R}_1$ and $\mathbb{R}_2$, $\mathbb{R}_3$, $\mathbb{R}_4$ and $\mathbb{R}_5$ use large conjunctively definable sets as building blocks of rough set approximations, which is preferred for deriving classification rules.

4   Conclusion

Pawlak rough set approximations use equivalence classes as the basic building blocks. By observing that equivalence classes are minimal nonempty conjunctively definable sets, we adopt conjunctively definable sets as the basic building blocks. The family of all nonempty conjunctively definable sets is a covering of the universe and is a generalization of the partition induced by an equivalence relation. Based on Bryniarski's definition, we propose a generalized definition of structured rough set approximations by using the family of conjunctively definable sets. In contrast to the partition-based formulation, there may exist redundant conjunctively definable sets in structured approximations. To resolve this problem, we propose two methods to reduce redundancy based on the notion of $\cup$-reducts. The first method finds an $\cup$-reduct directly. The second method first finds the set of all maximal conjunctively definable sets and then finds an $\cup$-reduct. The second method is preferred because it keeps large conjunctively definable sets, which are suitable for learning classification rules.

The proposed formulation of structured approximations is complementary to the formulation of LERS systems introduced by Grzymala-Busse [4, 5]. The former focuses on extensions of concepts and the latter focuses on intensions of concepts. Our formulation provides a new direction in studying covering-based rough sets [11, 17–19]. A main difference from some other studies is that we have a sound interpretation of subsets of objects in the covering in the sense that each set in the covering is a conjunctively definable set.

122   Y. Yao and M. Hu

References

1. Bryniarski, E.: A calculus of rough sets of the first order. Bulletin of the Polish Academy of Sciences, Mathematics 37, 71–78 (1989)
2. Buszkowski, W.: Approximation spaces and definability for incomplete information systems. In: Polkowski, L., Skowron, A. (eds.) RSCTC 1998. LNCS (LNAI), vol. 1424, pp. 115–122. Springer, Heidelberg (1998)
3. Dubois, D., Prade, H.: Rough fuzzy sets and fuzzy rough sets. International Journal of General Systems 17, 191–209 (1990)
4. Grzymala-Busse, J.W.: A new version of the rule induction system LERS. Fundamenta Informaticae 31, 27–39 (1997)
5. Grzymala-Busse, J.W.: LERS - A system for learning from examples based on rough sets. In: Słowiński, R. (ed.) Intelligent Decision Support: Handbook of Applications and Advances of the Rough Sets Theory, pp. 3–18. Kluwer Academic Publishers, Dordrecht (1992)
6. Hunt, E.B.: Concept Learning: An Information Processing Problem. John Wiley and Sons Inc., New York (1962)
7. Marek, W., Pawlak, Z.: Information storage and retrieval systems: Mathematical foundations. Theoretical Computer Science 1, 331–354 (1976)
8. Pawlak, Z.: Rough Sets, Theoretical Aspects of Reasoning About Data. Kluwer Academic Publishers, Dordrecht (1991)
9. Pawlak, Z.: Rough classification. International Journal of Man-Machine Studies 20, 469–483 (1984)
10. Pawlak, Z.: Rough sets. International Journal of Computer and Information Sciences 11, 341–356 (1982)
11. Restrepo, M., Cornelis, C., Gómez, J.: Duality, conjugacy and adjointness of approximation operators in covering-based rough sets. International Journal of Approximate Reasoning 55, 469–485 (2014)
12. van Mechelen, I., Hampton, J., Michalski, R., Thenus, P.: Categories and Concepts: Theoretical Views and Inductive Data Analysis. Academic Press, New York (1993)
13. Yao, Y.Y.: A note on definability and approximations. In: Peters, J.F., Skowron, A., Marek, V.W., Orłowska, E., Słowiński, R., Ziarko, W.P. (eds.) Transactions on Rough Sets VII. LNCS, vol. 4400, pp. 274–282. Springer, Heidelberg (2007)
14. Yao, Y.Y.: A partition model of granular computing. In: Peters, J.F., Skowron, A., Grzymała-Busse, J.W., Kostek, B., Swiniarski, R.W., Szczuka, M.S. (eds.) Transactions on Rough Sets I. LNCS, vol. 3100, pp. 232–253. Springer, Heidelberg (2004)
15. Yao, Y.Y., Deng, X.F.: A granular computing paradigm for concept learning. In: Ramanna, S., Jain, L.C., Howlett, R.J. (eds.) Emerging Paradigms in ML. SIST, vol. 13, pp. 307–326. Springer, Heidelberg (2013)
16. Yao, Y.Y., Fu, R.: The concept of reducts in Pawlak three-step rough set analysis. In: Peters, J.F., Skowron, A., Ramanna, S., Suraj, Z., Wang, X. (eds.) Transactions on Rough Sets XVI. LNCS, vol. 7736, pp. 53–72. Springer, Heidelberg (2013)
17. Yao, Y.Y., Yao, B.X.: Covering based rough set approximations. Information Sciences 200, 91–107 (2012)
18. Żakowski, W.F.: Approximations in the space (U, Π). Demonstratio Mathematica 16, 761–769 (1983)
19. Zhu, W., Wang, F.Y.: Reduction and axiomization of covering generalized rough sets. Information Sciences 152, 217–230 (2003)

Interactive Computations on Complex Granules

Andrzej Jankowski[1], Andrzej Skowron[2], and Roman Swiniarski[3,4]

[1] Institute of Computer Science, Warsaw University of Technology
Nowowiejska 15/19, 00-665 Warsaw, Poland
a.jankowski@ii.pw.edu.pl
[2] Institute of Mathematics, Warsaw University
Banacha 2, 02-097 Warsaw, Poland
skowron@mimuw.edu.pl
[3] Department of Computer Science, San Diego State University
5500 Campanile Drive San Diego, CA 92182, USA
[4] Institute of Computer Science Polish Academy of Sciences
Jana Kazimierza 5, 01-248 Warsaw, Poland
rswiniarski@mail.sdsu.edu

> *[...] a comprehensive theory of computation must reflect,*
> *in a stylized way aspects of the underlying physical world.*
> *– Tomasso Toffoli ([24])*

> *It seems that we have no choice but to recognize the dependence*
> *of our mathematical knowledge (...) on physics, and that being so,*
> *it is time to abandon the classical view of computation*
> *as a purely logical notion independent of that*
> *of computation as a physical process.*
> *– David Deutsch, Artur Ekert,and Rossella Lupacchini*
> *([3], p. 268)*

Abstract. Information granules (infogranules, for short) are widely discussed in the literature. In particular, let us mention here the rough granular computing approach based on the rough set approach and its combination with other approaches to soft computing. However, the issues related to interactions of infogranules with the physical world and to perception of interactions in the physical world represented by infogranules are not well elaborated yet. On the other hand, the understanding of interactions is the critical issue of complex systems. We propose to model complex systems by interactive computational systems (ICS) created by societies of agents. Computations in ICS are based on complex granules (c-granules, for short). In the paper we concentrate on some basic issues related to interactive computations based on c-granules performed by agents in the physical world.

Keywords: granular computing, rough set, interaction, information granule, physical object, hunk, complex granule, interactive computational system.

M. Kryszkiewicz et al. (Eds.): RSEISP 2014, LNAI 8537, pp. 123–134, 2014.

1 Introduction

Granular Computing (GC) is now an active area of research (see, e.g., [17]). Objects we are dealing with in GC are *information granules* (or *infogranules*, for short). Such granules are obtained as the result of information granulation [26, 29]. The concept of granulation is rooted in the concept of a linguistic variable introduced by Lotfi Zadeh in 1973 [25]. Information granules are constructed starting from some elementary ones. More compound granules are composed of finer granules that are drawn together by indistinguishability, similarity, or functionality [27].

Computations on granules should be interactive. This requirement is fundamental for modeling of complex systems [4]. For example, in [16] this is expressed as follows

> *[...] interaction is a critical issue in the understanding of complex systems of any sorts: as such, it has emerged in several well-established scientific areas other than computer science, like biology, physics, social and organizational sciences.*

Interactive Rough Granular Computing (IRGC) is an approach for modeling of interactive computations (see, *e.g.*, [18–23]). Computations in IRGC are progressing by interactions represented by interactive information granules. In particular, interactive information systems (IIS) are dynamic granules used for representing the results of the agent interaction with the environments. IIS can be also applied in modeling of more advanced forms of interactions such as hierarchical interactions in layered granular networks or generally in hierarchical modeling. The proposed approach [18–23] is based on rough sets but it can be combined with other soft computing paradigms such as fuzzy sets or evolutionary computing, and also with machine learning and data mining techniques. The notion of a highly interactive granular system is clarified as the system in which intrastep interactions [5] with the external as well as with the internal environments take place. Two kinds of interactive attributes are distinguished: perception attributes, including sensory ones and action attributes.

In this paper we extend the existing approach by introducing complex granules (c-granules) making it possible to model interactive computations performed by agents. Any c-granule consists of three components, namely soft_suit, link_suit and hard_suit. These components make it possible to deal with such abstract objects from soft_suit as infogranules as well as with physical objects from hard_suit. The link_suit of a given c-granule is used as a kind of c-granule interface for expressing interaction between soft_suit and and hard_suit. Any agent operates in a local world of c-granules. The agent control module aims at controlling computations performed by c-granules from this local world for achieving the target goals. Actions (sensors or plans) from link_suits of c-granules are used by the agent control in exploration and/or exploitation of the environment on the way to achieve their targets. C-granules are also used for representation of perception by agents of interactions in the physical world. Due to the bounds of the agent perception abilities usually only a partial information about the interactions from physical world may be available for agents. Hence, in particular the results of the actions performed by agents can not be predicted with certainty.

In Sect. 2, a general structure of c-granules is described and some illustrative examples are included explaining basic concepts related to c-granules. Moreover, some preliminary concepts related to agents performing computations on c-granules are discussed. Relationships of c-granules and satisfiability relations are discussed in Sect. 3.

This paper is a step toward the realization of the Wisdom Technology (WisTech) program [8, 9, 7]. The paper is an extension of [10].

2   Complex Granules and Physical World

We define the basic concepts related to c-granule with respect to a given agent ag. We assume that the agent ag has an access to a local clock making it possible to use the local time scale. In this paper we consider the discrete linear time scale.

We distinguish several kinds of objects in the environment in which the agent ag operates:

- *physical objects* are treated as four dimensional (spatio-temporal) *hunks of matter* (or *hunks*, for short) [6] such as physical parts of agents or robots, specific media for transmitting information; we distinguish hunks called as artifacts used for labeling other hunks or stigmergic markers used for indirect coordination between agents or actions [13]; note that hunks are spatio-temporal and are perceived by the agent ag as dynamic (systems) processes; any hunk h at the local time t of ag is represented by the state $st_h(t)$; the results of perception of hunk states by agent ag are represented by value vector of relevant attributes (features);
- *complex granules* (*c-granules*, for short) consisting of three parts: *soft_suit*, *link_suit*, and *hard_suit* (see Figure 1); c-granule at the local time t of ag is denoted by G_t (or G if this does not lead to misunderstanding); G receives some inputs and produces some outputs; inputs and outputs of c-granule G are c-granules of the specified admissible types; input admissible type is defined by some input preconditions and the output admissible type is defined by some output postconditions, there are distinguished input (output) admissible types which receive (send) c-granules from (to) the agent ag control;
    - G soft_suit consists of
        1. G name, describing the behavioral pattern description of the agent ag corresponding to the name used by agent for identification of the granule,
        2. G type consisting of the types of inputs and outputs of G c-granule,
        3. G status (*e.g.*, active, passive),
        4. G information granules (infogranules, for short) in mental imagination of the agent consisting, in particular of G specification, G implementation and manipulation method(s); any implementation distinguished

in an infogranule is a description in the agent ag language of transformation of input c-granules of relevant types into output c-granules of relevant types, $i.e.$, any implementation defines an interactive computation which takes as input c-granules (of some types) and produces some c-granules (of some types); inputs for c-granules can be delivered by the agent ag control (or by other c-granules), we also assume that the outputs produced by a given c-granule depend also on interactions of hunks pointed out by link_suite as well as some other hunks from the environment – in this way the semantics of c-granules is established;

- G link_suit consists of
  1. a representation of configuration of hunks at time t ($e.g.$, mereology of parts in the physical configurations perceived by the agent ag);
  2. links from different parts of the configuration to hunks;
  3. G links and G representations of elementary actions; using these links the agent ag may perform sensory measurement or/and actions on hunks; in particular, links point to the sensors or effectors in the physical world used by the considered c-granule; using links the agent ag may, $e.g.$, fix some parameters of sensors or/and actions, initiate sensory measurements or/and action performance; we also assume that using these links the agent ag may encode some information about the current states of the observed hunks by relevant information granules;
- G hard_suit is created by the environment of interacting hunks encoding G soft_suit, G link_suit and implementing G computations;
- soft_suit and link_suit of G are linked by G links for interactions between the G hunk configuration representation and G infogranules;
- link_suit and hard_suit are linked by G links for interactions between the G hunk configuration representation and hunks in the environment.

The interactive processes during transformation of inputs of c-granules into outputs of c-granules are influenced by

1. interaction of hunks pointed out by link_suit;
2. interaction of pointed hunks with relevant parts of configuration in link_suit.

Agent can establish, remember, recognize, process and modify relations between c-granules or/and hunks.

A general structure of c-granules is illustrated in Figure 1.

In Figure 2 we illustrate how the abstract definition of operation from soft_suit interacts with other suits of c-granule. It is necessary to distinguish two cases. In the first case, the results of operation $\otimes$ realized by interaction of hunks are consistent with the specification in the link_suit. In the second case, the result specified in the soft_suit can be treated only as an estimation of the real one which may be different due to the unpredictable interactions in the hard_suit.

In Figure 3 we explain the basic idea of interactions between hunks from the environment and hunks of the agent ($e.g.$, from the agent brain (or memory)).

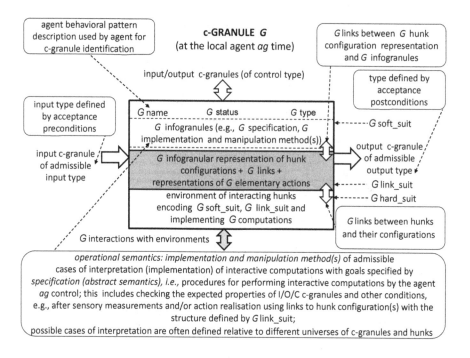

Fig. 1. General structure of c-granules

The idea of the link notion is illustrated in Figure 4. Parameters from p labeling the link are specifying a hunk configuration c necessary for creating the environment for the relevant interaction between h and h_0. Parameters from p are describing, *e.g.*, relationships between hunks in the configuration, actions to be performed, sensors to be used, time of initiation of actions and measurements by sensors, measurement method during the interaction process, time of interaction results measurement at h_0.

Figure 5 illustrates the meaning of aggregation of simple c-granules defined by links. The relation *Context* describes the relationships between parameter vectors of different links. For example, this may concern distance in time and/or space. $R(p_1, \ldots, p_k; u_1, \ldots, u_k)$ describes the expected result property of interactive aggregation of links. This idea is summarized in Eq. 1:

$$Context(p_1, \ldots, p_k) : \frac{g_i : h_i \longleftrightarrow_{p_i} u_i \ (i = 1, \ldots, k)}{R(p_1, \ldots, p_k; u_1, \ldots, u_k).} \tag{1}$$

The explanation of links joining interacting objects of different nature, *i.e.*, inforgranules and hunks is given in Fig. 6 The links (shown as thick arrows) are in fact a special 'composition' of links between hunks (shown as thin arrows) in the environment and the agent 'brain' (or memory) with the encoding between infogranule space and the space of hunks in brain (or memory).

In our approach, the agent can be also interpreted as a c-granule. However, this is a c-granule of higher order with embedded control. One can also consider

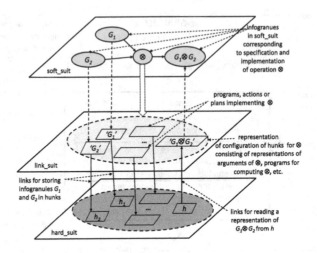

Fig. 2. Explanation of roles of different suits of a c-granule for operation $\otimes$

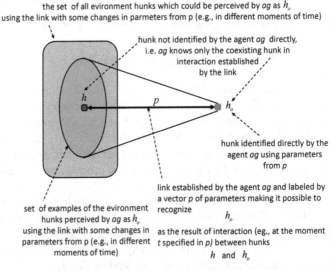

Fig. 3. Hunk interaction expressed by coexistence of hunks

another situation when the c-granules are autonomous but this is out of scope of this article. One can also consider interactions in societies of agents modeled by c-granules. We assume that for any agent ag there is distinguished a family of *elementary c-granules* and constructions on c-granules leading to more compound c-granules. The agent ag uses the constructed granules for modeling of attention and interaction with the environment. Note that for any new construction on elementary granules (such as network of c-granules) it should be defined the

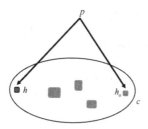

Fig. 4. Link interpretation

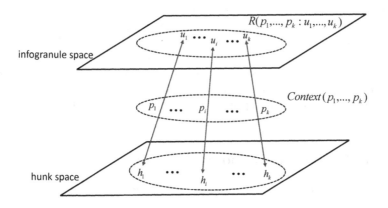

Fig. 5. Aggregation of c-granules

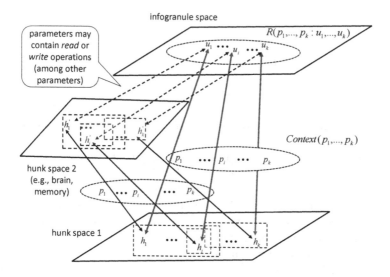

Fig. 6. Origin of links

corresponding c-granule. This c-granule should have appropriate soft_suit, link_suit, and hard_suit so that the constructed c-granule will satisfy the conditions of the new c-granule construction specification. Note that one of the constraints on such construction may follow from the interactions which the agent ag will have at the disposal in the uncertain environment.

The generalized c-granules corresponding to agents are defined using also the above classes of c-granules for defining corresponding suits of such generalized c-granules. The details of such construction will be presented in our next papers. Here, we would like to note only that there is a quite general approach for defining new c-granules from the simpler already defined.

The transition relation of a given agent ag is defined between configurations of ag at time t and the measurement time next to t. A configuration of ag at time t consists of all configurations of c-granules existing at time t. A configuration of c-granule G at time t consists of G itself as well as all c-granules selected on the basis of links in the link_suit of G at time t. These are, in particular all c-granules pointed by links corresponding to the c-granules stored in the computer memory during the computation process realised by c-granule as well as c-granules corresponding to perception at time t of the configuration of hunks at time t.

3   Complex Granules and Satisfiability

In this section, we discuss some examples of c-granules constructed over a family of satisfiability relations being at the disposal of a given agent. This discussion has some roots in intuitionism (see, *e.g.*, [14]). Let us consider a remark made by Per Martin-Löf in [14] about judgment presented in Figure 7.

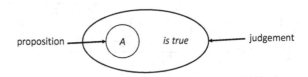

proposition →   A   is true   ← judgement

Fig. 7. "When we hold a proposition to be true, then we make a judgment" [14]

In the approach based on c-granules, the judgment for checking values of descriptors (or more compound formulas) pointed by links from simple c-granules is based on interactions of some physical parts considered over time and/or space (called hunks) and pointed by links of c-granules. The judgment for the more compound c-granules is defined by a relevant family of procedures also realized by means of interactions of physical parts.

Let us explain in more detail the above claims.

Let assume that a given agent ag has at the disposal a family of satisfiability relations

$$\{\models_i\}_{i \in I}, \tag{2}$$

where $\models_i \subseteq Tok(i) \times Typ(i)$, $Tok(i)$ is a set of tokens and $Type(i)$ is a set of types, respectively (using the terminology from [1]). The indices of satisfiability relations are vectors of parameters related, *e.g.*, to time, space, spatio-temporal features of physical parts represented by hunks or actions (plans) to be realized in the physical world.

In the discussed example of elementary c-granules, $Tok(i)$ is a set of hunks (more exactly, c-granules corresponding to hunks) and $Type(i)$ is a set of descriptors, *e.g.*, elementary infogranules (more exactly, c-granules corresponding to infogranules) pointed by link represented by $\models_i$ (more exactly, c-granules corresponding to $\models_i$). The procedure for computing the value of $h \models_i \alpha$, where h is a hunk and α is an infogranule (*e.g.*, descriptor or formula constructed over descriptors) is based on interaction of α with the physical world represented by hunk h.

The agent control can aggregate some simple c-granules into more compound c-granules, *e.g.*, by selecting some constraints on subsets of I making it possible to select relevant sets of simple c-granules and consider them as a new more compound c-granule. In constraints also values in descriptors pointed by links in elementary c-granules can be taken into account and sets of such more compound c-granules can be aggregated into new c-granules. Values of new descriptors pointed by links of these more compound granules are computed by new procedures. The computation process again is realized by interaction of the physical parts represented by hunks pointed by links of elementary c-granules included in the considered more compound c-granule as well as by using the procedure for computing of values of more compound descriptors from values of descriptors included in elementary c-granules of the considered more compound c-granule. Note that this procedure is also realized in the physical world thanks to relevant interactions.

In hierarchical modeling aiming at inducing of relevant c-granules (*e.g.*, for approximation of complex vague concepts), one can consider so far constructed c-granules as tokens. For example, they can be used to define structured objects representing corresponding hunks and link them using new satisfiability relations (from a given family) to relevant higher order descriptors together with the appropriate procedures (realized by interactions of hunks) for computing values of these descriptors. This approach generalizes hierarchical modeling developed for infogranules (see, *e.g.*, [15, 2]) to hierarchical modeling of c-granules which is important for many real-life projects.

We have assumed before that the agent ag is equipped with a family of satisfiability relations. However, in real-life projects the situation is more complicated. The agent ag should have strategies for discovery of new relevant satisfiability relations on the way of searching for solutions of target goals (problems). This is related to a question about the adaptive judgment relevant for agents performing computations based on configurations of c-granules. In the framework of granular computing based on c-granules, satisfiability relations are tools for constructing new c-granules. In fact, for a given satisfiability relation, the semantics of descriptors (and more compound formulas) relative to this relation can be defined. In this way candidates for new relevant c-granules are obtained. Hence, there arises a very important issue. The relevant satisfiability relations for the agent *ag* searching for solutions of problems are not given but they should be induced (discovered) on the basis of a partial information encoded in information (decision) systems including results of measurements of interaction of parts of physical world pointed by links of elementary c-granules as well as on the basis of c-granules representing domain knowledge. This problem is strongly related to reasoning from sensory measurement to perception [28, 29]. For real-life problems, it is often necessary to discover a hierarchy of satisfiability relations before the relevant target level will be obtained (see Figure 8) [11]. C-granules

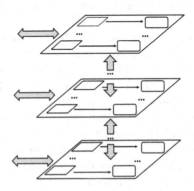

Fig. 8. Interactive hierarchical structures (gray arrows show interactions between hierarchical levels and the environment, arrows at hierarchical levels point from information (decision) systems representing partial specifications of satisfiability relations to induced from them theories consisting of rule sets)

constructed at different levels of this hierarchy finally lead to relevant c-granules (*e.g.*, for approximation of complex vague concepts) expressed very often in natural language. Concepts and relations from domain ontologies are used as hints in searching for relevant attributes (features) on different levels of the hierarchy. For details, the reader is referred to [15, 2].

4   Conclusions and Future Research

The outlined research on the nature of interactive computations is crucial for understanding complex systems. Our approach is based on complex granules (c-granules) performing computations through interaction with physical objects (hunks). Computations of c-granules are controlled by the agent control. More compound c-granules create agents and societies of agents. Other issues outlined in this paper such as interactive computations performed by societies of agents, especially communication language evolution and risk management in interactive computations will be discussed in more detail in our next papers. The discussed in the paper concept of computation based on interaction of c-granules differs from Turing computations and it seems to be a good starting point for developing computational models for Natural Computing too [12].

Acknowledgements. This work was supported by the Polish National Science Centre (NCN) grants DEC-2011/01/B/ ST6/03 867, DEC-2011/01/D /ST6/ 06981, DEC-2012/05/B/ ST6/03215, DEC-2013/09/B/ST6/01568 as well as by the Polish National Centre for Research and Development (NCBiR) under the grant SYNAT No. SP/I/1/ 77065/10 in frame of the strategic scientific research and experimental development program: "Interdisciplinary System for Interactive Scientific and Scientific-Technical Information", the grants O ROB/0010/03/001 as a part of Defence and Security Programmes and Projects and PBS2/B9/20/2013 in frame of Applied Research Programmes.

References

1. Barwise, J., Seligman, J.: Information Flow: The Logic of Distributed Systems. Cambridge University Press, Cambridge (1997)
2. Bazan, J.: Hierarchical classifiers for complex spatio-temporal concepts. In: Peters, J.F., Skowron, A., Rybiński, H. (eds.) Transactions on Rough Sets IX. LNCS, vol. 5390, pp. 474–750. Springer, Heidelberg (2008)
3. Deutsch, D., Ekert, A., Lupacchini, R.: Machines, logic and quantum physics. Neural Computation 6, 265–283 (2000)
4. Goldin, D., Smolka, S., Wegner, P. (eds.): Interactive Computation: The New Paradigm. Springer (2006)
5. Gurevich, Y.: Interactive algorithms 2005. In: Goldin, et al. (eds.) [4], pp. 165–181
6. Heller, M.: The Ontology of Physical Objects. Four Dimensional Hunks of Matter. Cambridge Studies in Philosophy. Cambridge University Press, Cambridge (1990)
7. Jankowski, A.: Practical Issues of Complex Systems Engineering: Wisdom Technology Approach. Springer, Heidelberg (in preparation, 2014)
8. Jankowski, A., Skowron, A.: A wistech paradigm for intelligent systems. In: Peters, J.F., Skowron, A., Düntsch, I., Grzymała-Busse, J., Orłowska, E., Polkowski, L. (eds.) Transactions on Rough Sets VI. LNCS, vol. 4374, pp. 94–132. Springer, Heidelberg (2007)
9. Jankowski, A., Skowron, A.: Wisdom technology: A rough-granular approach. In: Marciniak, M., Mykowiecka, A. (eds.) Bolc Festschrift. LNCS, vol. 5070, pp. 3–41. Springer, Heidelberg (2009)
10. Jankowski, A., Skowron, A., Swiniarski, R.: Interactive complex granules. In: Szczuka, M., Czaja, L., Kacprzak, M. (eds.) Proceedings of the 22nd International Workshop on Concurrency, Specification and Programming (CS&P 2013), Warsaw, Poland, September 25-27. CEUR Workshop Proceedings, vol. 1032, pp. 206–218. RWTH Aachen University (2013)
11. Jankowski, A., Skowron, A., Swiniarski, R.: Interactive computations: Toward risk management in interactive intelligent systems. In: Maji, P., Ghosh, A., Murty, M.N., Ghosh, K., Pal, S.K. (eds.) PReMI 2013. LNCS, vol. 8251, pp. 1–12. Springer, Heidelberg (2013)
12. Kari, L., Rozenberg, G.: The many facets of natural computing. Communications of the ACM 51, 72–83 (2008)

13. Marsh, L.: Stigmergic epistemology, stigmergic cognition. Journal Cognitive Systems 9, 136–149 (2008)
14. Martin-Löf, P.: Intuitionistic Type Theory (Notes by Giovanni Sambin of a series of lectures given in Padua, June 1980), Bibliopolis, Napoli, Italy (1984)
15. Nguyen, S.H., Bazan, J., Skowron, A., Nguyen, H.S.: Layered learning for concept synthesis. In: Peters, J.F., Skowron, A., Grzymała-Busse, J.W., Kostek, B., Swiniarski, R.W., Szczuka, M.S. (eds.) Transactions on Rough Sets I. LNCS, vol. 3100, pp. 187–208. Springer, Heidelberg (2004)
16. Omicini, A., Ricci, A., Viroli, M.: The multidisciplinary patterns of interaction from sciences to computer science. In: Goldin, et al. (eds.) [4], pp. 395–414
17. Pedrycz, W., Skowron, S., Kreinovich, V. (eds.): Handbook of Granular Computing. John Wiley & Sons, Hoboken (2008)
18. Skowron, A., Stepaniuk, J., Swiniarski, R.: Modeling rough granular computing based on approximation spaces. Information Sciences 184, 20–43 (2012)
19. Skowron, A., Szczuka, M.: Toward interactive computations: A rough-granular approach. In: Koronacki, J., Raś, Z.W., Wierzchoń, S.T., Kacprzyk, J. (eds.) Advances in Machine Learning II. SCI, vol. 263, pp. 23–42. Springer, Heidelberg (2010)
20. Skowron, A., Wasilewski, P.: Information systems in modeling interactive computations on granules. In: Szczuka, M., Kryszkiewicz, M., Ramanna, S., Jensen, R., Hu, Q. (eds.) RSCTC 2010. LNCS (LNAI), vol. 6086, pp. 730–739. Springer, Heidelberg (2010)
21. Skowron, A., Wasilewski, P.: Information systems in modeling interactive computations on granules. Theoretical Computer Science 412(42), 5939–5959 (2011)
22. Skowron, A., Wasilewski, P.: Toward interactive rough-granular computing. Control & Cybernetics 40(2), 1–23 (2011)
23. Skowron, A., Wasilewski, P.: Interactive information systems: Toward perception based computing. Theoretical Computer Science 454, 240–260 (2012)
24. Toffoli, T.: Physics and computation. International Journal of Theoretical Physics 21, 165–175 (1982)
25. Zadeh, L.A.: Outline of a new approach to the analysis of complex systems and decision processes. IEEE Trans. on Systems, Man and Cybernetics SMC-3, 28–44 (1973)
26. Zadeh, L.A.: Fuzzy sets and information granularity. In: Advances in Fuzzy Set Theory and Applications, pp. 3–18. North-Holland, Amsterdam (1979)
27. Zadeh, L.A.: Toward a theory of fuzzy information granulation and its centrality in human reasoning and fuzzy logic. Fuzzy Sets and Systems 90, 111–127 (1997)
28. Zadeh, L.A.: From computing with numbers to computing with words – from manipulation of measurements to manipulation of perceptions. IEEE Transactions on Circuits and Systems 45, 105–119 (1999)
29. Zadeh, L.A.: A new direction in AI: Toward a computational theory of perceptions. AI Magazine 22(1), 73–84 (2001)

Formulation and Simplification
of Multi-Granulation Covering Rough Sets

Tong-Jun Li, Xing-Xing Zhao, and Wei-Zhi Wu

School of Mathematics, Physics and Information Science, Zhejiang Ocean University,
Zhoushan, Zhejiang 316022, P.R. China
{ltj722,starzhao626}@163.com,
wuwz@zjou.edu.cn

Abstract. The theory of multi-granulation rough sets is one kind of effective methods for knowledge discovery in multiple granular structures. Based on rough sets on a single granular structure, various kinds of multi-granulation rough set models are proposed in the past decades. In this paper, according to two kinds of covering rough sets on single-granulation covering approximation spaces, four types of multi-granulation covering rough set models are defined. Properties of new models are examined in detail, comparison of multi-granulation covering approximation operators is done. Finally, simplification of four types of multi-granulation covering rough sets is investigated.

Keywords: Rough set, Multi-granulation, Covering rough set, Simplification.

1   Introduction

Rough set theory proposed by Pawlak [7] is an important tool to deal with inexact, uncertain and insufficient information. Pawlak rough set model has been extended for many real-world applications. Covering rough set is one type of the extensions of Pawlak rough sets. The lower and upper approximation operators in the earliest covering rough set models proposed by Zakowski [17] are not dual to each other [8,15]. In order to overcome this problem, many authors generalized one of them and defined the other by duality [2,15]. For covering rough sets, a lot of interesting results has been given [1,6]. By using a covering of the universe, some neighborhood operators, neighborhood systems, new coverings, and subsystems can be induced, thus various types of rough approximation operators can be constructed [16].

Partitions are basic knowledge structures in Pawlak rough set models. In practical applications, a lot of partitions of a universe may be considered at the same time. With respect to this situation, Qian et al [9] first proposed multi-granulation rough sets. In recent years, some generalized multi-granulation rough set models have been constructed [13,18]. For example, according to the decision-theoretic rough sets [14], Qian et al [11] established one kind of multi-granulation decision-theoretic rough sets. Furthermore, She and He [12] examine lattice and

M. Kryszkiewicz et al. (Eds.): RSEISP 2014, LNAI 8537, pp. 135–142, 2014.

136   T.-J. Li, X.-X. Zhao, and W.-Z. Wu

topology structure of the multi-granulation rough sets proposed by Qian et al [9].

A covering on a universe is a granulation of the universe. A family of coverings on the same universe is a multi-granulation structure on the universe. In multi-granulation covering approximation spaces, Liu et al [5] defined four kinds of multi-granulation covering rough sets by two induced neighborhood operators. Lin et al [3] proposed six pairs of multi-granulation covering approximation operators by using the subsets of the coverings of a multi-granulation covering approximation space. This paper focuses on the construction and comparison of rough approximation operators in multi-granulation approximation spaces. Meanwhile, simplification of four types of multi-granulation covering rough sets proposed in this paper is also discussed.

2   Preliminaries

In this section, we review some basic knowledge about covering rough sets and multi-granulation rough sets.

2.1   Covering Rough Sets

Let $\mathcal{C}$ be a covering on a universe of discourse U, that is, $\bigcup\{C|C \in \mathcal{C}\} = U$, then $(U,\mathcal{C})$ is called a covering approximation space. For any $X \subseteq U$, Zakowski [17] first constructed its lower and upper rough approximations as follows:

$$\underline{\mathcal{C}}(X) = \bigcup\{C \in \mathcal{C}|C \subseteq X\}, \quad \overline{\mathcal{C}}(X) = \bigcup\{C \in \mathcal{C}|C \cap X \neq \emptyset\}.$$

However, the lower and upper approximation operators are not dual to each other. Thus, the following modified models were proposed [2,15]: $\forall X \subseteq U$,

(I) $\underline{\mathcal{C}'}(X) = \bigcup\{C \in \mathcal{C}|C \subseteq X\}, \quad \overline{\mathcal{C}'}(X) =\sim \underline{\mathcal{C}'}(\sim X);$
(II) $\overline{\mathcal{C}'}(X) = \bigcup\{C \in \mathcal{C}|C \cap X \neq \emptyset\}, \quad \underline{\mathcal{C}''}(X) =\sim \overline{\mathcal{C}''}(\sim X).$

Where $\sim X$ denotes the complement of X.
$\underline{\mathcal{C}'}$ and $\overline{\mathcal{C}'}$ satisfy the following properties: $\forall X, Y \subseteq U$,

(L0) $\underline{\mathcal{C}'}(X) =\sim \overline{\mathcal{C}'}(\sim X)$, (U0) $\overline{\mathcal{C}'}(X) =\sim \underline{\mathcal{C}'}(\sim X);$
(L1) $\underline{\mathcal{C}'}(U) = U$, (U1) $\overline{\mathcal{C}'}(\emptyset) = \emptyset;$
(L1') $\underline{\mathcal{C}'}(\emptyset) = \emptyset$, (U1') $\overline{\mathcal{C}'}(U) = U;$
(L2) $X \subseteq Y \Rightarrow \underline{\mathcal{C}'}(X) \subseteq \underline{\mathcal{C}'}(Y)$, (U2) $X \subseteq Y \Rightarrow \overline{\mathcal{C}'}(X) \subseteq \overline{\mathcal{C}'}(Y);$
(L3) $\underline{\mathcal{C}'}(X) \subseteq X$, (U3) $X \subseteq \overline{\mathcal{C}'}(X);$
(L4') $\underline{\mathcal{C}'}(X \cap Y) \subseteq \underline{\mathcal{C}'}(X) \cap \underline{\mathcal{C}'}(Y)$, (U4') $\overline{\mathcal{C}'}(X \cup Y) \supseteq \overline{\mathcal{C}'}(X) \cup \overline{\mathcal{C}'}(Y).$
(L5) $\underline{\mathcal{C}'}(X) \subseteq \underline{\mathcal{C}'}(\underline{\mathcal{C}'}(X))$, (U5) $\overline{\mathcal{C}'}(\overline{\mathcal{C}'}(X)) \subseteq \overline{\mathcal{C}'}(X).$

The following properties do not hold for $\underline{\mathcal{C}'}$ and $\overline{\mathcal{C}'}$: (L4) $\underline{\mathcal{C}'}(X \cap Y) = \underline{\mathcal{C}'}(X) \cap \underline{\mathcal{C}'}(Y)$, (U4) $\overline{\mathcal{C}'}(X \cup Y) = \overline{\mathcal{C}'}(X) \cup \overline{\mathcal{C}'}(Y).$

As for $\underline{\mathcal{C}''}$ and $\overline{\mathcal{C}''}$, properties (L0)-(L4), (U0)-(U4), (L1') and (L4'), and (U1') and (U4') hold. In addition, $\forall X \subseteq U$, (L6) $X \subseteq \underline{\mathcal{C}''}\left(\overline{\mathcal{C}''}(X)\right)$, (U6) $\overline{\mathcal{C}''}\left(\underline{\mathcal{C}''}(X)\right) \subseteq X$.

Let $\mathcal{C}_1, \mathcal{C}_2$ be two coverings on U. Then $\mathcal{C}_1$ is called to be finer than $\mathcal{C}_2$, denoted as $\mathcal{C}_1 \preceq \mathcal{C}_2$, if for any $C \in \mathcal{C}_2$, there are a $\mathcal{D} \subseteq \mathcal{C}_1$ such that $\bigcup \mathcal{D} = C$, and $\mathcal{C}_1$ is called to be quasi-finer than $\mathcal{C}_2$, denoted as $\mathcal{C}_1 \sqsubseteq \mathcal{C}_2$, if for any $C_1 \in \mathcal{C}_1$, there exists a $C_2 \in \mathcal{C}_2$ such that $C_1 \subseteq C_2$.

Proposition 1. ([2]) *Let $\mathcal{C}_1, \mathcal{C}_2$ be two coverings on U. Then $\underline{\mathcal{C}_2'}(X) \subseteq \underline{\mathcal{C}_1'}(X)$, or equivalently $\overline{\mathcal{C}_1'}(X) \subseteq \overline{\mathcal{C}_2'}(X)$, $\forall X \subseteq U$, if and only if $\mathcal{C}_1 \preceq \mathcal{C}_2$.*

Proposition 2. *Let $\mathcal{C}_1, \mathcal{C}_2$ be two coverings on U. If $\mathcal{C}_1 \sqsubseteq \mathcal{C}_2$, then $\forall X \subseteq U$, $\underline{\mathcal{C}_2''}(X) \subseteq \underline{\mathcal{C}_1''}(X)$, $\overline{\mathcal{C}_1''}(X) \subseteq \overline{\mathcal{C}_2''}(X)$.*

2.2 Multi-Granulation Rough Sets

Let $R_1, R_2, \cdots, R_m$ be equivalence relations on U, denote $\mathcal{A} = \{R_1, R_2, \cdots, R_m\}$. Then $(U, \mathcal{A})$ is called a multi-granulation approximation space, and in this case, every $(U, R_i)(i = 1, 2, \cdots, m)$ is called a single-granulation approximation space.

The optimistic lower approximation and the optimistic upper approximation of X ($X \subseteq U$) in $(U, \mathcal{A})$, denoted by $\underline{\mathcal{A}^{op}}(X)$ and $\overline{\mathcal{A}^{op}}(X)$, respectively, are defined as follows [9]:

$$\underline{\mathcal{A}^{op}}(X) = \{x \in U | \exists i (1 \leq i \leq m)(R_i(x) \subseteq X)\}, \quad \overline{\mathcal{A}^{op}}(X) = \sim \underline{\mathcal{A}^{op}}(\sim X);$$

The pessimistic lower approximation and the pessimistic upper approximation of X ($X \subseteq U$) in $(U, \mathcal{A})$, denoted by $\underline{\mathcal{A}^{pe}}(X)$ and $\overline{\mathcal{A}^{pe}}(X)$, respectively, are defined as follows [10]:

$$\underline{\mathcal{A}^{pe}}(X) = \{x \in U | R_i(x) \subseteq X, i = 1, \cdots, m\}, \quad \overline{\mathcal{A}^{pe}}(X) = \sim \underline{\mathcal{A}^{pe}}(\sim X).$$

It is not difficult to check that

$$\overline{\mathcal{A}^{op}}(X) = \{x \in U | \forall i (1 \leq i \leq m)(R_i(x) \cap X \neq \emptyset)\},$$
$$\overline{\mathcal{A}^{pe}}(X) = \{x \in U | \exists i (1 \leq i \leq m)(R_i(x) \cap X \neq \emptyset)\}.$$

The following proposition shows the relationships between Pawlak rough sets and the multi-granulation rough sets.

Proposition 3. ([4]) *Let $(U, \mathcal{A})$ be a multi-granulation approximation space, where $\mathcal{A} = \{R_1, R_2, \cdots, R_m\}$. Then $\forall X \subseteq U$,*

(1) $\underline{\mathcal{A}^{op}}(X) = \bigcup\limits_{i=1}^{m} \underline{R_i}(X)$, $\overline{\mathcal{A}^{op}}(X) = \bigcap\limits_{i=1}^{m} \overline{R_i}(X)$,

(2) $\underline{\mathcal{A}^{pe}}(X) = \bigcap\limits_{i=1}^{m} \underline{R_i}(X)$, $\overline{\mathcal{A}^{pe}}(X) = \bigcup\limits_{i=1}^{m} \overline{R_i}(X)$.

3   Multi-Granulation Covering Rough Sets

Let U be a universe of discourse and $\mathcal{C} = \{C_1, \cdots, C_m\}$ a family of coverings on U. Then $(U, \mathcal{C})$ is called a multi-granulation covering approximation space, and every $(U, C_i)(i = 1, 2, \cdots, m)$ is called a (single-granulation) covering approximation space.

Example 1. Let $U = \{1, 2, 4, 5, 6\}$ and $\mathcal{C} = \{C_1, C_2, C_3\}$, where

$$C_1 = \{\{1, 2\}, \{3, 4\}, \{5, 6\}\}, C_2 = \{\{1, 2\}, \{2, 3\}, \{3, 5\}\}, \{4, 6\},$$

$$C_3 = \{\{1, 2, 3\}, \{2, 3\}, \{4, 5\}, \{4, 6\}\}.$$

Then $(U, \mathcal{C})$ is a multi-granulation covering approximation space.

By the relationship between the multi-granulation rough sets and the Pawlak rough sets shown in Proposition 3 and Formulations (I) and (II), we can define multi-granulation covering rough sets.

Definition 1. *Let $(U, \mathcal{C})$ be a multi-granulation covering approximation space, where $\mathcal{C} = \{C_1, \cdots, C_m\}$. For any $X \subseteq U$, type-1 optimistic lower approximation and type-1 optimistic upper approximation of X in $(U, \mathcal{C})$, denoted by $\underline{\mathcal{C}^{op_1}}(X)$ and $\overline{\mathcal{C}^{op_1}}(X)$, respectively, are defined as follows:*

$$\underline{\mathcal{C}^{op_1}}(X) = \bigcup_{i=1}^{m} \underline{C_i'}(X), \quad \overline{\mathcal{C}^{op_1}}(X) = \bigcap_{i=1}^{m} \overline{C_i'}(X).$$

Example 2. Let $(U, \mathcal{C})$ be the multi-granulation covering approximation space in Example 1 and $X = \{1, 3, 5\}$. Then by Definition 1 we have

$$\underline{\mathcal{C}^{op_1}}(X) = \{1, 3\}, \quad \overline{\mathcal{C}^{op_1}}(X) = \{1, 2, 3, 5\}.$$

The type-1 optimistic approximation operators can be represented as follows:

$$\underline{\mathcal{C}^{op_1}}(X) = \{x \in U | \exists C_i \in \mathcal{C}(1 \leq i \leq m)(\exists C \in C_i(x \in C, C \subseteq X))\},$$
$$\overline{\mathcal{C}^{op_1}}(X) = \{x \in U | \forall C_i \in \mathcal{C}(1 \leq i \leq m)(\forall C \in C_i(x \in C \Rightarrow C \cap X \neq \emptyset))\}.$$

Theorem 1. $\underline{\mathcal{C}^{op_1}}$ *satisfies the properties (L0)-(L3), (L5), (L1') and (L4'). Dually, $\overline{\mathcal{C}^{op_1}}$ satisfies (U0)-(U3), (U5), (U1') and (U4').*

Definition 2. *Let $(U, \mathcal{C})$ be a multi-granulation covering approximation space, where $\mathcal{C} = \{C_1, \cdots, C_m\}$. For any $X \subseteq U$, type-1 pessimistic lower approximation and type-1 pessimistic upper approximation of X in $(U, \mathcal{C})$, denoted by $\underline{\mathcal{C}^{pe_1}}(X)$ and $\overline{\mathcal{C}^{pe_1}}(X)$, respectively, are defined as follows:*

$$\underline{\mathcal{C}^{pe_1}}(X) = \bigcap_{i=1}^{m} \underline{C_i'}(X), \quad \overline{\mathcal{C}^{pe_1}}(X) = \bigcup_{i=1}^{m} \overline{C_i'}(X).$$

Example 3. Let $(U, \mathcal{C})$ be the multi-granulation covering approximation space in Example 1 and $X = \{1, 2, 3\}$. Then by Definition 2 we calculate

$$\underline{\mathcal{C}^{pe_1}}(X) = \{1, 2\}, \ \overline{\mathcal{C}^{ep_1}}(X) = \{1, 2, 3, 4, 5\}.$$

The type-1 pessimistic r approximation operators have the following formulations.

$$\underline{\mathcal{C}^{pe_1}}(X) = \{x \in U | \forall C_i \in \mathcal{C}(1 \le i \le m)(\exists C \in C_i(x \in C, C \subseteq X))\},$$
$$\overline{\mathcal{C}^{op_1}}(X) = \{x \in U | \exists C_i \in \mathcal{C}(1 \le i \le m)(\forall C \in C_i(x \in C \Rightarrow C \cap X \ne \emptyset))\}.$$

Theorem 2. $\underline{\mathcal{C}^{pe_1}}$ *satisfies the properties* (L0)-(L3), (L1') *and* (L4'). *Dually,* $\overline{\mathcal{C}^{op_1}}$ *satisfies* (U0)-(U3), (U1') *and* (U4').

By the similar way, according to Formulation (II), other two types of multi-granulation covering rough sets can be defined.

Definition 3. *Let* $(U, \mathcal{C})$ *be a multi-granulation covering approximation space, where* $\mathcal{C} = \{C_1, \cdots, C_m\}$. *For any* $X \subseteq U$, *type-2 optimistic lower approximation and type-2 optimistic upper approximation of* X *in* $(U, \mathcal{C})$, *denoted by* $\underline{\mathcal{C}^{op_2}}(X)$ *and* $\overline{\mathcal{C}^{op_2}}(X)$, *respectively, are defined as follows:*

$$\underline{\mathcal{C}^{op_2}}(X) = \bigcup_{i=1}^{m} \underline{C_i''}(X), \ \overline{\mathcal{C}^{op_2}}(X) = \bigcap_{i=1}^{m} \overline{C_i''}(X).$$

Example 4. Let $(U, \mathcal{C})$ be the multi-granulation covering approximation space in Example 1 and $X = \{1, 2, 4\}$. Then by Definition 3 we compute

$$\underline{\mathcal{C}^{op_2}}(X) = \{1, 2\}, \ \overline{\mathcal{C}^{op_2}}(X) = \{1, 2, 3, 4\}.$$

Another form of the operators in Definition 3 can be described as

$$\underline{\mathcal{C}^{op_2}}(X) = \{x \in U | \exists C_i \in \mathcal{C}(1 \le i \le m)(\forall C \in C_i(x \in C \Rightarrow C \subseteq X))\},$$
$$\overline{\mathcal{C}^{op_2}}(X) = \{x \in U | \forall C_i \in \mathcal{C}(1 \le i \le m)(\exists C \in C_i(x \in C, C \cap X \ne \emptyset))\}.$$

Theorem 3. $\underline{\mathcal{C}^{op_2}}$ *satisfies the properties* (L0)-(L3), (L6), (L1') *and* (L4'). *Dually,* $\overline{\mathcal{C}^{op_2}}$ *satisfies* (U0)-(U3), (U6), (U1') *and* (U4').

Definition 4. *Let* $(U, \mathcal{C})$ *be a multi-granulation covering approximation space, where* $\mathcal{C} = \{C_1, \cdots, C_m\}$. *For any* $X \subseteq U$, *type-2 pessimistic lower approximation and type-2 pessimistic upper approximation of* X *in* $(U, \mathcal{C})$, *denoted by* $\underline{\mathcal{C}^{pe_2}}(X)$ *and* $\overline{\mathcal{C}^{pe_2}}(X)$, *respectively, are defined as follows:*

$$\underline{\mathcal{C}^{pe_2}}(X) = \bigcap_{i=1}^{m} \underline{C_i''}(X), \ \overline{\mathcal{C}^{pe_2}}(X) = \bigcup_{i=1}^{m} \overline{C_i''}(X).$$

Example 5. Let $(U, \mathcal{C})$ be the multi-granulation covering approximation space in Example 1 and $X = \{1, 2, 3\}$. Then by Definition 4 we figure out

$$\underline{\mathcal{C}^{pe_2}}(X) = \{1, 2\}, \ \overline{\mathcal{C}^{pe_2}}(X) = \{1, 2, 3, 5\}.$$

The operators in Definition 4 have the following representations:

$$\underline{C}^{pe2}(X) = \{x \in U | \forall C_i \in C(1 \leq i \leq m)(\forall C \in C_i(x \in C \Rightarrow C \subseteq X))\},$$
$$\overline{C}^{pe2}(X) = \{x \in U | \exists C_i \in C(1 \leq i \leq m)(\exists C \in C_i(x \in C, C \cap X \neq \emptyset))\}.$$

Theorem 4. $\underline{C}^{pe2}$ *satisfies the properties* (L0)-(L4), (L6), (L1') *and* (L4'). *Dually,* $\overline{C}^{pe2}$ *satisfies* (U0)-(U4), (U6), (U1') *and* (U4').

4   Comparison of Multi-Granulation Covering Rough Sets

Proposition 4. [2] *Let* (U, C) *be a covering approximation space and* $X \subseteq U$. *Then*

$$\underline{C}''(X) \subseteq \underline{C}'(X) \subseteq X \subseteq \overline{C}'(X) \subseteq \overline{C}''(X).$$

By Proposition 4 and Definitions 1-4, we have

Theorem 5. *Let* (U, C) *be a multi-granulation covering approximation space, where* $C = \{C_1, \cdots, C_m\}$, *and* $X \subseteq U$. *Then*

$$\underline{C}^{pe2}(X) \subseteq \underline{C}^{op2}(X)(\text{ or } \underline{C}^{pe1}(X)) \subseteq \underline{C}^{op1}(X) \subseteq X$$
$$\subseteq \overline{C}^{op1}(X) \subseteq \overline{C}^{op2}(X)(\text{ or } \overline{C}^{pe1}(X)) \subseteq \overline{C}^{pe2}(X).$$

For a multi-granulation covering approximation space (U, C), if $C = \{C_1, \cdots, C_m\}$, then $\bigcup C = \overset{m}{\underset{i=1}{\bigcup}} C_i$ is a covering of U. The following conclusions hold.

Theorem 6. *Let* (U, C) *be a multi-granulation covering approximation space, where* $C = \{C_1, \cdots, C_m\}$. *Then*

$$\left(\bigcup C\right)' = \underline{C}^{op1}, \quad \overline{\left(\bigcup C\right)'} = \overline{C}^{op1}; \quad \left(\bigcup C\right)'' = \underline{C}^{pe2}, \quad \overline{\left(\bigcup C\right)''} = \overline{C}^{pe2}.$$

Theorem 6 shows that the type-1 optimistic and type-2 pessimistic multi-granulation covering rough sets are two types of covering rough sets on single-granulation covering approximation spaces.

5   Simplification of Multi-Granulation Covering Rough Sets

By Definitions 1-4, we gain the following theorem.

Theorem 7. *Let* (U, C_1) *and* (U, C_2) *be two multi-granulation covering approximation spaces. If* $C_1 \subseteq C_2$, *then* $\forall X \subseteq U$,

(1) $\underline{C}_1^{op1}(X) \subseteq \underline{C}_2^{op1}(X), \overline{C}_1^{op1}(X) \supseteq \overline{C}_2^{op1}(X)$;

(2) $\underline{C}_1^{pe1}(X) \supseteq \underline{C}_2^{pe1}(X), \overline{C}_1^{pe1}(X) \subseteq \overline{C}_2^{pe1}(X)$;

(3) $\underline{C}_1^{op2}(X) \subseteq \underline{C}_2^{op2}(X), \overline{C}_1^{op2}(X) \supseteq \overline{C}_2^{op2}(X)$;

(4) $\underline{C}_1^{pe2}(X) \supseteq \underline{C}_2^{pe2}(X), \overline{C}_1^{pe2}(X) \subseteq \overline{C}_2^{pe2}(X)$.

The below conclusions follows from Propositions 1 and 2.

Theorem 8. *Let $(U, \mathcal{C})$ be a multi-granulation covering approximation space, where $\mathcal{C} = \{C_1, \cdots, C_m\}$. If $C_1 \preceq \cdots \preceq C_m$, then*

$$\underline{\mathcal{C}}^{op_1} = \underline{C'_1}, \ \overline{\mathcal{C}}^{op_1} = \overline{C'_1}; \ \underline{\mathcal{C}}^{pe_1} = \underline{C'_m}, \ \overline{\mathcal{C}}^{pe_1} = \overline{C'_m}.$$

Theorem 9. *Let $(U, \mathcal{C})$ be a multi-granulation covering approximation space, where $\mathcal{C} = \{C_1, \cdots, C_m\}$. If $C_1 \sqsubseteq \cdots \sqsubseteq C_m$, then*

$$\underline{\mathcal{C}}^{op_2} = \underline{C''_1}, \ \overline{\mathcal{C}}^{op_2} = \overline{C''_1}; \ \underline{\mathcal{C}}^{pe_2} = \underline{C''_m}, \ \overline{\mathcal{C}}^{pe_2} = \overline{C''_m}.$$

From Theorems 8 and 9 we can see that when $\mathcal{C}$ is a totally ordered set under the relation $\preceq$ ($\sqsubseteq$, respectively), the optimistic (pessimistic, respectively) multi-granulation covering rough sets are identical to the corresponding covering rough sets on the finest (the coarsest, respectively) covering.

Theorem 10. *Let $(U, \mathcal{C})$ be a multi-granulation covering approximation space, where $\mathcal{C} = \{C_1, \cdots, C_m\}$. Then*

(1) $\underline{\mathcal{C}}^{op_1} = \underline{C'_1}$, *or* $\overline{\mathcal{C}}^{op_1} = \overline{C'_1}$, *if and only if* $C_1 \preceq C_i (i \neq 1)$;

(2) *If* $C_i \preceq C_1 (i \neq 1)$, *then* $\underline{\mathcal{C}}^{pe_1} = \underline{C'_1}$, $\overline{\mathcal{C}}^{pe_1} = \overline{C'_1}$;

(3) *If* $C_1 \sqsubseteq C_i (i \neq 1)$, *then* $\underline{\mathcal{C}}^{op_2} = \underline{C''_1}$, $\overline{\mathcal{C}}^{op_2} = \overline{C''_1}$;

(4) *If* $C_i \sqsubseteq C_1 (i \neq 1)$, *then* $\underline{\mathcal{C}}^{pe_2} = \underline{C''_1}$, $\overline{\mathcal{C}}^{pe_2} = \overline{C''_1}$.

Theorem 11. *Let $(U, \mathcal{C})$ be a multi-granulation covering approximation space and $\emptyset \neq \mathcal{D} \subset \mathcal{C}$. Then*

(1) $\underline{\mathcal{D}}^{op_1} = \underline{\mathcal{C}}^{op_1}$, *or* $\overline{\mathcal{D}}^{op_1} = \overline{\mathcal{C}}^{op_1}$, *if and only if* $\forall C \in \mathcal{C} - \mathcal{D}, \bigcup \mathcal{D} \preceq C$;

(2) *If* $\forall C \in \mathcal{C} - \mathcal{D}, \exists D \in \mathcal{D}(C \preceq D)$, *then* $\underline{\mathcal{D}}^{pe_1} = \underline{\mathcal{C}}^{pe_1}$, $\overline{\mathcal{D}}^{pe_1} = \overline{\mathcal{C}}^{pe_1}$;

(3) *If* $\forall C \in \mathcal{C} - \mathcal{D}, \exists D \in \mathcal{D}(D \sqsubseteq C)$, *then* $\underline{\mathcal{D}}^{op_2} = \underline{\mathcal{C}}^{op_2}$, $\overline{\mathcal{D}}^{op_2} = \overline{\mathcal{C}}^{op_2}$;

(4) *If* $\forall C \in \mathcal{C} - \mathcal{D}, \exists D \in \mathcal{D}(C \sqsubseteq D)$, *then* $\underline{\mathcal{D}}^{pe_2} = \underline{\mathcal{C}}^{pe_2}$, $\overline{\mathcal{D}}^{pe_2} = \overline{\mathcal{C}}^{pe_2}$.

6   Conclusions

Knowledge discovery in many granular structures is an important issue of granular computing theory. Covering rough sets are good method for data mining in single granulation of universe. In this paper, a new method for constructing multi-granulation covering rough sets in multi-granulation covering approximation space is proposed, so that the obtained four pairs of multi-granulation covering approximation operators could be understood clearly. Properties and comparison of the multi-granulation covering approximation operators are also investigated. Finally, some conditions are obtained, under which the multi-granulation covering rough sets defined in this paper can be simplified.

Acknowledgements. This work was supported by grants from the National Natural Science Foundation of China (Nos. 11071284, 61075120, 61272021, 61202206) and the Zhejiang Provincial Natural Science Foundation of China (Nos. LY14F030001, LZ12F03002, LY12F02021).

References

1. Chen, D.G., Wang, C.Z., Hu, Q.H.: A new approach to attribute reduction of consistent and inconsistent covering decision systems with covering rough sets. Information Sciences 177, 3500–3518 (2007)
2. Li, T.-J.: Rough approximation operators in covering approximation spaces. In: Greco, S., Hata, Y., Hirano, S., Inuiguchi, M., Miyamoto, S., Nguyen, H.S., Słowiński, R. (eds.) RSCTC 2006. LNCS (LNAI), vol. 4259, pp. 174–182. Springer, Heidelberg (2006)
3. Lin, G., Liang, J., Qian, Y.: Multigranulation rough sets: From partition to covering. Information Sciences 241, 101–118 (2013)
4. Lin, G., Qian, Y., Li, J.: NMGRS: Neighborhood-based multigranulation rough sets. International Journal of Approximate Reasoning 53, 1080–1093 (2012)
5. Liu, C., Miao, D., Qian, J.: On multi-granulation covering rough sets. International Journal of Approximate Reasoning (2014), http://dx.doi.org/10.1016/j.ijar.2014.01.002
6. Ma, L.W.: On some types of neighborhood-related covering rough sets. International Journal of Approximate Reasoning 53, 901–911 (2012)
7. Pawlak, Z.: Rough sets. International Journal of Computer and Information Sciences 11, 341–356 (1982)
8. Pomykala, J.A.: Approximation Operations in Approximation Space. Bulletin of the Polish Academy of Sciences: Mathematics 35, 653–662 (1987)
9. Qian, Y.H., Liang, J.Y., Yao, Y.Y., Dang, C.Y.: MGRS: A multi-granulation rough set. Information Sciences 180, 949–970 (2010)
10. Qian, Y.H., Liang, J.Y., Wei, W.: Pessimistic rough decision. In: Second International Workshop on Rough Sets Theory, pp. 440–449 (2010)
11. Qian, Y., Zhang, H., Sang, Y., Liang, J.: Multigranulation decision-theoretic rough sets. International Journal of Approximate Reasoning 55, 225–237 (2014)
12. She, Y., He, X.: On the structure of the multigranulation rough set model. Knowledge-Based Systems 36, 81–92 (2012)
13. Yang, X., Dou, H., Yang, J.: Hybrid multigranulation rough sets based on equivalence relation. Computer Science 39, 165–169 (2012) (A Chinese journal)
14. Yao, Y.Y., Wong, S.K.M.: A decision theoretic framework for approximating concepts. International Journal of Man-machine Studies 37, 793–809 (1992)
15. Yao, Y.Y.: Relational interpretations of neighborhood operators and rough set approximation operators. Information Sciences 101, 239–259 (1998)
16. Yao, Y.Y., Yao, B.: Covering based rough set approximations. Information Sciences 200, 91–107 (2012)
17. Zakowski, W.: Approximations in the space (U, P). Demonstratio Mathematica IXV, 761–769 (1983)
18. Zhang, M., Tang, Z., Xu, W., Yang, X.: Variable multigranulation rough set model. Pattern Recognition and Artificial Intelligence 25, 709–720 (2012) (A Chinese journal)

Covering Based Rough Sets
and Relation Based Rough Sets

Mauricio Restrepo and Jonatan Gómez

Universidad Nacional de Colombia,
Bogotá, Colombia
{mrestrepol,jgomezpe}@unal.edu.co
http://www.alife.unal.edu.co/grupo/index.html

Abstract. Relation based rough sets and covering based rough sets are two important extensions of the classical rough sets. This paper investigates relationships between relation based rough sets and the covering based rough sets in a particular framework of approximation operators, presents a new group of approximation operators obtained by combining coverings and neighborhood operators and establishes some relationships between covering based rough sets and relation based rough sets.

Keywords: rough sets, covering based rough sets, relation based rough sets.

1 Introduction

Rough set theory was introduced by Z. Pawlak in 1982, as a tool to deal with vagueness and granularity in information systems [8]. An equivalence relation on a set U is defined to discern its elements and therefore a partition of U is defined. A first generalization of rough sets is to replace the equivalence relation by a general binary relation. In this case, the binary relation determines collections of sets that no longer form a partition of U. Tolerance and similarity relations have been considered to investigate some attribute relation problems [11, 12]. This generalization has been used in applications with incomplete information systems and tables with continuous attributes [4, 5, 16, 17]. A second generalization is to replace the partition obtained by the equivalence relation with a covering; i.e., a collection of nonempty sets with union equal to U. Many definitions for lower and upper approximation operators have been proposed [10, 13, 15, 21, 22, 24, 26–29, 31]. There are many works in these two directions and some connections between the two generalizations have been established, for example in [17, 30, 32, 33]. Yao and Yao introduce a framework of twenty pairs of dual approximation operators [20] and Restrepo et al, establish some equivalences between these approximation operators [9].

W. Zhu, established an equivalence between a type of covering-based rough sets and a type of binary relation based rough sets [30]. Y. L. Zhang and M. K. Luo established the equivalence between four types of covering-based rough sets and a type of relation-based rough sets, respectively [23]. Covering based rough sets and tolerance relation based rough sets are used in information systems with missing and numerical data [2, 3, 7].

M. Kryszkiewicz et al. (Eds.): RSEISP 2014, LNAI 8537, pp. 143–152, 2014.

In this paper, we will extend the connections established by Y. L. Zhang and M. K. Luo to the Yao and Yao's framework. In section 2, we present preliminary concepts about rough set theory, covering based and relation based rough sets. In section 3, we review some equivalences between covering and relation based rough sets. We also present a new group of approximation operators, combining some coverings with neighborhood operators. Finally, we establish some equivalences of these operators and relation based rough sets. Section 4, presents some conclusions and outlines future work.

2 Preliminaries

Although some results are valid for infinite sets, we will assume that U is a finite and non-empty set. Rough set theory is based on an equivalence relation, called indiscernibility relation. Lower and upper approximations are two fundamental concepts in this theory. If $[x]_E$ represents the equivalence class of x and $A \subseteq U$, then:

$$\underline{apr}(A) = \{x \in U : [x]_E \subseteq A\} = \bigcup\{[x]_E \in U/E : [x]_E \subseteq A\} \tag{1}$$

$$\overline{apr}(A) = \{x \in U : [x]_E \cap A \neq \emptyset\} = \bigcup\{[x]_E \in U/E : [x]_E \cap A \neq \emptyset\} \tag{2}$$

are called the lower and upper approximation of A. U/E represents the equivalence classes, defined from the equivalence relation E.

The pair $(\underline{apr}, \overline{apr})$ is a dual pair of approximation operators, that is, for $A \subseteq U$, $\underline{apr}(\sim A) = \sim \overline{apr}(A)$, where $\sim A$ represents the complement of A, i.e., $\sim A = U \setminus A$.

2.1 Covering Based Rough Sets

An important generalization of rough sets theory can be obtained by changing U/E for a covering of U. Covering based rough sets are an extension of Pawlak's theory for data with missing values [2, 4, 5].

Definition 1. *[25] Let $\mathbb{C} = \{K_i\}$ be a family of nonempty subsets of U. $\mathbb{C}$ is called a covering of U, if $\cup K_i = U$. The ordered pair $(U, \mathbb{C})$ is called a covering approximation space.*

It is clear that a partition generated by an equivalence relation is a special case of a covering of U, so the concept of covering is a generalization of a partition.

In [20], Yao and Yao proposed a general framework for the study of covering based rough sets. The first part in Equations (1) and (2) is called element based definition and the second one is called granule based definition. Below, we briefly review the generalizations of the element, granule and system based definitions.

Element Based Definition.

Definition 2. *[20] A neighborhood operator is a mapping $N : U \to \mathscr{P}(U)$, where $\mathscr{P}(U)$ represents the collection of subsets of U. If $N(x) \neq \emptyset$ for all $x \in U$, N is called a serial neighborhood operator. If $x \in N(x)$ for all $x \in U$, N is called a reflexive neighborhood operator.*

Each neighborhood operator defines an ordered pair $(\underline{apr}_N, \overline{apr}_N)$ of dual approximation operators:

$$\underline{apr}_N(A) = \{x \in U : N(x) \subseteq A\} \tag{3}$$

$$\overline{apr}_N(A) = \{x \in U : N(x) \cap A \neq \emptyset\} \tag{4}$$

Different neighborhood operators, and hence different element based definitions of covering based rough sets, can be obtained from a covering $\mathbb{C}$. In general, we are interested in the sets K in $\mathbb{C}$ such that $x \in K$:

Definition 3. *[20] If $\mathbb{C}$ is a covering of U and $x \in U$, a neighborhood system $\mathscr{C}(\mathbb{C}, x)$ is defined by:*

$$\mathscr{C}(\mathbb{C}, x) = \{K \in \mathbb{C} : x \in K\} \tag{5}$$

In a neighborhood system $\mathscr{C}(\mathbb{C}, x)$, the minimal and maximal sets that contain an element $x \in U$ are particularly important.

Definition 4. *Let $(U, \mathbb{C})$ be a covering approximation space and x in U. The set*

$$md(\mathbb{C}, x) = \{K \in \mathscr{C}(\mathbb{C}, x) : (\forall S \in \mathscr{C}(\mathbb{C}, x), S \subseteq K) \Rightarrow K = S)\} \tag{6}$$

is called the minimal description of x [1]. On the other hand, the set

$$MD(\mathbb{C}, x) = \{K \in \mathscr{C}(\mathbb{C}, x) : (\forall S \in \mathscr{C}(\mathbb{C}, x), S \supseteq K) \Rightarrow K = S\} \tag{7}$$

is called the maximal description of x [29].

From $md(\mathbb{C}, x)$ and $MD(\mathbb{C}, x)$, Yao and Yao [20] presented the following neighborhood operators:

1. $N_1^{\mathbb{C}}(x) = \cap\{K : K \in md(\mathbb{C}, x)\}$
2. $N_2^{\mathbb{C}}(x) = \cup\{K : K \in md(\mathbb{C}, x)\}$
3. $N_3^{\mathbb{C}}(x) = \cap\{K : K \in MD(\mathbb{C}, x)\}$
4. $N_4^{\mathbb{C}}(x) = \cup\{K : K \in MD(\mathbb{C}, x)\}$

Therefore for each $N_i^{\mathbb{C}}$ with $i = 1, 2, 3, 4$, we have a pair of dual approximation operators $(\underline{apr}_{N_i}^{\mathbb{C}}, \overline{apr}_{N_i}^{\mathbb{C}})$ defined in Equations (3) and (4).

Granule Based Definition. For a covering approximation space, the second part of Equations (1) and (2) are not dual, so we can use each one with its dual operator. Generalizing the definitions given in Equations (1) and (2), the following dual pairs of approximation operators based on a covering $\mathbb{C}$ were considered in [20]:

$$\underline{apr}'_{\mathbb{C}}(A) = \bigcup\{K \in \mathbb{C} : K \subseteq A\} \tag{8}$$

$$\overline{apr}'_{\mathbb{C}}(A) = \sim \underline{apr}'_{\mathbb{C}}(\sim A) \tag{9}$$

$$\underline{apr}''_{\mathbb{C}}(A) = \sim \overline{apr}''_{\mathbb{C}}(\sim A) \tag{10}$$

$$\overline{apr}''_{\mathbb{C}}(A) = \bigcup\{K \in \mathbb{C} : K \cap A \neq \emptyset\} \tag{11}$$

Furthermore, Yao and Yao introduced six coverings derived from a covering $\mathbb{C}$:

1. $\mathbb{C}_1 = \cup\{md(\mathbb{C},x) : x \in U\}$
2. $\mathbb{C}_2 = \cup\{MD(\mathbb{C},x) : x \in U\}$
3. $\mathbb{C}_3 = \{\cap(md(\mathbb{C},x)) : x \in U\} = \{\cap(\mathscr{C}(\mathbb{C},x)) : x \in U\}$
4. $\mathbb{C}_4 = \{\cup(MD(\mathbb{C},x)) : x \in U\} = \{\cup(\mathscr{C}(\mathbb{C},x)) : x \in U\}$
5. $\mathbb{C}_\cap = \mathbb{C} \setminus \{K \in \mathbb{C} : (\exists \mathbb{K} \subseteq \mathbb{C} \setminus \{K\})(K = \cap \mathbb{K})\}$
6. $\mathbb{C}_\cup = \mathbb{C} \setminus \{K \in \mathbb{C} : (\exists \mathbb{K} \subseteq \mathbb{C} \setminus \{K\})(K = \cup \mathbb{K})\}$

The covering $\mathbb{C}_\cap$ is called intersection reduct and the covering $\mathbb{C}_\cup$ union reduct. These reducts eliminate intersection (respectively, union) reducible elements from the covering.

Each of the above six coverings determines two pairs of dual approximations given by Equations (8) and (9) and Equations (10) and (11), respectively.

So, for each covering $\mathbb{C}$ of U there exist fourteen pairs of dual approximation operators, obtained from granule based definition.

System Based Definition. Yao and Yao used the notion of a closure system over U, i.e., a family of subsets of U that contains U and is closed under set intersection [20]. Given a closure system $\mathbb{S}$ over U, one can construct its dual system $\mathbb{S}'$, containing the complements of each K in $\mathbb{S}$, as follows:

$$\mathbb{S}' = \{\sim K : K \in \mathbb{S}\} \tag{12}$$

The system $\mathbb{S}'$ contains $\emptyset$ and it is closed under set union. Given $S = (\mathbb{S}',\mathbb{S})$, a pair of dual lower and upper approximations can be defined as follows:

$$\underline{apr}_S(A) = \bigcup\{K \in \mathbb{S}' : K \subseteq A\} \tag{13}$$

$$\overline{apr}_S(A) = \bigcap\{K \in \mathbb{S} : K \supseteq A\} \tag{14}$$

As a particular example of a closure system, [20] considered the so-called intersection closure $S_{\cap,\mathbb{C}}$ of a covering $\mathbb{C}$, i.e., the minimal subset of $\mathscr{P}(U)$ that contains $\mathbb{C}$, $\emptyset$ and U, and is closed under set intersection. On the other hand, the union closure of $\mathbb{C}$, denoted by $S_{\cup,\mathbb{C}}$, is the minimal subset of $\mathscr{P}(U)$ that contains $\mathbb{C}$, $\emptyset$ and U, and is closed under set union. It can be shown that the dual system $S'_{\cup,\mathbb{C}}$ forms a closure system. Both $S_\cap = ((S_{\cap,\mathbb{C}})', S_{\cap,\mathbb{C}})$ and $S_\cup = (S_{\cup,\mathbb{C}}, (S_{\cup,\mathbb{C}})')$ can be used to obtain two pairs of dual approximation operations, according to Equations (13) and (14).

Twenty pairs of dual approximation operators were defined in this framework, four from the element based definition, fourteen from granule based definition and two from the system based definition based on $\cap$-closure and $\cup$-closure.

2.2 Relation Based Rough Sets

If R is a binary relation on U and $x \in U$, the sets:

$$S(x) = \{y \in U : xRy\} \text{ and } P(x) = \{y \in U : yRx\} \tag{15}$$

are called successor and predecessor neighborhood of x, respectively.

The lower and upper approximation of $A \subseteq U$, from a binary relation R on U, is given by:

$$\underline{apr}_R(A) = \{x \in U : R(x) \subseteq A\} \tag{16}$$
$$\overline{apr}_R(A) = \{x \in U : R(x) \cap A \neq \emptyset\} \tag{17}$$

where $R(x)$ can be replaced by $S(x)$ or $P(x)$. The pair (U,R) is called a relation approximation space. Therefore, each binary relation defines two pairs of approximation operators $(\underline{apr}_S, \overline{apr}_S)$ and $(\underline{apr}_P, \overline{apr}_P)$.

Järvinen shows in [6] that the ordered pairs $(\underline{apr}_S, \overline{apr}_S)$ and $(\underline{apr}_P, \overline{apr}_P)$, defined using the element based definitions, with $S(x)$ and $P(x)$, are dual pairs.

The main properties of the approximation operators can be summarized in the Proposition 1.

Proposition 1. *[30]. The operators defined in Equations (16) and (17) satisfy:*

a. $\overline{apr}_R(\emptyset) = \emptyset$
b. $\underline{apr}_R(U) = U$
c. $\overline{apr}_R(A \cup B) = \overline{apr}_R(A) \cup \overline{apr}_R(B)$
d. $\underline{apr}_R(A \cap B) = \underline{apr}_R(A) \cap \underline{apr}_R(B)$
e. $\underline{apr}_R(\sim A) = \sim \overline{apr}_R(A)$

3 Coverings and Relations

In this section we will establish an equivalence between relation based rough sets and some types of covering based rough sets from Yao and Yao's framework.

3.1 Coverings and Neighborhood

Definition 5. *Let $\mathbb{C}$ be a covering of U, for each neighborhood operator $N_i^{\mathbb{C}}$ with $i = 1,2,3,4$, two relations on U are defined by means of:*

$$xS_i^{\mathbb{C}}y \Leftrightarrow y \in N_i^{\mathbb{C}}(x) \tag{18}$$
$$xP_i^{\mathbb{C}}y \Leftrightarrow x \in N_i^{\mathbb{C}}(y) \tag{19}$$

It is easy to see that $S_i^{\mathbb{C}}$ and $P_i^{\mathbb{C}}$ for $i = 1,2,3,4$ are reflexive relations and $S_1^{\mathbb{C}}$ and $P_1^{\mathbb{C}}$ are transitive relations.

Definition 6. *Let N be a neighborhood operator. Two upper approximation operators are defined as follows:*

$$G_5^N(A) = \cup\{N(x) : x \in A\} \tag{20}$$
$$G_6^N(A) = \{x \in U : N(x) \cap A \neq \emptyset\}. \tag{21}$$

In [9] the conjugacy relation between G_5^N and G_6^N was established, i. e., $A \cap G_5^N(B) \neq \emptyset$ if and only if $B \cap G_6^N(A) \neq \emptyset$, for $A, B \subseteq U$.

The next propositions establish a connection between element based definition operators and relation based rough sets. Proposition (2) can be seen as a generalization of Theorem 6 in [30].

Proposition 2. *If $\mathbb{C}$ is a covering of U, $N_i^{\mathbb{C}}$ is any neighborhood operator and $\overline{apr}_{P_i}$ the upper approximation operator defined by the relation in Equation 19 then $\overline{apr}_{P_i} = G_5^{N_i}$.*

Proof. We can see that $x \in \overline{apr}_{P_i}(A)$ if and only if $P_i(x) \cap A \neq \emptyset$. So, there exist $w \in U$, such that xP_iw with $w \in A$, then $x \in N_i(w)$ and $w \in A$ if and only if $x \in G_5^{N_i}(A)$. □

Proposition 3. *Let $\mathbb{C}$ be a covering of U and $N_i^{\mathbb{C}}$ a neighborhood operator. If S_i is the relation defined in Equation (18), then $\underline{apr}_{S_i} = \underline{apr}_{N_i}^{\mathbb{C}}$ and $\overline{apr}_{S_i} = \overline{apr}_{N_i}^{\mathbb{C}}$.*

Proof. We can see that $w \in N_i(x)$ if and only if xS_iw, if and only if $w \in S_i(x)$. So, for a covering $\mathbb{C}$, $N_i(x) = S_i(x)$ and therefore $\underline{apr}_{S_i} = \underline{apr}_{N_i}^{\mathbb{C}}$ and $\overline{apr}_{S_i} = \overline{apr}_{N_i}^{\mathbb{C}}$. □

From Proposition (3), we know that each pair of covering-based approximation operators in element based definition, $(\underline{apr}_{N_i}, \overline{apr}_{N_i})$ can be treated as the relation based approximation operators, generated by a reflexive relation. Also, we know that: $G_6^{N_i} = \overline{apr}_{N_i}^{\mathbb{C}}$.

Proposition 4. *The operators $\overline{apr}_{S_i}$ and $\overline{apr}_{P_i}$ are conjugate.*

Proof. The conjugate relation between $G_5^{N_i}$ and $G_6^{N_i}$ was established in [9]. From Propositions (2) and (3), we have that $\overline{apr}_{S_i}$ and $\overline{apr}_{P_i}$ are conjugate. □

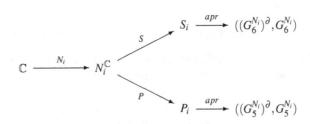

Fig. 1. Arrow diagram for element based definition approximation operators

In the arrow diagram of Figure 1, the arrow N_i represents the neighborhood operators given in Section 2.1, S and P arrows represent the successor and the predecessor neighborhood and *apr* the operator defining the dual pair of approximations, obtained from a neighborhood operator. The notation ∂ is used for representing the dual operator.

Proposition 5. *Let $\mathbb{C}$ be a covering of U and $N_4^{\mathbb{C}}$ the neighborhood system associated with the covering $\mathbb{C}$. If $S_4^{\mathbb{C}}$ is the relation defined in Equation (18) using $N_4^{\mathbb{C}}$, then $\underline{apr}_{\mathbb{C}}'' = \underline{apr}_{S_4}^{\mathbb{C}}$ and $\overline{apr}_{\mathbb{C}}'' = \overline{apr}_{S_4}^{\mathbb{C}}$.*

Proof. We will see that $\underline{apr}''_C = \underline{apr}_{S_4}$. In [9] the equivalence $\underline{apr}''_C = \underline{apr}^C_{N_4}$ was established. According to Proposition (3), $\underline{apr}^C_{N_4} = \underline{apr}_{S_4}$ then $\underline{apr}''_C = \underline{apr}^C_{S_4}$.   □

According to Proposition (5), we know that any approximation operator $\underline{apr}''_C$ from Yao and Yao's framework, can be seen as a relation based approximation.

The approximation operators $\underline{apr}'_C$, $\underline{apr}_{S_\cap}$ and $\underline{apr}_{S_\cup}$ do not have a relation based equivalence, because they do not satisfy the property (d) in Proposition 1.

3.2 New Approximation Operators

It is interesting to note that we only need the neighborhood system N_4^C to describe all the approximation operators $\underline{apr}''_C$, then it seems reasonable to ask about other approximation operators related with neighborhood systems N_1, N_2 and N_3. For each covering $\mathbb{C}$ and each neighborhood operator N_i^C, we have the operators $\underline{apr}^C_{N_i}$.

Combining each covering with the four neighborhood operators we get new approximation operators, for example $\underline{apr}^{C_3}_{N_2}$ is the lower approximation for covering $\mathbb{C}_3$ using the neighborhood operator N_2.

Example 1. For the covering $\mathbb{C} = \{\{1\}, \{1,2\}, \{1,3\}, \{3,4\}, \{1,2,3\}, \{2,4\}, \{2,3,4\}\}$ of $U = \{1,2,3,4\}$ the Table (1) shows the upper approximation operators $\overline{apr}^{C_j}_{N_i}$ for each singleton: $\{1\}$, $\{2\}$, $\{3\}$ and $\{4\}$.

The approximations for $A \subseteq U$ can be calculated using the property $\overline{apr}(A) = \cup_{x \in A}\overline{apr}(x)$.

Table 1. Upper approximation for singletons of U

Operator	{1}	{2}	{3}	{4}	Operator	{1}	{2}	{3}	{4}
$\overline{apr}^{C}_{N_1}$	{1}	{2}	{3}	{4}	$\overline{apr}^{C_1}_{N_1}$	{1}	{2}	{3}	{4}
$\overline{apr}^{C}_{N_2}$	{1,2,3}	{2,4}	{3,4}	{2,3,4}	$\overline{apr}^{C_1}_{N_2}$	{1,2,3}	{2,4}	{3,4}	{2,3,4}
$\overline{apr}^{C}_{N_3}$	{1}	{1,2,3,4}	{1,2,3,4}	{4}	$\overline{apr}^{C_1}_{N_3}$	{1}	{2}	{3}	{4}
$\overline{apr}^{C}_{N_4}$	{1,2,3}	{1,2,3,4}	{1,2,3,4}	{2,3,4}	$\overline{apr}^{C_1}_{N_4}$	{1,2,3}	{1,2,4}	{1,3,4}	{2,3,4}
$\overline{apr}^{C_2}_{N_1}$	{1}	{1,2,3,4}	{1,2,3,4}	{4}	$\overline{apr}^{C_3}_{N_1}$	{1}	{2}	{3}	{4}
$\overline{apr}^{C_2}_{N_2}$	{1,2,3}	{1,2,3,4}	{1,2,3,4}	{2,3,4}	$\overline{apr}^{C_3}_{N_2}$	{1}	{2}	{3}	{4}
$\overline{apr}^{C_2}_{N_3}$	{1}	{1,2,3,4}	{1,2,3,4}	{4}	$\overline{apr}^{C_3}_{N_3}$	{1}	{2}	{3}	{4}
$\overline{apr}^{C_2}_{N_4}$	{1,2,3}	{1,2,3,4}	{1,2,3,4}	{2,3,4}	$\overline{apr}^{C_3}_{N_4}$	{1}	{2}	{3}	{4}
$\overline{apr}^{C_4}_{N_1}$	{1}	{1,2,3,4}	{1,2,3,4}	{4}	$\overline{apr}^{C_n}_{N_1}$	{1}	{2}	{3}	{4}
$\overline{apr}^{C_4}_{N_2}$	{1,2,3}	{1,2,3,4}	{1,2,3,4}	{2,3,4}	$\overline{apr}^{C_n}_{N_2}$	{1,2,3}	{1,2,4}	{1,3,4}	{2,3,4}
$\overline{apr}^{C_4}_{N_3}$	{1,2,3,4}	{1,2,3,4}	{1,2,3,4}	{1,2,3,4}	$\overline{apr}^{C_n}_{N_3}$	{1}	{1,2,3,4}	{1,2,3,4}	{4}
$\overline{apr}^{C_4}_{N_4}$	{1,2,3,4}	{1,2,3,4}	{1,2,3,4}	{1,2,3,4}	$\overline{apr}^{C_n}_{N_4}$	{1,2,3}	{1,2,3,4}	{1,2,3,4}	{2,3,4}

We can use the results in Table (1) to establish the difference between some operators. For example, since $\overline{apr}^{\mathbb{C}}_{N_2}(3) \neq \overline{apr}^{\mathbb{C}}_{N_4}(3)$ then $\overline{apr}^{\mathbb{C}}_{N_2} \neq \overline{apr}^{\mathbb{C}}_{N_4}$.

The concepts of minimal and maximal description are related with a covering. If we consider other coverings, we have different minimal and maximal descriptions. We have some preliminary results about the operators defined in Table (1), as follows.

Proposition 6. *Let $\mathbb{C}$ be a covering of U, then $\mathbb{C}_3$ is an unary covering and $\underline{apr}^{\mathbb{C}_3}_{N_1} = \underline{apr}^{\mathbb{C}_3}_{N_2}$.*

Proof. The elements in $\mathbb{C}_3$ are the neighborhoods $N_1^{\mathbb{C}}(x)$, so $\mathbb{C}_3 = \{N_1^{\mathbb{C}}(x)\}$ and it is unary, because the minimal description of each $x \in U$ has only an element, $|md(\mathbb{C},x)| = 1$. In this case the neighborhood $N_1^{\mathbb{C}_3}$ and $N_2^{\mathbb{C}_3}$ are the same and therefore the approximation operators $\underline{apr}^{\mathbb{C}_3}_{N_1}$ and $\underline{apr}^{\mathbb{C}_3}_{N_2}$ are equal. □

In the same way the equality $N_3^{\mathbb{C}_3} = N_4^{\mathbb{C}_3}$ can be established and therefore $\underline{apr}^{\mathbb{C}_3}_{N_3} = \underline{apr}^{\mathbb{C}_3}_{N_4}$.

Proposition 7. *For a covering approximation space and the coverings defined above, we have:*

(i) $md(\mathbb{C},x) = md(\mathbb{C}_1,x)$.
(ii) $MD(\mathbb{C},x) = MD(\mathbb{C}_2,x) = MD(\mathbb{C}_\cap,x)$.

Proof. (i). Let us recall that $\mathbb{C}_1$ is the covering with the minimal sets of $\mathbb{C}$, so the minimal sets of $\mathbb{C}_1$ are equals to $\mathbb{C}_1$. (ii). Similarly for maximal description. □

Corollary 1. *The relations among neighborhood system are:*

(i) $N_1^{\mathbb{C}} = N_1^{\mathbb{C}_1}$ and $N_2^{\mathbb{C}} = N_2^{\mathbb{C}_1}$.
(ii) $N_3^{\mathbb{C}} = N_3^{\mathbb{C}_2} = N_3^{\mathbb{C}_\cap}$ and $N_4^{\mathbb{C}} = N_4^{\mathbb{C}_2} = N_4^{\mathbb{C}_\cap}$.

Proof. Let us recall that N_1 and N_2 are defined from $md(\mathbb{C},x)$ and N_3 and N_4 are defined from $MD(\mathbb{C},x)$. □

Corollary 2. *The relations among approximation operators are:*

(i) $\overline{apr}^{\mathbb{C}}_{N_1} = \overline{apr}^{\mathbb{C}_1}_{N_1}$ and $\overline{apr}^{\mathbb{C}}_{N_2} = \overline{apr}^{\mathbb{C}_1}_{N_2}$.
(ii) $\overline{apr}^{\mathbb{C}}_{N_3} = \overline{apr}^{\mathbb{C}_2}_{N_3} = \overline{apr}^{\mathbb{C}_\cap}_{N_3}$ and $\overline{apr}^{\mathbb{C}}_{N_4} = \overline{apr}^{\mathbb{C}_2}_{N_4} = \overline{apr}^{\mathbb{C}_\cap}_{N_4}$.

□

Finally we present an example to show that minimal and maximal description for different coverings are not the same.

Example 2. If $\mathbb{C} = \{\{1\},\{1,2\},\{1,2,3\},\{2,4\},\{2,3\},\{2,3,4\}\}$ is a covering of $U = \{1,2,3,4\}$. The coverings obtained from $\mathbb{C}$ are: $\mathbb{C}_1 = \{\{1\},\{1,2\},\{2,3\},\{2,4\}\}$, $\mathbb{C}_2 = \{\{1,2,3\},\{2,3,4\}\}$, $\mathbb{C}_3 = \{\{1\},\{2\},\{2,3\},\{2,4\}\}$, $\mathbb{C}_4 = \{\{1,2,3\},\{2,3,4\},\{1,2,3,4\}\}$ and $\mathbb{C}_\cap = \{\{1\},\{1,2\},\{1,2,3\},\{2,4\},\{2,3,4\}\}$.

The elements in the minimal description $md(\mathbb{C},x)$ for the coverings are listed in Table (2).

We can use the results in Table (2) to show some differences between approximation operators.

Table 2. Minimal descriptions for the coverings obtained from $\mathbb{C}$

Covering	$md(\mathbb{C},1)$	$md(\mathbb{C},2)$	$md(\mathbb{C},3)$	$md(\mathbb{C},4)$
$\mathbb{C}$	$\{1\}$	$\{1,2\},\{2,4\},\{2,3\}$	$\{2,3\}$	$\{2,4\}$
$\mathbb{C}_1$	$\{1\}$	$\{1,2\},\{2,4\},\{2,3\}$	$\{2,3\}$	$\{2,4\}$
$\mathbb{C}_2$	$\{1,2,3\}$	$\{1,2,3\},\{2,3,4\}$	$\{1,2,3\},\{2,3,4\}$	$\{2,3,4\}$
$\mathbb{C}_3$	$\{1\}$	$\{2\}$	$\{2,3\}$	$\{2,4\}$
$\mathbb{C}_4$	$\{1,2,3\}$	$\{1,2,3\},\{2,3,4\}$	$\{1,2,3\},\{2,3,4\}$	$\{2,3,4\}$
$\mathbb{C}_\cap$	$\{1\}$	$\{1,2\},\{2,4\}$	$\{1,2,3\},\{2,3,4\}$	$\{2,4\}$

4 Conclusions

We have studied relationships between relation based and covering based rough sets. In particular we have showed that each pair $(\underline{apr}_{N_i}, \overline{apr}_{N_i})$ of element based definition approximation and each pair in granule based definition approximation $(\underline{apr}''_{\mathbb{C}}, \overline{apr}''_{\mathbb{C}})$ can be treated as a relation based approximation operator. We also defined new approximation operators combining coverings and neighborhood operators. As a future work, we will establish other relationships between the new approximation operators defined Section 3.2.

References

1. Bonikowski, Z., Brynarski, E.: Extensions and Intensions in rough set theory. Information Science 107, 149–167 (1998)
2. Couso, I., Dubois, D.: Rough sets, coverings and incomplete information. Fundamenta Informaticae 108, 223–247 (2011)
3. Dai, J.: Rough sets approach to incomplete numerical data. Information Sciences 214, 43–57 (2013)
4. Grzymala-Busse, J., Siddhaye, S.: Rough Sets approach to rule induction from incomplete data. In: Proceedings of 10th International Conference on Information Processing and Management of Uncertainty in Knowledge-Based Systems, pp. 923–930 (2004)
5. Grzymała-Busse, J.W.: Data with Missing Attribute Values: Generalization of Indiscernibility Relation and Rule Induction. In: Peters, J.F., Skowron, A., Grzymała-Busse, J.W., Kostek, B., Swiniarski, R.W., Szczuka, M.S. (eds.) Transactions on Rough Sets I. LNCS, vol. 3100, pp. 78–95. Springer, Heidelberg (2004)
6. Järvinen, J.: Lattice Theory for Rough Sets. In: Peters, J.F., Skowron, A., Düntsch, I., Grzymała-Busse, J., Orłowska, E., Polkowski, L. (eds.) Transactions on Rough Sets VI. LNCS, vol. 4374, pp. 400–498. Springer, Heidelberg (2007)
7. Järvinen, J., Radeleczki, S.: Rough Sets determined by tolerance. International Journal of Approximate Reasoning (2013), http://dx.doi.org/10.1016/j.ijar.2013.12.005
8. Pawlak, Z.: Rough sets. International Journal of Computer and Information Sciences 11, 341–356 (1982)
9. Restrepo, M., Cornelis, C., Gómez, J.: Duality, Conjugacy and Adjointness of Approximation Operators in Covering-based Rough Sets. International Journal of Approximate Reasoning 55, 469–485 (2014)

10. Samanta, P., Chakraborty, M.K.: Covering based approaches to rough sets and implication lattices. In: Sakai, H., Chakraborty, M.K., Hassanien, A.E., Ślęzak, D., Zhu, W. (eds.) RSFD-GrC 2009. LNCS (LNAI), vol. 5908, pp. 127–134. Springer, Heidelberg (2009)

11. Skowron, A., Stepaniuk, J.: Tolerance Approximation Spaces. Fundamenta Informaticae 55, 245–253 (1996)

12. Slowinski, R., Venderpooten, D.: A generalized definition of rough approximation based on similarity. IEEE Transaction on Knowledge and Data Engineering 12, 331–336 (2000)

13. Wu, M., Wu, X., Shen, T.: A New Type of Covering Approximation Operators. In: International Conference on Electronic Computer Technology, pp. 334–338. IEEE (2009)

14. Xu, W., Zhang, W.: Measuring roughness of generalized rough sets induced by a covering. Fuzzy Sets and Systems 158, 2443–2455 (2007)

15. Yang, T., Li, Q.: Reduction about Approximation Spaces of Covering Generalized Rough Sets. International Journal of Approximate Reasoning 51, 335–345 (2010)

16. Yao, Y.Y.: Constructive and Algebraic Methods of the Theory of Rough Sets. Information Sciences 109, 21–47 (1998)

17. Yao, Y.Y.: Generalized Rough Sets Models. In: Polkowski, L., Skowron, A. (eds.) Knowledge Discovery, pp. 286–318. Physica-Verlag (1998)

18. Yao, Y.Y.: On generalizing Rough Sets Theory. In: Wang, G., Liu, Q., Yao, Y., Skowron, A. (eds.) RSFDGrC 2003. LNCS (LNAI), vol. 2639, pp. 44–51. Springer, Heidelberg (2003)

19. Yao, Y.Y., Yao, B.: Two views of the theory of rough sets in finite universes. International Journal of Approximate Reasoning 15, 299–317 (1996)

20. Yao, Y.Y., Yao, B.: Covering Based Rough Sets Approximations. Information Sciences 200, 91–107 (2012)

21. Zakowski, W.: Approximations in the Space (u, π). Demonstratio Mathematica 16, 761–769 (1983)

22. Zhang, Y.L., Li, J., Wu, W.: On Axiomatic Characterizations of Three Pairs of Covering Based Approximation Operators. Information Sciences 180, 274–287 (2010)

23. Zhang, Y.L., Luo, M.K.: Relationships between covering-based rough sets and relation-based rough sets. Information Sciences 225, 55–71 (2013)

24. Zhu, W.: Basics concepts in Covering-Based Rough Sets. In: Third International Conference on Natural Computation, vol. 5, pp. 283–286 (2007)

25. Zhu, W.: Properties of the First Type of Covering-Based Rough Sets. In: Proceedings of Sixth IEEE International Conference on Data Mining - Workshops, pp. 407–411 (2006)

26. Zhu, W.: Properties of the Second Type of Covering-Based Rough Sets. In: Proceedings of the IEEE/WIC/ACM International Conference on Web Intelligence and Intelligent Agent Technology, pp. 494–497 (2006)

27. Zhu, W.: Properties of the Third Type of Covering-Based Rough Sets. In: Proceedings of International Conference on Machine Learning and Cybernetics, pp. 3746–3751 (2007)

28. Zhu, W.: Properties of the Fourth Type of Covering-Based Rough Sets. In: Proceedings of Sixth International Conference on Hybrid Intelligence Systems, p. 43 (2006)

29. Zhu, W., Wang, F.: On Three Types of Covering Based Rough Sets. IEEE Transactions on Knowledge and Data Engineering 19, 1131–1144 (2007)

30. Zhu, W.: Relationship Between Generalized Rough Sets Based on Binary Relation and Covering. Information Sciences 179, 210–225 (2009)

31. Zhu, W., Wang, F.: A New Type of Covering Rough Set. In: Proceedings of Third International IEEE Conference on Intelligence Systems, pp. 444–449 (2006)

32. Zhu, W., Wang, F.: Reduction and Axiomatization of Covering Generalized Rough Sets. Information Sciences 152, 217–230 (2003)

33. Zhu, W., Wang, F.-Y.: Binary relation based Rough Sets. In: Wang, L., Jiao, L., Shi, G., Li, X., Liu, J. (eds.) FSKD 2006. LNCS (LNAI), vol. 4223, pp. 276–285. Springer, Heidelberg (2006)

A Rough Set Approach to Novel Compounds Activity Prediction Based on Surface Active Properties and Molecular Descriptors

Jerzy Błaszczyński[1], Łukasz Pałkowski[2], Andrzej Skrzypczak[3], Jan Błaszczak[3], Alicja Nowaczyk[4], Roman Słowiński[1], and Jerzy Krysiński[2]

[1] Poznań University of Technology, Institute of Computing Science,
Piotrowo 3a, 60-965 Poznań, Poland
{jerzy.blaszczynski,roman.slowinski}@cs.put.poznan.pl
[2] Nicolaus Copernicus University, Collegium Medicum,
Department of Pharmaceutical Technology, Jurasza 2, 85-089 Bydgoszcz, Poland
{lukaszpalkowski,jerzy.krysinski}@cm.umk.pl
[3] Poznań University of Technology, Institute of Chemical Technology,
Skłodowskiej-Curie 2, 60-965 Poznań, Poland
[4] Nicolaus Copernicus University, Collegium Medicum,
Department of Organic Chemistry, Jurasza 2, 85-094 Bydgoszcz, Poland

Abstract. The aim of this paper is to study relationship between biological activity of a group of 140 gemini-imidazolium chlorides and three types of parameters: structure, surface active, and molecular ones. Dominance-based rough set approach is applied to obtain decision rules, which describe dependencies between analyzed parameters and allow to create a model of chemical structure with best biological activity. Moreover, presented study allowed to identify attributes relevant with respect to high antimicrobial activity of compounds. Finally, we have shown that decision rules that involve only structure and surface active attributes are sufficient to plan effective synthesis pathways of active molecules.

Keywords: DRSA, structure-activity relationship, molecular properties, surfactant.

1 Introduction

In the analysis of structure-activity relationship (SAR) [7] of chemical compounds one may consider surface active and structure parameters, which can be controlled directly in synthesis. The second type of parameters are molecular descriptors, which are determined by calculation with a specific software. Molecular descriptors may be used to a molecular modelling of compounds (techniques used to model or mimic the behaviour of molecules). In this paper we present dependencies between both types of parameters, and antimicrobial activity for a group of 140 gemini-imidazolium chlorides, which formed 10 homologous series with a different n and R substituents. The type of compounds being analyzed is a group of gemini-imidazolium chlorides with a good antimicrobial activity [8].

M. Kryszkiewicz et al. (Eds.): RSEISP 2014, LNAI 8537, pp. 153–160, 2014.

First, we analyzed, separately, dependencies between the controlled parameters and antimicrobial activity. Then, the dependencies between molecular descriptors and antimicrobial activity. Finally, we performed an analysis with both types of parameters. Decision rules and attributes relevant in these analyses were discovered. The results allow to draw conclusions for synthesis of new compounds. The paper is organized as follows. In the next section, material and methods are presented, including description of analyzed compounds, transformation of the information table, rule discovery, and attribute relevance measurement. The last section is devoted to presentation of results and discussion.

2   Material and Methods

2.1   Material

For the sake of the analyses, 140 objects that represent new n-alkyl-bis- N-alkoxy-N-alkyl imidazolium chlorides were examined. 70 of those objects were previously described in [9,10]. Surface active properties of analyzed chlorides were described by the following parameters: CMC - critical micelle concentration (mol/L), γCMC - value of surface tension at critical micelle concentration (mN/m), $\Gamma \cdot 10^6$ - value of surface excess (mol/m^2), $A \cdot 10^{20}$ - molecular area of a single particle (m^2), ΔG_{ads} - free energy of adsorption of molecule (kJ/mol). Structure properties of chlorides were described by parameters: n - number of carbon atoms in n-substituent, and R - number of carbon atoms in R-substituent.

Domains of structure attributes, are presented in Table 1.

Table 1. Numerical coding of the structure condition attributes

Code	n	R	Code	R
1		CH_3	8	C_8H_{17}
2	C_2H_5	C_2H_5	9	C_9H_{19}
3	C_3H_7	C_3H_7	10	$C_{10}H_{21}$
4	C_4H_9	C_4H_9	11	$C_{11}H_{23}$
5	C_5H_{11}	C_5H_{11}	12	$C_{12}H_{25}$
6	C_6H_{13}	C_6H_{13}	14	$C_{14}H_{29}$
7		C_7H_{15}	16	$C_{16}H_{33}$

We also considered molecular parameters of analyzed compounds, which were calculated with Dragon (molecular descriptors calculation software)[1] and Gaussian (computational chemistry program)[2]. Those parameters were: MLOGP - Moriguchi octanol-water partition coefficient, Balaban index (BI), Narumi topological index (NTI), total structure connectivity index (TSCI), Wiener index (WI) - numerical parameters characterizing compounds' topology), HOMO - Highest Occupied Molecular Orbital, LUMO - Lowest Unoccupied Molecular Orbital, HOMO-LUMO gap (HL gap) - the energy difference between the HOMO and LUMO, dipole (dip) - electric dipole moment, radius of gyration (ROG) -

---

[1] http://www.talete.mi.it/index.htm

[2] http://www.gaussian.com/

the root mean square distance of the entities' parts from either its center of gravity or a given axis, molecular weight (MW) of compounds.

Staphylococcus aureus ATCC 25213 microorganisms were used to evaluate antibacterial activity of compounds by minimal inhibitory concentration (MIC). According to the value of MIC objects were sorted into three decision classes:

- class good - good antimicrobial properties: MIC ≤ 0.02 μM/L,
- class medium - medium antimicrobial properties: $0.02 <$ MIC < 1 μM/L,
- class weak - weak antimicrobial properties: MIC ≥ 1 μM/L.

2.2   Discovery of Relevant Decision Rules and Attributes

Data that concern antibacterial activity of n-alkyl-bis- N-alkoxy- N-alkyl imidazolium chlorides can be seen as classification data. Surface active properties and molecular parameters of chlorides are condition attributes (independent variables). Class labels: good, medium, and weak are assigned to chlorides by a decision attribute (dependent variable). To explain the class assignment in terms of condition attributes, we used the rough set concept [11], and its particular extension called Dominance-based Rough Set Approach (DRSA) [4,12]. In the classical rough-sets it is necessary to perform discretization procedure for all attributes with continuous values. In the case of this analysis continuous attributes are not discretized. DRSA proved to be an effective tool in analysis of classification data which are partially inconsistent [5,6]. In the context presented in this work, inconsistency means that two compounds have similar surface active properties and/or molecular parameters, while they are in different antimicrobial activity class. The rough sets representing classes discern between consistent and inconsistent compounds and prepare the ground for induction of decision rules. DRSA assumes that the value sets of condition attributes are ordered and monotonically dependent on the order of decision classes. In consequence, the rules induced from data structured using dominance-based rough sets are monotonic, which means that they have the following syntax:

$$\text{if } a_i(ch) \geq v_i \wedge a_j(ch) \geq v_j \wedge \ldots \wedge a_p(ch) \geq v_p$$
$$\text{then } ch \text{ belongs to a class }, \tag{1}$$
$$\text{if } a_k(ch) \leq v_k \wedge a_l(ch) \leq v_l \wedge \ldots \wedge a_s(ch) \leq v_s$$
$$\text{then } ch \text{ does not belong to a class}, \tag{2}$$

where a_h is an h-th condition attribute, v_h is a threshold value of this attribute, and ch stands for chloride. All of these pieces make an elementary condition $a_h(ch) \geq v_h$ or $a_h(ch) \leq v_h$ entering the condition part of a rule indicating the assignment of a compound to a given class or not. In the above syntax of the rules, it is assumed that value sets of all condition attributes are numerical and ordered such that the greater the value, the more likely is that the compound belongs to a given class; analogously, it is assumed that the smaller the value, the more likely is that a compound does not belong to a given class. Attributes ordered in this way are called gain-type. Cost-type attributes have value sets ordered in the opposite way, so elementary conditions on these attributes have opposite relation signs. In case of the type of data that we analyze, it is impossible to assume a priori if attributes are gain or cost attributes, thus we proceed as described in [1], i.e., we are considering each original attribute in two copies, and for the first copy we assume it is gain-type, while for the second copy we assume it is cost-type. The applied transformation of data is non-invasive, i.e., it does not bias the matter of discovered relationships between condition attributes and the decision attribute. Then, the induction algorithm constructs decision rules involving elementary conditions on one or both copies of particular attributes. For example, in a rule indicating the assignment of a compound to class good there may appear the following elementary conditions concerning attribute a_i:

- $\uparrow a_i(ch) \geq v_{i_1}$,
- $\downarrow a_i(ch) \leq v_{i_2}$,
- $\uparrow a_i(ch) \geq v_{i_1} \wedge \downarrow a_i(ch) \leq v_{i_2}$, which boils down to $a_i(ch) \in [v_{i_1}, v_{i_2}]$, provided that $v_{i_1} \leq v_{i_2}$,

where $\uparrow a_i$ and $\downarrow a_i$ are gain-type and cost-type copies of attribute a_i, respectively. Remark the transformation of attributes permits discovering global and local monotonic relationships between condition attributes and class assignment.

Sets of decision rules, which are essential for the analysis, were induced from data transformed in the way described above and structured using the concept of dominance-based rough sets. The induction algorithm is called VC-DomLEM [2]; it has been implemented as software package called jMAF[3], based on java Rough Set (jRS) library. The induced sets of rules were used to construct component classifiers in variable consistency bagging [3]. This type of bagging was applied to increase the accuracy of rule classifiers. Estimation of both, rule relevance and attribute relevance in rules, was made by measuring Bayesian confirmation [9,10]. In this process, decision rules are induced repetitively on bootstrap samples and tested on compounds, which are not included in the samples. Let us interpret a rule as a consequence relation "if E, then H", where E denotes rule premise, and H rule conclusion. As to relevance of a rule, the Bayesian confirmation measure quantifies the contribution of rule premise E to correct classification of unseen chlorides. From among many Bayesian confirmation measures proposed in the literature, we use measure $s(H, E)$ for its clear interpretation in terms of a difference of conditional probabilities involving H and E, i.e., $s(H, E) = Pr(H|E) - Pr(H|\neg E)$, where probability $Pr(\cdot)$ is estimated on the testing samples of compounds. Rules are, moreover, characterized by their strength defined as a ratio of the number of compounds matching the condition part a rule and the number of all compounds in the sample. Relevant rules are selected from a list of all rules induced for a class, and ordered by Algorithm 1.

As to relevance of single attributes, the Bayesian confirmation measure quantifies the degree to which the presence of attribute a_i in premise E, denoted by $a_i \triangleright E$, provides evidence for or against conclusion H of the rule. Here, we use again measure $s(H, a_i \triangleright E)$, which, in this case, is defined as follows,

---

Input : set of rules **R**,
  normalized values of confirmation of rule c_r, strength s_r,
  and confirmation of attributes in rule c_{r_a}, for each rule r in **R**.
Output: list of rules $\mathbf{R}^R$ ordered according to relevance.

1   $\mathbf{R}^R := \emptyset$;
2   $\mathbf{R}' :=$ list of rules from **R** sorted according to $c_r \times s_r \times c_{r_a}$;
3   **while** $\mathbf{R}' \neq \emptyset$ **do**
4     $r :=$ best rule from $\mathbf{R}'$;
5   remove r from $\mathbf{R}'$;
6   add r at the last position in $\mathbf{R}^R$;
7   increase normalized penalty $p_{r'}$ of each r' in $\mathbf{R}'$ according to number of
     objects supporting r' that are already covered by r;
8   sort rules in $\mathbf{R}'$ according to $p_{r'} \times c_{r'} \times s_{r'} \times c_{r'_a}$;

Algorithm 1. Greedy ordering of rules by relevance

$s(H, a_i \rhd E) = Pr(H|a_i \rhd E) - Pr(H|a_i \not\rhd E)$. In this way, attributes present in the premise of a rule that assigns compounds correctly, or the ones absent in the condition part of a rule that assigns incorrectly, are getting more relevant.

3 Results and Discussion

As it was mentioned in the introduction we divided our study into a few steps. We calculated dependencies between structure, surface active properties and antimicrobial activity, and then we made similar calculations for dependencies between molecular descriptors and antimicrobial activity. It should be emphasized that condition attributes in the first analysis were determined experimentally (measured in chemical laboratory), which was time consuming and expensive. For this reason, we also determined molecular descriptors for analyzed compounds by means of Dragon and Gaussian software. A molecular descriptor is the final result of a logic and mathematical procedure which transforms chemical information encoded within a symbolic representation of a molecule into a useful number or a result of a standardized experiment [13]. This is computationally expensive, but does not require work in a chemical laboratory. Molecular descriptors were chosen in such a way, that they should have structural interpretation, should have good correlation with at least one property of a molecule, should be possible to apply to compound's structure, and should not be based on experimental properties. Moreover, those descriptors were invariant. The invariance properties of molecular descriptors can be defined as the ability of the calculation algorithm to give a descriptor value that is independent of the particular characteristics of the molecular representation, such as atom numbering or labeling, spatial reference frame, molecular conformations, etc. Invariance to molecular numbering or labeling is assumed as a minimal basic requirement for any descriptor.

Results of estimation of confirmation of all attributes (structure, surface active and molecular) in rules induced for class good and weak are presented on Figure 1. We can observe that attributes: surface excess and Wiener index are the most relevant when the good class of activity is considered. On the other

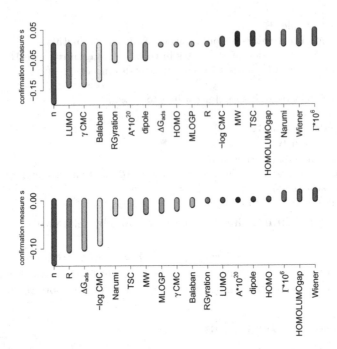

Fig. 1. Predictive confirmation of attributes for classes: good (top), and weak (bottom)

hand, the most relevant attributes for weak decision class are: Wiener index, HOMO-LUMO gap and surface excess. These results suggest that both: surface active and molecular parameters might be helpful in assigning new chemical compounds to a specific class of activity.

In Table 2, we presented decision rules obtained for structure and surface properties, and microbiological activity dependencies. All of the presented rules were obtained in the procedure described in Section 2.2. Compounds belonging to class good are characterized by R-substituent possessing preferably from 7 to 11 atoms of carbon in a chain, value of -log CMC ranging from 2.71 to 2.81, and value of surface excess in the range of 2.05 and 2.47. In contrast, compounds from the weak class are described by values of surface tension at critical micelle concentration more than 50.3, and values of surface excess higher than 2.57. Information about R-substituent length is absent in this group of decision rules. According to these results, we may conclude that, for the compounds that we have analyzed, the length of R-substituent, values of CMC and values of surface excess determine the best antimicrobial activity. On the other hand, values of surface tension at critical micelle concentration and of surface excess higher than 2.57 influence the most compounds with the lowest biological action.

In Table 3, we present predictive accuracy results obtained in repeated cross-validation, which has been conducted for: structure, surface condition attributes only, molecular condition attributes only, and all (structure, surface, and molecular) condition attributes. The results are similar regardless of the type of

Table 2. Decision rules, supporting examples, strength, and value of s measure - structure and surface active parameters

ID			Condition attributes				Support	Strength(%)	s
	n	R	-logCMC	γCMC	$\Gamma \cdot 10^6$	$A \cdot 10^{-20}$	ΔG_{ads}		
			class good						
1	$\langle 7, 11\rangle$			≤ 2.47				41	29.28	0.7894
2	$\langle 7, 10\rangle$			≤ 2.47				33	23.57	0.7500
3	$\langle 7, 9\rangle$			≤ 2.47				25	17.85	0.6956
4	≤ 11	≥ 2.71		≥ 2.71				24	17.14	0.7500
5	$\langle 7, 9\rangle$			≤ 2.46				24	17.14	0.6956
6	≤ 11	≥ 2.87						22	15.71	0.6956
7	≤ 11	≥ 2.88						21	15.00	0.6956
8		≥ 2.73		≥ 2.05		≥ 31.9		20	14.28	0.7272
9	≥ 7	≤ 2.81		≤ 2.47				20	14.28	0.7142
10	≥ 7	≤ 2.79		≤ 2.47				19	13.57	0.6956
			class weak						
11			≥ 50.3	≥ 2.57				18	12.85	0.9230
12				≥ 2.57		≤ 23.0		17	12.14	0.9583
13			≥ 52.1	≥ 2.57				17	12.14	0.9230
14			≥ 53.4	≥ 2.57				17	12.14	0.9200
15		≥ 2.6						17	12.14	0.8888
16			≥ 54.4	≥ 2.58				16	11.42	0.9200
17				≥ 2.61				16	11.42	0.8888
18	≤ 5			≥ 2.6				15	10.71	0.8888
19			≥ 55.7	≥ 2.6				14	10.00	0.9200
20		≥ 2.62						14	10.00	0.8888

descriptors that was set as condition attributes. This leads to conclusion that molecular descriptors might be used instead of structure and surface active parameters. At it was shown in [9,10] ranges of values of structure and surface active attributes in decision rules are important from the point of view of the synthesis of new chemical entities with good antimicrobial properties. This study revealed that rules involving molecular descriptors also provide some directions. However, structure and surface active parameters are explicit and easy to interpret, and might be processed into guidelines which compounds with preferable antimicrobial properties should be synthesized (class good), and which compounds should be abandoned (class weak). Furthermore, decision rules involving molecular descriptors, while interesting, are difficult to consider in projecting chemical synthesis, because they cannot be directly translated into directions of synthesis. Therefore, we consider decision rules involving structure and surface active attributes to be the more appropriate to plan synthesis pathways of new, active molecules with high biological activity against selected bacterial strains.

Table 3. Predictive accuracy of rule classifiers (results of repeated cross-validation)

	structure, surface		molecular		all	
	good	weak	good	weak	good	weak
% correctly classified chlorides	78.92	94.93	74.75	94.96	77.65	95.25
% incorrectly classified chlorides	21.07	5.06	25.24	5.03	22.34	4.75

Acknowledgment. The first and the sixth author wish to acknowledge financial support from the Polish National Science Centre, grant no. DEC-2011/01/B/ST6/07318.

References

1. Błaszczyński, J., Greco, S., Słowiński, R.: Inductive discovery of laws using monotonic rules. Eng. Appl. Artif. Intel. 25, 284–294 (2012)
2. Błaszczyński, J., Słowiński, R., Szeląg, M.: Sequential covering rule induction algorithm for variable consistency rough set approaches. Inf. Sciences 181(5), 987–1002 (2011)
3. Błaszczyński, J., Słowiński, R., Stefanowski, J.: Variable consistency bagging ensembles. In: Peters, J.F., Skowron, A. (eds.) Transactions on Rough Sets XI. LNCS, vol. 5946, pp. 40–52. Springer, Heidelberg (2010)
4. Greco, S., Matarazzo, B., Słowiński, R.: Rough sets theory for multicriteria decision analysis. European Journal of Operational Research 129, 1–47 (2001)
5. Greco, S., Matarazzo, B., Słowiński, R.: Multicriteria classification. In: Kloesgen, W., Żytkow, J. (eds.) Handbook of Data Mining and Knowledge Discovery, pp. 318–328. Oxford University Press, New York (2002)
6. Greco, S., Matarazzo, B., Słowiński, R.: Rough sets methodology for sorting problems in presence of multiple attributes and criteria. European J. of Operational Research 138(2), 247–259 (2002)
7. Krysiński, J., Płaczek, J., Skrzypczak, A., Błaszczak, J., Prędki, B.: Analysis of Relationships Between Structure, Surface Properties and Antimicrobial Activity of Quaternary Ammonium Chlorides. QSAR Comb. Sci. 28, 995–1002 (2009)
8. McBain, A.J., Ledder, R.G., Moore, L.E., Catrenich, C.E.: Effects of Quaternary-Ammonium-Based Formulations on Bacterial Community Dynamics and Antimicrobial Susceptibility. Appl. Environ. Microbiol. 70, 3449–3456 (2004)
9. Pałkowski, Ł., Błaszczyński, J., Krysiński, J., Słowiński, R., Skrzypczak, A., Błaszczak, J., Gospodarek, E., Wróblewska, J.: Application of Rough Set Theory to Prediction of Antimicrobial Activity of Bis-quaternary Ammonium Chlorides. In: Li, T., Nguyen, H.S., Wang, G., Grzymala-Busse, J., Janicki, R., Hassanien, A.E., Yu, H. (eds.) RSKT 2012. LNCS, vol. 7414, pp. 107–116. Springer, Heidelberg (2012)
10. Pałkowski, Ł., Błaszczyński, J., Skrzypczak, A., Błaszczak, J., Kozakowska, K., Wróblewska, J., Kozuszko, S., Gospodarek, E., Krysiński, J., Słowiński, R.: Antimicrobial activity and SAR study of new gemini imidazolium-based chlorides. Chem. Biol. Drug Des. 83(3), 278–288 (2014)
11. Pawlak, Z.: Rough sets. Theoretical aspects of reasoning about data. Kluwer, Dordrecht (1991)
12. Słowiński, R., Greco, S., Matarazzo, B.: Rough Sets in Decision Making. In: Meyers, R.A. (ed.) Encyclopedia of Complexity and Systems Science, pp. 7753–7786 (2009)
13. Todeschini, R., Consonni, V.: Handbook of Molecular Descriptors. Wiley-VCH (2000)

Rough Sets in Ortholog Gene Detection

Selection of Feature Subsets and Case Reduction Considering Imbalance

Deborah Galpert Cañizares, Reinier Millo Sánchez, María Matilde García Lorenzo,
Gladys Casas Cardoso, Ricardo Grau Abalo, and Leticia Arco García

Computer Science Department, Universidad Central "Marta Abreu" de Las Villas,
Carretera a Camajuaní km 5 ½, Santa Clara, Cuba
{deborah,mmgarcia,gcasas,rgrau,leticiaa}@uclv.edu.cu,
rmillo@uclv.cu

Abstract. Ortholog detection should be improved because of the real value of ortholog genes in the prediction of protein functions. Datasets in the binary classification problem can be represented as information systems. We use a gene pair extended similarity relation based on an extension of the Rough Set Theory and aggregated gene similarity measures as gene features, to select feature subsets with the aid of quality measures that take imbalance into account. The proposed procedure can be useful for datasets with few features and discrete parameters. The case reduction obtained from the approximation of ortholog and non-ortholog concepts might be an effective method to cope with extremely high imbalance in supervised classification.

Keywords: Ortholog Detection, Rough Sets, Classification.

1 Introduction

The successful applications of the Rough Set Theory (RST) [1] in classification problems have demonstrated its usefulness and versatility. Many solutions have arisen for the preprocessing of data in the feature and case dimensions. Specifically, many approaches have emerged to cope with class imbalance [2,3,4,5], that may be applied to gene classification problems, such as ortholog detection, since this branch of genomics has been deeply studied but even more accurate algorithms are still required because of the ortholog real value in the prediction of protein functions [6].

We aim at algorithms that compare two genomes to build gene groups but we classify pairs of genes in ortholog or non-ortholog pairs and we use curated reference classifications. The graph-based non-supervised solutions consulted [7,8,9,10,11,12] follow the gene pairwise comparison starting from the protein alignment to build bipartite graphs. They apply pruning processes to delete less similar relations in the graph and apply heurists or clustering to group orthologs.

Some algorithms [13,14] merge gene or protein alignment results with global genome rearrangements and synteny data to estimate the evolution distance, due to the fact that some proteins with sequence similarity under 20% show high similarity in

M. Kryszkiewicz et al. (Eds.): RSEISP 2014, LNAI 8537, pp. 161–168, 2014.
© Springer International Publishing Switzerland 2014

their function and structure. In this paper, we include a measure to compare the physi-cochemical properties of proteins by analyzing the periodicity of the contact energies of the amino acids in the primary structure of proteins. This analysis allows for the detection of similarity in the spectral dimension of sequences [15].

In order to compare gene pairs from the different perspectives mentioned above, we can combine gene pair local measures, into a global measure by means of aggrega-tion functions as the one we proposed in [16] to aggregate dissimilarity functions. In [17] we defined similarity measures (that we justify in this paper) and an aggregated gene pair global similarity function to underlie an extended gene pair similarity rela-tion for an ortholog detection information system. This relation and the selected Rough F-Measure validation measure are the bases of the contributions in this paper.

Here, we propose a procedure to select subsets of gene pair features with good classification results, having a fitness function that handle class imbalance. In order to build a model from a reference classification, we reduce cases considering the ex-tremely high class imbalance by selecting pairs of genes in the ortholog and non-ortholog positive regions. The new proposals are tested in a high dimension study case dataset of Saccharomyces Cerevisiae and Schizosaccharomyces Pombe genomes.

2   Local and Global Gene Pair Similarity Measures

Starting from the two genome representations being $G_1 = (X_1, X_2, ..., X_n)$ and $G_2 = (Y_1, Y_2, ..., Y_m)$, with n and m sequences respectively, in [17] we defined four normalized local similarity measures as features of the gene pair binary classification. The sequence alignment measure S_1 aggregates the local and global protein align-ment scores to combine their functional and structural relationship. Measure S_2 is obtained from the length of the sequences using the normalized difference [18] for continuous values. The similarity measure S_3 is calculated from the distance between pairs of sequences in regards to their membership to locally collinear blocks (LCBs). These blocks obtained with the Mauve software [19] are conserved truly homologous regions that do not seem to be altered by global genome rearrangements or inversions. The normalized difference is selected to compare the membership continuous values.

Based on the spectral representation of sequences from the global pairwise align-ment, measure S_4 uses the Linear Predictive Coding (LPC) [18]. In each matching region without "gaps", each amino acid is replaced by its contact energy, and the moving average of each spectrum is calculated with window size W. The similarity measure in a matching region is calculated with the Pearson correlation.

Each local function has continuous real values in [0..1] thus, the aggregated global si-milarity $S(x_i, y_j)$ between gene pairs can be calculated with Ruzicka, Roberts, Motyka, Bray-Curtis, Kulczynski 1 and 2, Baroni-Urbani-Buser, Covariance or Pearson correla-tion similarities in [18]. For now, we have selected the simplest average expression.

3   The Rough Set Framework

Considering the ortholog detection problem as a binary classification, we can define the information system [20] (U, A, Z) where $U = \{(x_i, y_j)\}, \forall x_i \in V_1, \forall y_j \in V_2$ is the universe of the gene pairs of the unpruned gene sets of both genomes $V_1 \subseteq G_1$ and $V_2 \subseteq G_2$. $A = \{S(x_i, y_j)\}$ is the attribute set defined as the values of the global aggregated gene similarity measure and $Z = \{Z^1, Z^0\}$, the set of concepts, Z^1 of ortholog pairs and Z^0 of non-ortholog pairs that can be defined from groups of a clustering process or from classes of a reference classification.

First, we define a similarity relation (1) between gene pairs of U from the global similarity measure S between genes. Then, we define an extended similarity relation R'_A (2) of a pair of U, by using a RST extension in [21], where ξ is a threshold value for the similarity between pairs of genes. It can be defined as the average of the maximum similarity values among pairs (3) following the approach in [22].

$$S_p\big((x_a, y_b), (x_c, y_d)\big) = \frac{\min(S(x_a,y_b),S(x_c,y_d))}{\max(S(x_a,y_b),S(x_c,y_d))} \tag{1}$$

$$R'_A((x_a, y_b)) = \big\{(x_c, y_d): (x_a = x_c \vee y_b = y_d) \wedge S_p\big((x_a, y_b),(x_c, y_d)\big) \geq \xi\big\} \tag{2}$$

$$\xi = \frac{1}{n}\Sigma_{i=1}^n \max_{\substack{j=1..n \\ i \neq j}}\big\{S_p\big(O_i, O_j\big)\big\}, \ O_i, \ O_j \in U, \ n = |U| \tag{3}$$

The approximation of a concept $Z^i \subseteq U$ is introduced from the extended inseparability relation by means of the R-lower (R'_*) (4) and R-upper (R'^*) (5) approximation sets. R'_* contains all the ortholog gene pairs related only with ortholog gene pairs while R'^* approximation contains all the gene pairs related with ortholog gene pairs.

$$R'_*(Z^i) = \{x \in Z^i : R'_A(x) \subseteq Z^i\} \tag{4}$$

$$R'^*(Z^i) = U_{x \in Z^i} R'_A(x) \tag{5}$$

4   Selection of Feature Subsets

In order to select combinations of features supposed to yield effective classification results, our fitness function is the Rough F-Measure (RFM) applied in [17] to take into account the imbalance of the dataset. The definition of this measure is based on the fact that both Quality and Precision of each concept approximation bring about a local assessment of each group [23], but due to imbalance, it is necessary to weight classes with an expression like the one in (6) to obtain measures in (7) and (8).

$$w(Z^i) = \frac{|U|-|Z^i|}{|U|} \tag{6}$$

$$\Gamma_G = \frac{\Sigma_{i=0}^1\big(|R'_*(Z^i)|\times w(Z^i)\big)}{|U|} \tag{7}$$

$$A_G = \frac{\Sigma_{i=0}^1\big(|R'_*(Z^i)|\times w(Z^i)\big)}{\Sigma_{i=0}^1\big(|R'^*(Z^i)|\times w(Z^i)\big)} \tag{8}$$

$$RFM = \frac{1}{\phi\times(1/\Gamma_G)+(1-\phi)\times(1/A_G)} \tag{9}$$

We evaluate the quality of classification with expression (9) where the value $\phi \in [0,1]$ is used to weight Precision or Quality depending on the problem to be solved. We decided to overweight precision with $\phi = 0.2$. The feature subset selection can be carried out with the steps below. This greedy procedure can be performed over datasets with few features and it is useful when similarity measures have few parameters with few discrete values as the window size W of the physicochemical profile. Parameter values can be combined into different models, for example, the substitution matrices and gap penalties.

```
maxRFM=0.
Repeat
    1 For each pair in U calculate S.
    2 Classify the pairs in U.
    3 Calculate R•A.
    4 Calculate R'*(Zⁱ) and R'*(Zⁱ) i=0, 1.
    5 Calculate RFM.
    If RFM >maxRFM then maxRFM=RFM
Until all combinations of features are tested.
```

The time complexity of step 1 is affected by the global alignment procedure in the calculation of the physicochemical profile. It is $O(k * z^2)$ where k is number of non-pruned gene pairs and z is the maximum length of the global alignment, which is N+M in the worst case, with N and M representing the maximum sequence length in genomes G_1 and G_2, respectively. The complexity of step 2 depends on the complexity of the classification algorithm, for example, $O(k^3)$. Steps 3 and 4 are $O(k^2)$. The overall procedure assumes the maximum complexity of steps 1 to 5 and it is to be multiplied by 2^t where t is the number of features. Dealing with space complexity, it is about $O(p * m * (N + M))$ where $p \in [1, n]$ is the total amount of alignments to be saved in a partial step of the calculation of the gene pair similarity measures.

5   Case Reduction Based on RST Considering Imbalance

Dealing with supervised classification, many solutions to manage imbalance can be found in literature. This problem refers to the fact that learning algorithms should learn to classify a minority class with fewer cases than the majority class. Traditional algorithms tend to produce high accuracy values for the majority class and low values for the minority one [3]. This author mentions some approaches from the data perspective: 1) under-sampling, in which the minority class is kept and the majority class is reduced by sampling; 2) over-sampling in which the minority class is over-sampled so that a desired distribution for each class can be obtained; 3) clustering-based sampling in which representative cases are randomly sampled from groups. Some disadvantages are also mentioned as the increased computational load and the overtraining due to replicated cases. Reduction by sampling does not include all the information in the learning dataset so it might lead to the missing of information. Reduction based on RST may include relevant information of the approximate sets [4]. A weighted approach is presented in [5] to improve the original RST approach.

We propose the use of rough sets for case reduction in the high dimension study case dataset with a reference classification. We calculate the lower approximation $R'_*(Z^0)$ of the majority class and apply the union to the lower approximation $R'_*(Z^1)$ of the minority class to build a reduced learning dataset for supervised algorithms. This reduction significantly reduce the data dimensionality and the extremely high imbalance of the training set while keeping the gene pairs that can be certainly classified as orthologs or non-orthologs.

For the selection of the evaluation metrics for imbalanced datasets we follow the statements and equations in [24] and [25] where the combination of Precision, Recall, F-Measure with global G-Mean measures (10) and (11) is recommended. Precision is sensitive to data distributions, while Recall is not. On the other hand, Recall provides no insight to how many examples are incorrectly labeled as positive. Similarly, Precision cannot assert how many positive examples are labeled incorrectly. Nevertheless, if we use measures based on both Precision and Recall as F-Measure and G-Mean1 (10) we can measure the effectiveness of classification remaining sensitive to data distributions. The G-Mean2 metric (11), evaluates the degree of inductive bias in terms of a ratio of positive accuracy and negative accuracy. However, F-Measure and G-Mean 1 and 2 are still ineffective in answering more questions related to the performance of classifiers over a range sample distributions. The ROC curve and the AUC measure are intended to overcome those limitations.

In the case of highly skewed data sets, it is observed that the ROC curve may provide an overly optimistic view of an algorithm's performance. It cannot capture the effect of a large change in the number of false positives (FP), since this change will not significantly change the FP_rate given that the number of negative examples is very large. Under such situations, the Precision-Recall curve (PRC) can provide a more informative representation of performance assessments since it is defined by plotting Precision rate over the Recall rate and the Precision metric considers the ratio of true positives (TP) with respect to TP+FP.

$$G - \text{Mean1} = \sqrt{\text{Precision} * \text{Recall}} \qquad (10)$$

$$G - \text{Mean2} = \sqrt{\text{Recall}_0 * \text{Recall}_1} \qquad (11)$$

6   The Study Case

The genome comparison dataset was constructed from 29340166 gene pairs of Saccharomyces Cerevisiae and Schizosaccharomyces Pombe genomes in NCBI. The classifications of gene pairs are drawn from the additional files in [26]. The imbalance ratios in each classification are both exceedingly high: for Inparanoid7.0 is 29335077/5089 and for GeneDB is 29336337/3829.

The comparison started from the global alignment of each pair of proteins. We defined four alignment models with recommended parameter values. Table 1 shows the identification of the models and the corresponding parameter values.

Table 1. Alignment models

Model id	Substitution matrix	Gap open	Gap extended
B50	BLOSUM50	15	8
B62_1	BLOSUM62	8	7
B62_2	BLOSUM62	12	6
P250	PAM250	10	8

Some previous experiments we presented in [17] with feature subsets and combination of parameter values have shown that the highest values of RFM are obtained with combinations of features with the physicochemical profile, emphasizing its influence in a non-supervised approach with some pruning policies and the Markov clustering followed by an assignment policy. The alignment feature, the most common feature in ortholog detection algorithms, is also included in combinations that yield the highest results. Table 2 shows the highest RFM values obtained with window size W=3. The results of combinations S_1, S_4 and S_1, S_3, S_4 for the B50 alignment model are highlighted since they slightly improve the RFM values.

For the case reduction experiments, we aggregate features S_1, S_2, S_3 and S_4 with W=3 in order to keep all the information. Starting from the original dataset of the B50 alignment model, the pairs with less than 40% of similarity in the global alignment were pruned taking into account the homology concept that two sequences are homologous if they are highly similar. This process built the DS dataset with 8092274/3633 imbalance ratio having the intersection of Inparanoid7.0 and GeneDB classifications. The dataset was split into 75% for training and 25% for testing. We use four classifiers of Weka software to show the imbalance effect and the possible improvement to be achieved with the case reduction based on RST.

While experimenting with DS dataset, we focused on the results of the minority class. The J48 (C4.5) classifier yielded the best results for the global measures, highlighted in bold in Table 3, mainly for the best result in the Recall$_1$ with a very limited deterioration in Precision. As shown in Table 4, the reduction policy improved the Recall$_1$, and hence the G-Mean2 of the four classifiers. Precision, however, was significantly reduced due to the increase in the false positives. The best results were obtained with RepTree and with RandomForest when we focused on Precision. Further research would be needed on the reduction policy to improve the results.

Table 2. Highest RFM values for different combinations of features using W=3 in S_4

Aggregated features	B50	B62_1	B62_2	P250
S_1, S_4	**0.816**	0.805	0.813	0.770
S_1, S_2, S_4	0.807	0.805	0.807	0.775
S_1, S_3, S_4	**0.817**	0.807	0.814	0.777
S_1, S_2, S_3, S_4	0.805	0.800	0.802	0.772

Table 3. Imbalance effect in four Weka classifiers in the DS dataset

Classifiers	Precision	Recall	F-Measure	PRC	G-Mean1	G-Mean2
J48	0.724	0.597	**0.655**	**0.682**	**0.657**	**0.773**
RandomForest	0.770	0.529	0.627	0.662	0.638	0.727
Logistic	0.759	0.469	0.579	0.578	0.597	0.685
RepTree	0.737	0.553	0.632	0.668	0.638	0.744

Table 4. The effect of case reduction for four Weka classifiers

Classifiers	Precision	Recall	F-Measure	PRC	G-Mean1	G-Mean2
J48	0.395	0.789	0.527	0.345	0.558	0.888
RandomForest	0.516	0.729	0.604	0.527	0.613	0.854
Logistic	0.262	0.656	0.375	0.502	0.415	0.810
RepTree	0.480	0.792	**0.598**	**0.579**	**0.617**	**0.890**

7   Conclusions

The proposed procedure to select subsets of adequate features can be useful for data-sets with few features and discrete parameters. In the study case, some experiments with a non-supervised algorithm have shown that the physicochemical profile feature, with window size 3, brings about valuable information for classification as well as the alignment feature. The highest results of the RFM measure were obtained for the B50 alignment model by aggregating the alignment and physicochemical profile features and by aggregating both of them to the membership of gene pairs to LCBs.

Other experiments with four Weka classifiers have shown that the case reduction based on RST, useful for high dimension datasets, should be improved to obtain better classification results, although the highest results were obtained with RepTree and with RandomForest when we try to diminish the false positives. Some other classifiers and reduction policies may be used in future experiments.

References

1. Pawlak, Z.: Rough sets. International Journal of Computer and Information Sciences 11(5), 341–356 (1982)
2. Liu, J., Hu, Q., Yu, D.: A comparative study on rough set based class imbalance learning. Knowledge-Based Systems 21, 753–763 (2008)
3. Chen, M.-C., et al.: An information granulation based data mining approach for classifying imbalanced data. Information Sciences 178, 3214–3227 (2008)
4. Stefanowski, J., Wilk, S.: Combining rough sets and rule based classifiers for handling imbalanced data. Fundamenta Informaticae (2006)
5. Liu, J., Hu, Q., Yu, D.: A weighted rough set based method developed for class imbalance learning. Information Sciences 178, 1235–1256 (2008)
6. Salichos, L., Rokas, A.: Evaluating Ortholog Prediction Algorithms in a Yeast Model Clade. PLoS ONE 6(4), 1–11 (2011)

7. Östlund, G., Schmitt, T., Forslund, K., Köstler, T., Messina, D.N., Frings, O., Sonnhammer, E.L.L., Roopra, S.: InParanoid 7: new algorithms and tools for eukaryotic orthology analysis. Nucleic Acids Research (2010)
8. Linard, B., et al.: OrthoInspector: comprehensive orthology analysis and visual exploration. BMC Bioinformatics 12(11), 1471–2105 (2011)
9. Muller, J., et al.: eggNOG v2.0: extending the evolutionary genealogy of genes with enhanced non-supervised orthologous groups, species and functional annotations. Nucleic Acids Res. 38, D190–D195 (2010)
10. Dessimoz, C., Cannarozzi, G.M., Gil, M., Margadant, D., Roth, A., Schneider, A., Gonnet, G.H.: OMA, A comprehensive, automated project for the identification of orthologs from complete genome data: Introduction and first achievements. In: McLysaght, A., Huson, D.H. (eds.) RECOMB 2005. LNCS (LNBI), vol. 3678, pp. 61–72. Springer, Heidelberg (2005)
11. Li, L., Stoeckert, C.J., Roos, D.S.: OrthoMCL: Identification of Ortholog Groups for Eukaryotic Genomes. Genome Research 13, 2178–2189 (2003)
12. Deluca, T.F., et al.: Roundup: a multi-genome repository of orthologs and evolutionary distances. Bioinformatics 22, 2044–2046 (2006)
13. Kamvysselis, M.K.: Computational comparative genomics: genes, regulation, evolution. In: Electrical Engineering and Computer Science, p. 100, Massachusetts Institute of Technology, Massachusetts (2003)
14. Fu, Z., et al.: MSOAR: A High-Throughput Ortholog Assignment System Based on Genome Rearrangement. Journal of Computational Biology 14, 16 (2007)
15. del Carpio-Muñoz, C.A., Carbajal, J.C.: Folding Pattern Recognition in Proteins Using Spectral Analysis Methods. Genome Informatics 13, 163–172 (2002)
16. Galpert, D.: A local-global gene comparison for ortholog detection in two closely related eukaryotes species. Investigación de Operaciones 33(2), 130–140 (2012)
17. Millo, R., et al.: Agregación de medidas de similitud para la detección de ortólogos, validación con medidas basadas en la teoría de conjuntos aproximados. Computación y Sistemas 18(1) (2014)
18. Deza, E.: Dictionary of Distances. Elsevier (2006)
19. Darling, A.E., Mau, B., Perna, N.T.: progressiveMauve: Multiple Genome Alignment with Gene Gain, Loss and Rearrangement. PLOS One 5(6) (2010)
20. Komorowski, J., Pawlak, Z., Polkowski, L.: Rough sets: a tutorial. In: Pal, S.K., Skowron, A. (eds.) Rough-Fuzzy Hybridization: A New Trend in Decision Making. Springer, Singapore (1999)
21. Slowinski, R., Vanderpooten, D.: Similarity relation as a basis for rough approximations. In: Wang, P.P. (ed.) Advances in Machine Intelligence & Soft-Computing, pp. 17–33 (1997)
22. Shulcloper, J.R., Arenas, A.G., Trinidad, J.F.M.: Enfoque lógico combinatorio al reconocimiento de patrones: Selección de variables y clasificación supervisada. Instituto Politécnico Nacional (1995)
23. Pawlak, Z.: Vagueness and uncertainty: a rough set perspective. Computational Intelligence: an International Journal 11, 227–232 (1995)
24. Kubat, M., Matwin, S.: Addressing the curse of imbalanced data sets: One-sided sampling. In: 14th International Conference on Machine Learning (1997)
25. He, H., Garcia, E.A.: Learning from Imbalanced Data. IEEE Transactions on Knowledge and Data Engineering 21(9), 1263–1284 (2009)
26. Koch, E.N., et al.: Conserved rules govern genetic interaction degree across species. Genome Biology 13(7) (2012)

Hybrid Model Based on Rough Sets Theory and Fuzzy Cognitive Maps for Decision-Making

Gonzalo Nápoles[1,2], Isel Grau[1], Koen Vanhoof[2], and Rafael Bello[1]

[1] Universidad Central "Marta Abreu" de Las Villas, Santa Clara, Cuba
[2] Hasselt University, Diepenbeek, Belgium
gnapoles@uclv.edu.cu

Abstract. Decision-making could be defined as the process to choose a suitable decision among a set of possible alternatives in a given activity. It is a relevant subject in numerous disciplines such as engineering, psychology, risk analysis, operations research, etc. However, most real-life problems are unstructured in nature, often involving vagueness and uncertainty features. It makes difficult to apply exact models, being necessary to adopt approximate algorithms based on Artificial Intelligence and Soft Computing techniques. In this paper we present a novel decision-making model called Rough Cognitive Networks. It combines the capability of Rough Sets Theory for handling inconsistent patterns, with the modeling and simulation features of Fuzzy Cognitive Maps. Towards the end, we obtain an accurate hybrid model that allows to solve non-trivial continuous, discrete, or mixed-variable decision-making problems.

Keywords: Decision-making, Rough Set Theory, Fuzzy Cognitive Maps.

1 Introduction

In recent years decision-making problems have become an active research area due to their impacts in solving real-world problems. Concisely speaking, decision-making process could be defined as the task of determining and selecting the most adequate action that allows solving a specific problem. This task is supported by the knowledge concerning the problem domain allowing justifying the selected decision. However, the knowledge obtained from experts regularly shows inconsistent patterns that could affect the inference results (e.g. different perception for the same observation).

The Rough Set Theory (RST) is a well-defined technique for handling uncertainty arising from inconsistency [1]. This theory adopts two approximations to describe a set, which are entirely based on the collected data [2] and does not require any further knowledge. Let us assume a decision system $S = (U, A \cup \{d\})$, where U is a non-empty finite set of objects (the universe of discourse), A is a non-empty finite set of attributes, whereas $d \notin A$ is the decision class. Any subset $X \subseteq U$ can be approximated by using two exact sets $B_*X = \{x \in U : [x]_B \subseteq X\}$ and $B^*X = \{x \in U : [x]_B \cap X \neq \emptyset\}$ called lower and upper approximation respectively. In this formulation $[x]_B$ denotes the set of inseparable objects associated to the instance x (equivalence class) using an indiscernibility relation defined by a subset of attributes $B \subseteq A$. The reader may notice that this indiscernibility relation is reflexive, transitive and symmetric.

M. Kryszkiewicz et al. (Eds.): RSEISP 2014, LNAI 8537, pp. 169–178, 2014.

The objects in B_*X are categorically members of X, whereas the objects in B^*X are possibly members of the set X. Notice that this model does not consider any tolerance of errors [3]: if two inseparable objects belong to different classes then the decision system will be inconsistent. The lower and upper approximations divide the universe into three pair-wise disjoint regions: the lower approximation as the positive region, the complement of the upper approximation as the negative region, and the difference between the upper and lower approximations as the boundary region [4]. Being more precise, objects belonging to the positive region $POS(X) = B_*X$ are certainly contained in X, objects that belong to the negative region $NEG(X) = U - B^*X$ are not confidently contained in the set X, whereas the boundary region $BND(X) = B^*X - B_*X$ represents uncertainty about the membership of related objects to the set X.

Such knowledge comprises an opposite knowledge when facing decision-making problems. For example, in reference [5] the author introduces the three-way decisions model. Rules constructed from the three regions are associated to different actions and decisions (see following equations). A positive rule makes a decision of acceptance, a negative rule makes a decision of rejection, and a boundary rule makes a decision of abstaining [4]. Observe that the model interpretation is not so critical in the classical rough set model since it does not involve any uncertainty. In an attempt to overcome this drawback, Wong and Ziarko [6] considered a probabilistic relationship between equivalence classes and X leading to the probabilistic three-way decisions. An object in the probabilistic positive region does not certainly belong to the decision class, but with a high probability. It is important in the probabilistic model, where acceptance and rejection are made with certain levels of tolerance for errors.

- $Des([x]) \rightarrow_P Des(d)$, for $[x] \subseteq POS(d)$
- $Des([x]) \rightarrow_B Des(d)$, for $[x] \subseteq BND(d)$
- $Des([x]) \rightarrow_N Des(d)$, for $[x] \subseteq NEG(d)$

The probabilistic three-way decisions showed to be superior with respect to the original algorithm [7], however, such models are mainly oriented to discrete decision-making problems. In this paper we present a novel hybrid model that combines three-way decisions rules, with the simulation aptitude of Fuzzy Cognitive Maps [8] using a sigmoid threshold function. This hybrid model not just allows to solve mixed-attribute or continuous problems, but also provides accurate inferences. The main idea consists in replacing the equivalence classes by similarity classes to define positive, negative, and boundaries regions. After that, we build a Sigmoid Fuzzy Cognitive Maps (which are a kind of recurrent neural network for modeling and simulation) using computed regions and the domain knowledge. Finally, a recurrent inference process is triggered allowing to the map to converge to a desired decision.

The rest of the paper is organized as follows: in following Section 2 the theoretical background of FCM is described. Here we point out some aspects concerning the map inference process using continuous threshold functions. In Section 3 we introduce the proposed hybrid model consisting in three main steps: (i) the computation of positive, negative and boundary regions, (ii) the construction of the map topology, and (iii) the map exploitation using the similarity class of the target instance. Section 4 provides numerical simulations illustrating the behavior of our algorithm. Finally, conclusions and further research aspects are discussed in Section 5.

2   Fuzzy Cognitive Maps

Fuzzy Cognitive Maps (FCM) are a suitable knowledge-based tool for modeling and simulation [9]. From a connectionist perspective, FCM are recurrent networks with learning capabilities, consisting of nodes and weighted arcs. Nodes are equivalent to neurons in connectionist models and represent variables, entities or objects; whereas weights associated to connections denote the *causality* among such nodes. Each link takes values in the range $[-1,1]$, denoting the causality degree between two concepts as a result of the quantification of a fuzzy linguistic variable, which is often assigned by experts during the modeling phase [10]. The activation value of neurons is also fuzzy in nature and regularly takes values in the range $[0,1]$. Therefore, the higher the activation value of a neuron, the stronger its influence over the investigated system, offering to decision-makers an overall picture of the systems behavior.

Without loss of generality, a FCM can be defined using a 4-tuple (C, W, A, f) where $C = \{C_1, C_2, C_3, ..., C_M\}$ is a set of M neurons, $W: (C_i, C_l) \rightarrow w_{il}$ is a function which associates a causal value $w_{il} \in [-1,1]$ to each pair of nodes (C_i, C_l), denoting the weight of the directed edge from C_i to C_l. The weigh matrix $W_{M \times M}$ gathers the system causality which is often determined by experts, although may be computed using a learning algorithm. Similarly, $A: (C_i) \rightarrow A_i$ is a function that associates the activation degree $A_i \in \mathbb{R}$ to each concept C_i at the moment t $(t = 1, 2, ..., T)$. Finally, a transformation function $f: \mathbb{R} \rightarrow [0,1]$ is used to keep the neuron's activation value in the interval $[0,1]$. Following Equation (1) portrays the inference mechanism using the vector A^0 as the initial configuration. This inference stage is iteratively repeated until a hidden pattern or a maximum number of iterations T is reached.

$$A_i^{t+1} = f\left(\phi_1 \sum_{j=1}^{M} w_{ji} A_j^t + \phi_2 w_{ii} A_i^t\right), i \neq j \qquad (1)$$

In the above equation ϕ_1 represents the influence from the interconnected concepts in the configuration of the new activation value, whereas ϕ_2 regulates the contribution of the neuron memory over its own state. In all experiments conducted in this paper we use $\phi_1 = 0.95$ and $\phi_2 = 1 - \phi_1$ since the new evidence is often desirable.

The most used threshold functions are: the bivalent function, the trivalent function, and the sigmoid variants. It should be stated that authors will be focused on Sigmoid FCM, instead of discrete ones. It is motivated by the benchmarking analysis discussed in reference [11] where results revealed that the sigmoid function outperformed the other functions by the same decision model. Therefore, the proper selection of this threshold function may be crucial for the system behavior. From [12] some important observations were concluded and summarized as follows:

- Binary and trivalent FCM cannot represent the degree of an increase or a decrease of a concept. Such discrete maps always converge to a fixed-point attractor or limit cycle since FCM are deterministic models.
- Sigmoid FCM, by allowing neuron's activation level, can also represent the neuron's activation degree. They are suitable for qualitative and quantitative tasks, however, may additionally show chaotic behaviors.

3   Rough Cognitive Networks

In this section we introduce a hybrid model for addressing decision-making problems called Rough Cognitive Networks (RCN). It combines the ability of RST for handling uncertainty and the simulation strength of FCM. The aim of this model is the mapping of an input vector to a feasible decision, using the knowledge obtained from historical data. Let us consider a set of decisions $D = \{d_1, ..., d_k, ..., d_n\}$ for some decision task, a decision system $DS = (U, A, \{d\})$ where problem attributes are mainly continuous, and an unlabeled problem instance O_i. Next steps describe how to design a RCN which is capable of computing the most fitting decision for the new instance O_i by ranking the activation value of decision concepts (i.e. map neurons).

3.1   Determining Positive, Negative and Boundary Regions

The first step of our proposal is oriented to determine positive, negative and boundary regions. It should be stated that we need to use weaker inseparability relations among objects in the universe U, since we assume decision-making problems with continuous or mixed variables. It could be achieved by extending the concept of inseparability, so that they are grouped together in the same class of not identical objects, according to a similarity relationship R. Hence, by replacing the equivalence relation with a weaker binary relation, an extension of the classical RST approach is achieved [2].

Equations (2) and (3) summarize how to compute lower and upper approximations respectively by using this scheme, where $R'(x)$ denotes the similarity class associated to the object x (that is, the set of objects which are similar to the instance x according to some similarity relation R). It suggests that an object can simultaneously belong to different similarity classes, so the covering induced by $R'(x)$ over U is not necessarily a partition [13]. Being more explicit, similarity relations do not provoke a partition of the universe U, but rather generate classes of similarity.

$$B_*X = \{x \in U : R'(x) \subseteq X\} \tag{2}$$

$$B^*X = \bigcup_{x \in X} R'(x) \tag{3}$$

While constructing an equivalence relation is trivial, constructing the apt similarity relation for a problem could be more complex. What is more, the global performance of our model will be reliant on the quality of such similarity relation, however, in the literature several methods for facing this challenge have been proposed. For example, Filiberto et al. [14] describe an optimization procedure for building accurate similarity relations using a population-based metaheuristic.

Another aspect to be considered when designing a similarity relation is the opposite selection of the similarity (or distance) function. It is used for measuring the similarity (or difference) degree between two objects. In reference [15] the authors widely study the properties of several distance functions, where problem attributes are grouped into two large groups: continuous or discrete (both nominal and ordinal). This subject will be detailed in the Section 5, during numerical simulations.

3.2   Designing the Map Topology

During this step, positive, negative and boundary regions are denoted as map neurons denoting input variables of the system. Following a similar reasoning, we use $|D|$ new concepts for measuring the activation degree of each decision. It should be mentioned that a boundaries region will be considered or not (this point will be clarified in next sub-sections). Afterwards, once concepts have been defined, we establish connections among all map neurons where causal values are computed as follows:

- R_1: if C_i is P_k and C_l is d_k then $w_{il} = 1.0$
- R_2: if C_i is P_k and C_l is $d_{v \neq k}$ then $w_{il} = -1.0$
- R_3: if C_i is P_k and C_l is $P_{v \neq k}$ then $w_{il} = -1.0$
- R_4: if C_i is N_k and C_l is d_k then $w_{il} = -1.0$

In rules detailed above C_i and C_l denote two neurons, P_k and N_k are the positive and negative region related to the kth decision respectively, whereas w_{il} denote the causal weight between the cause C_i and the effect C_l. If the positive region P_k is activated, then the FCM stimulates the kth decision node since we confidently know that objects belonging to the positive region will be categorically members of X_k. On the contrary, if the negative region N_k is activated, then the map will inhibit the corresponding decision, but we cannot conclude about other decisions.

Boundary regions usually report uncertain information about the acceptance of the investigated decision, however, an unlabeled object $O_i \in BND(X_k)$ could be correctly associated to decision d_k. Being more specific, let us suppose a problem having three decisions where $O_i \in BND(X_1)$, $O_i \in BND(X_2)$ and $O_i \notin BND(X_3)$. It means that the instance could be labeled as d_1 or d_2 as well, but it provides no evidence supporting decision class d_3. Encouraged by this remark we include another rule where further knowledge about boundary regions is considered.

- R_5: if C_i is B_k and C_l is d_v and $(BND(X_k) \cap BND(X_v) \neq \emptyset)$ then $w_{il} = 0.5$

Observe that rules $R_1 - R_4$ are independent of the upper and lower approximations, while the rule R_5 requires information about boundary regions. Hence, the final map could have at most $4|D|$ neurons and $3|D|(1 + |D|)$ causal links since the number of boundary concepts will depend on the upper and lower approximations.

3.3   Inferring the Most Fitting Decision

The final phase of this hybrid model is related to the FCM exploitation, so we need to compute the activation value of input concepts (neurons that denote positive, negative and boundary regions). Being more explicit, the excitation vector will be calculated using objects belonging to the similarity class $R'(O_i)$ and their relation to each region. For example, let us suppose that $|POS(X_1)| = 20$, $|R'(O_i)| = 10$, whereas the number of similar objects that belong to the positive region is given by the following expression: $|R'(O_i) \cap POS(X_k)| = 7$. Then the activation degree of the positive neuron associated to the first decision will be $7/20 = 0.35$. Following three rules generalize this scheme for all map concepts (regions), thus complementing the proposal.

- R_6: $if C_i$ is P_k then $A_i^0 = \frac{|Y|}{|POS(X_k)|}$ where $Y = R'(O_i) \cap POS(X_k)$
- R_7: $if C_i$ is N_k then $A_i^0 = \frac{|Y|}{|NEG(X_k)|}$ where $Y = R'(O_i) \cap NEG(X_k)$
- R_8: $if C_i$ is B_k then $A_i^0 = \frac{|Y|}{|BND(X_k)|}$ where $Y = R'(O_i) \cap BND(X_k)$

Figure 1 illustrates the FCM resulting from a decision-making problem having two decisions, assuming inconsistencies on the information system. Notice that each input neuron has a self-reinforcement connection with causal weight $w_{ii} = 1$ which partially preserves the initial knowledge during the simulation (exploitation) stage.

The main reason for adopting FCM as inference mechanism is that they can handle incomplete or conflicting information. Besides, decision-making problems are usually characterized by various concepts or facts interrelated in complex ways, so the system feedback plays a prominent role by efficiently propagating causal influences in non-trivial pathways. Formulating a precise mathematical model for such systems may be difficult or even impossible due to lack of numerical data, its unstructured nature, and dependence on imprecise verbal expressions. Observe that the performance of FCM is dependent on the initial weight setting and architecture [16], but our model provides a general framework for facing problems having different features.

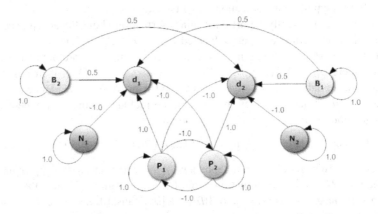

Fig. 1. Resultant FCM model for decision-making problems for two decisions

It should be stated that our algorithm only requires the estimation of a similarity threshold, which is used for building lower and upper approximations. It means that we need to build precise similarity relations to ensure high-quality results. Being more explicit, if this similarity value is excessively small then positive regions will be small as well, leading to poor excitation of related neurons. It could be essential to select the most fitting decision when a new scenario is observed: the more active the positive region, the more desirable the decision (although the algorithm will compute the final decision taking into account all the evidence). If the similarity threshold is too large then boundary regions will be also large, increasing the global uncertainty during the model inference phase, decreasing the overall algorithm performance.

4   Numerical Simulations

In the present section we study the behavior of the proposed RCN approach by using both synthetic and real-life data. In all simulations we adopt a Sigmoid RCN resulting in a FCM having a sigmoid threshold function. The sigmoid function uses a constant parameter $\lambda > 0$ to adjust its inclination. In this paper we use $\lambda = 1$, because this value showed best results in previous studies [16]. During synthetic simulations a decision-making problem having three outcomes d_1, d_2 or d_3 is assumed. With this purpose in mind we evaluate the system response for the following scenarios:

   i.   the set $R'(O_i)$ activates a single positive region
  ii.   the set $R'(O_i)$ activates two positive regions
 iii.   the set $R'(O_i)$ activates boundary regions

To reach the first case we use the excitation values $P_1 = 0.23$ and $N_2 = N_3 = 0.12$ leading to the output $(1,0,0)$ which is the desired solution. It should be stated that we know the expert preference for each target object O_i in advance, but it is only used for measuring the algorithm performance. The second scenario could be more challenging and clarifies the real contribution of our algorithm. For example, the excitation values $P_1 = 0.045$, $P_2 = 0.044$, $N_1 = 0.0136$, $N_2 = 0.0138$, $N_3 = 0.4$, $B_1 = B_2 = 0.685$ has the output vector $(1.0,0.68,0)$ which is a right state. Observe that activation values of positive regions P_1 and P_2 are quite similar, however, the overall evidence against decision d_2 suggests accepting d_1. It could be also possible that two dominant positive regions have the same activation value being difficult to take a decision. In such cases the map computes the decision by using negative and boundary regions.

The last scenario takes place when multiple boundary regions are noted at the same time, but no positive region is activated (e.g. $N_1 = 0.4$ and $B_2 = B_3 = 0.6$). This initial state leads to the output $(0,0.54,0.54)$ where choices d_2 and d_3 are equally adequate, but it definitely suggest rejecting d_1. In such cases the decision-maker could adjust the similarity threshold with the goal of reducing the cardinality of boundary regions, and so reducing the overall uncertainty over the system. But if no change is observed then both decisions should be equally considered by experts. Following figure 2 illustrates the network behavior for all synthetic scenarios discussed above, where the activation degree of each decision neuron (over the time) is plotted.

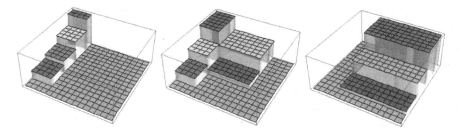

Fig. 2. Activation value of decision neurons for different scenarios. i) a single positive region is activated, (ii) two positive regions are activated, (iii) only boundaries regions are activated.

4.1   Decision-Making in Travel Behavior Problems

The transport management appears in all modern societies due to the cost that implies and because of the importance for social and economic process in a country. In such situations users intuitively select the most convenient transport mode mainly based on their expertise about a wide range of context variables such as: temperature, saving money, bus frequency, precipitation, physical comfort, etc.

Recently, M. León et al. [17] proposed a modeling based on Cognitive Mapping theory to characterize mental representations of users. Moreover, a learning algorithm based on Swarm Intelligence for tuning the system causality was developed, allowing the simulation of observed patterns. It facilitates the formulation of new strategies that efficiently manage existing resources according to the customer's preferences. Being more explicitly, each variable is evaluated by experts and then the best transport mode is selected (i.e. bus, car or bike). In order to evaluate the prediction capability of our model, we compare the accuracy on predicting the most adequate choice against other well-known algorithms such as: Multilayer Perceptron (MLP), Decision Trees (DT), Bayesian Networks (BN), and a Fuzzy Cognitive Map (FCM).

Before presenting experimental results a similarity relation to determine when two objects belong to the same similarity class is required. Following equation shows the symmetric relation R used in this work, where x and y are two objects of the universe of discourse, A is the set of attributes describing the task, m_i and M_i are the lower and upper values for the ith variable respectively, whereas the factor $0 < \varepsilon \leq 1$ represents the similarity threshold. Notice that, if $\varepsilon = 1$ then the relation R is reflexive, transitive and symmetric leading to the Pawlak's model for discrete problems. It is also possible to incorporate a weight for each attribute denoting the relevance degree of each variable which provides accurate approximations. However, in all simulations performed next, we assume that attributes variables have the same importance.

$$R: xRy \iff \frac{1}{|A|} \sum_{i=1}^{|A|} \left(1 - \frac{|x_i - y_i|}{M_i - m_i} \right) \geq \varepsilon \tag{4}$$

Table 1 shows the averaged prediction accuracy of RCN against four models taken from references [18]. It involves the accuracy for both optimistic and pessimistic trials over 220 knowledge bases concerning travel behavior problems. Toward this goal two studies were designed. In the first study (E1) stored scenarios serve either for training and validation (empirical error). The optimistic test is necessary because it reflects the self-consistency of the method, a prediction algorithm certainly cannot be deemed as a good one if its self-consistency is poor. In the second study (E2) testing cases were used to obtain a pessimist estimation (real error) using a cross- validation process with 10-folds. In such experiments "predicting the most important decision" means to find the transport mode having better expected utility, while "predicting the correct order in decisions" means to establish a proper ranking among decisions.

In case of the RCN model the similarity threshold ε is fixed to 0.9. This process is performed by "trial and error" although we could use a learning method as was stated in the previous section. However, in the present paper the authors prefer to be focused on the methodology to deal with decision-making problems.

Table 1. Prediction accuracy achieved by selected algorithms for both studies

	Predicting the most important decision					Predicting the correct order in decisions				
Study	MLP	BN	DT	FCM	RCN	MLP	BN	DT	FCM	RCN
E1	97.38	95.63	94.26	99.47	99.50	94.38	93.12	87.29	96.27	95.11
E2	92.06	91.37	89.39	93.74	93.29	82.40	80.25	77.59	88.72	90.32
Av	94.72	93.50	91.82	96.60	95.89	88.39	86.68	82.44	92.45	92.71

From Table 1 we can conclude that the RCN model is an appropriate alternative for addressing decision-making tasks having continuous (or mixed) features. It showed good prediction ability, outperforming traditional approaches such as MLP, BN or DT which have proved to be very competent classifiers. What is more, for this study case, our network performs comparable regarding the FCM model. It should remarked that the FCM topology introduced in [17] is problem-dependent, designed for addressing decision-making problems concerning public transportation issues, hence it cannot be generalized to other application domains. On the contrary, the RCN topology does not require specific information about the decision problem.

5   Conclusions

Decision-making problems have become an interesting research direction mainly due to their complexity and practical applications. In such problems experts express their preferences about multiple feasible alternatives according to the problem attributes or descriptors; the decision-making method must to derive a solution.

A relevant issue in decision-making tasks is related to the knowledge quality, since experts frequently have different perceptions for the same situation. To deal with this drawback some RST-based algorithms have been developed (e.g. three-way decision rules). This paper proposed a novel approach which combines the capability of Rough Sets for handling inconsistent patterns, with the simulation features of Sigmoid Fuzzy Cognitive Maps. It results in a new hybrid model, called Rough Cognitive Networks, which allows to solve difficult decision-making problems having discrete, continuous or mixed attributes. This model comprises three stages: (i) the estimation of positive, negative and boundary regions, (ii) the construction of the map topology, and (iii) the exploitation phase using the similarity class of the target instance.

During numerical simulations we observed that RCN are capable of computing the expected decision in scenarios where positive or boundary regions have similar (even identical) activation value. The main reason behind this positive result arises from the map inference process, which iteratively accentuates a pattern. The sigmoid threshold function also plays an important role in the simulation phase because it quantifies the activation degree of decision neurons. Similarly, using all computed regions allows to select the appropriate output from decisions having similar activation.

In a second study we compare the performance of our algorithm against traditional classifiers by using a real-life study case. From this study we noticed that RCM have high prediction capability, even without using specific knowledge about the decision problem. The future work will be focused on exploring the accuracy of such networks when facing more challenging classification problems.

References

1. Pawlak, Z.: Rough sets. Int. J. of Information and Computer Sciences 11, 341–356 (1982)
2. Bello, R., Verdegay, J.: Rough sets in the Soft Computing environment. Information Science 212, 1–14 (2012)
3. Pérez, R.B., Garcia, M.M.: Probabilistic approaches to the Rough Set Theory and their applications in decisionmaking. In: Espin, R., Pérez, R.B., Cobo, A., Marx, J., Valdés Olmos, R.A. (eds.) Soft Computing for Business Intelligence. SCI, vol. 537, pp. 67–80. Springer, Heidelberg (2014)
4. Yao, Y.: Three-way decisions with probabilistic rough sets. Information Science 180, 341–353 (2010)
5. Yao, Y.: Three-way decision: An interpretation of rules in rough set theory. In: Wen, P., Li, Y., Polkowski, L., Yao, Y., Tsumoto, S., Wang, G. (eds.) RSKT 2009. LNCS (LNAI), vol. 5589, pp. 642–649. Springer, Heidelberg (2009)
6. Wong, S., Ziarko, W.: Algorithm for inductive learning. Bulletin of the Polish Academy of Sciences Technical Sciences 34, 271–276 (1986)
7. Yao, Y.: The superiority of three-way decisions in probabilistic rough set models. Information Science 180, 1080–1096 (2011)
8. Kosko, B.: Fuzzy Cognitive Maps. Int. Journal of Man-Machine Studies 24, 65–75 (1986)
9. Kosko, B.: Hidden patterns in combined and adaptive knowledge networks. International Journal of Approximate Reasoning 2, 377–393 (1988)
10. Kosko, B.: Fuzzy Engineering. Prentice-Hall Inc., New York (1997)
11. Bueno, S., Salmeron, J.L.: Benchmarking main activation functions in Fuzzy cognitive maps. Expert Syst. Appl. 36, 5221–5229 (2009)
12. Tsadiras, A.K.: Comparing the inference capabilities of binary, trivalent and sigmoid fuzzy cognitive maps. Information Science 178, 3880–3894 (2008)
13. Slowinski, R., Vanderpooten, D.: A generalized definition of rough approximations based on similarity. IEEE Transactions on Data and Knowledge Engineering 12, 331–336 (2000)
14. Filiberto, Y., Bello, R., et al.: A method to build similarity relations into extended rough set theory. In: Proceedings of the 10th International Conference on Intelligent Systems Design and Applications, ISDA 2010, pp. 1314–1319. IEEE (2010)
15. Wilson, D.R., Martínez, T.R.: Improved heterogeneous distance functions. Journal of Artificial Intelligence Research 6, 1–34 (1997)
16. Miao, Y., Liu, Z.Q.: On causal inference in fuzzy cognitive maps. IEEE Transactions on Fuzzy Systems 8, 107–119 (2000)
17. León, M., Nápoles, G., García, M.M., Bello, R., Vanhoof, K.: Two Steps Individuals Travel Behavior Modeling through Fuzzy Cognitive Maps Pre-definition and Learning. In: Batyrshin, I., Sidorov, G. (eds.) MICAI 2011, Part II. LNCS (LNAI), vol. 7095, pp. 82–94. Springer, Heidelberg (2011)
18. León, M., Nápoles, G., Bello, R., Mkrtchyan, L., Depaire, B., Vanhoof, K.: Tackling Travel Behaviour: An approach based on Fuzzy Cognitive Maps. International Journal of Computational Intelligence Systems 6, 1012–1039 (2013)

Decision Rules-Based Probabilistic MCDM Evaluation Method – An Empirical Case from Semiconductor Industry

Kao-Yi Shen[1,*] and Gwo-Hshiung Tzeng[2]

[1] Department of Banking and Finance, Chinese Culture University (SCE), Taipei, Taiwan
kyshen@sce.pccu.edu.tw
[2] Graduate Institute of Urban Planning, College of Public Affairs, National Taipei University,
151, University Rd., San Shia District, New Taipei City, 23741, Taiwan
ghtzeng@mail.ntpu.edu.tw

Abstract. Dominance-based rough set approach has been widely applied in multiple criteria classification problems, and its major advantage is the inducted decision rules that can consider multiple attributes in different contexts. However, if decision makers need to make ranking/selection among the alternatives that belong to the same decision class—a typical multiple criteria decision making problem, the obtained decision rules are not enough to resolve the ranking problem. Using a group of semiconductor companies in Taiwan, this study proposes a decision rules-based probabilistic evaluation method, transforms the strong decision rules into a probabilistic weighted model—to explore the performance gaps of each alternative on each criterion—to make improvement and selection. Five example companies were tested and illustrated by the transformed evaluation model, and the result indicates the effectiveness of the proposed method. The proposed evaluation method may act as a bridge to transform decision rules (from data-mining approach) into a decision model for practical applications.

Keywords: Dominance-based rough set approach (DRSA), multiple-criteria decision making (MCDM), VIKOR, financial performance (FP), performance gap.

1 Introduction

The complexity of commercial data nowadays has caused increasing interests in exploring their implicit patterns or rules, to find potential business opportunities or to make better decisions. Also, the knowledge discovering process has gained supports from data-mining techniques [1] and artificial intelligence (AI)[2, 3] recently. While all seem to agree that complex or vague data could be analyzed by certain computational techniques to acquire useful patterns, opinions differ as to how data should be processed and the ways of modeling. Some have argued that the nonlinearity of data

---

* Corresponding author.

M. Kryszkiewicz et al. (Eds.): RSEISP 2014, LNAI 8537, pp. 179–190, 2014.
© Springer International Publishing Switzerland 2014

set is common in real world problem, the analytical methods based on linear assumption of data set is unpersuasive [4]. A more flexible analytical method that may resolve the imprecise property in data set and interdependence among variables should be more realistic, such as fuzzy set theory, grey relational analysis, artificial neural network (ANN)[5], and support vector machine (SVM) [6]. Other researchers, however, argue that the process of numerical calculation of certain AI techniques (e.g. ANNs) is not transparent, and the obtained result from ANNs is like a black-box, difficult for decision maker to gain useful knowledge [7]. Therefore, the methods that may obtain understandable rules—such as decision tree, fuzzy neural network, dominance-based rough set approach (DRSA)—are more meaningful for decision makers. Therefore, the present study also chooses DRSA to obtain decision rules for resolving classification problem.

Although the use of DRSA can generate understandable decision rules for complex data sets, the classification of objects is based on the approximation of dominance relationship in attributes and decision classes, also termed as granules of knowledge [8]. DRSA may assign objects into different classes; however, the classified result from DRSA can only provide ordinal scale of decision classes. For alternatives classified in the same decision class, DRSA cannot provide ratio scale ranking for alternatives. Thus, to overcome the aforementioned limitation, the present study attempts to propose a decision rule-based probabilistic weighted model to solve MCDM ranking/selection or improvement problems. To demonstrate the proposed model, a group of semiconductor companies in Taiwan stock market is examined, and the data of five semiconductor stocks in 2012 are validated.

The remainder of this study is structured as below. Section 2 provides a discussion regarding the involved research methods, including DRSA and VIKOR. In Section 3, the proposed two-stage model is explained with needed steps. Section 4 takes the public-listed semiconductor stocks in Taiwan as an empirical case for examining the proposed model with results. Section 5 concludes this study with the research limitations of the proposed model.

2   Preliminary

The section briefly reviews the concept of RSA, DRSA and the mainstream MCDM evaluation methods applied in finance. Also, the aggregation method VIKOR—applied in this study—is also discussed.

2.1   Dominance-Based Rough Set Approach (DRSA)

The rough set approach (RSA) was proposed by Pawlak [9], which is regarded capable in solving complex classification problem. However, the classical RSA does not consider the preferential property of attributes, which is inevitable in solving many multi-criteria decision making problems. To overcome this limitation, Greco et al. [8] proposed DRSA to consider the preferential characteristic of attributes while making classification. Based on the dominance relationship among attributes, DRSA can generate a set of decision rules to categorize objects, which has been applied in various fields, such as marketing [10], finance [11], and the evaluation of national competitiveness [12]. To extend the multiple criteria classification capability of DRSA, Blaszczynski et al. [13] enhanced assignments of objects evaluated by a set of preference-ordered attributes by proposing a new classification score, which may calculate the closeness of an object to the covering rules. Although the proposed score may support to assign objects into ordinal decision classes more accurately, the proposed method still has difficulty in ranking within a specific decision class. Therefore, inspired by the idea of Greco et al. [8] and Blaszczynski et al. [13], the present study proposes a decision rules-based probabilistic weighted model to resolve the ranking or selection problem in MCDM field.

2.2   VIKOR and MCDM Evaluation Methods

The compromised ranking method VIKOR was proposed to select the ideal solution (alternative) with the shortest performance gap to the aspiration level [14]. To introduce VIKOR briefly, assume that a group of alternatives $A_1, A_2, ..., A_m$ are to be evaluated, and the strength of the jth criterion is expressed by f_{kj} for alternative A_k, and w_j is the influential weight of the jth variable/criterion, where $j=1,2,...,N$. The compromised alternatives will be chosen while the group utility (average gap) and individual regret (gap) be considered according to different settings. The VIKOR method has been combined with other MCDM evaluation methods—such as AHP [15], ANP [16], DEMATEL—to form hybrid evaluation models for solving various problems in practice, such as the selection of glamor stocks [7], suppliers/vendors [17], and information security risk evaluation [18]. The integration of VIKOR may extend the ranking capability of the proposed model into improvement planning.

3   Research Model

This section provides the background information of the incorporated methods in the proposed model. How to obtain decision rules from DRSA is discussed at first, and the decision rules can be transformed as the inputs for constructing an MCDM evaluation model.

3.1   Dominance-Based Rough Set Approach

Originated from the classical RSA, DRSA considers the preferential characteristic of both conditional attributes and decision classes, which also considers multiple criteria for making classification (decision). A typical DRSA model starts from a 4-tuple information system, i.e., $IS = (U, Q, V, f)$, which represents the data for learning decision rules from DRSA. In an IS of DRSA, U is a finite set of universe, Q is a finite set of n attributes (i.e., $Q = \{q_1, q_2 ..., q_n\}$), V is the value domain of attribute (i.e., $V = \bigcup_{q \in Q} V_q$), and f denotes a total function (i.e. $f : U \times Q \to V$). The set Q comprises of two main parts: conditional attributes (criteria) and a decision attribute. Let's assume that there are g objects in U, a complete outranking relation on U

can be defined as $\succeq_q$ with respect to a criterion $q \in Q$. For any two objects x and y (alternatives) in U, a complete outranking relation with regard to a criterion q can be defined as $\succeq_q$; if $x \succeq_q y$, then it denotes that "x is at least as good as y with respect to criterion q". For the decision attribute d that belongs to D, which divides U into a finite number of decision classes, denoted as $Cl = \{Cl_t : Cl_1, Cl_2, ..., Cl_m\}$ for m decision classes with $t=1,...,m$. For each $x \in U$, object x belongs to only one class Cl_t ($Cl_t \in Cl$). For Cl with predefined preferential order (i.e. for all $r, s = 1, ..., m$, if $r \succ s$, the decision class Cl_r is preferred to Cl_s), an downward union $Cl_t^{\leq}$ and upward union $Cl_t^{\geq}$ of classes can be defined as Eq. (1):

$$Cl_t^{\leq} = \bigcup_{s \leq t} Cl_s \text{ and } Cl_t^{\geq} = \bigcup_{s \geq t} Cl_s \tag{1}$$

Because the upward union is demonstrated in this study for transforming decision rules into a probabilistic weighing model, the discussion hereafter will focus on the upward union. On the conditional attribute side, a subset $P \subseteq C$ can be applied to classify decision classes Cl by the so-called dominance relation. For $x, y \in U$, if x dominates y with respect to the attribute set P, it can be denoted as $x D_p y$ to represent x P-dominates y. Then, for a set of objects that dominate x with regard to P, it can be denoted as $D_P^+(x) = \{y \in U : y D_p x\}$, the P-dominating set. On the contrary, a set of objects that are dominated by x with regard to P can be denoted as $D_P^-(x) = \{y \in U : x D_p y\}$, the P-dominated set.

The P-dominating set $D_P^+(x)$ and P-dominated set $D_P^-(x)$ can be used to representing a collection of upward and download unions of decision classes. The P-lower and P-upper approximation of an upward union with respect to $P \subseteq C$ can be define by $\underline{P}(Cl_t^{\geq}) = \{x \in U : D_P^+(x) \subseteq Cl_t^{\geq}\}$ and $\overline{P}(Cl_t^{\geq}) = \{x \in U : D_P^-(x) \cap Cl_t^{\geq} \neq \varnothing\}$ respectively. The P-lower approximation $\underline{P}(Cl_t^{\geq})$ denotes all of the objects $x \in U$ that are for sure to be included in the upward union $Cl_t^{\geq}$, which represents the certain knowledge. The P-upper approximation $\overline{P}(Cl_t^{\geq})$ can be interpreted as all of the objects possibly belongs to $Cl_t^{\geq}$ (i.e., not certain). With the P-upper approximation and P-lower approximation of $Cl_t^{\geq}$, the P-boundary of $Cl_t^{\geq}$ is defined as Eq. (2):

$$Bn_P = \overline{P}(Cl_t^{\geq}) - \underline{P}(Cl_t^{\geq}), t=2,..., m \tag{2}$$

For every subset $P \subseteq C$, the quality of approximation of Cl (by using a set of conditional attributes P) is defined as Eq. (3). If a full quality of approximation (fully consistent) is reached, which means that if an object x dominating object y on all considered criteria $P \subseteq C$, then the object x should also dominate y on the decision attribute.

$$\gamma_P(Cl) = \frac{\left| U - \left(\bigcup_{t \in \{2,\ldots,m\}} Bn_P \left(Cl_t^{\geq} \right) \right) \right|}{|U|} \tag{3}$$

A set of decision rules can be obtained by using the dominance-based approximation discussed above, in the form of "**if** *antecedents* **then** *consequence*". Although the inconsistency of approximation was also discussed, this study mainly adopts the consistent decision rules to construct a probabilistic weighted model. For each decision rule, the number of the fully consistent (i.e., all of the requirements on each attribute in each rule are satisfied) objects denotes the strength of a rule, and each dominance-consistent (i.e., fully consistent) object is called a support of the decision rule. The required steps to construct the proposed model are as below:

Step 1: Discretize the raw figures of conditional attributes, and assign the used objects into a 4-tuple *IS* table. In this study, only three decision classes are used to denote *Good* (improvement), *Middle* (not much change), and *Bad* (deterioration) of future FP respectively. The used discretization method for the conditional attributes will be explained in Section 4.

Step 2: Obtain the strong decision rules (only keep the decision rules with supports more than an assigned target threshold) associated with *Good* and *Bad* decision classes.

3.2   Transformed Probabilistic Weight of Decision Rule

In a typical MCDM model, the performance score of each alternative on each criterion should be measured before calculating the final scores. Compared with conventional MCDM model, the proposed probabilistic weighted model regards each strong decision rule as a transformed criterion. The transformation is explained in **Step 3** and **Step 4**.

Step 3: Calculate the probabilistic weight of each decision rule associated with *Good* and *Bad* decision classes separately. The number of supports divided by the total number of a decision class denotes the probabilistic weight of a decision rule, shown in Eq. (4):

$$\Pr(Rule_i) = \frac{card\left(\text{supp}_d\left(Rules_i\right)\right)}{card\left(Cl_d\right)}, \ d=Good \text{ or } Bad \tag{4}$$

Take the *h*th decision rule associated with *Good* decision class with 14 supports for example, if the total number of objects that belong to the *Good* decision class is 28, then its probabilistic weight $\Pr(Rule_h) = {}^{14}\!/_{28} = 50\%$. This definition is originated from the idea of coverage in DRSA.

Step 4: Evaluate the performance score of each target alternative on each strong decision rule. Suppose that there were *n* conditional attributes in the *i*th decision rule, and the alternative *y* satisfies three conditional criteria ($z \geq 3$) among the

z criteria, then the performance score of y equals to $3/z$ on the ith decision rule, which represents the consistent percentage of an alternative with regard to a decision rule.

After obtaining the probabilistic weight of each decision rule, based on VIKOR, we may further evaluate alternatives based on their performance gap in each decision rule.

3.3   VIKOR Method

The compromising outranking method VIKOR is incorporated with the probabilistic weighted model obtained in **Step 3** and **Step 4**. The VIKOR technique can synthesize the performance gaps on the considered criteria (i.e., strong decision rules in this stage) to rank or select alternatives.

Assume that there are l alternatives, expressed as $A_1, A_2, ..., A_l$. The performance of the jth criterion is denoted by h_{kj} for alternative k, w_j is the probabilistic weight of the jth criterion, where $j=1,2,...,N$, and N is the number of the criteria (the criteria in this stage denotes the strong decision rules). The VIKOR begins with an L_p-metric, in

which the $L_k^P = \left\{ \sum_{j=1}^{N} \left[w_j \left(\left| h_j^* - h_{kj} \right| \right) / \left(h_j^* - h_j^- \right) \right]^P \right\}^{1/p}$, the compromised alternative

based on $\min_k = L_k^P$ could be chosen to minimize the total performance gap.

Step 5: Form the original rating matrix by placing criteria (i.e., strong decision rules, comprise of rules associated with both *Good* and *Bad* decision classes) in column and alternatives in row.

Step 6: Normalize the original rating matrix. Choose the best h_j^* and the worst h_j^- for all criteria, $j=1,2,...,N$, where j represents the jth criterion. While the jth criterion represents a benefit, and the best implies $h_j^* = \max h_{kj}$ and $h_j^- = \min h_{kj}$ respectively; our new approach sets the best as the aspiration level and the worst as the worst value. An original rating matrix ($V = [v_{ij}]_{m \times n}$) can be transformed into a normalized matrix by Eq. (5); therefore, we may avoid "choose the best among inferior alternatives" [19]-[20], i.e., avoid "pick the best apple among a barrel of rotten apples".

$$r_{kj} = \left(\left| h_j^* - h_{kj} \right| \right) / \left(\left| h_j^* - h_j^- \right| \right) \tag{5}$$

In Eq. (5), r_{kj} denotes the normalized performance gap to the best (aspired) performance on the criterion j.

Step 7: *Compute the rating indexes* S_k *and* R_k *by Eq. (6):*

$$S_k = \sum_{j=1}^{N} w_j r_{kj} \text{ and } R_k = \max_j \left\{ w_j r_{kj} \mid j = 1, 2, ..., N \right\} \tag{6}$$

Where S_k is the synthesized total performance gap of alternative k in the evaluation model, R_k is the maximal performance gap of alternative k on the criteria; in the conventional terms of VIKOR method, these two indexes also represents the mean of group utility (i.e., average performance gap) and maximal regret (i.e., top priority performance gap) respectively. Details of the two indices S_k, R_k could be found in [14].

Step 8: *Compute the index values* Q_k, *k=1,2,..., l.*

$$Q_k = v \times \frac{\left(S_k - S^*\right)}{\left(S^- - S^*\right)} + (1-v) \times \frac{\left(R_k - R^*\right)}{\left(R^- - R^*\right)} \qquad (7)$$

Where S^* =min S_k, S^- =max S_k and R^* =min R_k, R^- =max R_k in traditional approach. In which, v is introduced as a weight for the strategy of maximal group utility (minimal gap), and $(1-v)$ is the weight of the individual regret (maximal gap). In the new approach, where S^*=0, S^- =1, and R^* =0, R^- =1, and Eq. (7) can be re-written as $Q_k = v \times S_k + (1-v) \times R_k$.

Step 9: Rank the alternatives. By sorting the values S_k, R_k and Q_k, for k=1, 2,..., l, the ranking of alternatives can be obtained by the ratio scale index Q_k with certain value in v.

4   Empirical Case of Semiconductor Industry in Taiwan

To illustrate the proposed model, a group of semiconductor stocks listed in the Taiwan stock market was examined. The data from 2007 to 2011 were trained to obtain DRSA decision rules, and the data in 2012 were tested to validate the model.

4.1   Data

All of the public-listed semiconductor companies in Taiwan were included for analysis, and the data from 2007 to 2011 were kept as the training set, the data in 2012 the testing set. After excluding incomplete data, 145 objects were kept as for training, and 37 for validation. The data were retrieved from the website MOPS [21], which is hosted by Taiwan stock exchange. This study chose the improvement of return on assets in the subsequent year (ΔROA) as the decision attribute, and divided the decision attribute into three decision classes—*Good, Middle, Bad*—by ranking the top 1/3, middle 1/3, and bottom 1/3 ΔROA in each year.

MOPS reports the summarized financial result of each public-listed stock by 20 key financial indicators (Table 1), and this study included those 20 indicators as the conditional attributes for DRSA. Furthermore, the discretization of the 20 attributes took the similar approach as the decision attribute, and the 20 attributes of each alternative were ranked in each year and discretized as "H", "M", and "L". Moreover, the financial data of five semiconductor stocks in 2012 were validated, and the names of the five companies are: 1) TSMC; 2) UMC; 3) KINSUS; 4) GUC; 5) SPIL.

Table 1. Definitions of the financial indicators

Financial ratios	Symbols	Definitions and brief explanations
Debt to total asset	*Debt*	total debt/ total asset
Long capital to total asset	*LongCapital*	Long-term capital/total asset
Liquidity ratio	*Liquidity*	current asset/ current liability
Speed ratio	*Speed*	(current asset-inventory)/current liability
Interest coverage ratio	*InterestCoverage*	(net profit before tax+interest expense) /interest expense
Accounts receivable ratio	*AR_turnover*	Net credit sales/average AR
Days for collecting AR	*AR_days*	(days*AR)/credit sales
Inventory turnover rate	*InvTurnover*	total operational cost/ average inventory
Average days for sales	*DAYs*	(average ending inventory/ operational cost)*365 days
Fixed asset turnover rate	*FAssetTurnover*	total revenue/ total fixed asset
Return on total asset	*ROA*	net profit before tax/ average total asset
Return on equity	*ROE*	net profit before tax/ average total equity
Operational profit to total capital	*OP_capital*	operational profit/total capital
Net profit before tax to total capital	*NP_capital*	net profit before tax/total capital
Net profit ratio	*NetProfit*	net profit/net sales
Earnings per share	*EPS*	(Net income-dividends on preferred stocks)/total outstanding shares
Cash-flow ratio	*CashFlow*	(operational cashflow-cash dividend for preferred stocks)/weighted average equity
Cash-flow adequacy ratio	*CashFlow_adq*	cash flow from operation/annual current maturities
Cash-flow reinvestment ratio	*CashFlow_inv*	(increase in fixed asset+increase in working capital)/net income+noncash expense-non cash sales-dividends)

4.2   Decision Rules from DRSA

The classification accuracy (CA) and mean absolute error (MAE) were both measured for DRSA model. The software jMAF [22] for conducting DRSA analysis (by setting consistency level=1) is developed by the laboratory of Intelligent Decision System (IDSS). At first, the training set was examined by conducting 5-fold cross validation and repeated for 5 times. Also, discriminant analysis was also conducted in similar approach for comparison in Table 2.

The data for the training of DRSA model generated 78.04% classification accuracy in average; therefore, this study used the whole training set to generate decision rules, and the untouched testing set was validated by the obtained decision rules. The validation yielded 88.24% classification accuracy for the testing set, which is regarded as acceptable to use the obtained decision rules. To adopt the strong decision rules as criteria in the next stage, this example selected the decision rules with the top four highest supports (supports ≥ 12) for the *Good* (i.e., positive rules) and *Bad* (i.e., nega-tive rules) decision classes separately, and the obtained rules are listed in Table 3.

Table 2. The 5-fold cross validation of the training set

		DRSA		Discriminant Analysis	
		CA	MAE	CA	MAE
		78.51%	28.97%	76.55%	33.79%
		77.31%	30.34%	72.55%	37.93%
		76.55%	29.66%	70.95%	40.00%
		78.62%	30.34%	74.27%	37.24%
		79.21%	28.97%	71.55%	36.55%
Average		78.04%	29.66%	73.17%	37.01%
SD*		(1.08%)	(0.69%)	(2.27%)	(2.26%)

*Standard deviation of the five experiments.

Table 3. The strong rules used for probabilistic weighted model

	Supp	Prob. Weight	**Positive Rules**
R1	17	0.3263	(*LongCapital* $\geq$ M) & (*Speed* $\geq$ M) & (*AR_turnover* $\geq$ M) & (*AR_days* $\geq$ H) & (*ROE* $\geq$ M) & (*CashFlow_adq* $\geq$ M)
R2	16	0.3077	(*Speed* $\geq$ M) & (*AR_turnover* $\geq$ H) & (*EPS* $\geq$ M) & (*CashFlow* $\geq$ H) & (*Cash-Flow_adq* $\geq$ M)
R3	14	0.2692	(*Debt* $\geq$ M) & (*InterestCoverage* $\geq$ M) & (*AR_turnover* $\geq$ H) & (*Inventory* $\geq$ M) & (*EPS* $\geq$ H) & (*CashFlow* $\geq$ M)
R4	13	0.2500	(*LongCapital* $\geq$ M) & (*AR_days* $\geq$ H) & (*ROA* $\geq$ M) & (*CashFlow* $\geq$ H)
			Negative Rules
R5	30	0.3226	(*LongCapital* $\leq$ L) & (*Liquidity* $\leq$ M)
R6	29	0.3118	(*AR_turnover* $\leq$ M) & (*AR_days* $\leq$ L)
R7	29	0.3118	(*Liquidity* $\leq$ M) & (*InterestCoverage* $\leq$ L)
R8	28	0.3011	(*AR_days* $\leq$ M) & (*Inventory* $\leq$ L)

The probabilistic weights in Table 3 need further explanation. Take the probabilis-tic weight of R1 for example, as there were 52 objects in the *Good* decision class, the probabilistic weight of R1 equals $^{17}\!/_{52} = 0.3263$. The probabilistic weight could be interpreted as the total objects fully covered by a decision rules within a decision class. In the next stage, the financial data of the five sample stocks in 2012 were ex-amined by the eight strong decision rules (rule with only one conditional attribute is not included). To calculate the performance scores of each sample stock on each crite-rion (i.e., decision rules in this stage), the conditional attributes in each rule were as-sumed to be equally weighted to calculate the performance scores. Take the data of TSMC on decision rule R1 for example, five conditional attributes are satisfied out of the six conditional attributes (i.e., (*LongCapital* ≥ M) & (*Speed* ≥ M) & (*AR_turnover* ≥ M) & (*AR_days* ≥ H) & (*ROE* ≥ M) & (*CashFlow_adq* ≥ M); therefore, the performance score of TSMC on R1 is 5/6=0.8333. The performance score of each stock were measured on all of the eight strong decision rules (Table 3), which positioned a stock by the strong positive and negative decision rules. The performance scores of the five example stocks on the eight decision rules are in Table 4.

Table 4. Performance scores of the five example stocks on the eight rules (V)

	R1	R2	R3	R4	R5	R6	R7	R8
TSMC	**0.8333**	1.0000	1.0000	0.6667	0.0000	0.0000	0.5000	0.0000
UMC	0.8333	0.8000	0.8333	0.6667	0.5000	0.5000	0.5000	0.0000
KINSUS	0.8333	0.8000	0.8333	0.6667	0.5000	0.5000	0.5000	0.5000
GUC	0.5000	0.2000	0.5000	0.3333	0.0000	1.0000	0.0000	0.5000
SPIL	0.8333	0.8000	0.6667	0.3333	0.5000	0.5000	0.5000	0.0000

With the performance scores of each stock on the eight decision rules (criteria), the performance scores of the example stocks were synthesized by the VIKOR method. The original weighting matrix V and the normalized weighting matrix V^N are shown in Table 4 and Table 5 respectively (refer **Step 6** in subsection 3.2).

Table 5. Normalized weighting matrix V^N

	R1	R2	R3	R4	R5	R6	R7	R8
TSMC	0.3263	0.3077	0.2692	0.2500	0.3226	0.3118	0.0000	0.3011
UMC	0.3263	0.2308	0.179449	0.2500	0.0000	0.1559	0.0000	0.3011
KINSUS	0.3263	0.2308	0.179449	0.2500	0.0000	0.1559	0.0000	0.0000
GUC	0.0000	0.0000	0.0000	0.0000	0.3226	0.0000	0.3118	0.0000
SPIL	0.3263	0.2308	0.0898	0.0000	0.0000	0.1559	0.0000	0.3011

By setting v=0.7 and 0.5 respectively, S_k, R_k, and Q_k of the five sample stocks are summarized in Table 6 based on Eq. (6)-(7).

Table 6. Final performance scores of the five sample stocks based on VIKOR

	R1	R2	R3	R4	R5	R6	R7	R8	S_k	R_k	Q_k v=0.7	Q_k v=0.5
TSMC	0.00	0.00	0.00	0.00	0.00	0.00	1.00	0.00	0.31	0.31	0.31(1)	0.31(1)
UMC	0.00	0.25	0.33	0.00	1.00	0.50	1.00	0.00	0.95	0.32	0.42(5)	0.64(5)
KINSUS	0.00	0.25	0.33	0.00	1.00	0.50	1.00	1.00	0.66	0.32	0.37(2)	0.49(2)
GUC	1.00	1.00	1.00	1.00	0.00	1.00	0.00	1.00	0.66	0.33	0.38(3)	0.50(3)
SPIL	0.00	0.25	0.66	1.00	1.00	0.50	1.00	0.00	0.79	0.32	0.39(4)	0.56(4)

In Table 6, the two columns of Q_k (v=0.7 and v=0.5) both indicate the same ranking sequence of the five stocks: TSMC ≻ KINSUS ≻ GUC ≻ SPIL ≻ UMC. The actual ROA performances of the five example stocks were 19.56% (TSMC), 10.06% (KINSUS), 12.51% (GUC), 6.83% (SPIL), and 3.07% (UMC) respectively. Although the ranking is not fully consistent with the actual ROA performance in 2012 (i.e., the actual ROA performance ranking was TSMC ≻ GUC ≻ KINSUS ≻ SPIL ≻ UMC), the top performed TSMC were ranked as the best choice by both Q_k (v=0.7 and v=0.5) and SPIL and UMC were ranked in the bottom, which indicated the effectiveness of the proposed model.

5   Conclusion and Remarks

To summarize, the present study proposes a probabilistic-weighted model to transform DRSA decision rules into a MCDM model. Compared with conventional MCDM models, the proposed model does not need to acquire opinions from decision makers or domain experts; therefore, the obtained result is relatively more objective. By using a group of semiconductor stocks, this study demonstrated how to measure the performance score of each alternative on each criterion and adopted the VIKOR method to explore the performance gap of each stock. The final result suggested that TSMC is the top performer among the five sample stocks, which is consistent with the actual performances of the five alternatives in 2012. The proposed approach extends the application of DRSA decision rules into a MCDM model with a proposed transformation method, which is the main novelty and contribution of this study.

The proposed approach in the present study is inspired by previous works; nevertheless, there are at least two main differences compared with previous methods. First, the proposed classification score of Blaszczynski et al. (2006) calculates the closeness of an alternative to the covered rules, to assign an alternative to a specific decision class more accurately, which might encounter obstacles in making further ranking or selection in the same decision class. However, the present study used positive and negative rules with strong supports to measure all alternatives, and calculated the final performance scores for each alternative. The obtained result may rank alternatives in the same decision class. Second, this study proposes the idea of probabilistic-weighted evaluation model, which assigns the probabilistic weights to each decision rule based on the number of supported objects in historical data. This approach only considers the decision rules that happened with high frequency in the past, which is similar to the process of making decisions in human beings. Though the present study demonstrated its effectiveness in making selection, it is still in the experimental stage with some limitations, such as the requirements of attributes in each decision rule were assumed to be equally important; future research is suggested to resolve this limitation.

References

1. Ngai, E., Hu, Y., Wong, Y.-H., Chen, Y., Sun, X.: The application of data mining techniques in financial fraud detection: A classification framework and an academic review of literature. Decision Support Systems 50(3), 559–569 (2011)
2. Derelioğlu, G., Gürgen, F.: Knowledge discovery using neural approach for SME's credit risk analysis problem in Turkey. Expert Systems with Applications 38(8), 9313–9318 (2011)
3. Ho, G.-T., Ip, W.-H., Wu, C.-H., Tze, Y.-K.: Using a fuzzy association rule mining approach to identify the financial data association. Expert Systems with Applications 39(10), 9054–9063 (2012)

4. Liou, J.-J., Tzeng, G.-H.: Comments on "Multiple criteria decision making (MCDM) methods in economics: an overview". Technological and Economic Development of Economy 18(4), 672–695 (2012)
5. Shen, K.-Y.: Implementing value investing strategy by artificial neural network. International Journal of Business and Information Technology 1(1), 12–22 (2011)
6. Bahrammirzaee, A.: A comparative survey of artificial intelligence applications in finance: artificial neural networks, expert system and hybrid intelligent systems. Neural Computing & Applications 19(8), 1165–1195 (2010)
7. Shen, K.-Y., Yan, M.-R., Tzeng, G.-H.: Combining VIKOR-DANP model for glamor stock selection and stock performance improvement. Knowledge-Based Systems 58, 86–97 (2013)
8. Greco, S., Matarazzo, B., Slowinski, R.: Multicriteria classification by dominance-based rough set approach. In: Handbook of Data Mining and Knowledge Discovery. Oxford University Press, New York (2002)
9. Pawlak, Z.: Rough sets. International Journal of Computer & Information Sciences 11(5), 341–356 (1982)
10. Liou, J.-J., Tzeng, G.-H.: A dominance-based rough set approach to customer behavior in the airline market. Information Sciences 180(11), 2230–2238 (2010)
11. Zaras, K.: The Dominance-based rough set approach (DRSA) applied to bankruptcy prediction modeling for small and medium businesses. In: Multiple Criteria Decision Making/The University of Economics in Katowice, pp. 287–295 (2011)
12. Ko, Y.-C., Tzeng, G.-H.: A Dominance-based rough set approach of mathematical programming for inducing national competitiveness. In: Watada, J., Phillips-Wren, G., Jain, L.C., Howlett, R.J. (eds.) Intelligent Decision Technologies. SIST, vol. 10, pp. 23–36. Springer, Heidelberg (2011)
13. Błaszczyński, J., Greco, S., Słowiński, R.: Multi-criteria classification – A new scheme for application of dominance-based decision rules. European Journal of Operational Research 181(3), 1030–1044 (2007)
14. Opricovic, S., Tzeng, G.-H.: Compromise solution by MCDM methods: A comparative analysis of VIKOR and TOPSIS. European Journal of Operational Research 156(2), 445–455 (2004)
15. Kaya, T., Kahraman, C.: Multicriteria renewable energy planning using an integrated fuzzy VIKOR & AHP methodology: The case of Istanbul. Energy 35(6), 2517–2527 (2010)
16. Wang, Y.-L., Tzeng, G.-H.: Brand marketing for creating brand value based on a MCDM model combining DEMATEL with ANP and VIKOR methods. Expert Systems with Applications 39(5), 5600–5615 (2012)
17. Hsu, C.-H., Wang, F.-K., Tzeng, G.-H.: The best vendor selection for conducting the recycled material based on a hybrid MCDM model combining DANP with VIKOR. Resources, Conservation and Recycling 66, 95–111 (2012)
18. Ou Yang, Y.-P., Shieh, H.-M., Tzeng, G.-H.: A VIKOR technique based on DEMATEL and ANP for information security risk control assessment. Information Sciences 232, 482–500 (2013)
19. Liu, C.-H., Tzeng, G.-H., Lee, M.-H.: Improving tourism policy implementation – The use of hybrid MCDM models. Tourism Management 33(2), 413–426 (2012)
20. Hu, S.-K., Lu, M.-T., Tzeng, G.-H.: Exploring smart phone improvements based on a hybrid MCDM model. Expert Systems with Applications 41(9), 4401–4413 (2014)
21. MOPS, http://emops.twse.com.tw/emops_all.htm (accessed in 2013)
22. Błaszczyński, J., Greco, S., Matarazzo, B., Slowinski, R., Szelag, M.: jMAF-Dominance-based rough set data analysis framework. In: Skowron, A., Suraj, Z. (eds.) Rough Sets and Intelligent Systems-Professor Zdzisław Pawlak in Memoriam. ISRL, vol. 42, pp. 185–209. Springer, Heidelberg (2013)

Decision Rule Classifiers
for Multi-label Decision Tables

Fawaz Alsolami[1,2], Mohammad Azad[1], Igor Chikalov[1], and Mikhail Moshkov[1]

[1] Computer, Electrical and Mathematical Sciences and Engineering Division
King Abdullah University of Science and Technology
Thuwal 23955-6900, Saudi Arabia
[2] Computer Science Department, King Abdulaziz University, Saudi Arabia

Abstract. Recently, multi-label classification problem has received significant attention in the research community. This paper is devoted to study the effect of the considered rule heuristic parameters on the generalization error. The results of experiments for decision tables from UCI Machine Learning Repository and KEEL Repository show that rule heuristics taking into account both coverage and uncertainty perform better than the strategies taking into account a single criterion.

Keywords: decision rules, rule heuristics, classification.

1 Introduction

Multi-label data sets appear in important problems of discrete optimization, pattern recognition, computational geometry, decision making, inconsistent decision tables in rough set theory, etc. [8–10]. In real life applications we can meet multi-label data when we study, e.g., problem of semantic annotation of images [5], music categorization into emotions [13], functional genomics [4], and text categorization [14]. In contrast with single-label data sets, rows (objects) of multi-label data sets are associated with a set of decisions (labels).

Two main approaches have been proposed for constructing multi-label classifiers. The first approach is based on algorithm adaptation methods which means to extend existing learning algorithms to handle multi-label data sets such as decision trees [3, 6]. The second approach is called problem transformation methods where a given multi-label data set transforms into the corresponding single-label data set(s), for example, binary relevance method (BR) [11, 12]. In this paper, we consider the former approach. A model which is expressed as a set of decision rules is built for a given multi-label data set using some rule heuristic algorithms [2]. Each decision rule is associated with a single decision (label).

This work is devoted to study statistically the influence of the considered rule heuristic algorithm parameters to accuracy of produced classifiers. Particularly, a post-hoc Nemenyi test is used in order to rank the considered algorithms based on the achieved accuracy over multiple data sets and draw out differences between classifiers that are statistically significant [7]. The test recognizes two well-performed rule heuristics in terms of accuracy.

M. Kryszkiewicz et al. (Eds.): RSEISP 2014, LNAI 8537, pp. 191–197, 2014.

The remainder of this paper consists of four sections. Section 2 contains definitions of main notions that employ throughout this paper. We present procedures of a variety of greedy algorithms in Sections 3. The experiments setup and the results of the experiments are described in Section 4. Finally, Section 5 concludes the paper.

2   Main Definitions

A decision table with many-valued decisions T is a rectangular table consisting of rows and columns. Rows are filled with numbers or strings (values of attributes). Columns of this table are labeled with attributes $f_1, \ldots, f_n$. Rows of the table are pairwise different, and each row is labeled with a nonempty finite set of decisions. Note that each decision table with one-valued decisions can be interpreted also as a decision table with many-valued decisions. In such table, each row is labeled with a set of decisions which has one element. Table 1 shows an example of a decision table with many-valued decisions. We denote by $N(T)$ the number of

Table 1. A decision table with many-valued decisions

	f_1	f_2	
r_1	1	0	{1,2}
r_2	0	0	{1}
r_3	0	1	{3,2}

rows in table T, and by $D(r)$ the set of decisions attached to the row r. A decision d is a common decision of T if $d \in D(r)$ for any row r of T. We will say that T is a *degenerate* table if T does not have rows or it has a common decision. A table obtained from T by removing some rows is called a subtable of T. We denote by $T(f_{i_1}, b_1), \ldots, (f_{i_m}, b_m)$ a *subtable* of T which consists of rows that at the intersection with columns $f_{i_1}, \ldots, f_{i_m}$ have values $b_1, \ldots, b_m$.

The expression

$$(f_{i_1} = a_1) \wedge \ldots \wedge (f_{i_m} = a_m) \to t \tag{1}$$

is called a decision rule over T, where $f_{i_1}, \ldots, f_{i_m} \in \{f_1, \ldots, f_n\}, a_1, \ldots, a_m$ are the values of the corresponding attributes, and t is a decision. The rule is called realizable for a row $r = (b_1, \ldots, b_n)$ if

$$b_{i_1} = a_1, \ldots, b_{i_m} = a_m.$$

The rule is called *true* for T, if for any row r of T for which the rule is realizable, $t \in D(r)$. We will say that the considered rule is a *rule for T and r*, if this rule is true for T and realizable for r. The number m is called the *length* of the rule. The *coverage* of the rule is the number of rows r from T for which the rule is realizable and for which $t \in D(r)$.

3 Greedy Algorithms

A greedy algorithm constructs decision rules in a sequential manner by adding conditions in left-hand side by minimizing some heuristic function at each step. Let $r = (b_1, \ldots, b_n)$ be a row of T. For the considered heuristics, we will fix a decision $d \in D(r)$ to be on the right-hand side of the rule.

In this paper, we describe five different heuristics. Generally they work as follows: Let $\mathcal{H}$ is a heuristic that works with a fixed decision $d \in D(r)$ for a row r. This heuristic starts initially with a rule whose left-hand side is empty $\to d$, and then sequentially conditions get added to the left-hand side of such rule.

Let during the work of the heuristic $\mathcal{H}_1$, we already constructed the following left-hand side:

$$(f_{i_1} = b_{i_1}) \wedge \ldots \wedge (f_{i_m} = b_{i_m}) \tag{2}$$

Obviously, this left-hand side of the rule describes a subtable T' of T:

$$T' = T(f_{i_1}, b_{i_1}), \ldots, (f_{i_m}, b_{i_m}).$$

If a stopping condition is not met for the heuristic $\mathcal{H}_1$, a new attribute $f_{i_{m+1}}$ is chosen to be added to the constructed rule that satisfy some conditions. Different heuristics have different conditions to build left side conditions of the rules and finally we get:

$$(f_{i_1} = b_{i_1}) \wedge \ldots \wedge (f_{i_m} = b_{i_m}) \wedge (f_{i_{m+1}} = b_{i_{m+1}}) \to d \tag{3}$$

Below we describe all such conditions:

Miscoverage_heuristic (M) $\mathcal{H}_1$. For this heuristic $\mathcal{H}_1$, a new attribute $f_{i_{m+1}}$ is chosen to be added to the constructed rule that results in minimizing number of miscoverage. In other words, the heuristic $\mathcal{H}_1$ chooses an attribute $f_{i_{m+1}}$ if the corresponding subtable $T'' = T'(f_{i_{m+1}}, b_{i_{m+1}})$ that has a minimum value of M'', where we denote by M'' the number of rows from the subtable T'' for which the decision d is not belong to any row r'' of this subtable $d \notin D(r'')$. This heuristic takes into account a single criterion minimizing uncertainty.

Relative_miscoverage_heuristic (RM) $\mathcal{H}_2$. For the heuristic $\mathcal{H}_2$, a new attribute $f_{i_{m+1}}$ is chosen to be added to the constructed rule that results in minimizing number of relative miscoverage: $\frac{M''}{N(T'')}$. Here we denote by M'' the number of rows from the subtable T'' for which the decision d is not belong to any row r'' of this subtable $d \notin D(r'')$, and $N(T'')$ is the number of rows in T''.

$\{\frac{b}{(a+1)}\}$_heuristic (Div_Poly) $\mathcal{H}_3$. We denote by C' the number of rows r' from T' such that $d \in D(r')$ and by M' the number of rows r' from T' for which $d \notin D(r')$. We also denote by C'' the number of rows r' from T'' such that $d \in D(r')$ and by M'' the number of rows r' from T'' for which $d \notin D(r')$. Finally, we denote two parameters : $a = C' - C''$ and $b = M' - M''$. This heuristic takes into account two criteria maximizing coverage a and minimizing uncertainty b.

If a stopping condition is not met for the heuristic $\mathcal{H}_3$, a new attribute $f_{i_{m+1}}$ is chosen to be added to the constructed rule that results in minimizing the value of $\frac{b}{(a+1)}$. Our intuition here is that minimizing b will minimize miscoverage and maximizing a will maximize coverage of the constructed rule. Therefore, if we consider the fractional value $\frac{b}{(a+1)}$, we can achieve our goal.

$\{\frac{b}{\log_2(a+2)}\}$_heuristic (Div_Log) $\mathcal{H}_4$. We denote by C' the number of rows r' from T' such that $d \in D(r')$ and by M' the number of rows r' from T' for which $d \notin D(r')$. We also denote by C'' the number of rows r' from T'' such that $d \in D(r')$ and by M'' the number of rows r' from T'' for which $d \notin D(r')$. Finally, we denote two parameters : $a = C' - C''$ and $b = M' - M''$. This heuristic takes into account two criteria maximizing coverage a and minimizing uncertainty b.

If a stopping condition is not met for the heuristic $\mathcal{H}_4$, a new attribute $f_{i_{m+1}}$ is chosen to be added to the constructed rule that results in minimizing the value of $\frac{b}{\log_2(a+2)}$. This heuristic follows the same intuition as $\mathcal{H}_3$ except we introduce logarithmic form for denominator.

Conditional_maximum_coverage_heuristics (Max_Cov) $\mathcal{H}_5$. We denote by C' the number of rows r' from T' such that $d \in D(r')$ and by M' the number of rows r' from T' for which $d \notin D(r')$. We also denote by C'' the number of rows r' from T'' such that $d \in D(r')$ and by M'' the number of rows r' from T'' for which $d \notin D(r')$. Finally, we denote two parameters : $a = C' - C''$ and $b = M' - M''$. This heuristic takes into account a single criterion which is maximizing coverage a.

If a stopping condition is not met for the heuristic $\mathcal{H}_5$, a new attribute $f_{i_{m+1}}$ is chosen to be added to the constructed rule that results in maximizing the value of a provided that $b > 0$. Our goal is to keep at least one nonzero value for b, and maximize the value of a which will eventually maximize the coverage of the considered rule.

4   Experimental Results for Classifiers Based on Sets of Decision Rules

We performed the experiments on two modified decision tables from UCI Machine Learning Repository [1] as shown in Table 2 and three decision tables from KEEL data Repository[1] shown in Table 3.

4.1   Data Sets: UCI ML Repository

In some tables there were missing values. Each such value was replaced with the most common value of the corresponding attribute. Some decision tables contain conditional attributes that take unique value for each row. Such attributes were removed. We removed from these tables more conditional attributes. As a

---

[1] http://sci2s.ugr.es/keel/multilabel.php

result we obtained inconsistent decision tables which contained equal rows with different decisions.

Each group of identical rows was replaced with a single row from the group which is labeled with the set of decisions attached to rows from the group. The information about obtained decision tables with many-valued decisions can be found in Table 2. This table contains the name of initial table from [1] with an index equal to the number of removed conditional attributes, number of rows (column "Rows"), number of attributes (column "Attributes"), and spectrum of this table (column "Spectrum"). Spectrum of a decision table with many-valued decisions is a sequence #1, #2,..., where #i, $i = 1, 2, \ldots$, is the number of rows labeled with sets of decisions with the cardinality equal to i.

4.2   Classification Method

We evaluate the accuracy of considered classifiers using two-fold cross validation approach where a given decision table is randomly partitioned into two folds of equal size. In all five rule heuristics algorithms, 30% of samples of the first fold are reserved for validation. Train and validation subtables (samples) are passed to the considered rule heuristic algorithm that builds a model as rulesets. The model is applied to predict decisions for all samples from the second fold. Then the folds are swapped: the model is constructed from the samples of the second fold and makes prediction for all samples from the first fold. Finally, misclassification error is estimated as the average of misclassification error of the first and the second fold. In order to reduce variation of the estimate, 15 experiments for each model and each data set were performed.

A post-pruning technique is applied to a set of rules $\mathcal{R}$ generated by one of the present algorithms. For each rule in $\mathcal{R}$ that has uncertainty less than the predefined threshold t, we keep removing constraints from the end of the rule until the uncertainty of the resulted rule is greater than or equal the threshold t. The threshold that shows the minimim misclassification error on the validation subtables (samples) is selected.

4.3   Results

In this section, we compare prediction power of classifiers based on sets of decision rules constructed by the presented algorithms. Let now $\mathcal{H}$ be one of the heuristics. Then, for a given decision table T, this heuristic to construct a decision rule for T, each row r of T, and each decision $d \in D(r)$. The constructed set of rules is considered as a classifier that uses a voting scheme procedure where each rule votes for the decision in its right-hand side. For an unseen row r, the classifier assigns the decision with maximum votes to the row r. The row r is correctly classified if the assigned decision belongs to the set of decisions $D(r)$.

Figure 1 shows the critical difference diagram at the significance level of $\alpha = 0.05$ for the considered classifiers where a classifier with a small average rank has a low error rate. We can see from the critical diagram that the worst performing classifier in terms of error rates is $\mathcal{H}_5$ (Max_Cov). For the other classifiers based

Table 2. Characteristics of modified decision tables from UCI ML Repository

Decision table T	Rows	Attributes	Spectrum			
			#1	#2	#3	#4
cars-1	432	5	258	161	13	
mushroom-5	4078	17	4048	30		

Table 3. Decision tables from KEEL Multi-label data sets Repository

Decision table	Rows	Attributes	Decisions
corel5k	5000	499	374
enron	1702	1001	53
genbase	662	1186	27

on $\mathcal{H}_2$ (RM), $\mathcal{H}_2$ (RM),$\mathcal{H}_3$ (Div_Log), and $\mathcal{H}_4$ (Div_Poly) the experimental data is not sufficient to enough to distinguish between them. Note that the two well-performed heuristics algorithms $\mathcal{H}_3$ (Div_Poly) and $\mathcal{H}_4$ (Div_Log) are able to generate rules with coverage on average for real datasets close to optimal [2] (at most 6.23%,8.83% respectively).

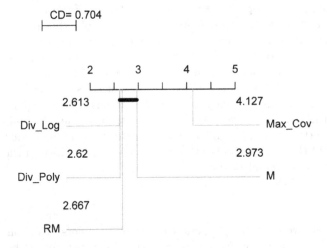

Fig. 1. The critical diagram for the considered classifiers

5   Conclusion

This paper presented a comparative study for rule heuristics algorithms for multi-label decision tables. Specifically, we compared statistically the performance of classifiers constructed by the considered rule heuristics algorithms. The obtained results show that rule heuristics taking into account both coverage and uncertainty perform better than the strategies taking into account a single criterion (maximum coverage or minimum uncertainty).

References

1. Asuncion, A., Newman, D.J.: UCI Machine Learning Repository (2007), http://www.ics.uci.edu/~mlearn/
2. Azad, M., Chikalov, I., Moshkov, M.: Optimization of decision rule complexity for decision tables with many-valued decisions. In: 2013 IEEE International Conference on Systems, Man, and Cybernetics (SMC), pp. 444–448 (October 2013)
3. Blockeel, H., De Raedt, L., Ramon, J.: Top-down induction of clustering trees. In: Proceedings of the Fifteenth International Conference on Machine Learning, pp. 55–63. Morgan Kaufmann Publishers Inc. (1998)
4. Blockeel, H., Schietgat, L., Struyf, J., Džeroski, S., Clare, A.: Decision trees for hierarchical multilabel classification: A case study in functional genomics. In: Fürnkranz, J., Scheffer, T., Spiliopoulou, M. (eds.) PKDD 2006. LNCS (LNAI), vol. 4213, pp. 18–29. Springer, Heidelberg (2006)
5. Boutell, M.R., Luo, J., Shen, X., Brown, C.M.: Learning multi-label scene classification. Pattern Recognition 37(9), 1757–1771 (2004)
6. Clare, A., King, R.D.: Knowledge discovery in multi-label phenotype data. In: Siebes, A., De Raedt, L. (eds.) PKDD 2001. LNCS (LNAI), vol. 2168, pp. 42–53. Springer, Heidelberg (2001)
7. Demšar, J.: Statistical comparisons of classifiers over multiple data sets. Journal of Machine Learning Research 7, 1–30 (2006)
8. Greco, S., Matarazzo, B., Słowiński, R.: Rough sets theory for multicriteria decision analysis. European Journal of Operational Research 129(1), 1–47 (2001)
9. Moshkov, M., Zielosko, B.: Combinatorial Machine Learning–A Rough Set Approach. SCI, vol. 360. Springer, Heidelberg (2011)
10. Pawlak, Z.: Rough Sets: Theoretical Aspects of Reasoning About Data. Kluwer Academic Publishers, Bosten (1991)
11. Tsoumakas, G., Katakis, I.: Multi-label classification: An overview. International Journal of Data Warehouse and Mining 3(3), 1–13 (2007)
12. Tsoumakas, G., Katakis, I., Vlahavas, I.: Mining multi-label data. In: Data Mining and Knowledge Discovery Handbook, pp. 667–685. Springer US (2010)
13. Wieczorkowska, A., Synak, P., Lewis, R., Raś, Z.W.: Extracting emotions from music data. In: Hacid, M.-S., Murray, N.V., Raś, Z.W., Tsumoto, S. (eds.) ISMIS 2005. LNCS (LNAI), vol. 3488, pp. 456–465. Springer, Heidelberg (2005)
14. Zhou, Z.H., Jiang, K., Li, M.: Multi-instance learning based web mining. Applied Intelligence 22(2), 135–147 (2005)

Considerations on Rule Induction Procedures by STRIM and Their Relationship to VPRS

Yuichi Kato[1], Tetsuro Saeki[2], and Shoutarou Mizuno[1]

[1] Shimane University,
1060 Nishikawatsu-cho, Matsue City, Shimane 690-8504, Japan
ykato@cis.shimane-u.ac.jp
[2] Yamaguchi University,
2-16-1 Tokiwadai, Ube City, Yamaguchi 755-8611, Japan
tsaeki@yamaguchi-u.ac.jp

Abstract. STRIM (Statistical Test Rule Induction Method) has been proposed as a method to effectively induct if-then rules from the decision table. The method was studied independently of the conventional rough sets methods. This paper summarizes the basic notion of STRIM and the conventional rule induction methods, considers the relationship between STRIM and their conventional methods, especially VPRS (Variable Precision Rough Set), and shows that STRIM develops the notion of VPRS into a statistical principle. In a simulation experiment, we also consider the condition that STRIM inducts the true rules specified in advance. This condition has not yet been studied, even in VPRS. Examination of the condition is very important if STRIM is properly applied to a set of real-world data set.

1 Introduction

Rough Sets theory as introduced by Pawlak [1] provides a database called the decision table, with various methods of inducting if-then rules and determining the structure of rating and/or knowledge in the database. Such rule induction methods are needed for disease diagnosis systems, discrimination problems, decision problems, and other aspects, and consequently many effective algorithms for rule induction by rough sets have been reported to date [2–7]. However, these methods and algorithms have paid little attention to mechanisms of generating the database, and have generally focused on logical analysis of the given database.

In a previous studies [8, 9] we (1) devised a model of data generation for the decision table with if-then rules specified in advance, and proposed a statistical rule induction method and an algorithm named STRIM; (2) In a simulation experiment based on the model of the data generation, STRIM was confirmed to successfully induct the if-then true rules from different databases generated from the same specified rules[8]; (3) found that, when conventional methods [4, 6, 7] were used, significant rules could barely be inducted, and different rule sets were inducted from different sample data sets with the same rules; i.e.

M. Kryszkiewicz et al. (Eds.): RSEISP 2014, LNAI 8537, pp. 198–208, 2014.
© Springer International Publishing Switzerland 2014

Table 1. An example of a decision table

U	$C(1)$	$C(2)$	$C(3)$	$C(4)$	$C(5)$	$C(6)$	D
1	5	6	3	2	4	2	3
2	2	5	6	1	2	4	6
3	1	1	6	2	2	6	1
4	4	1	6	6	4	6	6
5	4	4	5	5	4	1	4
...	...	...	...	...	...	...	...
$N-1$	1	5	1	2	5	2	2
N	5	1	3	1	3	5	4

when using these methods, results were highly dependent on the sample set; (4) considered the data size of the decision table needed to induct the true rules with the probability of w [9], since conventional methods have not examined this problem.

This paper briefly summarizes STRIM and the conventional methods and investigates the principle of STRIM in more depth by using an example simulation experiment. Specifically, this paper summarizes STRIM as a two stage process. The first stage finds rule candidates, and the second arranges these candidates. We then consider the relationship between the principle of rule induction in the first stage and that in the conventional method, especially in VPRS (Variable Precision Rough Set) [5]. STRIM was developed independently from the conventional rough set theory. Our considerations show that STRIM can develop the notion of VPRS into a statistical principle, and the admissible classification error in VPRS corresponds to the significance level of the statistical test by STRIM. Consideration of the statistacl approach by Jaworski [10] is complex and difficult to understand, since he studies the confident intervals of accuracy and coverage of the rules inducted by VPRS.

We further examine the validity of the second process and the arrangement by STRIM, based on a statistical model which clearly shows the standard of the arrangement. In contrast, that achieved by VPRS is shown to be to not so clear, and the analyst studying the matter is required to make a decision. Both considerations in the two processes are also illustrated by the results of the simulation experiment, and seem to be useful for understanding and/or interpreting the results when analyzing real-world data sets.

2   Data Generation Model and Decision Table

Rough Sets theory is used for inducting if-then rules hidden in the decision table S. S is conventionally denoted $S = (U, A = C \cup \{D\}, V, \rho)$. Here, $U = \{u(i)|i = 1, ..., |U| = N\}$ is a sample set, A is an attribute set, $C = \{C(j)|j = 1, .., |C|\}$ is a condition attribute set, $C(j)$ is a member of C and a condition attribute, and D is a decision attribute. V is a set of attribute values denoted by $V = \bigcup_{a \in A} V_a$ and is characterized by an information function $\rho: U \times A \to V$. Table 1 shows the example where $|C| = 6$, $|V_{a=C(J)}| = M_{C(j)} = 6$, $|V_{a=D}| = M_D = 6$, $\rho(x = u(1), a = C(1)) = 5$, $\rho(x = u(2), a = C(2)) = 5$, and so on.

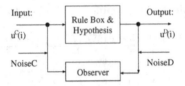

Fig. 1. Relationship between the condition attributes' value and the decision attribute's value

Table 2. Hypothesis with regard to the decision attribute value

Hypothesis 1 $u^C(i)$ coincides with $R(k)$, and $u^D(i)$ is uniquely determined as $D = d(k)$ (uniquely determined data).
Hypothesis 2 $u^C(i)$ does not coincide with any $R(d)$, and $u^D(i)$ can only be determined randomly (indifferent data).
Hypothesis 3 $u^C(i)$ coincides with several $R(d)$ $(d = d1, d2, ...)$, and their outputs of $u^C(i)$ conflict with each other. Accordingly, the output of $u^C(i)$ must be randomly determined from the conflicted outputs (conflicted data).

STRIM considers the decision table to be a sample data set obtained from an input-output system including a rule box (Fig. 1) and a hypothesis regarding the decision attribute values (Table 2). A sample $u(i)$ consists of its condition attributes values of $|C|$-tuple $u^C(i)$ and its decision attribute $u^D(i)$. $u^C(i)$ is the input into the rule box, and is transformed into the output $u^D(i)$ using the rules contained in the rule box and the hypothesis. For example, specify the following rules in the rule box as true rules to be inducted by STRIM introduced in 3:

$R(d)$: if Rd then $D = d$, $(d = 1, ..., M_D = 6)$,

where $Rd = (C(1) = d) \wedge (C(2) = d) \vee (C(3) = d) \wedge (C(4) = d)$. Generate $u^C(i) = (v_{C(1)}(i), v_{C(2)}(i), ..., v_{C(|C|)}(i))$ of $u(i)$ $(i = 1, ..., |U| = N)$ by use of random numbers with a uniform distribution, and then $u^D(i)$ is determined using the rules specified in the rule box and the hypothesis. For example, Hypothesis 1 is applied to $u(3)$ and $u(4)$, Hypothesis 2 is to $u(1)$, $u(2)$, $u(N - 1)$, $u(N)$ and Hypothesis 3 is to $u(5)$ in Table 1. In contrast, $u(i) = (u^C(i), u^D(i))$ is measured by an observer, as shown in Fig. 1. Existence of *NoiseC* and *NoiseD* makes missing values in $u^C(i)$, and changes $u^D(i)$ to create another values of $u^D(i)$, respectively. This model is closer to the real-world system. However, Table 1 is an example generated by this specification without both noises, for a plain explanation of the system. Inducting if-then rules from the decision table then identifies the rules in the rule box, by use of the set of inputs-output $(u^C(i), u^D(i))$ $(i = 1, ..., |U| = N)$.

3   Summaries of Rule Induction Procedures by STRIM

STRIM inducts if-then rules from the decision table through two processes in separate stages. The first stage process is that of statistically discriminating and

Table 3. An example of a condition part and corresponding frequency of their decision attribute values

trying $CP(k)$	$C(1)$	$C(2)$	$C(3)$	$C(4)$	$C(5)$	$C(6)$	$f = (n_1, n_2, ..., n_6)$	z
1	1	0	0	0	0	0	(469, 240, 275, 238, 224, 226)	12.52
2	2	0	0	0	0	0	(238, 454, 245, 244, 232, 219)	12.12
3	3	0	0	0	0	0	(236, 213, 477, 271, 232, 222)	13.36
4	0	0	0	0	0	6	(289, 277, 300, 255, 296, 296)	0.97
5	1	1	0	0	0	0	(235, 10, 12, 2, 7, 8)	30.77
6	1	2	0	0	0	0	(41, 47, 41, 41, 40, 49)	1.06
7	1	3	0	0	0	0	(46, 52, 57, 60, 51, 40)	1.46
8	1	0	0	0	0	6	(84, 36, 52, 38, 34, 39)	5.95
9	2	1	0	0	0	0	(46, 35, 42, 51, 37, 37)	1.73
10	2	2	0	0	0	0	(8, 227, 6, 11, 4, 6)	30.47
11	2	3	0	0	0	0	(49, 43, 44, 55, 49, 44)	1.30
12	0	0	0	0	6	6	(52, 50, 60, 46, 52, 43)	1.54
13	1	1	0	0	1	0	(38, 1, 3, 1, 0, 1)	12.61
14	1	1	0	0	2	0	(38, 3, 1, 0, 3, 3)	11.81
15	1	1	0	0	5	0	(49, 3, 2, 1, 1, 1)	14.22
16	1	1	0	0	6	0	(46, 0, 3, 0, 0, 1)	14.48
17	1	1	0	0	0	3	(45, 4, 3, 0, 3, 1)	12.97
18	2	2	0	5	0	0	(1, 45, 1, 3, 1, 0)	13.90
19	2	2	0	0	1	0	(0, 55, 0, 2, 0, 1)	16.15
20	2	2	0	0	2	0	(0, 38, 1, 1, 2, 2)	12.61
21	2	2	0	0	4	0	(4, 37, 1, 2, 0, 1)	12.00

separating the set of indifferent data from the set of uniquely determined or conflicted data in the decision table (See Table 2). Specifically, assume $CP(k) = \bigwedge_j (C(j_k) = v_j) (\in V_{C(j_k)})$ as the condition part of the if-then rule, and derive the set $U(CP(k)) = \{u(i)|u^C(i)$ satisfies $CP(k)$, which is denoted by $u^{C=CP(k)}(i)$ hereafter $\}$. Also derive $U(m) = \{u(i)|u^{D=m}(i)\}$ $(m = 1, ..., M_D)$. Calculate the distribution $f = (n_1, n_2, ..., n_{M_D})$ of the decision attributes of $U(CP(k))$, where $n_m = |U(CP(k)) \cap U(m)|$ $(m = 1, ..., M_D)$. If the assumed $CP(k)$ does not satisfy the condition $U(Rd) \supseteq U(CP(k))$ (sufficient condition of specified rule Rd) or $U(CP(k)) \supseteq U(Rd)$ (necessary condition), $CP(k)$ only generates the indifferent data set based on Hypothesis 2 in Table 2, and the distribution f does not have partiality of decisions. Conversely, if $CP(k)$ satisfies either condition, f has partiality of the distribution, since $u^D(i)$ is determined by Hypothesis 1 or 3. Accordingly, whether f has partiality or not determines whether the assumed $CP(k)$ is a neither necessary nor sufficient condition. Whether f has partiality or not can be determined objectively by a statistical test of the following null hypothesis $H0$ and its alternative hypothesis $H1$:

$H0$: f does not have partiality. $H1$: f has partiality.

In order to illustrate this concept, Table 3 shows the number of examples of $CP(k)$, $(n_1, n_2, ..., n_{M_D})$ and an index of the partiality by z derived from Table 1 with $N = 10000$. For example, the first row means the following: 100000 denotes $CP(k = 1) = (C(1) = 1)$ (the rule length is $RL = 1$) and its corresponding $f = (496, 240, 275, 238, 224, 226)$ and $z = 12.52$, where,

$$z = \frac{n_d + 0.5 - np_d}{(np_d(1 - p_d))^{0.5}}, \tag{1}$$

```
int main(void) {
int rule[|C|]={0,...,0}; //initialize trying rules
int tail=-1; //initial vale set
input data; // set decision table
rule_check(tail,rule); // 1)-5) strategies
make Pyramid(l) (l=1,2,...) so that every r(k) belongs to
one Pyramid at least; // strategy 6)
make rePyramid(l) (l=1,2,...); // strategy 7)
reduce rePyramid; // strategy 8)
} // end of main

int rule_check(int tail,int rule[|C|]) {
  for (ci=tail+1; cj<|C|; ci++) {
    for (cj=1; cj<=|C[ci]|; cj++) {
      rule[ci]=cj; // a trying rule sets for test
      count frequency of the trying rule; // count n1 n2
      if (frequency>=N0) { //sufficient frequency ?
        if (|z|>3.0) { //sufficient evidence ?
          store necessary data such as rule, frequency of n1
          and n2, and z
        } // end of if |z|
        rule_check(ci,rule);
      } // end of if frequency
    } // end of for cj
    rule[ci]=0; // trying rules reset
  } // end of for ci
} // end of rule_check
```

Fig. 2. An algorithm for STRIM (Statistical Test Rule Induction Method)

$n_d = \max(n_1, n_2, ..., n_{M_D} = n_6)$, $(d \in \{1, 2, ..., M_{D=6}\})$, $p_d = P(D = d)$, $n = \sum_{m=1}^{M_D} n_m$. In principle, $(n_1, n_2, ..., n_{M_D})$ under $H0$ obeys a multinomial distribution which is sufficiently approximated by the standard normal distribution by use of n_d under the condition[11]: $p_d n \geq 5$ and $n(1 - p_d) \geq 5$. In the same way, the fifth row 110000 denotes $CP(k = 5) = (C(1) = 1 \wedge C(2) = 1)$ $(RL = 2)$, the 13-th row 110010 denotes $C(1) = 1 \wedge C(2) = 1 \wedge C(5) = 1$ $(RL = 3)$, and so on. Here, if we specify a standard of the significance level such as $z \geq z_\alpha = 3.0$ and reject $H0$, then the the assumed $CP(k)$ becomes a candidate for the rules in the rule box. For example, see $CP(1)$ having $z = 12.52 \geq z_\alpha = 3.0$ in Table 3 and confirm the partiality of f that n_1 is much greater than n_l $(l = 2, ..., 6)$.

The second stage process is that of arranging the set of rule candidates derived from the first process, and finally estimating the rules in the rule box, since some candidates may satisfy the relationship: $CP(ki) \supseteq CP(kj) \supseteq CP(kl)$ ..., for example, in the case $100000 \supset 110000 \supset 110010$ (see Table 3). The basic notion is to represent the $CP(k)$ of the maximum z, that is, the maximum partiality. In the above example, STRIM selects the $CP(k)$ of 110000, which by chance coincides with the rule specified in advance. Figure 2 shows the STRIM algorithm[8].

Table 4 shows the estimated results for Table 1 with $N = 10000$. STRIM inducts all of twelve rules specified in advance, and also one extra rule. However, there are clear differences between them in the indexes of accuracy and coverage.

Table 4. Results of estimated rules for the decision table in Table 1 by STRIM

esti-mated rule $R(i)$	$C(1)$	$C(2)$	$C(3)$	$C(4)$	$C(5)$	$C(6)$	D	$f = (n_1, ..., n_6)$	p-value (z)	accuracy	coverage
1	5	5	0	0	0	0	5	(7,8,5,7,271,4)	0(34.15)	0.897	0.162
2	0	0	1	1	0	0	1	(243,6,5,6,4,3)	0(32.68)	0.910	0.148
3	4	4	0	0	0	0	4	(10,2,8,252,7,6)	0(32.58)	0.884	0.150
4	0	0	5	5	0	0	5	(5,5,6,11,249,7)	0(32.27)	0.880	0.149
5	6	6	0	0	0	0	6	(10,12,4,7,6,253)	0(32.16)	0.866	0.154
6	3	3	0	0	0	0	3	(6,3,254,13,8,12)	0(31.00)	0.858	0.150
7	0	0	2	2	0	0	2	(4,243,2,8,5,14)	0(31.90)	0.880	0.146
8	0	0	3	3	0	0	3	(11,8,243,5,7,7)	0(31.48)	0.865	0.143
9	0	0	6	6	0	0	6	(7,2,8,10,9,240)	0(31.41)	0.870	0.146
10	0	0	4	4	0	0	4	(8,12,13,245,7,7)	0(30.91)	0.839	0.146
11	1	1	0	0	0	0	1	(235,10,12,2,7,8)	0(30.77)	0.858	0.143
12	2	2	0	0	0	0	2	(8,227,6,11,4,6)	0(30.47)	0.866	0.136
13	0	0	0	0	1	1	2	(39,61,44,44,35,31)	6.26e-4(3.23)	0.240	0.037

4 Studies of the Conventional Methods and Their Problems

The most basic strategy to induct the rules from a decision table is to use the inclusion relationship between the set derived by the condition attributes and the set by the decision attribute. Many methods of achieving this have been proposed [4–7]. Figure 3, for example, shows the well-known LEM2 algorithm[4]. In this B at LN (Line No.) $= 0$ is specified like $B = U(d) = \{u(i)|u^{D=d}(i)\}$ removing the conflicted data set. LEM2 with lower approximation derives $CP(k)$, satisfying $U(d) \supseteq U(CP(k))$. In the figure, t corresponds to $C(j_k) = v_{C(j_k)}$, $[t]$ to $U(t) = \{u(i)|u^{C=t}(i)\}$, T to $CP(k)$, and the final result τ to $\bigvee_k CP(k)$ is obtained by repeating from $LN = 3$ to $LN = 16$ until the condition $U(d) = U(\bigvee_k CP(k))$ is satisfied.

However, as previously shown [8], LEM2 is likely to induct many sub-rules of their true rules with longer rule length, since it executes the algorithm until the condition $U(d) = U(\bigvee_k CP(k))$ is satisfied. In 1993 Ziarko [5] introduced the variable precision rough set, which inducts $CP(k)$ satisfying the following conditions:

$$C_\varepsilon(U(d)) = \{u(i)|acc \geq acc0,$$
$$acc = |U(d) \cap U(CP(k))|/|U(CP(k))| = n_d/n\}, \qquad (2)$$

where acc is accuracy of the rule and acc0 is a constant depending on ε. Ziarko further defined (2) in two cases as follows:

$$\underline{C}_\varepsilon(U(d)) = \{u(i)|acc \geq 1 - \varepsilon\}, \qquad (2a)$$
$$\overline{C}_\varepsilon(U(d)) = \{u(i)|acc \geq \varepsilon\}, \qquad (2b)$$

where $\varepsilon \in [0, 0.5)$ is an admissible classification error. $\underline{C}_\varepsilon(U(d))$ and $\overline{C}_\varepsilon(U(d))$ are respectively called a ε-lower and ε-upper approximation of VPRS, and coincides with the ordinary lower and upper approximation by $\varepsilon = 0$. Their difference

```
Line Procedure LEM2
No.
  0 (input: a set B
     output: a single local covering τ of set B);
  1 begin
  2   G := B;
      τ := φ;
  3   while G ≠ φ
  4   begin
  5     T := φ;
        T(G) := {t|[t] ∩ G ≠ φ};
  6     while T = φ or not([T] ⊆ B)
  7     begin
  8       select a pair t ∈ T(G)
          such that |[t] ∩ G| is maximum;
          if a tie occurs,
          select a pair t ∈ T(G)
          with the smallest cardinality of [t];
          if another tie occurs,
          select first pair;
  9       T := T ∪ {t};
 10       G := [t] ∩ G;
 11       T(G) := {t|[t] ∩ G ≠ φ};
 12       T(G) := T(G) - T;
 13     end {while}
 14     for each t ∈ T do
          if [T - {t}] ⊆ B then T := T - {t};
 15     τ := τ ∪ {T};
        G := B - ∪_{T∈τ}[T];
 16   end {while};
 17   for each T ∈ τ do
        if ∪_{S∈τ-{T}}[S] = B then τ := τ - {T};
 18 end {procedure}.
```

Fig. 3. An algorithm for LEM2

is in the range of their accuracy; that is the accuracy of $\underline{C}_\varepsilon(U(d)) \in (0.5, 1.0]$ and that of $\overline{C}_\varepsilon(U(d)) \in (0.0, 0.5]$. VPRS adopts the rules with the high index of coverage defined by $cov = |U(d) \cap U(CP(k))| / |U(d)|$; this can squeeze the above sub-rules. VPRS has been widely used for solving real-world problems, and a variety of modified VPRSs have been proposed [12–14]. However, the standard of adopting rules is not so clear. For example, (acc, cov) of $CP(k = 1)$, $CP(k = 5)$ and $CP(13)$ in Table 3 are $(0.281, 0.285)$, $(0.857, 0.143)$ and $(0.864, 0.0231)$ respectively. $CP(k = 13)$ should be adopted as the most accurate, $CP(k = 1)$ as the widest coverage, and $CP(k = 5)$ as the moderate index of both; this requires a decision by the analyst studying the matter. This unclearness and lack of the standard lead the making of an algorithm such as LEM2 to difficulty.

Jaworski [10] further pointed out the problem in VPRS that the standard of adopting rules by using (acc, cov) is highly dependent on the decision table; that is the sample set, as (acc, cov) will change in each sample set. He then extended the decision table in the sample set to that in the population, and proposed a type of confidence interval for each index. For example, for the index of accuracy,

$$P(acc|population) \geq acc|_{sample} - \sqrt{\frac{\ln(1 - \gamma_n)}{-2n}}, \qquad (3)$$

where, γ_n is a degree of confidence. In $R(i = 1)$ of Table 4, $acc|_{sample} = 0.897$, $n = 302$ and let specify $\gamma_n = 0.85$, then $P(acc|population) \geq 0.897 - 0.037 = 0.860$. This means that the accuracy in the population is greater than 0.860, with the degree of confidence of 0.85. However, two-story uncertainty $acc \in [0.0, 1.0]^{[0.0, 1.0]}$ is very complicated, and hard to understand.

5   Studies of the Relationship between VPRS and STRIM

Let us consider a $CP(k)$ satisfying (2a). The greater part of the decision attribute value of the $U(CP(k))$ is now included in $U(d)$, since the $CP(k)$ satisfies (2a). Accordingly, the distribution of the decision attribute value of the $U(CP(k))$ has partiality in $D = d$, which coincides with the basic concept of STRIM. As shown in 3, STRIM statistically tests whether $(n_1, n_2, ..., n_{M_D})$ is partial or not, and gives the decision with a significance level z_α. In (2) n_d satisfies the event $n_d \geq n \cdot acc0$, and the probability of the event is evaluated thus:

$$P(n_d \geq n - n \cdot \varepsilon) = P\left(\frac{n_d + 0.5 - np_d}{(np_d(1 - p_d))^{0.5}} \geq \frac{n + 0.5 - n \cdot \varepsilon - np_d}{(np_d(1 - p_d))^{0.5}}\right). \quad (4)$$

Accordingly, $z_\alpha = \dfrac{n + 0.5 - n \cdot \varepsilon - np_d}{(np_d(1 - p_d))^{0.5}}$ is obtained comparing (4) with (1) and then the following relationship between $acc0$ and z_α holds:

$$acc0 = 1 - p_d + 0.5/n - z_\alpha \left(\frac{p_d}{n}(1 - p_d)\right)^{0.5} \quad (5)$$

In $R(i = 1)$ in Table 4, substituting $n = 302$, $p_d = 1/6$ for (5), $acc0 = 0.229 \in (0.0, 0.5]$ and then $\varepsilon = 0.229$. Let specify $z_\alpha = 30.0$ since $z = 34.5$ in $R(i = 1)$, then $acc0 = 0.808 \in (0.5, 1.0]$ and $\varepsilon = 1 - acc0 = 0.192$.

The covering index $cov = n_d/|U(d)|$ in VPRS is considered to reflect the degree of support and/or sufficient evidence of the inducted rule. However, the standard of the degree of support has not yet been clearly expressed. On the other hand, STRIM requested the number of data as the evidence satisfying the conditions [11]: $p_d n \geq 5$ and $n(1 - p_d) \geq 5$, which are needed for testing $H0$. Accordingly, the covering index corresponds to the testing condition in STRIM and the condition is clearly given by statistics, whether the covering index is somewhat higher or lower (See Table 4).

As the relationships considered above between VPRS and STRIM, STRIM can yield the validity of inducting rules; that is, the clear meaning and standard of the index of accuracy and coverage for the inducted rules from statistical viewpoints.

6   Studies of Arrangement of Rule Candidates

There may be rule candidates satisfying relationships $CP(ki) \supset CP(kj) \supset CP(kl)$ ... after the first stage process. STRIM selects the candidate with the

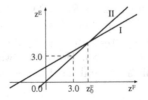

Fig. 4. Relationship between z^F and z^E

maximum partiality in the relationship. Let us consider whether STRIM is assured of selecting the true rule in the rule box. Hereafter, $CP(F)$ denotes the true rule in the rule box, and we assume that it satisfies the relationship $CP(E) \supset CP(F) \supset CP(G)$ and has the maximum n_d at $d = 1$ like $CP(E) = "100000"$, $CP(F) = "110000"$ and $CP(G) = "110010"$ in Table 3 and the distribution of $f = (n_1^F, n_2^F, ..., n_{M_D}^F)$. Then: $z^F = \dfrac{n_1^F + 0.5 - n^F p_1}{(n^F p_1(1 - p_1))^{0.5}} \simeq \dfrac{n^F(a^F - p_1)}{\sigma^F}$. Here, 0.5

$<< n^F p_1$, $n_1^F = a^F n^F$ $(0 < a^F \le 1)$, $n^F = \displaystyle\sum_{j=1}^{M_D} n_j^F$ and $\sigma^F = (n^F p_1(1 - p_1))^{0.5}$.

In the same way with regard to G, $z^G = \dfrac{n_1^G + 0.5 - n^G p_1}{(n^G p_1(1 - p_1))^{0.5}} \simeq \dfrac{n^G(a^G - p_1)}{\sigma^G}$.

Here $U(F) \supset U(G)$, $a^G \simeq a^F$ and $n^F > n^G$ lead to $n^G = rn^F$ $(0 < r < 1)$

Accordingly, $z^G \simeq \dfrac{n^G(a^G - p_1)}{\sigma^G} = \dfrac{r^{0.5} n^F(a^F - p_1)}{\sigma^F} = r^{0.5} z^F$, which means z^G

$< z^F$ and STRIM selects not $CP(G)$ but $CP(F)$. In the same way with regard to E, $z^E = \dfrac{n_1^E + 0.5 - n^E p_1}{(n^E p_1(1 - p_1))^{0.5}}$. Here, $n_1^E = n_1^F + n_1^{EF} = a^E n^E + a^{EF} n^{EF}$,

$n^{EF} = |U(EF)|$ and $U(EF) = U(E) - U(F)$. $U(EF)$ is an indifferent data set (See Hypothesis 2). Taking their relationships into consideration, the following

equation holds: $z^E \simeq \dfrac{n^F(a^F - p_1)}{\sigma^E} + \dfrac{n^{EF}(a^{EF} - p_1)}{\sigma^E} = \dfrac{\sigma^F}{\sigma^E} z^F + \dfrac{\sigma^{EF}}{\sigma^E} z^{EF} =$

$s_E^F z^F + s_E^{EF} z^{EF}$. Here $n^F < n^E$ and $n^{EF} < n^E$ lead to $\dfrac{\sigma^F}{\sigma^E} = s_E^F < 1$ and $\dfrac{\sigma^{EF}}{\sigma^E}$

$= s_E^{EF} < 1$. Figure 4 shows the relationship between z^F and z^E by use of the following two lines: $z^E = s_E^F z^F + s_E^{EF} z^{EF}$ (I), $z^E = z^F$ (II). The cross point

of the two lines is $\left(z_0^F = \dfrac{s_E^{EF}}{1 - s_E^F} z^{EF}, z_0^F \right)$. Accordingly, STRIM always selects

$CP(F)$ if the inequality $z^F \ge z_0^F$ holds. As conclusion in this section, STRIM necessarily selects $CP(F)$ of the true rule only if $z^F \ge z_0^F$ holds. In Table 1, $s_E^F = \left(\dfrac{1}{M_C} \right)^{0.5}$, $s_E^{EF} = \left(1 - \dfrac{1}{M_C} \right)^{0.5}$ and $M_C = 6$, and $|z^{EF}| < z_\alpha = 3.0$ holds with

less than 1 [%] error. Accordingly, if $z^F \ge z_0^F = \dfrac{s_E^{EF}}{1 - s_E^F} \simeq 5$ holds, then STRIM selects not $CP(E)$ but $CP(F)$. We can confirm the validity of the consideration of this section in Table 3 and 4. Especially, note that $R(i = 13)$ in Table 4 does not satisfy the condition $z \geq 5.0$, and doubt of the inducted result thus arises.

7   Conclusions

This paper summarized the basic concept of the rule induction method by STRIM [8, 9], and the conventional methods, especially VPRS [5] and their problems, using a simulation experiment. We illustrated the following features and relationships between results from the STRIM model and those from conventional methods, especially VPRS[5]:

1) VPRS uses the indexes of accuracy and coverage with an admissible error when it selects the rule candidates. The accuracy can be recognized as the index of the partiality of the distribution of the decision attribute values for the trying rule, which coincides with the idea of STRIM. The corresponds to the significance level by z_α in STRIM. The coverage corresponds to the applicable condition for a statistical test by STRIM. However, VPRS does not have the objective standard of both indexes to select rule candidates, since to date the conventional methods do not view the decision table as a sample data set obtained from its population.
2) STRIM provides assurance for an analyst searching for the true rules under the proper conditions studied in 6, as the results show whether those rules inducted are true or not. In contrast, VPRS provides no such assurance, since it is not a method based on a data generating model, as is STRIM.

Focus for future studies:

1) To consider relationship to Variable Consistency Rough Sets Approaches (VC-IRSA and VC-DRSA) [15].
2) To see how good are rules found by STRIM in a accuracy cross-validation experiment comparing them to the ones found by LEM2 [4] and/or other classifiers.
3) To consider relationship to rule quality measures which seek to find a trade-off between the rule precision (rule accuracy) and coverage [16].

References

1. Pawlak, Z.: Rough sets. Internat. J. Inform. Comput. Sci. 11(5), 341–356 (1982)
2. Skowron, A., Rauszer, C.M.: The Discernibility Matrix and Functions in Information Systems. In: Slowinski, R. (ed.) Intelligent Decision Support, Handbook of Application and Advances of Rough Set Theory, pp. 331–362. Kluwer Academic Publishers (1992)
3. Bao, Y.G., Du, X.Y., Deng, M.G., Ishii, N.: An Efficient Method for Computing All Reducts. Transaction of the Japanese Society for Artificial Intelligence 19(3), 166–173 (2004)

4. Grzymala-Busse, J.W.: LERS- A system for learning from examples based on rough sets. In: Słowński, R. (ed.) Intelligent Decision Support. Handbook of Applications and Advances of the Rough Sets Theory, pp. 3–18. Kluwer Academic Publishers (1992)

5. Ziarko, W.: Variable precision rough set model. Journal of Computer and System Science 46, 39–59 (1993)

6. Shan, N., Ziarko, W.: Data-based acquisition and incremental modification of classification rules. Computational Intelligence 11(2), 357–370 (1995)

7. Nishimura, T., Kato, Y., Saeki, T.: Studies on an Effective Algorithm to Reduce the Decision Matrix. In: Kuznetsov, S.O., Ślęzak, D., Hepting, D.H., Mirkin, B.G. (eds.) RSFDGrC 2011. LNCS (LNAI), vol. 6743, pp. 240–243. Springer, Heidelberg (2011)

8. Matsubayashi, T., Kato, Y., Saeki, T.: A new rule induction method from a decision table using a statistical test. In: Li, T., Nguyen, H.S., Wang, G., Grzymala-Busse, J., Janicki, R., Hassanien, A.E., Yu, H. (eds.) RSKT 2012. LNCS (LNAI), vol. 7414, pp. 81–90. Springer, Heidelberg (2012)

9. Kato, Y., Saeki, T., Mizuno, S.: Studies on the Necessary Data Size for Rule Induction by STRIM. In: Lingras, P., Wolski, M., Cornelis, C., Mitra, S., Wasilewski, P. (eds.) RSKT 2013. LNCS (LNAI), vol. 8171, pp. 213–220. Springer, Heidelberg (2013)

10. Jaworski, W.: Rule Induction: Combining Rough Set and Statistical Approaches. In: Chan, C.-C., Grzymala-Busse, J.W., Ziarko, W.P. (eds.) RSCTC 2008. LNCS (LNAI), vol. 5306, pp. 170–180. Springer, Heidelberg (2008)

11. Walpole, R.E., Myers, R.H., Myers, S.L., Ye, K.: Probability & Statistics for Engineers & Scientists, 8th edn., pp. 187–194. Pearson Prentice Hall (2007)

12. Xiw, G., Zhang, J., Lai, K.K., Yu, L.: Variable precision rough set group decision-making, An application. International Journal of Approximate Reasoning 49, 331–343 (2008)

13. Inuiguchi, M., Yoshioka, Y., Kusunoki, Y.: Variable-precision dominance-based rough set approach and attribute reduction. International Journal of Approximate Reasoning 50, 1199–1214 (2009)

14. Huang, K.Y., Chang, T.-H., Chang, T.-C.: Determination of the threshold β of variable precision rough set by fuzzy algorithms. International Journal of Approximate Reasoning 52, 1056–1072 (2011)

15. Greco, S., Matarazzo, B., Słowiński, R., Stefanowski, J.: Variable Consistency Model of Dominance-Based Rough Sets Approach. In: Ziarko, W., Yao, Y. (eds.) RSCTC 2000. LNCS (LNAI), vol. 2005, pp. 170–181. Springer, Heidelberg (2001)

16. Janssen, F., Fürnkranz, J.: On the quest for optimal rule learning heuristics. Machine Learning 78, 343–379 (2010)

Generating Core in Rough Set Theory: Design and Implementation on FPGA

Maciej Kopczynski, Tomasz Grzes, and Jaroslaw Stepaniuk

Faculty of Computer Science
Bialystok University of Technology
Wiejska 45A, 15-351 Bialystok, Poland
{m.kopczynski,t.grzes,j.stepaniuk}@pb.edu.pl
http://www.wi.pb.edu.pl

Abstract. In this paper we propose the FPGA based device for data processing using rough set methods. Presented architecture has been tested on a real-world data. Obtained results confirm the huge acceleration of the computation time using hardware supporting core generation in comparison to software implementation.

Keywords: Rough sets, FPGA, hardware, core.

1 Introduction

The theory of rough sets has been developed in the eighties of the twentieth century by Prof. Z. Pawlak. Rough sets are used as a tool for data analysis and classification as well as for the extraction of important characteristics that describe the objects.

There exist many software implementations of rough set methods and algorithms. However, they require significant amount of resources of a computer system and a lot of time for algorithms to complete, especially during processing large amount of data.

Field Programmable Gate Arrays (FPGAs) are a group of integrated circuits, whose functionality is not defined by the manufacturer, but by the user using a hardware description language, such as VHDL, which allows describing architecture and the functional properties of the digital system [1].

At the moment there is no comprehensive hardware implementation of rough set methods. In the literature one can find descriptions of concepts or partial rough set methods hardware implementations. The idea of sample processor generating decision rules from decision tables was described in [5]. In [4] was presented architecture of rough set processor based on cellur networks. More detailed summary of the existing ideas and hardware implementations of rough set methods can be found in [3]. Previous authors' research results focused on this subject can be found in [2,9].

The paper is organized as follows. In Section 2 some information about the notion of core and dataset used during research are provided. The Section 3 focuses on description of hardware solution, while Section 4 is devoted to the experimental results.

M. Kryszkiewicz et al. (Eds.): RSEISP 2014, LNAI 8537, pp. 209–216, 2014.

2   Introductory Information

2.1   The Notion of Core in the Rough Set Theory

In decision table some of the condition attributes may be superfluous. This means that their removal cannot worsen the classification. The set of all indispensable condition attributes is called the core. One can also observe that the core is the intersection of all decision reducts - each element of the core belongs to every reduct. Thus, in a sense, the core is the most important subset of condition attributes. In order to compute the core we can use discernibility matrix. Below one can find pseudocode for calculating core using discernibility matrix.

INPUT: discernibility matrix DM
OUTPUT: core C

```
1. C ← ∅
2. for x ∈ U do
3.     for y ∈ U do
4.         if |DM(x, y)| = 1 then
5.             C ← C ∪ DM(x, y)
6.         end if
7.     end for
8. end for
```

Core is initialized as empty set in line 1. Two loops in lines 2 and 3 iterates over all objects (denoted as U) in discernibility matrix. Condition instruction in line 4 checks if matrix cell contains only one attribute. If so, then this attribute is added to the core C.

A much more detailed description of the concept of the core can be found, for example, in the article [6] or in the book [8].

2.2   Data to Conduct Experimental Research

In this paper, we conduct experimental studies using data about children with insulin-dependent diabetes mellitus (type 1). Insulin-dependent diabetes mellitus is a chronic disease of the body's metabolism characterized by an inability to produce enough insulin to process carbohydrates, fat, and protein efficiently. Treatment requires injections of insulin. Twelve condition attributes, which include the results of physical and laboratory examinations and one decision attribute (microalbuminuria) describe the database used in our experiments. The data collection so far consists of 107 cases. All this information is collected during treatment of diabetes mellitus.

The database is shown at the end of the paper [7]. A detailed analysis of the above data (only with the use of software systems) is in chapter 6 of the book [8].

Dataset had to be transformed to binary version. Exemplary transformation of three first objects is presented below. Numerical values were discretized and

Table 1. Exmaple of data encoding for first three objects in dataset

Sex	Age	Time	Family	Type	Infect.	Remis.	HbA1c	HT	BM	Ch	TG	d
f	12	5	no	KITIIT	yes	no	7.28	yes	3-97	no	no	yes
m	1	4	no	KIT	yes	no	10.00	no	3-97	no	no	no
m	15	5	yes	KIT	yes	no	6.65	no	3-97	no	no	no
01	01	00	00	01	01	00	00	01	01	00	00	01
00	00	00	00	00	01	00	10	00	01	00	00	00
00	10	00	01	00	01	00	00	00	01	00	00	00

each attributes' value was encoded using two bits. Table 1 presents objects with original and encoded values.

Transformed data is stored in wide-word FPGA internal memory, thus prepared hardware units doesn't have to reconfigured for different datasets until these datasets fit into configured and compiled unit.

3 Hardware Implementation

Authors prepared three implementations of the discernibility matrix related methods of the core calculation:

- combinational circuit with "OR-cascade" (referred as "cascade"),
- combinational circuit with "big OR" (referred as "big OR"),
- mixed combinational/sequential circuit (referred as "mixed").

The architecture of the first solution is shown on Fig. 1 in part a). Input of this block is decision table. Circuit consists of three functionally separated blocks:

1. **Comparators** – block of identical comparators which calculate the entries of discernibility matrix. Each comparator has two inputs which are connected to two objects of decision table.
2. **OR-gates cascade** – block of OR-gates connected in a cascade. Every gate calculates logical OR operation on two elements: one from previous gate in a cascade and second from comparator. Result goes to next OR gate and finally to **CORE** register which stores the result of calculations. Calculation of OR operation can be blocked by input from the **Singleton Detector**.
3. **Singleton Detector** – block for checking if entry in discernibility matrix is a singleton, ie. consists of only one logical '1'. Outputs from this block are connected to OR-gates cascade.

Discernibility matrix entries are calculated by comparators very quickly, mostly because of simplicity of each comparator architecture. Then all entries goes to OR-gates cascade. The time to calculate the result depends on the size of discernibility matrix, increasing with its size. Last gate in cascade stores the result of calculations in the **CORE** register.

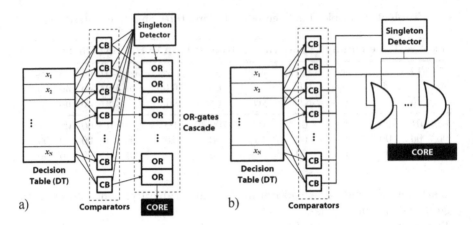

Fig. 1. Hardware implementation block diagram of the (a) "cascade" and (b) "big OR" core calculation unit

The architecture of the second solution is shown on Fig. 1 in part b). The difference between these two solutions is the implementation of the OR-gates. In the second solution we have one OR-gate for one bit of **CORE** register. Each gate has as many inputs as the number of comparators. Additionally every input can be gated by a signal from the **Singleton Detector** block.

The time to calculate the result also depends on the size of discernibility matrix as in first solution, but the dependence is much smaller because of elimination of cascade.

Two solutions described previously were pure combinational circuits. The main disadvantage of these circuits is very big resources utilization of the FPGA. Thus to limit the utilization of LEs (Logical Elements) the third solution was prepared which is mixed combinational/sequential unit. Block diagram of the third solution is shown on Fig. 2.

Main differences to previous solutions are:

1. The number of comparators is equal to number of objects in decision table, not the number of elements in discernibility matrix.
2. The number of OR-gates in a cascade is equal to number of objects in decision table, not the number of elements in discernibility matrix.
3. Multiplexer MUX in every turn selects the following object from decision table and puts it into the comparators.
4. Control Logic block counts the objects in decision table, and sends latching signal for temporary register (TEMP).

Third implementation is much smaller in terms of resources utilization than previous two, but needs more time to calculate the final result. Number of cycles needed to complete the calculation is equal to the number of objects in the decision table.

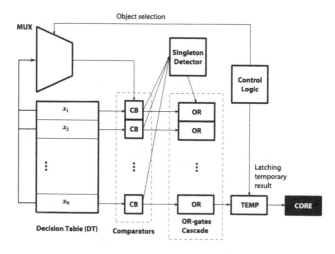

Fig. 2. Hardware implementation block diagram of mixed combinational/sequential solution

4   Experimental Results

For the research purpose the same rough set methods were implemented in C language. Results of the software implementation were obtained using a PC equipped with an 8 GB RAM and 4-core Intel Core i7 3632QM with maximum 3.2 GHz in Turbo mode clock speed running Windows 7 Professional operational system. The source code of application was compiled using the GNU GCC 4.8.q compiler. Given times are averaged for 10 000 runs of each algorithm with the same set of data.

The hardware implementation used single execution unit module covering the entire decision table. Quartus II 13.1 was used for creation, compilation, synthesis and verifying simulation of the hardware implementation in VHDL language. Sythesized hardware blocks were downloaded and run on TeraSIC DE-3 equipped with Stratix III EP3SL150F1152C2N FPGA chip. FPGA clock running at 50 MHz for the sequential parts of the project was derived from development board oscillator. Timing results were obtained using Tektronix TDS3052B (500 MHz bandwidth, 5 GS/s) oscilloscope.

It should be noticed, that PCs clock is 64 times faster than development boards clock source.

All calculations were performed using diabetes dataset. Presented results have been obtained using full set and smaller subsets (in terms of objects).

Two out of three versions of core solutions use the discernibility matrix to return the final result. For the sake of clarity, columns labeled with w/DM include time required for discerniblity matrix and core calculation. Columns labeled with $w/o\ DM$ present time for performing pure core operation.

Table 2 presents the results of the time elapsed for software and hardware solutions for the core calculation (hardware version "cascade").

Table 2. Comparision of execution time for calculating core (hardware version "cascade")

Objects	Software - t_S		Hardware - t_H		$\frac{t_S}{t_H}$	
	w/DM	w/o DM	w/DM	w/o DM	w/DM	w/o DM
—	[μs]	[μs]	[μs]	[μs]	—	
15	112.28	2.28	0.0084	0.0052	13 366	438
30	420.22	10.22	0.0104	0.0054	40 405	1 892
45	983.42	33.42	0.0216	0.0132	45 528	2 531
60	1 736.67	66.67	0.0310	0.0196	56 021	3 401
90	4 074.80	154.8	0.0584	0.0392	69 773	3 948
107	5 990.00	210.00	0.0683	0.0437	87 701	4 805

Table 3 presents the results of the time elapsed for software and hardware solutions for the core calculation (hardware version "big OR").

Table 3. Comparision of execution time for calculating core (hardware version "big OR")

Objects	Software - t_S		Hardware - t_H		$\frac{t_S}{t_H}$	
	w/DM	w/o DM	w/DM	w/o DM	w/DM	w/o DM
—	[μs]	[μs]	[μs]	[μs]	—	
15	112.28	2.28	0.0080	0.0048	14 035	475
30	420.22	10.22	0.0110	0.0060	38 201	1 703
45	983.42	33.42	0.0188	0.0104	52 309	3 213
60	1 736.67	66.67	0.0242	0.0128	71 763	5 208
90	4 074.80	154.8	0.0392	0.0200	103 948	7 740
107	5 990.00	210.00	0.0498	0.0252	120 281	8 333

Table 4 presents the results of the time elapsed for software and hardware solutions for the core calculation (hardware version "mixed"). Table doesn't include previously introduced discerniblity matrix labels because this version of core calculation derives its result from decision table (using discerniblity matrix explicitly). Results for software solution include time needed for discernibility matrix calculation.

Table 5 presents the FPGA structure resources utilization in LEs (Logical Elements) basis for each implemented type and version of block.

Presented results show a huge increase in the speed of data processing. Hardware module execution time compared to the software implementation is at least 1 order of magnitude shorter for sequential core calclulation hardware units what is shown in Table 4 in columns $\frac{t_S}{t_H}$ and is increasing with larger data sets. For the combinational versions of hardware units, times are 2 to 3 orders of magnitude shorter. After taking calculation time for creating discernibility matrix under consideration, these results are getting even better.

Table 4. Comparision of execution time for calculating core (hardware version "mixed")

Objects	Software - t_S	Hardware - t_H	$\frac{t_S}{t_H}$
—	$[\mu s]$	$[\mu s]$	—
15	112.28	0.57	196
30	420.22	1.17	359
45	983.42	1.77	555
60	1 736.67	2.40	723
90	4 074.80	3.58	1 138
107	5 990.00	4.26	1 406

Table 5. FPGA structure resources utilization in LEs

Objects	DMB	Core "cascade"	"big OR"	"mixed"
15	1 365	1 374	1 374	1 252
30	5 655	5 675	5 675	1 979
45	12 870	12 666	12 666	3 418
60	23 010	22 863	22 863	3 541
90	52 065	52 339	52 339	5 335
107	73 723	74 916	74 916	7 171

Let comparison of objects' attribute value in the decision table or getting an attribute from the data container (e.g. linked list) be an elementary operation. Let's assume that hardware module is big enough to store the entire decision table. k in the table denotes number of conditional attributes and n is the number of objects in decision table.

First step of implemented algorithms is discernibility matrix calculation. Using this matrix, the cores are calculated. Computational complexity of software implementation for the creating discernibility matrix is $\Theta(n^2 k)$ and using hardware implementation, complexity is $\Theta(1)$. Computational complexity of software implementation for the core calculation is $\Theta(n^2)$. Using hardware implementation, complexity of core calculation is $\Theta(1)$ for both "cascade" and "big OR" hardware versions and $\Theta(n)$ for "mixed" hardware version.

Table 5 shows FPGA structure utilization. Combinational solutions consumes much more resources than sequential types, but these are few orders of magnitude faster. For practical solutions sequential units are preferred because of their resources saving implementation and FPGA limited LEs. Even sequential implementation with 64 times slower clock than PCs clock is much faster than software solution.

Of course, for most real data sets it is impossible to create a single hardware structure capacious enough to store the entire data set. In this case, the input data set must be divided into a number of subsets, where each of them will be separately processed by a single hardware unit.

216 M. Kopczynski, T. Grzes, and J. Stepaniuk

5 Conclusions and Future Research

The hardware implementation is the main direction of using scalable rough sets methods in real time solutions. Software implementations are universal, but rather slow. Hardware realizations are deprived of this universality, however, allow us performing specific calculations in substantially shorter time.

Further research will focus on developing methods for transforming sequential rough sets algorithms into mixed combinational-sequential hardware solutions. The effort will be put also towards developing methods for explicitly dividing large input dataset into smaller subsets.

Acknowledgements. The research is supported by the Polish National Science Centre under the grant 2012/07/B/ST6/01504 and by the scientific grant S/WI/3/2013. Maciej Kopczynski is a beneficiary of the project "Scholarships for PhD students of Podlaskie Voivodeship". The project is co-financed by European Social Fund, Polish Government and Podlaskie Voivodeship.

References

1. Athanas, P., Pnevmatikatos, D., Sklavos, N. (eds.): Embedded Systems Design with FPGAs. Springer (2013)
2. Grześ, T., Kopczyński, M., Stepaniuk, J.: FPGA in rough set based core and reduct computation. In: Lingras, P., Wolski, M., Cornelis, C., Mitra, S., Wasilewski, P. (eds.) RSKT 2013. LNCS (LNAI), vol. 8171, pp. 263–270. Springer, Heidelberg (2013)
3. Kopczyński, M., Stepaniuk, J.: Hardware Implementations of Rough Set Methods in Programmable Logic Devices. In: Skowron, A., Suraj, Z. (eds.) Rough Sets and Intelligent Systems - Professor Zdzisław Pawlak in Memoriam. ISRL, vol. 43, pp. 309–321. Springer, Heidelberg (2013)
4. Lewis, T., Perkowski, M., Jozwiak, L.: Learning in Hardware: Architecture and Implementation of an FPGA-Based Rough Set Machine. In: 25th Euromicro Conference (EUROMICRO 1999), vol. 1, p. 1326 (1999)
5. Pawlak, Z.: Elementary rough set granules: Toward a rough set processor. In: Pal, S.K., Polkowski, L., Skowron, A. (eds.) Rough-Neurocomputing: Techniques for Computing with Words, Cognitive Technologies, pp. 5–14. Springer, Berlin (2004)
6. Pawlak, Z., Skowron, A.: Rudiments of rough sets. Information Sciences 177(1), 3–27 (2007)
7. Stepaniuk, J.: Knowledge discovery by application of rough set models. In: Polkowski, L., Tsumoto, S., Lin, T.Y. (eds.) Rough Set Methods and Applications. New Developments in Knowledge Discovery in Information Systems, pp. 137–233. Physica–Verlag, Heidelberg (2000)
8. Stepaniuk, J.: Rough–Granular Computing in Knowledge Discovery and Data Mining. Springer (2008)
9. Stepaniuk, J., Kopczyński, M., Grześ, T.: The First Step Toward Processor for Rough Set Methods. Fundamenta Informaticae 127, 429–443 (2013)

Attribute Ranking Driven
Filtering of Decision Rules

Urszula Stańczyk

Institute of Informatics, Silesian University of Technology,
Akademicka 16, 44-100 Gliwice, Poland

Abstract. In decision rule induction approaches either minimal, complete, or satisfying sets of constituent rules are inferred, with an aim of providing predictive properties while offering descriptive capabilities for the learned concepts. Instead of limiting rules at their induction phase we can also execute post-processing of the set of generated decision rules (whether it is complete or not) by filtering out those that meet some constraints. The paper presents the research on rule filtering while following a ranking of conditional attributes, obtained in the process of sequential forward selection of input features for ANN classifiers.

Keywords: Decision Algorithm, Decision Rule, DRSA, Feature Selection, Ranking, Rule Filtering, ANN, Stylometry, Authorship Attribution.

1 Introduction

A decision algorithm is a form of a classification system that allows for very explicit and direct representation of knowledge inferred from available data by its constituent rules. The premise part of each rule specifies conditions which need to be met in order for the rule to be applicable and to classify to the class indicated in the decision part of the rule. Classification accuracy of the algorithm depends on the quality of rules, that, in turn, depend on available data from which they are inferred, but also on the applied approach to induction.

We can calculate only a minimal subset of rules, all rules for provided learning examples, or, using some additional search criteria or constraints, a set that is neither minimal nor complete, but satisfying and containing only some interesting rules [1]. Yet another approach is to execute post-processing of the already induced set of rules, by tailoring it to some specific purposes. The rules are filtered by exploiting some weights or measures of importance [13], quality [9], or re-defined to provide a better fit for a given task [8].

The paper presents a methodology where firstly a ranking of the considered attributes is obtained through the procedures of sequential forward feature selection [5] for artificial neural networks (ANN) [3] working in the wrapper model. Next, the complete decision algorithm is generated within Dominance-Based Rough Set Approach (DRSA) [10] and filtering of its rules is executed while following the previously found ranking of variables. In the process there are

M. Kryszkiewicz et al. (Eds.): RSEISP 2014, LNAI 8537, pp. 217–224, 2014.

discarded all rules that include conditions on specific attributes, regardless of presence or absence of other conditions on other attributes.

A ranking typically organises elements in the decreasing order, firstly most important, then less and less important elements, hence ranking driven reduction of rules causes significant lowering of the number of rules but at the cost of decreased classifier performance, while the reversed ranking enables keeping the rules with more important variables and good predictive properties.

The procedures described were employed in the domain of stylometry, which is a branch of science dedicated to analysis of linguistic styles [2]. The main task of authorship attribution can be perceived as classification, so both connectionist [14] and rule classifiers are often exploited in processing [11,12].

The paper is organised as follows. Section 2 presents briefly some search procedures aiming at feature selection. Section 3 explains conducted experiments, with comments on domain of application, employed techniques, and obtained results. Section 4 contains concluding remarks.

2   Feature Selection and Ranking

A search procedure for characteristic features can start with the empty set to which elements are added one after another in sequential forward selection. As the search begins with many inducers of low dimensions, the features are studied in the limited context where observing their importance and the interdependencies among them is more difficult and can even be impossible. With just few inputs the performance of the system can be so low, the differences between them so small, that it is hard to make some informed choice.

The opposite starting point for the search procedure is that of a complete set of all available features, some of which are next rejected in backward elimination. This approach offers the widest possible view on variable interactions yet with their multitude some individual relevancy can be obscured, and irrelevant or repetitive features not so easy to detect. What is more, high dimensionality of inducers can result in higher computational complexity of the learning phase.

Algorithms dedicated to feature selection are typically grouped into three main categories: filers, wrappers, and embedded solutions [5]. Filters work as kind of pre-processing engines, independently on the classification stage. They assign some measures to variables, using such concepts as entropy, information gain, consistency. In wrappers the choice of attributes is conditioned by the performance of employed classification systems, and embedded approaches exploit mechanisms dedicated to feature selection that are inherent to data mining techniques used, such as input pruning for artificial neural networks, or the concept of relative reducts for rough set processing.

Feature selection can refer to a notion of a ranking of features, where an evaluating function returns values basing on which the variables are ordered, typically in a decreasing manner — firstly the most important attributes, then gradually less and less important. The ranking can cover all available features or only their subset. By the formal definition ranking belongs with filters.

3   Experimental Setting

Since a ranking of features reflects their relevance, and while filtering decision rules we want to keep the rules that refer to important attributes, the ranking can be used to drive the process of selection of rules. The methodology proposed and illustrated in this paper consists of three subsequent stages: (i) Pre-processing stage for preparation of inputs datasets, (ii) Ranking of attributes with ordering of variables based on the process of sequential forward selection of input features for ANN classifiers, (iii) Filtering of decision rules with reduction of rules from the generated all rules on examples DRSA decision algorithm (with conditions on specific attributes), while following the obtained ranking and its reverse.

3.1   Input Datasets

The rule filtering methodology is employed in a task of authorship attribution that belongs with stylometry. Stylometry performs an analysis of texts with respect to their styles rather than the subject content [2]. To achieve the most important goal of author style recognition we need to detect and describe patterns present in individual writing habits and linguistic preferences.

To make use of modern computing technologies we need quantitative textual descriptors, such as term frequencies, averages, and distributions [7]. Disregarding the elements of formatting makes the analysis more general and the textual markers most widely exploited refer to the usage of common parts of speech or syntactic elements rather than some specific vocabulary.

For reliable stylometric analysis an access to sufficient number of representative text samples is necessary. The individual samples should be neither too short nor too long. Characteristics of very small text parts can be unique and incomparable to others, while in some general statistics for huge documents local variations of styles can cease to be visible. Therefore it is more advantageous to divide long manuscripts into smaller parts where we can observe all subtleties and at the same time increase the number of available samples.

To construct the learning and testing datasets for experiments the frequencies of usage of selected function words and punctuation marks were calculated over groups of text samples for two writers, Jane Austen and Edith Wharton. The novels of respective authors were divided into smaller parts of comparable length.

Variations of linguistic styles are less likely to appear within one piece of writing, even when it is long. Therefore just random selection of testing samples could lead to falsely high predictive accuracies. To prevent that the testing samples were based on completely separate texts. In the training set there were 100 samples per author and in the testing set 50 samples per author, corresponding to usage frequencies for 25 attributes combined from two groups:

- lexical (17): and, as, at, but, by, for, from, not, if, in, of, on, that, this, what, with, to,
- syntactic (8): a comma, a fullstop, a colon, a semicolon, a question mark, an exclamation mark, a bracket, a hyphen.

With two authors to recognise, the classification is binary and for both writers there was the same number of samples in both learning and testing sets providing balanced datasets. In case of many authors to recognise we can either transform the task into several binary classifications, or increase the number of classes but we need to remember that when some class is under- or overrepresented it may result in classification bias and significantly lower performance.

3.2 Sequential Forward Selection for ANNs

Multilayer Perceptron is a unidirectional feed-forward network that consists of interconnected layers comprised of neurons, widely employed in classification tasks due to its good generalisation properties [3]. A learning stage (typically exploiting backpropagation training algorithm) begins with randomisation of weights associated with connections which are then adjusted to minimise the error on the network output, equal to the difference between the expected and generated value. The initial values of weights can greatly influence the learning process and resulting performance, hence to counteract it the multi-starting approach can be employed, with multiple training phases.

Sequential forward selection starts with construction of N (N being the number of available features) single-input ANN classifiers, which are trained and tested with multi-starting approach. From all tested networks the one with the highest average performance is selected and to its single input another is added, and $(N-1)$ networks with two inputs are constructed. The process continues as long as there are some features to be added to the input set. The total number of trained networks equals to:

$$\binom{N}{1} + \binom{N-1}{1} + \cdots + \binom{2}{1} + \binom{1}{1} = \sum_{i=0}^{N-1}\binom{N-i}{1} = \frac{(N+1)N}{2}. \quad (1)$$

Artificial neural networks typically deal better with excessive rather than with insufficient numbers of input features. When the network has only a single input, or just few of them, the training requires more time and the network can have trouble converging. In such cases the training is stopped after some pre-defined number of runs is reached or when some other limit is achieved, for example the network learns some specific percentage of facts.

The performance of the classifier is displayed in Fig. 1. For each step the maximal average classification accuracy that led to selection of features is shown along with the minimal and averaged performance. In the first 3–4 steps the level of predictive accuracy rises steeply, with each added feature significant gain can be observed, then it kind of levels once the networks have more than 5 inputs, but still some increase can be noted. The maximum of 97.78% is reached for 13 variables, then the performance decreases slightly, to 91.5% for the complete set of 25 features. This drop indicates that not all variables are in fact needed for this classifier to achieve the maximum accuracy. Yet the fact that the maximum is not just local is only known when the entire set of attributes is processed.

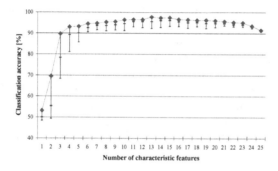

Fig. 1. Classification accuracy of ANN in relation to the number of characteristic features. For each maximum there are also indicated minimum and average levels.

If a goal of the procedure was to find only some number of relevant features, we could stop the search when some maximum performance was reached. As we want to detect the least ranking variables, the complete search process needs to be executed, starting with the empty set and ending with the entire set of features, and the resulting attribute ranking is given in Table 1.

3.3   Attribute Ranking Based Filtering of Rules

Before filtering of rules can be attempted, all rules on examples algorithm, using all available attributes, needs to be induced and this was executed in Dominance-Based Rough Set Approach (DRSA) [4,10], a modification of Classical Rough Set Approach (CRSA) invented by Z. Pawlak [6]. Unlike CRSA which allows only for nominal classification, DRSA observes orderings in value sets of attributes and employs the dominance principle which states that x should be classified at least as good as y if the values of considered attributes for x are at least as good as those for y, which enables also ordinal classification.

To induce rules, excessive and repetitive information present in the learning examples is eliminated and rough approximations calculated. Inferred rules classify to either at most or at least some decision class and are characterised by lengths equal to the number of included conditions and support values indicating for how many training samples each rule is valid.

The results of classification are given in three categories: correct decisions — when all matching rules classify correctly, incorrect decisions — when all matching rules classify incorrectly, or ambiguous decisions — when some of the matching rules classify correctly while others incorrectly, or when there are no rules matching. In cases of contradicting decisions voting or weighting of rules can be executed to arrive at some final decision, but it can significantly lengthen the whole process. In the research only correct decisions were taken under considerations and all ambiguous treated as incorrect. However, such attitude means resigning from the possible increase of classification accuracy.

The full decision algorithm generated contained 62,383 rules. Application of such long algorithm is not practical, also multitude of rules causes conflicting classification verdicts. We can impose some hard constraints on rules, for example on minimal support that rules must have in order to be considered, or their maximal length. Support equal at least 66 characterises just 17 decision rules which gave the highest correct classification accuracy of 86.67%.

Next the methodology of filtering rules by exploiting the ranking of attributes was employed. By general understanding a ranking is organised in decreasing order. If this order is used in filtering out rules it means discarding rules that include conditions on important attributes, which in turn results in the worsened performance of decision algorithms employed as classifiers, as plotted in Fig. 2.

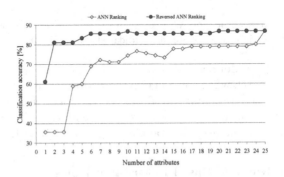

Fig. 2. Classification accuracy of rule classifiers in relation to the number of attributes involved, with filtering of decision rules driven by ANN ranking and its reverse

The typical goal of rule filtering is to obtain a limited set of rules with good predictive capabilities. To find it it is best to keep the rules with conditions on highly ranking variables and discard those irrelevant or less relevant. Therefore the reversed ranking needs to be employed (as shown in Fig. 2) and with such approach it is possible to reject significant number of attributes and filter out high number of rules, the details of which are listed in Table 1. Since at each step there is given the attribute that was rejected at this step, the last row of the table indicates the only remaining variable. When this element is rejected the set of rules is empty, hence no other parameters can be given.

Both the ranking and its reverse enable significant reduction in the number of rules, yet the decrease is greater for reversed ranking, while at the same time generated algorithms maintain satisfactory level of predictive accuracy, as displayed in Fig. 2. For ranking driven filtering the performance is immediately and irrecoverably worsened. As the particular values of predictive accuracies depend on the available input datasets, and they may or may not be representative, it is best to observe trends in performance rather than specific classification ratios.

The filtering procedure can be stopped when some criteria with respect to the number of rules or variables are met, or when the performance is worsened, however, the detected changes in accuracy are not necessarily strictly monotonic and only when the complete procedure is executed, with construction of N decision algorithms, we know with certainty whether extrema are local or global.

Table 1. Filtering of decision rules based on attribute ranking. Columns specify: a) Number of attributes still involved, b) the attribute rejected at this step, c) Number of remaining decision rules, d) Maximal support required of rules to arrive at the maximal classification accuracy, e) Number of decision rules meeting constrains on support.

ANN Ranking					Reversed ANN Ranking				
a)	b)	c)	d)	e)	a)	b)	c)	d)	e)
24	not	61,382	48	49	24	comma	55,418	62	21
23	semico	56,666	38	35	23	bracke	44,836	62	21
22	colon	52,470	38	35	22	at	34,878	62	21
21	hyphen	44,619	38	35	21	to	27,106	62	21
20	on	40,343	38	35	20	fullst	21,217	62	21
19	as	31,517	38	27	19	for	15,190	62	17
18	if	24,707	38	27	18	of	11,037	62	17
17	questi	20,055	38	27	17	in	7,947	62	17
16	by	15,434	38	19	16	from	5,769	62	16
15	exclam	11,138	38	18	15	what	4,184	62	16
14	this	8,110	21	110	14	with	3,125	62	16
13	and	5,810	24	63	13	but	2,182	62	16
12	that	4,420	20	107	12	that	1,659	62	16
11	but	2,867	20	66	11	and	939	62	16
10	with	2,040	20	59	10	this	601	55	23
9	what	1,217	20	47	9	exclam	407	52	23
8	from	825	11	98	8	by	195	52	20
7	in	582	11	88	7	questi	150	52	20
6	of	280	12	65	6	if	108	52	20
5	for	118	9	38	5	as	66	31	23
4	fullst	43	9	24	4	on	39	55	13
3	to	4	19	2	3	hyphen	36	55	13
2	at	3	19	2	2	colon	27	55	13
1	bracke	3	19	2	1	semico	10	55	4
	comma					not			

4 Conclusions

One of the possible ways to arrive at a decision algorithm comprised of rules with good predictive and at the same time descriptive properties is firstly to induce the complete set of decision rules, then filter some of these rules using some kind of weights, measures, or approaches. The paper presents a methodology within which filtering of rules is driven by a ranking of features.

Firstly, for all attributes considered their ranking is obtained by the procedures of sequential forward selection employed for artificial neural networks. Then, using Dominance-Based Rough Set Approach all rules on examples algorithm is calculated. In the next step the ranking of features and its reverse are exploited to reduce such rules from the algorithm that contain conditions on selected variables. The procedure enables to filter out significant amount of rules without worsening the power of the rule classifier. The system is employed in the binary classification task, to execute categorisation of text samples with respect to their authors, basing on their linguistic styles.

Acknowledgements. The research described in the paper was performed in the framework of the statutory project BK/215/RAu2/2013/502 at the Silesian University of Technology, Gliwice, Poland. 4eMka Software used in search for decision rules was developed at the Poznan University of Technology (idss.cs.put.poznan.pl). ANNs were simulated with California Scientific Brainmaker software. Texts exploited in experiments are available for on-line reading and download thanks to Project Gutenberg (www.gutenberg.org).

References

1. Bayardo Jr., R., Agrawal, R.: Mining the most interesting rules. In: Proceedings of the 5th ACM SIGKDD International Conference on Knowledge Discovery and Data Mining, pp. 145–154 (1999)
2. Craig, H.: Stylistic analysis and authorship studies. In: Schreibman, S., Siemens, R., Unsworth, J. (eds.) A Companion to Digital Humanities. Blackwell, Oxford (2004)
3. Fiesler, E., Beale, R.: Handbook of neural computation. Oxford University Press (1997)
4. Greco, S., Matarazzo, B., Słowiński, R.: Rough set theory for multicriteria decision analysis. European Journal of Operational Research 129(1), 1–47 (2001)
5. Jensen, R., Shen, Q.: Computational Intelligence and Feature Selection. John Wiley & Sons, Inc., Hoboken (2008)
6. Pawlak, Z.: Rough sets and intelligent data analysis. Information Sciences 147, 1–12 (2002)
7. Peng, R., Hengartner, H.: Quantitative analysis of literary styles. The American Statistician 56(3), 15–38 (2002)
8. Sikora, M.: Redefinition of classification rules by evaluation of elementary conditions occurring in the rule premises. Fundamenta Informaticae 123(2), 171–197 (2013)
9. Sikora, M., Wróbel, Ł.: Data-driven adaptive selection of rules quality measures for improving the rules induction algorithm. In: Kuznetsov, S.O., Ślęzak, D., Hepting, D.H., Mirkin, B.G. (eds.) RSFDGrC 2011. LNCS (LNAI), vol. 6743, pp. 278–285. Springer, Heidelberg (2011)
10. Słowiński, R., Greco, S., Matarazzo, B.: Dominance-based rough set approach to reasoning about ordinal data. In: Kryszkiewicz, M., Peters, J.F., Rybiński, H., Skowron, A. (eds.) RSEISP 2007. LNCS (LNAI), vol. 4585, pp. 5–11. Springer, Heidelberg (2007)
11. Stańczyk, U.: DRSA decision algorithm analysis in stylometric processing of literary texts. In: Szczuka, M., Kryszkiewicz, M., Ramanna, S., Jensen, R., Hu, Q. (eds.) RSCTC 2010. LNCS (LNAI), vol. 6086, pp. 600–609. Springer, Heidelberg (2010)
12. Stańczyk, U.: Rule-based approach to computational stylistics. In: Bouvry, P., Kłopotek, M.A., Leprévost, F., Marciniak, M., Mykowiecka, A., Rybiński, H. (eds.) SIIS 2011. LNCS (LNAI), vol. 7053, pp. 168–179. Springer, Heidelberg (2012)
13. Stańczyk, U.: Decision rule length as a basis for evaluation of attribute relevance. Journal of Intelligent and Fuzzy Systems 24(3), 429–445 (2013)
14. Stańczyk, U.: Establishing relevance of characteristic features for authorship attribution with ANN. In: Decker, H., Lhotská, L., Link, S., Basl, J., Tjoa, A.M. (eds.) DEXA 2013, Part II. LNCS, vol. 8056, pp. 1–8. Springer, Heidelberg (2013)

Evaluation of Leave-One Out Method
Based on Incremental Sampling Scheme

Shusaku Tsumoto and Shoji Hirano

Department of Medical Informatics, School of Medicine,
Faculty of Medicine Shimane University
89-1 Enya-cho Izumo 693-8501 Japan
{tsumoto,hirano}@med.shimane-u.ac.jp

Abstract. This paper proposes a new framework for evaluation of leave-out one methods based on incremental sampling scheme. Although incremental sampling scheme is used for incremental rule induction, this paper shows that the same idea can be used for deletion of exampling. Then, we applied this technique to the leave-one out method for rules defined by the propositions whose constraints were defined by inequalities of accuracy and coverage. The results show that the evaluation framework gives a powerful tool for evaluation of the leave-out method.

Keywords: incremental sampling scheme, rule induction, accuracy, coverage, leave-one-out.

1 Introduction

Tsumoto and Hirano proposed a incremental rule induction method by using a new framework of incremental sampling [3]. Since accuracy and coverage [2] are defined by the relations between conditional attributes (R) and decision attribute(D), incremental sampling gives four possible cases according to the update of accuracy coverage as shown in Table 1. Using this classification, the behavior of these indices can be evaluated for four cases. Furthermore, they conducted experimental evaluation of rule induction method based on this framework, which gave comparable results compared with conventional approaches.

Table 1. Incremental sampling scheme

	R	D	$R \wedge D$
1.	0	0	0
2.	0	+1	0
3.	+1	0	0
4.	+1	+1	+1

In this paper, we extend this scheme by including the negations of R and D to evaluate the set-based indices where the negated terms are used. The results show

M. Kryszkiewicz et al. (Eds.): RSEISP 2014, LNAI 8537, pp. 225–236, 2014.

that the evaluation framework gives a powerful tool for evaluation of set-based indices. Especially, it is found that the behavior of indices can be determined by a firstly given dataset.

This paper proposes a new framework for evaluation of leave-out one methods based on incremental sampling scheme. Although incremental sampling scheme is used for incremental rule induction, this paper shows that the same idea can be used for deletion of exampling. Then, we applied this technique to the leave-one out method for rules defined by the propositions whose constraints were defined by inequalities of accuracy and coverage. The results show that the evaluation framework gives a powerful tool for evaluation of the leave-out method.

The paper is organized as follows: Section 2 makes a brief description about rough set theory and the definition of probabilistic rules based on this theory. Section 3 introduces incremental sampling scheme. Section 4 disccusses estimation of removal of data by using incremental sampling scheme. Then, Section 5 show the anlaysis of leave-one out method. Next, Section 6 discusses the extension of this method to generalized framework. Finally, Section 7 concludes this paper.

2 Rough Sets and Probabilistic Rules

2.1 Rough Set Theory

Rough set theory clarifies set-theoretic characteristics of the classes over combinatorial patterns of the attributes, which are precisely discussed by Pawlak [1,4]. This theory can be used to acquire some sets of attributes for classification and can also evaluate how precisely the attributes of database are able to classify data. One of the main features of rough set theory is to evaluate the relationship between the conditional attributes and the decision attributes by using the hidden set-based relations. Let a conditional attribute or conjunctive formula of attributes a decision attribute be denoted by R and D. Then, a relation between R and D can be evaluated by each supporting sets ($[x]_R$ and $[x]_D$) and their overlapped region denoted by $R \wedge D$ ($[x]_R \cap [x]_D$). If $[x]_R \subset [x]_D$, then a proposition $R \rightarrow D$ will hold and R will be a part of lower approximation of D. Dually, D can be called a upper approximation of R. In this way, we can define the characteristics of classification in the set-theoretic framework. Let n_R, n_D and n_{RD} denote the cardinality of $[x]_R$, $[x]_D$ and $[x]_R \cap [x]_D$, respectively. Accuracy (true predictive value) and coverage (true positive rate) can be defined as:

$$\alpha_R(D) = \frac{n_{RD}}{n_R} \quad and \quad \kappa_R(D) = \frac{n_{RD}}{n_D}, \tag{1}$$

It is notable that $\alpha_R(D)$ measures the degree of the sufficiency of a proposition, $R \rightarrow D$, and that $\kappa_R(D)$ measures the degree of its necessity. For example, if $\alpha_R(D)$ is equal to 1.0, then $R \rightarrow D$ is true. On the other hand, if $\kappa_R(D)$ is equal to 1.0, then $D \rightarrow R$ is true. Thus, if both measures are 1.0, then $R \leftrightarrow D$.

For further information on rough set theory, readers could refer to [1,4,2].

2.2   Probabilistic Rules

The simplest probabilistic model is that which only uses classification rules which have high accuracy and high coverage.[1] This model is applicable when rules of high accuracy can be derived. Such rules can be defined as:

$$R \overset{\alpha,\kappa}{\to} d \text{ s.t.} \quad R = \vee_i R_i = \vee \wedge_j [a_j = v_k],$$
$$\alpha_{R_i}(D) > \delta_\alpha \text{ and } \kappa_{R_i}(D) > \delta_\kappa,$$

where δ_α and δ_κ denote given thresholds for accuracy and coverage, respectively. where $|A|$ denotes the cardinality of a set A, $\alpha_R(D)$ denotes an accuracy of R as to classification of D, and $\kappa_R(D)$ denotes a coverage, or a true positive rate of R to D, respectively. We call these two inequalities *rule selection inequalities*.

3   Theory for Incremental Rule Induction

3.1   Incremental Induction of Probabilistic Rules

Usually, datasets will monotonically increase. Let $n_R(t)$ and $n_D(t)$ denote cardinalities of a supporting set of a formula R in given data and a target concept d at time t.

$$n_R(t+1) = \begin{cases} n_R(t) + 1 & \text{an additional example satisfies } R \\ n_R(t) & \text{otherwise} \end{cases}$$

$$n_D(t+1) = \begin{cases} n_D(t) + 1 & \text{an additional example belongs} \\ & \text{to a target concept } d. \\ n_D(t) & \text{otherwise} \end{cases}$$

Let $\neg R$ and $\neg D$ be the negations of R and D, respectively. Then, the above two possiblities have the following two dual cases.

$$n_{\neg R}(t+1) = \begin{cases} n_{\neg R}(t) & \text{an additional example satisfies } R \\ n_{\neg R}(t) + 1 & \text{otherwise} \end{cases}$$

$$n_{\neg D}(t+1) = \begin{cases} n_{\neg D}(t) & \text{an additional example belongs} \\ & \text{to a target concept } d. \\ n_{\neg D}(t) + 1 & \text{otherwise} \end{cases}$$

Thus, from the definition of accuracy (Eqn.(1)) and coverage (Eqn. (1)), accuracy and coverage may nonmonotonically change due to the change of the intersection of R and D, n_{RD}. Since the above classification gives four additional patterns, we will consider accuracy and coverage for each case as shown in Table 2, called incremental sampling scheme, in which 0 and +1 denote stable and increase in each value.

---

[1] In this model, we assume that accuracy is dominant over coverage.

Table 2. Incremental Sampling Scheme

	R	D	$\neg R$	$\neg D$	$R \wedge D$	$\neg R \wedge D$	$R \wedge \neg D$	$\neg R \wedge \neg D$
1.	0	0	+1	+1	0	0	0	+1
2.	0	+1	+1	0	0	+1	0	0
3.	+1	0	0	+1	0	0	+1	0
4.	+1	+1	0	0	+1	0	0	0

Table 3. Four patterns for an additional example

t:	$[x]_R(t)$	$D(t)$	$[x]_R \cap D(t)$
original	n_R	n_D	n_{RD}

t+1:	$[x]_R(t+1)$	$D(t+1)$	$[x]_R \cap D(t+1)$
Both negative	n_R	n_D	n_{RD}
R: positive	$n_R + 1$	n_D	n_{RD}
d: positive	n_R	$n_D + 1$	n_{RD}
Both positive	$n_R + 1$	$n_D + 1$	$n_{RD} + 1$

Since accuracy and coverage use only the postivie sides of R and D, we will consider the following subtable for the updates of accuracy and coverage (Table 3).

Thus, in summary, Table 4 gives the classification of four cases of an additional example.

Table 4. Summary of change of accuracy and coverage

Mode				$\alpha(t+1)$	$\kappa(t+1)$
Both negative	n_R	n_D	n_{RD}	$\alpha(t)$	$\kappa(t)$
R: positive	$n_R + 1$	n_D	n_{RD}	$\frac{\alpha(t)n_R}{n_R+1}$	$\kappa(t)$
d: positive	n_R	$n_D + 1$	n_{RD}	$\alpha(t)$	$\frac{\kappa(t)n_D}{n_D+1}$
Both positive	$n_R + 1$	$n_D + 1$	$n_{RD} + 1$	$\frac{\alpha(t)n_R+1}{n_R+1}$	$\frac{\kappa(t)n_D+1}{n_D+1}$

4   Analysis of Leave-One Out Method

4.1   Leave-One Out

Leave-one out method is one of the methods for estimation of predictive values of statistical indices obtained by classification method as follows. First, one example is deleted from training examples. Second, a classification method is applied to the training examples. Third, the method is evaluated by a selected example. These three procedures are repeated for all the elements in training examples.

4.2   Framework

From the definition of accuracy and coverage, Equations(1) accuracy and coverage may nonmonotonically change. Since the above classification gives four additional patterns, we will consider accuracy and coverage for each case as shown in Table 5. in which $|[x]_R(t)|$, $|D(t)|$ and $|[x]_R \cap D(t)|$ are denoted by n_R, n_D and n_{RD}. Table 6 gives the classification of four cases of deletion of one example. Then, updates of accuracy and coverage are obtained as follows.

Table 5. Four patterns for deletion of one example

t:	$[x]_R(t)$	$D(t)$	$[x]_R \cap D(t)$
	n_R	n_D	n_{RD}

t+1:	$[x]_R(t+1)$	$D(t+1)$	$[x]_R \cap D(t+1)$
	$n_R - 1$	$n_D - 1$	$n_{RD} - 1$
	$n_R - 1$	n_D	n_{RD}
	n_R	$n_D - 1$	n_{RD}
	n_R	n_D	n_{RD}

Table 6. Summary of change of accuracy and coverage

				$\alpha(t+1)$	$\kappa(t+1)$
1. Both Negative (BN)	n_R	n_D	n_{RD}	$\alpha(t)$	$\kappa(t)$
2. d-Positive (dP)	n_R	$n_D - 1$	n_{RD}	$\alpha(t)$	$\frac{\kappa(t)n_D}{n_D-1}$
3. R-Positive (RP)	$n_R - 1$	n_D	n_{RD}	$\frac{\alpha(t)n_R}{n_R-1}$	$\kappa(t)$
4. Both Positive (BP)	$n_R - 1$	$n_D - 1$	$n_{RD} - 1$	$\frac{\alpha(t)n_R-1}{n_R-1}$	$\frac{\kappa(t)n_D-1}{n_D-1}$

4.3   Framework of Analysis

Table 7 shows the number of examples for each case of deletion. Then, by using these combination, expected value of updates of accuracy can be calculated as follows:

$$\bar{\alpha}_R(D) = \frac{N - (n_R + n_D - n_R D)}{N}\alpha_R(D)$$
$$+ \frac{n_R - n_{RD}}{N}\frac{\alpha_R(D)n_R}{n_R - 1}$$
$$+ \frac{n_D - n_{RD}}{N}\alpha_R(D)$$
$$+ \frac{n_{RD}}{N}\frac{\alpha_R(D)n_R - 1}{n_R - 1}$$
$$= \alpha_R(D)$$

Table 7. Number of examples for deletion of one example

			Number of Examples	
1. BN	n_R	n_D	n_{RD}	$N - (n_R + n_D - n_{RD})$
2. dP	n_R	$n_D - 1$	n_{RD}	$n_R - n_{RD}$
3. RP	$n_R - 1$	n_D	n_{RD}	$n_D - n_{RD}$
4. BP	$n_R - 1$	$n_D - 1$	$n_{RD} - 1$	n_{RD}

In the same way,

$$\bar{\kappa}_R(D) = \kappa_R(D)$$

Thus, simply, leave one out estimators for both accuracy and coverage are equal to their original values.

It is notable that these equations can be also applied to the estimation of error rate. Since any constraint is not given, this means that estimated values of accuracy and coverage of a set of rules without any constraints are equal to the original values. Thus, the difference between original and estimated values of accuracy and coverage will be caused by the nature of rule selection inequalities.

5 Updates of Probabilistic Rules

For simplicity, let the threshold of coverage be equal to 0. Then, for inclusion of R to a set of rules for d, updates of accuracy can be considered. Rule selection inequality will become: $\alpha(t+1) > \delta_\alpha$. Then, the conditions for updating can be calculated from the original datasets: when accuracy or coverage does not satisfy the constraint, the corresponding formula should be removed from the candidates. On the other hand, both accuracy and coverage satisfy both constraints, the formula should be included into the candidates. Thus, the following inequality are important for exclusion of R into the conditions of rules for D:

$$\alpha(t+1) = \frac{\alpha(t)n_R - 1}{n_R - 1} \leq \delta_\alpha$$

For its inclusion, the following inequality is important:

$$\alpha(t+1) = \frac{\alpha(t)n_R}{n_R - 1} > \delta_\alpha$$

Then, the follow four cases should be considered as shown in Table 8. For example, if R satisfies the inequality

$$\delta_\alpha < \alpha_R(D) < \frac{n_R - 1}{n_R}\delta_\alpha + 1,$$

then R will be removed from a set of Rules when a deleted example satisfied R and belongs to d (in case of BP).

Thus, for estimation of α will be obtained as follows.

Table 8. Four possible cases for inclusion and removal

A. $\alpha_R(D) \le \frac{n_R-1}{n_R}\delta_\alpha$	Not included in a Set of Rules
B. $\frac{n_R-1}{n_R}\delta_\alpha < \alpha_R(D) \le \delta_\alpha$	Included into a set of Rules in cases of RP
C. $\delta_\alpha < \alpha_R(D) \le \frac{n_R-1}{n_R}\delta_\alpha + 1$	Removed from a set of Rules in cases of BP
D. $\alpha_R(D) > \frac{n_R-1}{n_R}\delta_\alpha + 1$	Always included

5.1   When $\alpha_R(D) \le \frac{n_R-1}{n_R}\delta_\alpha$

Since this formula R will not be included in any case, the estimated value of $\alpha_R(D)$ in a set of rule is equal to 0.

5.2   When $\frac{n_R-1}{n_R}\delta_\alpha < \alpha_R(D) \le \delta_\alpha$

Since the formula will be included in a set of rules when a deleted example satisfies R but does not belong to d (RP). Thus, the estimated value of $\alpha_R(D)$ is equal to:

$$\frac{\alpha_R(D)}{n_R - 1}(n_R - n_{RD}) = \frac{n_{RD}(1 - \alpha_R(D))}{n_R - 1}.$$

5.3   When $\delta_\alpha < \alpha_R(D) \le \frac{n_R-1}{n_R}\delta_\alpha + 1$

Since the formula will be excluded from a set of rules when a deleted example satisfies R and belongs to d (BP). Thus, the estimate value of $\alpha_R(D)$ is equal to:

$$\bar{\alpha}_R(D) = \frac{N - (n_R + n_D - n_{RD})}{N}\alpha_R(D)$$
$$+ \frac{n_R - n_{RD}}{N}\frac{\alpha_R(D)n_R}{n_R - 1}$$
$$+ \frac{n_D - n_{RD}}{N}\alpha_R(D)$$
$$= \alpha_R(D) - \frac{n_{RD}}{N}\frac{n_D - 1}{n_R - 1}$$

5.4   When $\alpha_R(D) > \frac{n_R-1}{n_R}\delta_\alpha + 1$

In this case, a rule $R \to d$ will be included into any case, so the estimated values of accuracy and coverage are equal to the original values.

In summary, expected value of accuracy of a set of rules is formulated as follows.

$$\bar{\alpha} = \frac{1}{\#R}\sum_{\#R \in \{\frac{n_R-1}{n_R}\delta_\alpha < \alpha_R(D) \le \delta_\alpha\}} \frac{n_{RD}(1 - \alpha_R(D))}{n_R - 1}$$

$$+ \frac{1}{\#R} \sum_{\#R \in \{\delta_\alpha < \alpha_R(D) \le \frac{n_R-1}{n_R}\delta_\alpha + 1\}} \alpha_R(D) - \frac{n_{RD}}{N} \frac{n_D - 1}{n_R - 1}$$

$$+ \frac{1}{\#R} \sum_{\#R \in \{\alpha_R(D) > \frac{n_R-1}{n_R}\delta_\alpha + 1\}} \alpha_R(D) \qquad (2)$$

5.5   Estimation of Error Rate

The error rate of $R \to d$, denoted by $\varepsilon_R(D)$ is defined as:

$$\varepsilon_R(D) = 1 - \frac{n_{RD} + n_{\neg R \neg D}}{n_R + n_{\neg R}}$$

Thus, updates of error rate is obtained as:

$$\varepsilon_R(D, t+1) = 1 - \frac{n_{RD} + n_{\neg R \neg D} - 1}{n_R + n_{\neg R} - 1} = 1 - \frac{N(1 - \varepsilon_R(D))}{N - 1}$$

$$= \frac{N\varepsilon_R(D) - 1}{N - 1}.$$

Since the definition of error rate includes positive and negative values of R and D, the values will not be changed in any type of deletion of an example.

Thus, the estimated value of error rate of a set of rules will be obtained as:

$$\bar{\varepsilon} = \frac{1}{\#R} \sum_{\#R \in \{\frac{n_R-1}{n_R}\delta_\alpha < \alpha_R(D) \le \delta_\alpha\}} \frac{N\varepsilon_R(D) - 1}{N - 1}$$

$$+ \frac{1}{\#R} \sum_{\#R \in \{\delta_\alpha < \alpha_R(D) \le \frac{n_R-1}{n_R}\delta_\alpha + 1\}} \frac{N\varepsilon_R(D) - 1}{N - 1}$$

$$+ \frac{1}{\#R} \sum_{\#R \in \{\alpha_R(D) > \frac{n_R-1}{n_R}\delta_\alpha + 1\}} \frac{N\varepsilon_R(D) - 1}{N - 1}$$

$$= \frac{1}{\#R} \sum_{\#R \in \{\alpha_R(D) > \frac{n_R-1}{n_R}\delta_\alpha\}} \frac{N\varepsilon_R(D) - 1}{N - 1} \qquad (3)$$

Since the total number of sum in the above formula is smaller than $\#R$, the expected error rate will satisfy the following inequality:

$$\bar{\varepsilon} < \max_R \frac{N\varepsilon_R(D) - 1}{N - 1}$$

6   Updates of Probabilistic Rules (II)

In a general case, since we have two rule selection inequalities, the situation may be a little complicated. Since rules is defined as a probabilistic proposition with two inequalities, supporting sets should satisfy the following constraints:

$$\alpha(t+1) > \delta_\alpha \quad , \quad \kappa(t+1) > \delta_\kappa \qquad (4)$$

Then, the conditions for updating can be calculated from the original datasets: when accuracy or coverage does not satisfy the constraint, the corresponding formula should be removed from the candidates. On the other hand, both accuracy and coverage satisfy both constraints, the formula should be included into the candidates. Thus, the following inequalities are important for exclusion of R into the conditions of rules for D:

$$\alpha(t+1) = \frac{\alpha(t)n_R - 1}{n_R - 1} \leq \delta_\alpha \quad , \quad \kappa(t+1) = \frac{\kappa(t)n_D - 1}{n_D - 1} \leq \delta\kappa.$$

For its inclusion, the following inequalities are important:

$$\alpha(t+1) = \frac{\alpha(t)n_R}{n_R - 1} > \delta_\alpha \quad , \quad \kappa(t+1) = \frac{\kappa(t)n_D}{n_D - 1} > \delta\kappa.$$

Thus, in addition to four possible cases for inclusion and removal for accuracy shown in Table 8, coverage updates should be considered. For simplicirty, first we assume that accuracy satisfies the inequality for D in the above table. Table 9 shows the four possible cases.

Table 9. Four possible cases for inclusion and removal for coverage

A2. $\kappa_R(D) \leq \frac{n_R-1}{n_R}\delta_\kappa$	Not included in a Set of Rules
B2. $\frac{n_R-1}{n_R}\delta_\kappa < \kappa_R(D) \leq \delta_\kappa$	Included into a set of Rules in cases of dP
C2. $\delta_\kappa < \kappa_R(D) \leq \frac{n_R-1}{n_R}\delta_\kappa + 1$	Removed from a set of Rules in cases of BP
D2. $\kappa_R(D) > \frac{n_R-1}{n_R}\delta_\kappa + 1$	Always included

Updates of Rules can be considered with the combination of two tables: Table 8 and Table 9, which 16 combinations should be considered. A matrix (Table 10) gives a table for updates of rule inclusion.

Table 10. Matirx for updates of rule inclusion

	A	B	C	D
A2	Always Deleted	Always Deleted	Always Deleted	Always Deleted
B2	Always Deleted	Always Deleted	Always Deleted	**Included when dP**
C2	Always Deleted	Always Deleted	**Deleted when BP**	**Deleted when BP**
D2	Always Deleted	**Included when RP**	**Deleted when BP**	**Always Included**

Therefore, the following six cases will be candidates for rule updates, where (a, b) denotes the pair of the type of inequalities for accuracy and coverage.

(D,D2) Rules which satisfies D and D will be always included into a set of rules. Thus, accuracy, coverage, error rate are equal to ones obtained by original examples.

$$\bar{\alpha}_R(D) = \alpha_R(D), \bar{\kappa}_R(D) = \kappa_R(D), \bar{\varepsilon}_R(D) = \varepsilon_R(D).$$

(C,C2) In this case, coverageis estimated as:

$$\bar{\kappa}_R(D) = \frac{N - (n_R + n_D - n_{RD})}{N}\kappa_R(D)$$
$$+ \frac{n_R - n_{RD}}{N}\kappa_R(D)$$
$$+ \frac{n_D - n_{RD}}{N}\frac{\kappa_R(D)n_D}{n_D - 1}$$
$$= \kappa_R(D) - \frac{n_{RD}}{N}\frac{n_R - 1}{n_D - 1}$$

Thus, estimated values of accuracy and coverage will be:

$$\bar{\alpha}_R(D) = \alpha_R(D) - \frac{n_{RD}}{N}\frac{n_D - 1}{n_R - 1}$$
$$\bar{\kappa}_R(D) = \kappa_R(D) - \frac{n_{RD}}{N}\frac{n_R - 1}{n_D - 1}$$

(D,C2) Estimated values of accuracy and coverage will be:

$$\bar{\alpha}_R(D) = \alpha_R(D)$$
$$\bar{\kappa}_R(D) = \kappa_R(D) - \frac{n_{RD}}{N}\frac{n_R - 1}{n_D - 1}$$

(C,D2) This case can be regarded as a dual one of $(C, D2)$. Thus, estimated values of accuracy and coverage will be:

$$\bar{\alpha}_R(D) = \alpha_R(D) - \frac{n_{RD}}{N}\frac{n_D - 1}{n_R - 1}$$
$$\bar{\kappa}_R(D) = \kappa_R(D)$$

In the same way, the estimated values of accuracy and coverage for $(D, B2)$ and $(B, D2)$ are obtained as follows.

(D,B2)

$$\bar{\alpha}_R(D) = \alpha_R(D)$$
$$\bar{\kappa}_R(D) = \frac{n_{RD}(1 - \kappa_R(D))}{n_D - 1}$$

(B,D2)

$$\bar{\alpha}_R(D) = \frac{n_{RD}(1 - \alpha_R(D))}{n_R - 1}$$

$$\bar{\kappa}_R(D) = \kappa_R(D)$$

Thus, expected values of error rate of a set of rules will be:

$$\bar{\varepsilon} = \frac{1}{\#R} \sum_{\#R \in \{\frac{n_R - 1}{n_R}\delta_\alpha < \alpha_R(D) \le \delta_\alpha, \kappa_R(D) > \delta_\kappa\}} \frac{N\varepsilon_R(D) - 1}{N - 1}$$

$$= \frac{1}{\#R} \sum_{\#R \in \{\alpha_R(D) > \delta_\alpha, \frac{n_D - 1}{n_D}\delta_\kappa < \kappa_R(D) \le \delta_\kappa\}} \frac{N\varepsilon_R(D) - 1}{N - 1}$$

$$+ \frac{1}{\#R} \sum_{\#R \in \{\alpha_R(D) > \delta_\alpha, \kappa_R(D) > \delta_\kappa\}} \frac{N\varepsilon_R(D) - 1}{N - 1} \qquad (5)$$

$$(6)$$

7 Conclusion

This paper introduces a framework of analysis of leave-one out method based on incremental sampling scheme. The key idea is that the formulae obtained for incremental sampling can be reversed and the similar formulae for estimation of accuracy and coverage are obtained. Then, the leave-one out estimators are obtained as follows. For each case of deletion type, the updated values of accuracy and coverage are calculated. Then, sum up these values for each case, and the expected value will be obtained as the ratio of total sum of updated values to the total number of the formulae. Since leave-one out method gives the estimated values of accuracy and coverage equal to the original values, the estimation of probabilistic rules depends on the values of the thresholds for accuracy and coverage. Even in case of error rate, the dependence of the effects of thresholds is included into the summation of each possible case. This is a preliminary approach of formal studies on leave-one out method based on incremental sampling scheme. Since leave-one out method is a specific type of a cross-validation method, the next step will be to extend this approach to a generalized case.

Acknowledgments. This research is supported by Grant-in-Aid for Scientific Research (B) 24300058 from Japan Society for the Promotion of Science(JSPS).

References

1. Pawlak, Z.: Rough Sets. Kluwer Academic Publishers, Dordrecht (1991)
2. Tsumoto, S.: Automated induction of medical expert system rules from clinical databases based on rough set theory. Information Sciences 112, 67–84 (1998)
3. Tsumoto, S., Hirano, S.: Incremental rules induction based on rule layers. In: Li, T., Nguyen, H.S., Wang, G., Grzymala-Busse, J., Janicki, R., Hassanien, A.E., Yu, H. (eds.) RSKT 2012. LNCS (LNAI), vol. 7414, pp. 139–148. Springer, Heidelberg (2012)
4. Ziarko, W.: Variable precision rough set model. Journal of Computer and System Sciences 46, 39–59 (1993)

Optimization of Decision Rules Relative to Coverage - Comparative Study

Beata Zielosko

Institute of Computer Science, University of Silesia
39, Będzińska St., 41-200 Sosnowiec, Poland
beata.zielosko@us.edu.pl

Abstract. In the paper, we present a modification of the dynamic programming algorithm for optimization of decision rules relative to coverage. The aims of the paper are: (i) study of the coverage of decision rules, and (ii) study of the size of a directed acyclic graph (the number of nodes and edges), for a proposed algorithm. The paper contains experimental results with decision tables from UCI Machine Learning Repository.

Keywords: decision rules, coverage, dynamic programming.

1 Introduction

Decision rules are used in many areas connected with knowledge discovery and machine learning [9, 10, 14]. Exact decision rules can be overfitted, i.e., dependent essentially on the noise or adjusted too much to the existing examples. Classifires based on approximate decision rules often give better results than classifiers based on exact decision rules. If decision rules are considered as a way of knowledge representation then instead of exact decision rules with many attributes, it is more appropriate to work with partial decision rules which contain smaller number of attributes and have relatively good accuracy. Therefore, aproximate decision rules and closely connected with them aproximate reducts are studied intensively last years by H.S. Nguyen, A. Skowron, D. Ślęzak, Z. Pawlak and others [7, 9–12].

There are different approaches for construction of decision rules, for example, Boolean reasoning [10, 12], different kinds of greedy algorithms [8, 10], separate and conquer approach [6, 7], dynamic programming approach [2, 3]. Also, there are different rule quality measures that are used for induction or classification processes [4, 13].

In this paper, we study a modification of a dynamic programming algorithm for construction and optimization of decision rules relative to coverage [2]. We try to find a heuristic, modification of a dynamic programming algorithm that allows us to find the values of coverage of decision rules close to optimal ones. The rule coverage is a measure that allows to discover major patterns in the data. Construction and optimization of decision rules relative to coverage can be considered as important task for knowledge representation and knowledge discovery.

M. Kryszkiewicz et al. (Eds.): RSEISP 2014, LNAI 8537, pp. 237–247, 2014.

There are two aims for a proposed algorithm: (i) study the coverage of approximate rules and comparison with the coverage of decision rules constructed by the dynamic programming algorithm [2], (ii) study the size of a directed acyclic graph (the number of nodes and edges) and comparison with the size of a directed acyclic graph constructed by the dynamic programming algorithm.

To work with approximate decision rules, we use an uncertainty measure $J(T)$ that is the difference between number of rows in a given decision table and the number of rows labeled with the most common decision in this table. We fix a nonnegative threshold γ, and study so-called γ-decision rules that localize rows in subtables which uncertainty is at most γ.

We based on the dynamic programming algorithm for optimization of decision rules relative to coverage [2]. For a given decision table T a directed acyclic graph $\Delta_\gamma(T)$ is constructed. Nodes of this graph are subtables of a decision table T described by descriptors (pairs attribute = value). We finish the partition of a subtable when its uncertainty is at most γ. In [2] subtables of the directed acyclic graph were constructed for each value of each (not constant) attribute from T. In the presented approach, subtables of the graph $\Delta_\gamma(T)$ are constructed for one (not constant) attribute from T with the minimum number of values, and for the rest of (not constant) attributes from T - the most frequent value of each attribute (value of an attribute attached to the maximum number of rows) is chosen. So, the size of the graph $\Delta_\gamma(T)$ (the number of nodes and edges) is smaller than the size of the graph constructed by the dynamic programming algorithm. This fact is important from the point of view of scalability. Based on the graph $\Delta_\gamma(T)$ we can describe sets of γ-decision rules for rows of table T. Then, based on a procedure of optimization of the graph $\Delta_\gamma(T)$ relative to coverage we can find for each row r of T a γ-decision rule with the maximum coverage. We compare these values of coverage with the optimal ones obtained using dynamic programming algorithm. In [15] we studied modified dynamic programming algorithm and we used another uncertainty measure $R(T)$ which is the number of unordered pairs of rows with different decisions in the decision table T. It contains also comparison of the coverage of exact decision rules based on another modification of the dynamic programming algorithm.

The paper consists of six sections. Section 2 contains main notions connected with a decision table and decision rules. In Section 3, modified algorithm for construction of a directed acyclic graph is presented. Section 4 contains a description of a procedure of optimization relative to coverage. Section 5 contains experimental results with decision tables from UCI Machine Learning Repository, and Section 6 - conclusions.

2   Main Notions

In this section, we present notions corresponding to decision tables and decision rules.

A *decision table* T is a rectangular table with n columns labeled with conditional attributes $f_1, \ldots, f_n$. Rows of this table are filled with nonnegative integers that are interpreted as values of conditional attributes. Rows of T are pairwise different and each row is labeled with a nonnegative integer that is interpreted as a value of a decision attribute.

A minimum decision value which is attached to the maximum number of rows in T will be called the *most common decision for T*.

We denote by $N(T)$ the number of rows in table T and by $N_{mcd}(T)$ we denote the number of rows in the table T labeled with the most common decision for T. We will interpret the value $J(T) = N(T) - N_{mcd}(T)$ as *uncertainty* of the table T.

The table T is called *degenerate* if T is empty or all rows of T are labeled with the same decision. It is clear that $J(T) = 0$ if and only if T is a degenerate table.

A table obtained from T by the removal of some rows is called a *subtable* of the table T. Let T be nonempty, $f_{i_1}, \ldots, f_{i_k} \in \{f_1, \ldots, f_n\}$ and $a_1, \ldots, a_k$ be nonnegative integers. We denote by $T(f_{i_1}, a_1) \ldots (f_{i_k}, a_k)$ the subtable of the table T which contains only rows that have numbers $a_1, \ldots, a_k$ at the intersection with columns $f_{i_1}, \ldots, f_{i_k}$. Such nonempty subtables (including the table T) are called *separable subtables* of T.

We will say that an attribute $f_i \in \{f_1, \ldots, f_n\}$ is *not constant* on T if it has at least two different values. For the attribute that is not constant on T we can find *the most frequent value*. It is an attribute's value attached to the maximum number of rows in T. If there are two or more such values then we choose the most frequent value for which exists the most common decision for T.

We denote by $E(T)$ a set of attributes from $\{f_1, \ldots, f_n\}$ that are not constant on T. The set $E(T)$ contains one attribute with the minimum number of values and attributes with the most frequent value. For any $f_i \in E(T)$, we denote by $E(T, f_i)$ a set of values of the attribute f_i in T. Note, that if $f_i \in E(T)$ is the attribute with the most frequent value then $E(T, f_i)$ contains only one element.

The expression

$$f_{i_1} = a_1 \wedge \ldots \wedge f_{i_k} = a_k \rightarrow d \tag{1}$$

is called a *decision rule over T* if $f_{i_1}, \ldots, f_{i_k} \in \{f_1, \ldots, f_n\}$, and $a_1, \ldots a_k, d$ are nonnegative integers. It is possible that $k = 0$. In this case (1) is equal to the rule

$$\rightarrow d. \tag{2}$$

Let $r = (b_1, \ldots, b_n)$ be a row of T. We will say that the rule (1) is *realizable for r*, if $a_1 = b_{i_1}, \ldots, a_k = b_{i_k}$. If $k = 0$ then the rule (2) is realizable for any row from T.

Let γ be a nonnegative real number. We will say that the rule (1) is *γ-true for T* if d is the most common decision for $T' = T(f_{i_1}, a_1) \ldots (f_{i_k}, a_k)$ and $J(T') \leq \gamma$. If $k = 0$ then the rule (2) is γ-true for T if d is the most common decision for T and $J(T) \leq \gamma$.

If the rule (1) is γ-true for T and realizable for r, we will say that (1) is a *γ-decision rule for T and r*. Note that if $\gamma = 0$ we have an exact decision rule for T and r.

Let τ be a decision rule over T and τ be equal to (1). The *coverage* of τ is the number of rows in T for which τ is realizable and which are labeled with the decision d. We denote it by $c(\tau)$. If $k = 0$ then $c(\tau)$ is equal to the number of rows in T that are labeled with decision d.

3   Algorithm for Directed Acyclic Graph Construction

In this section, we present a modification of the dynamic programming algorithm that construct, for a given decision table T, a *directed acyclic graph* $\Delta_\gamma(T)$. Based on this graph we can describe set of decision rules for T and each row r of T. Nodes of the graph are separable subtables of the table T. At each step, the algorithm processes one node and marks it with the symbol *. At the first step, the algorithm constructs a graph containing a single node T that is not marked with *.

Let us assume that the algorithm has already performed p steps. We describe now the step $(p+1)$. If all nodes are marked with the symbol * as processed, the algorithm finishes its work and presents the resulting graph as $\Delta_\gamma(T)$. Otherwise, choose a node (table) Θ, that has not been processed yet. Let d be the most common decision for Θ. If $J(\Theta) \leq \gamma$ label the considered node with the decision d, mark it with symbol * and proceed to the step $(p + 2)$. If $J(\Theta) > \gamma$ then for each attribute $f_i \in E(\Theta)$, draw a bundle of edges from the node Θ if f_i is the attribute with the minimum number of values. If f_i is the attribute with the most frequent value draw one edge from the node Θ. Let f_i be the attribute with the minimum number of values and $E(\Theta, f_i) = \{b_1, \ldots, b_t\}$. Then draw t edges from Θ and label them with pairs $(f_i, b_1) \ldots (f_i, b_t)$ respectively. These edges enter to nodes $\Theta(f_i, b_1), \ldots, \Theta(f_i, b_t)$. For the rest of attributes from $E(\Theta)$ draw one edge, for each attribute, from the node Θ and label it with pair (f_i, b_1), where b_1 is the most frequent value of the attribute f_i. This edge enters to a node $\Theta(f_i, b_1)$. If some of nodes $\Theta(f_i, b_1), \ldots, \Theta(f_i, b_t)$ are absent in the graph then add these nodes to the graph. We label each row r of Θ with the set of attributes $E_{\Delta_\gamma(T)}(\Theta, r) \subseteq E(\Theta)$, some attributes from this set can be removed later during a procedure of optimization. Mark the node Θ with the symbol * and proceed to the step $(p + 2)$.

The graph $\Delta_\gamma(T)$ is a directed acyclic graph. A node of such graph will be called *terminal* if it does not have outgoing edges. Note that a node Θ of $\Delta_\gamma(T)$ is terminal if and only if $J(\Theta) \leq \gamma$.

In the next section, we will describe a procedure of optimization of the graph $\Delta_\gamma(T)$ relative to the coverage. As a result we will obtain a graph G with the same sets of nodes and edges as in $\Delta_\gamma(T)$. The only difference is that any row r of each nonterminal node Θ of G is labeled with a nonempty set of attributes $E_G(\Theta, r) \subseteq E(\Theta)$. It is possible also that $G = \Delta_\gamma(T)$.

Now, for each node Θ of G and for each row r of Θ, we describe a set of γ-decision rules $Rul_G(\Theta, r)$. We will move from terminal nodes of G to the node T.

Let Θ be a terminal node of G labeled with the most common decision d for Θ. Then

$$Rul_G(\Theta, r) = \{\to d\}.$$

Let now Θ be a nonterminal node of G such that for each child Θ' of Θ and for each row r' of Θ', the set of rules $Rul_G(\Theta', r')$ is already defined. Let $r = (b_1, \ldots, b_n)$ be a row of Θ. For any $f_i \in E_G(\Theta, r)$, we define the set of rules $Rul_G(\Theta, r, f_i)$ as follows:

$$Rul_G(\Theta, r, f_i) = \{f_i = b_i \wedge \sigma \to s : \sigma \to s \in Rul_G(\Theta(f_i, b_i), r)\}.$$

Then

$$Rul_G(\Theta, r) = \bigcup_{f_i \in E_G(\Theta, r)} Rul_G(\Theta, r, f_i).$$

To illustrate the presented algorithm we consider a simple decision table T depicted on the top of Fig. 1. We set $\gamma = 2$, so during the construction of the graph $\Delta_2(T)$ we stop the partitioning of a subtable Θ of T if $J(\Theta) \le 2$. We denote $G = \Delta_2(T)$.

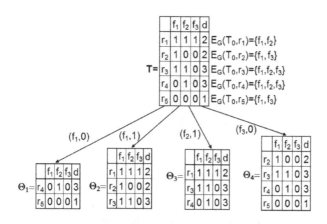

Fig. 1. Directed acyclic graph $G = \Delta_2(T)$

For each node Θ of the graph G and for each row r of Θ we describe the set $Rul_G(\Theta, r)$. We will move from terminal nodes of G to the node T. Terminal nodes of the graph G are Θ_1, Θ_2, Θ_3, Θ_4. For these nodes,

$Rul_G(\Theta_1, r_4) = Rul_G(\Theta_1, r_5) = \{\to 1\}$;
$Rul_G(\Theta_2, r_1) = Rul_G(\Theta_2, r_2) = Rul_G(\Theta_2, r_3) = \{\to 2\}$;
$Rul_G(\Theta_3, r_1) = Rul_G(\Theta_3, r_3) = Rul_G(\Theta_3, r_4) = \{\to 3\}$;
$Rul_G(\Theta_4, r_2) = Rul_G(\Theta_4, r_3) = Rul_G(\Theta_4, r_4) = Rul_G(\Theta_4, r_5) = \{\to 3\}$;

Now we describe the sets of rules attached to rows of T:

$Rul_G(T, r_1) = \{f_1 = 1 \to 2, f_2 = 1 \to 2\}$;
$Rul_G(T, r_2) = \{f_1 = 1 \to 2, f_3 = 0 \to 3\}$;

$Rul_G(T, r_3) = \{f_1 = 1 \to 2, f_2 = 1 \to 3, f_3 = 0 \to 3\};$
$Rul_G(T, r_4) = \{f_1 = 0 \to 1, f_2 = 1 \to 3, f_3 = 0 \to 3\};$
$Rul_G(T, r_5) = \{f_1 = 0 \to 1, f_3 = 0 \to 3\}.$

4   Procedure of Optimization Relative to Coverage

In this section, we describe a procedure of optimization of the graph G relative to the coverage c. For each node Θ in the graph G, this procedure assigns to each row r of Θ the set $Rul_G^c(\Theta, r)$ of γ-decision rules with the maximum coverage from $Rul_G(\Theta, r)$ and the number $Opt_G^c(\Theta, r)$ – the maximum coverage of a γ-decision rule from $Rul_G(\Theta, r)$.

Figure 2 presents the directed acyclic graph G^c obtained from the graph G (see Fig. 1) by the procedure of optimization relative to the coverage.

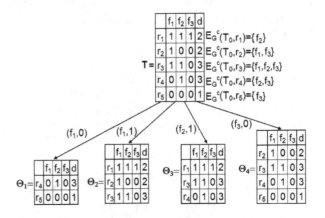

Fig. 2. Graph G^c

We will move from the terminal nodes of the graph G to the node T. We will assign to each row r of each table Θ the number $Opt_G^c(\Theta, r)$ and we will change the set $E_G(\Theta, r)$ attached to the row r in the nonterminal node Θ of G. We denote the obtained graph by G^c.

Let Θ be a terminal node of G and d be the most common decision for Θ. Then we assign to each row r of Θ the number $Opt_G^c(\Theta, r)$ that is equal to the number of rows in Θ which are labeled with the decision d.

Let Θ be a nonterminal node of G and all children of Θ have already been treated. Let $r = (b_1, \ldots, b_n)$ be a row of Θ. We assign the number

$$Opt_G^c(\Theta, r) = \max\{Opt_G^c(\Theta(f_i, b_i), r) : f_i \in E_G(\Theta, r)\}$$

to the row r in the table Θ and we set

$$E_{G^c}(\Theta, r) = \{f_i : f_i \in E_G(\Theta, r), Opt_G^c(\Theta(f_i, b_i), r) = Opt_G^c(\Theta, r)\}.$$

Using the graph G^c presented in Fig. 2 we can describe for each row r_i, $i = 1, \ldots, 5$, of the table T the set $Rul_G^c(T, r_i)$ of decision rules for T and r_i with the maximum coverage. We will give also the value $Opt_G^c(T, r_i)$ which is equal to the maximum coverage of decision rule for T and r_i. This value was obtained during the procedure of optimization of the graph G relative to the coverage. We have

$Rul_G(T, r_1) = \{f_1 = 1 \to 2\}$, $Opt_G^c(T, r_1) = 2$;
$Rul_G(T_0, r_2) = \{f_1 = 1 \to 2, f_3 = 0 \to 3\}$, $Opt_G^c(T, r_2) = 2$;
$Rul_G(T_0, r_3) = \{f_1 = 1 \to 2, f_2 = 1 \to 3, f_3 = 0 \to 3\}$, $Opt_G^c(T, r_3) = 2$;
$Rul_G(T_0, r_4) = \{f_2 = 1 \to 3, f_3 = 0 \to 3\}$, $Opt_G^c(T, r_4) = 2$;
$Rul_G(T_0, r_5) = \{f_3 = 0 \to 3\}$, $Opt_G^c(T, r_5) = 2$.

5 Experimental Results

We made experiments on decision tables from UCI Machine Learning Repository [5] using system Dagger [1] created in King Abdullah University of Science and Technology. Some decision tables contain conditional attributes that take unique value for each row. Such attributes were removed. In some tables there were equal rows with, possibly, different decisions. In this case each group of identical rows was replaced with a single row from the group with the most common decision for this group. In some tables there were missing values. Each such value was replaced with the most common value of the corresponding attribute.

Let T be one of these decision tables. We consider for this table values of γ from the set $\Gamma(T) = \{\lfloor J(T) \times 0.0 \rfloor, \lfloor J(T) \times 0.01 \rfloor, \lfloor J(T) \times 0.1 \rfloor, \lfloor J(T) \times 0.3 \rfloor, \lfloor J(T) \times 0.5 \rfloor\}$.

Table 1. Average coverage of γ-decision rules for $\gamma \in \{\lfloor J(T) \times 0.0 \rfloor, \lfloor J(T) \times 0.01 \rfloor\}$

Decision table	atr	rows	$\lfloor J(T) \times 0.0 \rfloor$			$\lfloor J(T) \times 0.01 \rfloor$		
			avg	avg-dp	rel dif	avg	avg-dp	rel dif
adult-stretch	4	16	6.25	7.00	11%	6.25	7.00	11%
balance-scale	4	625	3.07	4.21	27%	13.96	17.76	21%
breast-cancer	9	266	6.15	9.53	35%	6.15	9.53	35%
cars	6	1728	325.58	332.76	2%	327.36	335.30	2%
house-votes	16	279	73.08	73.52	1%	73.08	73.52	1%
lymphography	18	148	20.69	21.54	4%	20.69	21.54	4%
mushroom	22	8124	2121.84	2135.46	1%	2264.80	2273.95	0%
nursery	8	12960	1483.58	1531.04	3%	1513.36	1695.13	11%
soybean-small	35	47	12.53	12.53	0%	12.53	12.53	0%
teeth	8	23	1.00	1.00	0%	1.00	1.00	0%
tic-tac-toe	9	958	59.49	66.68	11%	60.66	67.17	10%
zoo-data	16	59	11.07	11.07	0%	11.07	11.07	0%
average					8%			8%

Tables 1 and 2 present the average values of the maximum coverage of γ-decision rules. Column *atr* contains the number of attributes, column *rows* - the number of rows in T. For each row r of T, we find the maximum coverage of a γ-decision rule for T and r. After that, we find for rows of T the average

Table 2. Average coverage of γ-decision rules for $\gamma \in \{\lfloor J(T) \times 0.1\rfloor, \lfloor J(T) \times 0.3\rfloor, \lfloor J(T) \times 0.5\rfloor\}$

Decision table	$\lfloor J(T) \times 0.1\rfloor$			$\lfloor J(T) \times 0.3\rfloor$			$\lfloor J(T) \times 0.5\rfloor$		
	avg	avg-dp	rel dif	avg	avg-dp	rel dif	avg	avg-dp	rel dif
adult-stretch	6.25	7.00	11%	6.75	7.00	4%	7.50	7.50	0%
balance-scale	65.13	66.30	2%	88.48	92.31	4%	88.48	92.31	4%
breast-cancer	34.74	36.15	4%	78.48	79.62	1%	138.59	140.39	1%
cars	354.76	371.08	4%	458.67	485.51	6%	473.04	499.26	5%
house-votes	141.46	141.79	0%	154.69	156.06	1%	159.22	159.96	0%
lymphography	40.61	41.22	1%	57.80	57.82	0%	58.13	58.13	0%
mushroom	3287.94	3287.94	0%	3328.17	3328.17	0%	3822.32	3822.32	0%
nursery	1816.65	2129.73	15%	2385.13	3066.10	22%	2917.33	3066.10	5%
soybean-small	13.34	13.34	0%	15.68	15.68	0%	17.00	17.00	0%
teeth	1.00	1.00	0%	1.00	1.00	0%	1.00	1.00	0%
tic-tac-toe	103.52	113.45	9%	244.42	259.90	6%	327.78	327.79	0%
zoo-data	13.88	14.07	1%	17.34	17.36	0%	18.93	18.93	0%
average			4%			4%			1%

coverage of rules with the maximum coverage - one for each row. The results can be found in a column *avg*. To make comparison with the average coverage of γ-decision rules constructed by the dynamic programming algorithm [2] we made experiments and present results in a column *avg-dp*. The last column *rel dif* presents a relative difference (as a percentage). This value is equal to $(Opt\_Coverage - Coverage)/Opt\_Coverage \times 100$, where *Coverage* denotes the average coverage of γ-decision rules constructed by the proposed algorithm, *Opt_Coverage* denotes the average coverage of γ-decision rules constructed by the dynamic programming algorithm. The last row in Tables 1 and 2 presents the average value of the relative difference for considered decision tables.

Based on the presesnted results we can see that the average relative difference is nonincreasing when the value of γ is increasing. The relative difference is equal to 0, for $\gamma \in \Gamma(T)$, for "soybean-small" and "teeth", and for "mushroom" (with the exception $\gamma = \lfloor J(T) \times 0.0\rfloor$), and "zoo-data" (with the exception $\gamma = \lfloor J(T) \times 0.1\rfloor$). The biggest relative difference exists for $\gamma = \lfloor J(T) \times 0.0\rfloor$ and $\gamma = \lfloor J(T) \times 0.01\rfloor$, for "breast-cancer" - 35%.

Table 3. Size of the directed acyclic graph for $\gamma \in \{\lfloor J(T) \times 0.0\rfloor, \lfloor J(T) \times 0.01\rfloor\}$

Decision table	$\lfloor J(T) \times 0.0\rfloor$				$\lfloor J(T) \times 0.01\rfloor$			
	nd	edg	nd-dp	edg-dp	nd	edg	nd-dp	edg-dp
adult-stretch	36	37	72	108	36	37	72	108
balance-scale	654	808	1212	3420	185	214	647	1460
breast-cancer	2483	9218	6001	60387	2483	9218	6001	60387
cars	799	1133	7007	19886	406	564	4669	11379
house-votes	123372	744034	176651	1981608	123372	744034	176651	1981608
lymphography	26844	209196	40928	814815	26844	209196	40928	814815
mushroom	75125	524986	149979	2145617	34988	193301	83959	947650
nursery	18620	27826	115200	434338	1631	2392	16952	42001
soybean-small	3023	38489	3592	103520	3023	38489	3592	103520
teeth	118	446	135	1075	118	446	135	1075
tic-tac-toe	14480	41214	42532	294771	7324	13753	32596	103015
zoo-data	4087	37800	4568	83043	4087	37800	4568	83043

Table 4. Size of the directed acyclic graph for $\gamma \in \{\lfloor J(T) \times 0.1 \rfloor, \lfloor J(T) \times 0.3 \rfloor, \lfloor J(T) \times 0.5 \rfloor\}$

Decision table	$\lfloor J(T) \times 0.1 \rfloor$ nd	edg	nd-dp	edg-dp	$\lfloor J(T) \times 0.3 \rfloor$ nd	edg	nd-dp	edg-dp	$\lfloor J(T) \times 0.5 \rfloor$ nd	edg	nd-dp	edg-dp
adult-stretch	36	37	72	108	16	17	52	76	6	5	20	20
balance-scale	37	36	165	260	9	8	21	20	9	8	21	20
breast-cancer	539	976	3241	7748	120	162	633	883	41	46	201	225
cars	76	92	737	1276	23	22	158	197	16	15	57	57
house-votes	21984	55578	48341	131306	2463	3945	5839	8688	450	575	993	1162
lymphography	8773	31180	22523	123280	1652	4168	4223	13397	681	1422	1653	4456
mushroom	6119	18663	18058	93661	953	2119	2894	10438	227	377	606	1770
nursery	177	226	1126	1748	42	43	217	245	10	9	28	27
soybean-small	2374	12449	3456	35333	724	1847	1473	4789	227	385	495	999
teeth	103	323	133	806	54	111	99	288	30	48	58	138
tic-tac-toe	421	697	2114	4035	83	101	308	435	12	11	28	27
zoo-data	3492	22226	4426	48009	1562	4986	2533	10760	506	984	913	2062

Table 5. Comparison of the size of the directed acyclic graph

Decision table	$\lfloor J(T) \times 0.0 \rfloor$ nd dif	edg dif	$\lfloor J(T) \times 0.01 \rfloor$ nd dif	edg dif	$\lfloor J(T) \times 0.1 \rfloor$ nd dif	edg dif	$\lfloor J(T) \times 0.3 \rfloor$ nd dif	edg dif	$\lfloor J(T) \times 0.5 \rfloor$ nd dif	edg dif
adult-stretch	2.00	2.92	2.00	2.92	2.00	2.92	3.25	4.47	3.33	4.00
balance-scale	1.85	4.23	3.50	6.82	4.46	7.22	2.33	2.50	2.33	2.50
breast-cancer	2.42	6.55	2.42	6.55	6.01	7.94	5.28	5.45	4.90	4.89
cars	8.77	17.55	11.50	20.18	9.70	13.87	6.87	8.95	3.56	3.80
house-votes	1.43	2.66	1.43	2.66	2.20	2.36	2.37	2.20	2.21	2.02
lymphography	1.52	3.89	1.52	3.89	2.57	3.95	2.56	3.21	2.43	3.13
mushroom	2.00	4.09	2.40	4.90	2.95	5.02	3.04	4.93	2.67	4.69
nursery	6.19	15.61	10.39	17.56	6.36	7.73	5.17	5.70	2.80	3.00
soybean-small	1.19	2.69	1.19	2.69	1.46	2.84	2.03	2.59	2.18	2.59
teeth	1.14	2.41	1.14	2.41	1.29	2.50	1.83	2.59	1.93	2.88
tic-tac-toe	2.94	7.15	4.45	7.49	5.02	5.79	3.71	4.31	2.33	2.45
zoo-data	1.12	2.20	1.12	2.20	1.27	2.16	1.62	2.16	1.80	2.10
average	2.71	6.00	3.59	6.69	3.77	5.36	3.34	4.09	2.71	3.17

Tables 3 and 4 present a size of the directed acyclic graph, i.e., number of nodes (column nd) and number of edges (column edg) in the graph constructed by the proposed algorithm and dynamic programming algorithm (columns nd-dp and edg-dp respectively).

Table 5 presents comparison of the number of nodes (column $nd\ dif$) and number of edges (column $edg\ dif$) of the directed acyclic graph. Values of these columns are equal to the number of nodes/edges in the directed acyclic graph constructed by the dynamic programming algorithm divided by the number of nodes/edges in the directed acyclic graph constructed by the proposed algorithm.

Presented results show that the size of the directed acyclic graph constructed by the proposed algorithm is smaller than the size of the directed acyclic graph constructed by the dynamic programming algorithm. In particular, for the data sets "soybean-small", "teeth", "mushroom" and "zoo-data", the results of the average coverage are almost the same (see Tables 1 and 2) but there exists a difference relative to the number of nodes (more than one time) and relative to the number of edges (more than two times).

The size of the directed acyclic graph is related in some way with the properties of a data set, e.g., the number of attributes, the distribution of attribute values, the number of rows. To understand these relationships we need to consider deeply the structure of the graph, e.g., the number of nodes in each layer of the graph, the number of subtables and the number of incoming edges, the number of subtables and the number of rows in T. However, these considerations are not the aim of the paper.

6   Conclusions

We presented a modification of the dynamic programming algorithm for optimization of γ-decision rules relative to the coverage. Experimental results show that the size of the directed acyclic graph constructed by the proposed algorithm is smaller than the size of the directed acyclic graph constructed by the dynamic programming algorithm, and in the case of edges, the difference is at least two times. The average coverage of γ-decision rules constructed by the proposed algorithm is equal to optimal values for two data sets, for $\gamma \in \Gamma(T)$. In the future works, we will compare accuracy of rule based classifiers using considered algorithms.

Acknowledgements. The author wishes to thanks the anonymous reviewers for useful comments.

Thanks also to Prof. Moshkov and Dr. Chikalov for possibility to work with Dagger software system.

References

1. Alkhalid, A., Amin, T., Chikalov, I., Hussain, S., Moshkov, M., Zielosko, B.: Dagger: A tool for analysis and optimization of decision trees and rules. In: Computational Informatics, Social Factors and New Information Technologies: Hypermedia Perspectives and Avant-Garde Experiences in the Era of Communicability Expansion, pp. 29–39. Blue Herons (2011)
2. Amin, T., Chikalov, I., Moshkov, M., Zielosko, B.: Dynamic programming approach for partial decision rule optimization. Fundam. Inform. 119(3-4), 233–248 (2012)
3. Amin, T., Chikalov, I., Moshkov, M., Zielosko, B.: Dynamic programming approach to optimization of approximate decision rules. Inf. Sci. 221, 403–418 (2013)
4. An, A., Cercone, N.J.: Rule quality measures improve the accuracy of rule induction: An experimental approach. In: Ohsuga, S., Raś, Z.W. (eds.) ISMIS 2000. LNCS (LNAI), vol. 1932, pp. 119–129. Springer, Heidelberg (2000)
5. Asuncion, A., Newman, D.J.: UCI Machine Learning Repository ((2007), http://www.ics.uci.edu/~mlearn/
6. Błaszczyński, J., Słowiński, R., Szeląg, M.: Sequential covering rule induction algorithm for variable consistency rough set approaches. Inf. Sci. 181(5), 987–1002 (2011)
7. Dembczyński, K., Kotłowski, W., Słowiński, R.: Ender: a statistical framework for boosting decision rules. Data Min. Knowl. Discov. 21(1), 52–90 (2010)
8. Moshkov, M., Piliszczuk, M., Zielosko, B.: Partial Covers, Reducts and Decision Rules in Rough Sets - Theory and Applications. SCI, vol. 145. Springer, Heidelberg (2008)

9. Moshkov, M., Zielosko, B.: Combinatorial Machine Learning - A Rough Set Approach. SCI, vol. 360. Springer, Heidelberg (2011)
10. Nguyen, H.S.: Approximate boolean reasoning: Foundations and applications in data mining. In: Peters, J.F., Skowron, A. (eds.) Transactions on Rough Sets V. LNCS, vol. 4100, pp. 334–506. Springer, Heidelberg (2006)
11. Nguyen, H.S., Ślęzak, D.: Approximate reducts and association rules - correspondence and complexity results. In: Zhong, N., Skowron, A., Ohsuga, S. (eds.) RSFD-GrC 1999. LNCS (LNAI), vol. 1711, pp. 137–145. Springer, Heidelberg (1999)
12. Pawlak, Z., Skowron, A.: Rough sets and boolean reasoning. Inf. Sci. 177(1), 41–73 (2007)
13. Sikora, M., Wróbel, L.: Data-driven adaptive selection of rule quality measures for improving rule induction and filtration algorithms. Int. J. General Systems 42(6), 594–613 (2013)
14. Stefanowski, J., Vanderpooten, D.: Induction of decision rules in classification and discovery-oriented perspectives. Int. J. Intell. Syst. 16(1), 13–27 (2001)
15. Zielosko, B.: Optimization of approximate decision rules relative to coverage. In: Kozielski, S., Mrozek, D., Kasprowski, P., Małysiak-Mrozek, B. (eds.) BDAS 2014. CCIS, vol. 424, pp. 170–179. Springer, Heidelberg (2014)

MedVir: An Interactive Representation System of Multidimensional Medical Data Applied to Traumatic Brain Injury's Rehabilitation Prediction

Santiago Gonzalez[1], Antonio Gracia[1], Pilar Herrero[1],
Nazareth Castellanos[2], and Nuria Paul[3]

[1] Computer Science School, Universidad Politécnica de Madrid, Madrid, Spain
{sgonzalez,pherrero}@fi.upm.es, antonio.gracia@upm.es
[2] Center for Biomedical Technology, Universidad Politécnica de Madrid, Spain
nazareth@pluri.ucm.es
[3] Department of Basic Psychology I, Complutense University of Madrid, Spain
napaul@med.ucm.es

Abstract. Clinicians could model the brain injury of a patient through his brain activity. However, how this model is defined and how it changes when the patient is recovering are questions yet unanswered. In this paper, the use of MedVir framework is proposed with the aim of answering these questions. Based on complex data mining techniques, this provides not only the differentiation between TBI patients and control subjects (with a 72% of accuracy using 0.632 Bootstrap validation), but also the ability to detect whether a patient may recover or not, and all of that in a quick and easy way through a visualization technique which allows interaction.

Keywords: dimensionality reduction, multivariate medical data, feature selection, data mining, visualization, interaction, virtual reality, TBI, MEG.

1 Introduction

The possibility of detecting if a patient with a traumatic brain injury (TBI) can be rehabilitated by means of a treatment is not an easy work, however it is interesting. This is because it would allow to adjust and personalize treatments (in time and economy) of TBI patients to the needs of each one. One of the possibilities of evaluating the impact of brain injury is using MagnetoEncephaloGraphic (MEG) recordings through the obtaining of the functional connectivity patterns. These recordings are performed during several minutes of brain activity per individual (that is, time series of 148 sensors included in MEG machine).

The data generated by MEG are very complex (multidimensional and multivariate) and require much time for analysis. Nowadays, the idea of getting a prediction of the evolution of a TBI patient is out of the reach of clinicians, even without taking into account the analysis time. But if the data gathering process through MEG happens repeatedly, then is further complicated, so it is necessary to carry out an analysis mechanism that allows to draw conclusions (and extract new knowledge) in a quick and easy way.

M. Kryszkiewicz et al. (Eds.): RSEISP 2014, LNAI 8537, pp. 248–257, 2014.

In this work we were aimed to design a 3D visual interface for medical analysis easy to be used by clinicians. MedVir is a robust and powerful 3D visual interface to analyze, in this case, MEG data. After several stages, MedVir represents the information in two and three dimensions. Furthermore, the interface allows the experts to interact with the data in order to provide a more exhaustive analysis in the shortest time possible.

The paper is organized as follows: in section 2 a short overview of related work is presented. Section 3 presents the MedVir framework. Section 4 describes the TBI data obtaining and the experiments carried out. Finally, the conclusions and future lines (section 5) are reported.

2 Related Work

Nowadays, from the Data Mining point of view, there is no researches about TBI analysis through MEG recordings. However, three of the pillars supporting the proposed framework are very known there: Feature Subset Selection, Dimensionality Reduction and Data Visualization. Thus, in this section a short overview about theses points is presented.

Feature Subset Selection (FSS) problem [1] deals with the search of the best subset of attributes to train a classifier. This is a very important issue in several areas of knowledge discovery such as machine learning, optimization, pattern recognition and statistics. The goal behind FSS is the appropriate selection of a relevant subset of features upon which to focus the attention of a classification algorithm, while ignoring the rest. The FSS problem is based on the fact that the inclusion of more attributes in a training dataset does not necessarily improve the performance of the model. Two different kinds of variables can be distinguished: irrelevant (variable has no relation with the target of the classifier) and redundant (variables whose information can be deduced from other variables) features.

The literature describes several approaches to tackle this problem. To achieve the best possible performance with a particular learning algorithm on a particular training set, a FSS method should consider how the algorithm and the training set interact. Thus, there are two alternatives to consider this interaction: Filter [2] (analytical and statistical information among the features to evaluate each available feature) and Wrapper methods [3] (they use the induction algorithm itself to evaluate the performance of each candidate feature selection).

Wrapper methods often achieve better feature selections but the computational cost is higher. There are two main aspects that influence deeply on the computational cost of these techniques: (i) the optimization algorithm could be more or less exhaustive. For example, forward selection, backward elimination, and their stepwise variants can be viewed as simple hill-climbing techniques in the space of feature subsets; (ii) the robustness of the validation method (for instance LOOCV, Bootstrap, ...) applied to evaluate the quality of the results obtained by each candidate selection. It includes the measure to use, but also the validation schema.

An accurate FSS technique based on wrapper approaches that combines both a powerful search method and a robust validation approach is still a challenge, particularly in high-dimensional datasets. An appropriate alternative is to use a hybrid approach [4,5].

The most common one is the use of a filter to reduce the number of features (features are ranked based on their representativeness and the worst ones are removed), and a wrapper to perform the final selection. This represents a balance between the number of features to make the wrapper technique reasonable in computational time and the number of features included in the optimal subset selection.

As regards Data Visualization (DV), and specifically Multidimensional (unknown relations between attributes) Multivariate Data Visualization (MMDV), there are four broad categories [6] according to the approaches taken to generate the resulting visualizations. The first, *Geometric projection*, includes techniques that aim to find informative projections and transformations of multidimensional datasets [7] such as the Scatterplot Matrix [8], the Prosection Matrix [9], Parallel Coordinates [10] and Star Coordinates [11]. The second category groups the *Pixel-oriented* techniques [7] that represent a feature value by a pixel based on a color scale. This group includes the Space Filling Curve [12], the Recursive Pattern [13] and Spiral and Axes Techniques [14], among others. The techniques of the third category, *Hierarchical techniques*, subdivide the data space and present sub-spaces in a hierarchical way [7], for example, the Hierarchical Axis [15] and Dimensional Stacking [16] methods. The last category, *Iconography*, represents icon-based techniques that map the multidimensional data to different icons, or glyphs [17]. Some of them are Chernoff Faces [18] and Star Glyph [19].

Another way of visualizing multidimensional and multivariate data is by carrying out a Dimensionality Reduction (DR) process, which is one of the usual operations in Data Analysis (DA) [20]. Historically, the main reasons for reducing the dimensionality of the data is to remove possible noise or redundancy in the data, and reducing the computational load in further processing. One of the fields in which DR techniques for DV are currently very useful, is the scientific interactive visualization field, or Visual Analytics (VA). For DV, one of the main applications of DR is to map a set of observations into a 2 or 3 dimensional space that preserves the intrinsic geometric structure of the data as much as possible [21]. More related work about DR is presented in [22].

3   MedVir

The *MedVir* framework has been devised to abstract the clinicians the slow and tedious task of extracting conclusions about patients, treatments and rehabilitation when they work with multidimensional multivariate data analysis. The idea is that the expert only has to select the data to work with, and MedVir carries out a pipeline containing the most important steps of the KDD (Knowledge Discovery in Databases) process. As a result, data can be easily visualized in a virtual environment allowing a complete interaction, in order to get more conclusions about the interests of the clinicians.

MedVir comprises the following stages, as illustrated in Figure 1: i) *data preprocessing*, in which a set of data transformations and formatting are carried out so that the data can be properly treated by the following steps; ii) *selection of a reduced number of attributes* that best describe the original nature of the dataset. This step is carried out by using an extensive and intensive FSS process, in which five filter methods, four wrapper methods and four classification algorithms are used to obtain the models that perform better in supervised learning tasks; iii) *dimensionality data reduction* up to 2 or 3 dimensions to correctly represent the data on the display, with a minimum loss of quality; iv) *visualization of the data* facilitating a quick data interpretation.

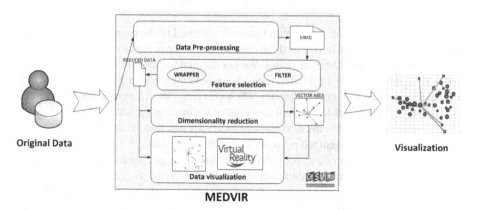

Fig. 1. The MedVir framework

Data Pre-processing. Real data often have a lot of redundancy, as well as incorrect or missing values, depending on different factors. Thus, it is usually necessary to perform some techniques in order to clean up and prepare the data. The algorithms included in this stage are replicated features handling, missing value handling and imputing missing values with *KNNImpute* algorithm [23]. Real data often have a lot of redundancy, as well as incorrect or missing values, depending on different factors. Thus, it is usually necessary to perform some techniques in order to clean up and prepare the data. The algorithms included in this stage are replicated features handling, missing value handling and imputing missing values with *KNNImpute* algorithm [23].

Feature Subset Selection. The second stage consists of a FSS process, which is responsible for selecting a reduced subset of attributes, from a very large number of initial attributes. The aim is to obtain a reduced dataset that retains or improves efficiency in many different Data Mining tasks. Thus, the main advantage of this stage is that the number of data attributes are strongly reduced from tens of thousands to a few dozens of attributes, thus reducing the computational cost and retaining or even improving their accuracy in different tasks, such as supervised or unsupervised classification. It is worth mentioning that the study presented here is limited to supervised classification tasks.

This step consists of two sub-stages: *filter* and *wrapper*. To implement the filter approach, five filter methods were used (Information gain, ReliefF, Symmetrical Uncertainty, Gain ratio and Chi squared) [24,25], and each one of these is executed *P times*, that is, for the different numbers of attributes to be filtered (eg, 500, 1000, 2000, ...). Once the filtered dataset is obtained, a wrapper process is carried out, using four search methods (Greedy, Best first, Genetic and Linear forward selection (LFS)) [24,25] and four classification algorithms (C4.5, SVM, Bayes Net and K-NN) [24,25] to obtain a reduced dataset containing, most of the cases, a few dozens of attributes. The combined use of wrapper and filter methods generate $80P$ different models and those that produces the best values, in terms of accuracy, are selected. To validate the results of each model, the *0.632 Bootstrap* [26] validation method has been used. Note that P can be set according to the number of attributes contained in the original data (e.g., if the dataset has 5000 attributes, P executions could be 6: 500, 1000, 2000, 3000, 4000 and 4500).

Dimensionality Reduction. The optimal dataset obtained in the previous stage can still not be directly visualized in two or three dimensions, since in many cases these data are supposed to have more than 3 attributes. We say optimal because, at this point, a dataset with a minimum number of attributes has been obtained, which always preserves or even improves (never worsens) efficacy when carrying out different tasks. Therefore, the third stage is responsible for obtaining a set of vector axes (generated by a particular DR algorithm) to be used in the next stage of MedVir's pipeline, so that the reduced data are transformed to be visualized properly in 2 or 3 dimensions.

   Different DR algorithms can be, indeed, used at this stage. For example, for clustering tasks, one might be interested in using PCA, since due to its great ability to obtain the directions of maximum variance of data, it produces minimum loss of quality of data [22], thus making more reliable the visualization of the real structure of data. Instead, LDA could be useful for supervised tasks, because even if the effectiveness in the preservation of the original geometry data is drastically reduced [22], the spatial directions of maximum discrimination between classes are easily obtained. This will facilitate the separation of different classes when the data are displayed. Therefore, depending on the used DR algorithm, a set of vectors are generated (as many vectors as attributes has the reduced dataset prior to this stage) to be used in the last stage of MedVir.

Data Visualization. The last stage generates the final visualization of the reduced data. To do so, the *star coordinates* (SC) algorithm is used [27]. SC algorithm works as follows: first, each attribute is represented as a vector radiating from the center of a circle to its circumference. Then the coordinate axes are arranged onto a flat (2-dimensional) surface forming equidistant angles between the axes. The mapping of an D-dimensional point to a 2-dimensional Cartesian coordinate is computed by means of the sum of all unit vectors on each coordinate, multiplied by the data value for that coordinate. In this paper, the input to the SC algorithm comprises two different elements: the *reduced data* and the set of *vector axes* generated in the previous stage. Thus, final visualization will be adjusted to the DR algoritm's requirements.

   The MedVir's *visualization* and *interaction* comprise, among many others, the carrying out of the following tasks: 1) if two points of different class (color) are very close or even overlapped in the visualization. This could strongly suggest that the expert might have made a mistake when originally labelling those instances. 2) if an attribute is selected, all points will be resized based on that attribute's value. This could represent the *importance* or *influence* of that attribute on a particular class. 3) if one or more attributes are selected and their lengths are modified, we would be giving them more or less weight on the representation, so all the instances will be reorganized based on those new weights. For example, if we give a greater weight to an attribute and a point with class A approaches another point with class B, this could suggest that a higher value of that attribute will mean a change in instance status from class A to B. 4) selected attributes can be *removed* to reduce their influence on the data representation. 5) if the instances are patients, their clinical information can be visualized *quickly* and *easily*. 6) a different visual *dispersion* among members of the same class and other classes may suggest different levels of cohesion between different instances. 7) display can be adjusted to achieve a comfortable interface, and the 2D and 3D visualization is represented by different colors, sizes, transparencies and shapes. Furthermore, navigation is *simple* and *intuitive*.

4   MedVir Applied to TBI

MedVir was applied to a real world case (figure 2), that is a Traumatic Brain Injury (TBI) rehabilitation prediction [28]. The study was performed by 12 control subjects and 14 patients with brain injury. All patients have completed a neurorehabilitation program, which was adapted specifically to each individual's requirements. This program was conducted in individual sessions attempting to offer an intensive neuropsychological-based rehabilitation, provided in 1h sessions for 3-4 days a week. In some cases, cognitive intervention was coupled with other types of neurorehabilitation therapies according to the patient's profile.

Patients had MEG recordings before and after the neuropsychological rehabilitation program. In this study control subjects were measured once, assuming that brain networks do not change in their structure in less than one year, as demonstrated previously in young.

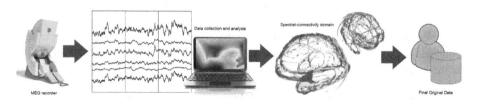

Fig. 2. MEG data obtaining process

The magnetic fields were recorded using a 148-channel whole-head magnetometer confined in 40 magnetically shielded room. MEG data were submitted to an interactive environmental noise reduction procedure. Fields were measured during a no task eyes-open condition. Time-segments containing eye movements or blinks or other myogenic or mechanical artefacts were rejected and time windows not containing artefacts were visually selected by experienced investigators, up to a segment length of 12s. By using a wavelet transformation [29], we perform a time-frequency analysis of rhythmic components in a MEG signal, and hence estimate the wavelet coherence for a pair of signals, a normalized measure of association between two time series [30]. Finally, MEG data were digitalized and transformed into a simple dataset of 26 instances x 10878 attributes, where each instance is a patient and each attribute is the relationship between each pair of channels.

4.1 Experiments

Experiments on the aforementioned dataset consists of a *FSS* process and *visualization* of the obtained data.

The first stage, *FSS*, is responsible for selecting, from among the 10,878 original attributes, a set of reduced data which improve accuracy when classifying new patients, compared to the original data. To classify new patients, a specific dataset consisting of 14 new instances is used. The FSS process consists of two parts: the first one uses filter methods, and the second one uses wrapper methods on the previous filtered attributes.

In total, 480 different models (5 *filter methods* x 6 *number of attributes to be filtered* x 4 *search methods* x 4 *classification algorithms*) have been obtained over the two parts, of which those who obtained the best accuracy were selected. The aim was to apply those models to the data to eventually visualize them. Note that the P value, described in Section 3 has been set to 6, since each filter method is carried out on the 500, 1000, 2000, 3000, 4000 and 5000 best attributes. The implementation of these models has been carried out in parallel and using the Magerit supercomputer, thus 480 nodes of the supercomputer have been used simultaneously to obtain the results. At the end of the process, a ranking of the 480 models was obtained, sorted by time spent to carry out the experiments (see Figure 3). Furthermore, the results have been validated using 0.632 Bootstrap method, as indicated in Section 3.

Model	% Accuracy (Original)	Accuracy (Filtered)	Accuracy (Wrappered)	N° of attributes	Time (s)
TBI_Relieff_500_Genetic_KNN	64.546	63.996	71.168	10	578.041
TBI_SymmetricalUncert_5000_Genetic_SVM	67.893	63.411	72.243	65	6126.37
...	...	...	...	...	...
...	...	...	...	...	...

Fig. 3. An example of the ranking of models obtained after the FSS stage

The criterion to select the best models is based on the highest values of accuracy achieved after the carrying out of the wrapper methods (fourth column from left). So, the models that have obtained the best accuracy are:

- **TBI_Relieff_500_Genetic_KNN (71.16%)**. The first model has used the *relieff* filter to obtain the best 500 attributes. After this, a *genetic* algorithm carried out an extensive search to select a subset of the 10 best attributes that best discriminate between the original classes, when classifying the instances by using the *K-NN* classification algorithm.
- **TBI_SymmetricalUncert_5000_Genetic_SVM (72.24 %)**. The second model has used the *symmetrical uncert* filter to rank the best 5000 attributes. Then, a *genetic* algorithm has selected a subset of the 65 best attributes that best discriminate between the original classes, when using the *SVM* classification algorithm.

Therefore, once the two reduced datasets were obtained, the classification of new patients was carried out. The results of the classification task are shown in Figure 4 (0 represents control subjects and 1 represents TBI patients). Except for patients 3 and 4 contained in the test dataset, there is a clear unanimity between the classification carried out by both models.

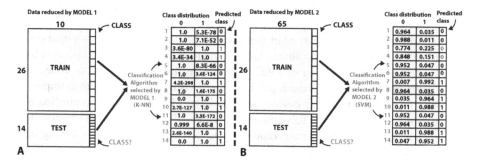

Fig. 4. Two models to classify the new patients

In the last step, *visualization*, MedVir represents the two datasets (figure 5). There, the blue color represents control subjects, whiles red means TBI patients and new classified patients are represented in magenta. The dotted line indicates the linear decision boundary in classification tasks. And it is at this time, when experts analyze the resulting work in order to extract the maximum possible information and draw relevant conclusions.

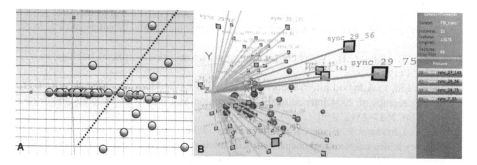

Fig. 5. Visualization in MedVir

5   Conclusions

MedVir can visualize multidimensional and multivariate medical data in 3D, so this allows conclusions to be obtained in a more simple, quick and intuitive way. Furthermore, the use of MedVir allows the clinicians to interact with the data they collect daily.

Specifically for this study, MedVir allows to effectively segment between control subjects and TBI patients with a 72% of accuracy. This is not a very high value, but it must be, indeed, taked into account that the validation mechanism used (Bootstrap) is certainly pessimistic due to the small number of instances in the data, so this strongly penalizes the final accuracy. In this paper, MedVir has been presented as a quick and easy tool to classify and visualize new subjects included in the TBI study. Visualization and interaction with the data can provide extra useful information to discern between uncertain class patients, after obtaining the results of a classification. In addition, MedVir could even be used to estimate if a TBI patient is in process of rehabilitation or not, so clinicians could be able to change the treatment or stop it.

However, there will be further research behind this work. In terms of data analysis, regression models and neuropsychological tests are going to be included to estimate the exact situation of a TBI patient in the recovery process, and how much treatment time he will need to fully rehabilitate. Another interesting future research point is the DR of data based on clustering of MEG sensors (creation of brain regions based on sensors locations and their relationships). In terms of visualization, we want to improve the user's interaction, using IO devices such as Leap Motion (MedVir controlled by gestual movements) and voice recognition (by means of expert orders). Furthermore, the interaction with the data visualization should be further studied by carrying out usability tests to test its reliability.

Concluding, MedVir, as a analysis tool, has successfully served for its purpose, allowing to know the status of rehabilitation of the TBI patients in an easy way. Of course, this tool could be applied to another interesting field, in which the number of attributes are too high that makes impossible a direct data analysis.

References

1. Blum, A.L., Langley, P.: Selection of relevant features and examples in machine learning. Artificial Intelligence 97, 245–271 (1997)
2. Jeffery, I.B., Higgins, D.G., Culhane, A.C.: Comparison and evaluation of methods for generating differentially expressed gene lists from microarray data. BMC Bioinformatics 7, 359+ (2006)
3. Kohavi, R., John, G.H.: Wrappers for feature subset selection. Artificial Intelligence 97, 273–324 (1997)
4. Ni, B., Liu, J.: A hybrid filter/wrapper gene selection method for microarray classification. In: Proceedings of 2004 International Conference on Machine Learning and Cybernetics, vol. 4, pp. 2537–2542. IEEE (2004)
5. Inza, I., Larrañaga, P., Blanco, R., Cerrolaza, A.J.: Filter versus wrapper gene selection approaches in dna microarray domains. Artificial Intelligence in Medicine 31, 91–103 (2004)
6. Keim, D.A.: Information Visualization and Visual Data Mining. IEEE Transactions on Visualization and Computer Graphics 8, 1–8 (2002)
7. Keim, D.A., Kriegel, H.P.: Visualization Techniques for Mining Large Databases: A Comparison. Transactions on Knowledge and Data Engineering, Special Issue on Data Mining 8, 923–938 (1996)
8. Hartigan, J.: Printer graphics for clustering. Journal of Statistical Computation and Simulation 4, 187–213 (1975)
9. Furnas, G.W., Buja, A.: Prosection Views: Dimensional Inference through Sections and Projections. Journal of Computational and Graphical Statistics 3, 323–385 (1994)
10. Inselberg, A.: Multidimensional Detective. In: Proceedings of the 1997 IEEE Symposium on Information Visualization, INFOVIS 1997, pp. 100–107. IEEE Computer Society, Washington, DC (1997)
11. Beddow, J.: Shape Coding of Multidimensional Data on a Microcomputer Display. In: IEEE Visualization, pp. 238–246 (1990)
12. Peano, G.: Sur une courbe, qui remplit toute une aire plane. Mathematische Annalen 36, 157–160 (1890)
13. Keim, D.A., Ankerst, M., Kriegel, H.P.: Recursive Pattern: A Technique for Visualizing Very Large Amounts of Data. In: Proceedings of the 6th Conference on Visualization 1995, VIS 1995, pp. 279–286. IEEE Computer Society, Washington, DC (1995)

14. Keim, D.A., Krigel, H.P.: VisDB: Database Exploration Using Multidimensional Visualization. IEEE Comput. Graph. Appl. 14, 40–49 (1994)
15. Mihalisin, T., Gawlinski, E., Timlin, J., Schwegler, J.: Visualizing a Scalar Field on an N-dimensional Lattice. In: Proceedings of the 1st Conference on Visualization 1990, VIS 1990, pp. 255–262. IEEE Computer Society Press, Los Alamitos (1990)
16. LeBlanc, J., Ward, M.O., Wittels, N.: Exploring N-dimensional Databases. In: Proceedings of the 1st Conference on Visualization 1990, VIS 1990, pp. 230–237. IEEE Computer Society Press, Los Alamitos (1990)
17. de Oliveira, M.C.F., Levkowitz, H.: From Visual Data Exploration to Visual Data Mining: A Survey. IEEE Trans. Vis. Comput. Graph. 9, 378–394 (2003)
18. Chernoff, H.: The Use of Faces to Represent Points in K-Dimensional Space Graphically. Journal of the American Statistical Association 68, 361–368 (1973)
19. Chambers, J., Cleveland, W., Kleiner, B., Tukey, P.: Graphical Methods for Data Analysis. The Wadsworth Statistics/Probability Series. Duxury, Boston (1983)
20. Lee, J.A., Verleysen, M.: Nonlinear dimensionality reduction. Springer, New York (2007)
21. Wang, J.: Geometric Structure of High-dimensional Data and Dimensionality Reduction. Higher Education Press (2012)
22. Gracia, A., González, S., Robles, V., Menasalvas, E.: A methodology to compare Dimensionality Reduction algorithms in terms of loss of quality. Information Sciences (2014)
23. Speed, T.: Statistical analysis of gene expression microarray data. CRC Press (2004)
24. Hall, M., Frank, E., Holmes, G., Pfahringer, B., Reutemann, P., Witten, I.H.: The weka data mining software: an update. ACM SIGKDD Explorations Newsletter 11, 10–18 (2009)
25. Witten, I.H., Frank, E.: Data Mining: Practical Machine Learning Tools and Techniques, 2nd edn. Morgan Kaufmann Series in Data Management Systems. Morgan Kaufmann Publishers Inc., San Francisco (2005)
26. Efron, B., Tibshirani, R.: Improvements on cross-validation: the 632+ bootstrap method. Journal of the American Statistical Association 92, 548–560 (1997)
27. Kandogan, E.: Visualizing multi-dimensional clusters, trends, and outliers using star coordinates. In: KDD 2001: Proceedings of the Seventh ACM SIGKDD International Conference on Knowledge Discovery and Data Mining, pp. 107–116. ACM, New York (2001)
28. Castellanos, N.P., Paul, N., Ordonez, V.E., Demuynck, O., Bajo, R., Campo, P., Bilbao, A., Ortiz, T., del Pozo, F., Maestu, F.: Reorganization of functional connectivity as a correlate of cognitive recovery in acquired brain injury. Brain Journal 133, 2365–2381 (2010)
29. Mallat, S.: A Wavelet Tour of Signal Processing, The Sparse Way, 3rd edn. Academic Press (2008)
30. Torrence, C., Compo, G.P.: A practical guide to wavelet analysis. Bull. Am. Meteorol. Soc. 79, 61–78 (1998)

SnS: A Novel Word Sense Induction Method*

Marek Kozłowski and Henryk Rybiński

Warsaw University of Technology
Warsaw, Poland
{M.Kozlowski,H.Rybinski}@ii.pw.edu.pl
http://www.ii.pw.edu.pl

Abstract. The paper is devoted to the word sense induction problem. We propose a knowledge-poor method, called SenseSearcher (SnS), which induces senses of words from text corpora, based on closed frequent sets. The algorithm discovers a hierarchy of senses, rather than a flat list of concepts, so the results are easier to comprehend. We have evaluated the SnS quality by performing experiments for web search result clustering task with the datasets from SemEval-2013 Task 11.

Keywords: word sense induction, sense based clustering, text mining.

1 Introduction

Huge mass of textual data available on Web makes a real challenge in developing automatic information and knowledge retrieval methods. Most of the current text mining methods are based on lexico-syntactic analysis of text. One of the main flaw of such approach is lack of sense-awareness, which in Information Retrieval systems leads to retrieving documents that are not pertinent to the user needs. Knowledge of an actual meaning of a polysemous word can significantly improve the quality of the information retrieval process by means of retrieving more relevant documents or extracting relevant information from texts. The importance of the polysemy problem becomes much clearer when we look at the statistics for the languages. As stated in [2], about 73% of words in common English are polysemous, and the average number of senses per word for all words found in English texts is about 6,5 [3].

The process of identifying meaning of a word in a given context is called word sense disambiguation (WSD). It is a computational process aiming at identifying the meaning of words in a context. Especially important role in word sense disambiguation is played by dedicated knowledge resources, however, a manual creation of such resources is an expensive and time-consuming task. Additionally, because of dynamic changes of the language in time, it is a never ending process. This problem is known as the knowledge acquisition bottleneck [4]. Clearly,

---

* This work was supported by the National Centre for Research and Development (NCBiR) under Grant No. SP/I/1/77065/10 devoted to the Strategic scientific research and experimental development program: "Interdisciplinary System for Interactive Scientific and Scientific-Technical Information".

M. Kryszkiewicz et al. (Eds.): RSEISP 2014, LNAI 8537, pp. 258–268, 2014.

because of the scale of language processing algorithms, the supervised machine learning algorithms, well-known in AI, cannot be effectively applied in solving language-related problems. Unsupervised methods seem to be the only chance to overcome the knowledge acquisition bottleneck. These methods are commonly known as Word Sense Induction ones (WSI). They consist in automatic identification of senses from textual repositories, without using hand-crafted resources or manually annotated data. The main advantage of the WSI methods is that they can detect new senses from the corpus, whereas the systems that reuse a man-made inventory are limited to the senses provided by lexicographers.

In the paper we present a novel approach to sense induction, called SenseSearcher (SnS). It is a knowledge-poor algorithm, using raw text corpus, and based only on text mining methods. The senses induced by SnS are easier to comprehend, mainly because the algorithm discovers a hierarchy of senses, rather than a flat list of concepts, thus showing important relationships between them. In other words, SnS builds a tree of senses, where coarse-grained senses contain related sub-senses (fine-grained senses).

The paper is organized as follows: in Section 2 we discuss related work. Section 3 presents the algorithm. Then we provide experimental results of the method in Section 4. Finally, in Section 5 we conclude the paper.

2   Related Work

WSI can be seen as a clustering problem, where words are grouped into non-disjoint clusters according to their meanings. Each cluster can be considered as a separate sense of the word. The methods are based on the distributional approach, which makes distinctions in word meanings based on the assumption that words that occur in similar contexts will have similar meanings. The WSI approaches can be divided into the following groups:

1. clustering context vectors;
2. extended clustering techniques;
3. bayesian methods;
4. graph-based techniques; and
5. frequent termsets based algorithms.

Clustering context vectors approach ([5], [6]) consists in grouping of the contexts, so that a given target word occurs. Thus, resulting clusters are built of contexts that use the target word in the same sense. The methods are based on the use of first- and second-order features. First-order features occur directly in a context being clustered, while second-order features are those that occur with a first order feature, but may not occur in the context being clustered. Initially in [5] automatic word sense discrimination was proposed. Pedersen and Bruce [6] presented clustering contexts using a small set of localized features and EM algorithm and Gibbs Sampling.

Referring to (2), several works aiming at improving the accuracy of WSI have been presented (e.g. [7] or [8]). They discover cluster of words that are related by virtue of their use in similar contexts. They rely on measuring similarity between word co-occurrence vectors or association between words. Given such information about a word, it is possible to identify other words that have a similar profile, which likely implies that they occur in related contexts and have similar meanings. In [7] a technique known as LSA has been presented. It lies in analyzing relationships between a set of contexts and the terms they contain by producing a set of likely topics/senses. It models the meaning of words and documents by projecting them into a vector space of reduced dimensionality, which is built up by applying singular value decomposition. Pantel and Lin [8] introduced a clustering algorithm called Clustering by Committee that automatically discovers word senses from text, and copes with the problem of infrequent meanings. The algorithm inertially discovers a set of tight clusters, called committees, that are well scattered in the similarity space. A cluster vector is a centroid of each committee. The words are assigned to their most similar clusters.

Bayesian methods (3) model the contexts of the ambiguous word as samples from a multinomial distribution over senses (characterized as distributions over words). The Bayesian framework provides a fundamental way to incorporate a wide range of features beyond lexical co-occurrences and to systematically assess their utility on the sense induction task. A recent example of this approach is presented by Brody and Lapata [9].

The WSI graph-based approaches ([10], [11], [12]) represent each word w_i co-occurring with the target word (within a context) as a vertex. Two vertices are connected via an edge if they co-occur in one or more contexts of the target word. Different graph clustering algorithms are used to induce senses from the graph. Each cluster, built of a set of vertices, represents a induced sense.

Frequent termsets based algorithms exploit classical data mining methods (as association rule mining) to induce senses. The first attempts of association mining application in sense discovery related area were conducted by Maedche and Staab [13]. The authors proposed an approach extending semi-automatic acquisition of taxonomies, by the discovery of non-taxonomic conceptual relations from textual data. Frequent sets mining, was applied directly as a method for sense discovery in papers [14], [15]. The method described in [14] consists in determining atomic contexts of terms of interest by means of maximal frequent sets of terms, which then are used for determining discriminant contexts. Distinct meanings of homonyms are indicated by various distinct contexts in which they appear frequently. Similar approach is proposed in [15], which is based on concise representations of frequent patterns. The method attempts to discover not only word senses that are dominating, but also senses that are dominated and infrequent in the repository.

3   The SnS Method

Semantically similar words tend to occur in similar contexts. In this Section we present SnS in more detail. The rest of the section is organized as follows. Subsection 3.1 describes the concepts, which are used by the method. Subsection

3.2 presents the algorithm, discusses its phases and finally provides it in the form of a pseudo-code.

3.1   Basic Concepts

Context is a normalized set of terms (termset) from a text unit (e.g. snippet). Precisely, it is a set of words surrounding a target term in the paragraph. Two assumptions are taken: all terms must be normalized (lemmatized, lower case, ascii encoded); all terms must be filtered by PoS rules and proper names detector (the context consists of nouns and proper names).

We define dictionary $\mathfrak{D} = \{t_1, t_2, ..., t_m\}$ as a set of distinct terms. Terms can be unigrams or n-grams. By $\mathcal{P}(t)$ we denote a set of all paragraphs retrieved for term t, where each paragraph is a text unit represented as a bag of words.

We also distinguish a set of proper names[1] $PN = \{pn_1, pn_2, ..., pn_n\}$, which are single or compound terms that are well defined in some external knowledge resource (e.g. Wikipedia), or mined by text mining methods, e.g. [17]. In this paper proper names are just extracted Wikipedia article's titles. By $pos(x)$ we denote a function determining a part of speech of x. It is used only for unigrams. If x is a unigram and is a noun the function returns the value $noun$, $pos(x) = noun$. Given the paragraph P and target term t we define context $C(P, t)$ as:

$$C(P,t) = \{x \mid x, t \in \mathfrak{D} \land x \neq t \land P \in \mathcal{P}(t) \land x \in P \land \\ (x \in PN \lor pos(x) = noun)\} \tag{1}$$

A family of contexts describing term t will be denoted as $\mathcal{C}(t) = \{C(P,t) \mid P \in \mathcal{P}(t)\}$. A termset is supported by a context if all its terms are contained in the context. We introduce the support function $sup(X) = n$, which means that X is supported by n contexts. X is frequent termset if $sup(X) > \epsilon$, where ϵ is a user defined threshold. Following [19] we say that a frequent termset is *maximal* if it has no frequent superset. We also say that a termset is *closed* iff none of its supersets has the same support. By $\mathcal{F}(t)$ we denote frequent termsets within contexts of a term t.

Given a set of contexts extracted from a corpus, a procedure for mining contextual patterns is performed. We aim at finding relevant patterns among the contexts. These patterns should maximize information (cover as many terms as possible without a loss of information), and minimize duplicating of information. To this end, looking for all frequent termsets is not justified, because they duplicate each other. Also maximal frequent termsets are not good - actually by representing contexts by maximal frequent termsets we discriminate some significant information, especially in the case where the support threshold is too low. Additionally with mined maximal termsets it is difficult (if possible at all) to create hierarchy of senses. On the other hand, closed frequent termsets This is why in our approach the contextual patterns are closed frequent termsets.

---

[1] Proper names are names of persons, places, or multiword concept names.

By $\mathcal{CP}(\mathcal{C}(t))$ we denote all contextual patterns of a term t, i.e. closed frequent termsets within contexts of a term t. In the sequel, whenever it is clear, the family of contextual patterns for t is written shorter as $\mathcal{CP}(t)$.

$$\mathcal{CP}(t) = \{X \mid X \in \mathcal{F}(t) \text{ and } \nexists Y \supset X \text{ s.t. } sup(X) = sup(Y)\} \qquad (2)$$

We also define a notion of a *sense frame*. Sense frame is a multi-hierarchical structure organizing contextual patterns. It builds tree structure with unlimited number of levels. The root is a main contextual pattern, which is a representative label for the sense frame, and it is typically discriminative against other sense frames. The main contextual pattern has sub-trees, which are sub-contexts trees. Sub-contexts are supersets of the main contextual pattern, but also they can be in subset relation among themselves. For a given term t and pattern $cp \in \mathcal{CP}(t) \cup \emptyset$, a sub-context tree is denoted by $stree(cp, t)$.

All sense frames are denoted as a sequence of tuples containing main contextual pattern (the shortest pattern, which is not a superset of other contextual patterns, further denoted as mp_i) and the tree $stree(mp_i, t)$:

$$\mathcal{SF}(t) = stree(\emptyset, t) = \langle (mp_1, stree(mp_1, t)), ..., (mp_n, stree(mp_n, t)) \rangle =$$
$$\langle sf_{1,t}, ..., sf_{n,t} \rangle, \text{ where } (mp_i, stree(mp_i, t)) \text{ is denoted by } sf_{i,t}, \text{ and } i = 1, .., n \qquad (3)$$

In most of the cases, especially when the repositories are representative sufficiently, sense frames correspond to distinctive senses. In the cases where the repositories are too small, two or more sense frames may refer to the same sense. Therefore within the SnS algorithm the obtained sense frames are compared, and the similar frames are grouped into the sense clusters. The similarity between the frames is expressed by a measure, which is based on how many terms of the sense frames are common. If for two frames the relative cardinality of their intersection is higher than a given threshold σ, then the two sense frames are grouped. Formally, the function $senseFrameTerms(x)$ returns a set of terms which are included in the sense frame x. Then the similarity measure between two $x, y \in \mathcal{SF}(t)$ is calculated as follows:

$$sim(x, y) = \frac{|senseFrameTerms(x) \cap senseFrameTerms(y)|}{min(|senseFrameTerms(x)|, |senseFrameTerms(y)|)} \qquad (4)$$

Let S be a subset of sense frames. By $seed(S)$ we denote a sense frame with the highest position in $\mathcal{SF}(t)$ (the sense frame's root has the highest support), compared to other sense frames in S. Finally, the set of senses is defined as below:

$$\mathcal{S}(t) = \{S \mid S \subseteq \mathcal{SF}(t) \wedge \forall x \in S \setminus seed(S) \; sim(seed(S), x) > \sigma\} \qquad (5)$$

3.2 Algorithm

The SnS algorithm consists of five phases, which we present below. The pseudocode of the whole algorithm is presented as Algorithm 1.

In Phase I, we build the index (e.g: full-text search index) for the corpus. In the SemEval Task 11 index contains terms with assigned set of snippets.

In Phase II, with a given term we run a query on the index, and find paragraphs (in the SemEval Task 11 snippets) related to the term. Then we convert them into context representations.

In Phase III contextual patterns are discovered from contexts generated in Phase II. The patterns are closed frequent termsets in the context space. The contexts are treated as transactions (itemsets are replaced by termsets) and the process of mining closed frequent termsets is performed with the use of the CHARM algorithm [18].

Phase IV is devoted to forming contextual patterns into sense frames, building a hierarchical structure of senses. In some exceptional states few sense frames may refer to one sense, it may result from the corpus limitations (lack of representativeness and high synonymity against descriptive terms).

In Phase V, sense frames are clustered. The clusters of sense frames are called senses. Optionally senses can be labelled with some descriptive terms.

4   Experiments

Evaluating of WSI methods is not an easy task, as there is no easy way to compare and rank various representations of senses. It can be actually seen as a special case of a more general and difficult problem of evaluating clustering algorithms. In order to find out more rigid ways to compare results of sense induction systems, Navigli and Vannella [1] organized Semeval-2013 Task 11[2]. The task is stated as follows: *given a target query, induction and disambiguation systems are requested to cluster and diversify the search results returned by a search engine for that query.*

Such a task enables evaluating and comparing various WSI methods. In order to perform comparisons with SemEval systems we customized SnS by adding the results clustering and diversification phase. The clustering is performed in two phases: (1) simultaneously during sense induction, and (2) after sense discovering clustering the results that remained not grouped before. Senses are grouped sense frames. Each sense frame has the main contextual pattern, so according to sense frames the snippets containing the main pattern are grouped in the corresponding result cluster. Non-grouped snippets are tested iteratively against each of the induced sense. Clustering of remaining snippets consists in using the similarity measure defined as intersection cardinality between the snippet and sense cluster's bag-of-words. Within each cluster the snippets are sorted using this measure. Clusters are sorted by the support of their sense frames.

4.1   Compared Systems

Within SemEval task 11 five teams submitted 10 systems: nine WSI-based and one WSD-based using wikipedia sense inventory. Below the systems are listed:

---

[2] http://www.cs.york.ac.uk/semeval-2013/task11

Algorithm 1. SnS algorithm

Input: $Corp$ as textual corpus, t as queried term and ϵ as minimal support, σ as sense frames similarity threshold

Output: $S(Corp, t, \epsilon, \sigma)$ as a set of senses

    Phase I - for each document $d \in Corp$ paragraphs are extracted and added to index $L(Corp)$

    Phase II - the index $L(Corp)$ is queried for a term t and retrieved $\mathcal{P}$ are converted into $\mathcal{C}$

       $\mathcal{C} = \{\}$

       for paragraph $p \in \mathcal{P}$ **do**

          $c = buildContext(p)$ {convert p into a context c, persisting only proper names and nouns}

          $\mathcal{C} = \mathcal{C} \cup c$

       end for

    Phase III - discovering contextual patterns $\mathcal{CP}$ from contexts $\mathcal{C}$, and sorting them by support descending

       $\mathcal{CP} = contextualPatternsMining(\mathcal{C}, \epsilon)$

    Phase IV - building sense frames $\mathcal{SF}$ from $\mathcal{CP}$

       $LC = \{\}, LT = \{\}, \mathcal{SF} = <>$

       for contextual pattern $cp \in \mathcal{CP}$ **do**

          if $cp \notin LC$ and $cp \cap LT = \emptyset$ **then**

              $mp = cp$ {cp becomes main contextual pattern mp}

              $sf = SenseFrame(mainContextPattern = mp, sTree = \emptyset)$

              $scp = findSubContextPatterns(mp)$ {find supersets of mp among $\mathcal{CP}$}

              $LC = LC \cup mp \cup scp; LT = terms(LC)$ {expand locked contexts and locked terms}

              $buildSenseFrame(sf, scp)$

              {organize scp in a tree hierarchy and assign to $sf.sTree$}

              $\mathcal{SF}.append(sf)$

          end if

       end for

    Phase V - clustering sense frames $\mathcal{SF}$ in order to find tight senses

       $clusters = senseFramesClustering(\mathcal{SF}, \sigma)$ {clusters are groups of similar sense frames}

       $labeledSenses = labelSenses(clusters)$ {clusters are labeled by the most common noun terms among grouped sense frames}

       $S(Corp, t, \epsilon, \sigma) = labeledSenses$ {labeled sense contains descriptive sense label, and set of incorporated sense frames}

---

1. HDP systems adopt WSI methodology based on a non-parametric model using Hierarchical Dirichlet Process. Systems are trained over extracts from the full text of English Wikipedia. In the contest two variants was submitted : HDP-CLS-LEMMA (search queries are lemmatized) and HDP-CLS-NOLEMMA (search queries are not lemmatized).
2. Satty-approach system implements the idea of monotone submodular function optimization, using a greedy algorithm.
3. UKP systems exploit graph-based WSI methods and external resources (Wikipedia or ukWaC). In the contest three variants of systems were submitted: UKP-WSI-WP-PMI (Wikipedia as a background corpus, and PMI as association metric were used), UKP-WSI-WP-LLR2 (Wikipedia as a background corpus and log-likelihood ratio as association metric were used), UKP-WSI-WACKY-LLR (ukWaC as a background corpus and log-likelihood ratio as association metric were used).
4. Duluth systems are based on second-order context clustering as provided in SenseClusters, a freely available open source software package. In the contest three variants were submitted: DULUTH.SYS9.PK2 (uses additionally paragraphs of Associated Press, apart task data), DULUTH.SYS1.PK2 (all the Web snippet results for all 100 queries were combined into a single corpus), DULUTH.SYS7.PK2 (treats the 64 snippets returned by each query as the corpus for creating a cooccurrence matrix).
5. Rakesh system exploits external sense inventories for performing the disambiguation task. It employs YAGO hierarchy and DBPedia in order to assign senses to the search results.

4.2 Scoring

Following [16] and [1], the systems were evaluated in terms of the clustering quality and the diversification quality. Clustering evaluation problem is a difficult issue, and one for which there exists no unequivocal solution. Many evaluation measures have been proposed in the literature so, in order to get exhaustive results, we calculated four distinct measures, namely: Rand Index (RI), Adjusted Rand Index (ARI), Jaccard Index (JI) and F1 measure. Diversification is a technique aimed at reranking search results on the basis of criteria that maximize their diversity. Quantifying the impact of web search result clustering methods on flat-list search engines is measured by $S\text{-}recall@K$ (sense recall at rank K) and $S\text{-}precision@r$ (Sense precision at recall r). The above mentioned measures are described in details in [16] and [1].

4.3 Results

We show the results for the measures RI, ARI, JI and F1 in Table 1. Additionally, the last two columns show the average number of clusters (cl.) and average cluster size (ACS). The best results in each class are in bold.

The SnS method outperforms the best systems taking part in SemEval - HDP based. SnS reports considerably higher values in RI and ARI. As one can see, SnS also performs best in both measures - JI and F1.

Table 1. The results of clustering experiments on SEMEVAL data set (in %)

Type	System	RI	ARI	JI	F1	cl.	ACS
	HDP-CLS-LEMMA	**65.22**	21.31	33.02	**68.30**	6.63	11.07
	HDP-CLS-NOLEMMA	64.86	**21.49**	33.75	68.03	6.54	11.68
	SATTY-APPROACH1	59.55	7.19	15.05	67.09	9.90	6.46
	DULUTH.SYS9.PK2	54.63	2.59	22.24	57.02	3.32	19.84
WSI	DULUTH.SYS1.PK2	52.18	5.74	31.79	56.83	2.53	26.45
	DULUTH.SYS7.PK2	52.04	6.78	31.03	58.78	3.01	25.15
	UKP-WSI-WP-LLR2	51.09	3.77	31.77	58.64	4.17	21.87
	UKP-WSI-WP-PMI	50.50	3.64	29.32	60.48	5.86	30.30
	UKP-WSI-WACKY-LLR	50.02	2.53	**33.94**	58.26	3.64	32.34
WSD	RAKESH	**58.76**	**8.11**	**30.52**	**39.49**	9.07	2.94
SNS	SNS	**65.84**	**22.19**	**34.26**	**70.16**	8.82	8.46

To get more insights into the performance of the various systems, we calculated the average number of clusters and the average cluster size per clustering produced by each system and compared it with the gold standard average. The best performing system in the case of all the above mentioned categories has clustering size and clusters size similar to the gold standard. SnS reports results similar to HDP ones (the best ones in the WSI class). The expected values of number of snippets per query is close to 64.

Finally, we come to the diversification performance, calculated in terms of S-recall@K and S-precision@r, which results are shown in Table 2. In these categories generally HDP systems obtain the best performance. The SnS methods is still near to the top systems, reporting results higher than the remaining verified WSI systems.

Table 2. The results for S-recall@K and S-precision@r on SEMEVAL data set (in %)

Type	System	S-recall				S-precision			
		K=5	K=10	K=20	K=40	r=50	r=60	r=70	r=80
	HDP-CLS-LEMMA	48.13	**65.51**	78.86	91.68	**48.85**	42.93	**35.19**	27.62
	HDP-CLS-NOLEMMA	**50.80**	63.21	**79.26**	**92.48**	48.18	**43.88**	34.85	**29.30**
	SATTY-APPROACH1	38.97	48.90	62.72	82.14	34.94	26.88	23.55	20.40
	DULUTH.SYS9.PK2	37.15	49.90	68.91	83.65	35.90	29.72	25.26	21.26
WSI	DULUTH.SYS1.PK2	37.11	53.29	71.24	88.48	40.08	31.31	26.73	24.51
	DULUTH.SYS7.PK2	38.88	53.79	70.38	86.23	39.11	30.42	26.54	23.43
	UKP-WSI-WP-LLR2	41.07	53.76	68.87	85.87	42.06	32.04	26.57	22.41
	UKP-WSI-WP-PMI	40.45	56.25	68.70	84.92	42.83	33.40	26.63	22.92
	UKP-WSI-WACKY-LLR	41.19	55.41	68.61	83.90	42.47	31.73	25.39	22.71
WSD	RAKESH	**46.48**	**62.36**	**78.66**	**90.72**	**48.00**	**39.04**	**32.72**	**27.92**
SNS	SNS	**47.36**	**62.96**	**74.10**	**87.79**	**47.95**	**37.99**	**31.68**	**24.10**

5   Conclusion

In this paper we propose a novel WSI knowledge-poor algorithm SnS, which is based on text mining methods. It uses closed frequent termsets as contextual patterns, identifying senses.

The algorithm converts simple contexts (based on bag-of-words paragraph representation) into relevant contextual patterns (which are much more concise, and representative). Using significant patterns SnS builds a hierarchical structure composed of so called sense frames. The discovered sense frames usually are independent senses, but sometimes (e.g. because of too small corpus) few can point the same sense. In order to reduce redundancy of discovered senses SnS uses a simple clustering method to group similar sense frames, referring to the same main sense.

We have shown that small to medium size text corpus can be used to identify efficiently senses. Additional interesting feature of the algorithm is discovering of the dominated meanings. An extensive set of experiments, corresponding to SemEval 2013 Task 11, confirms that SnS provides significant improvements over the existing methods in the domain of clustering quality. Using a broad scope of measures and the SEMEVAL data set we have experimentally shown that SnS outperforms others WSI systems, and it can be efficiently used as a clustering engine for end-users.

References

1. Navigli, R., Vannella, D.: SemEval-2013 Task 11: Word Sense Induction and Disambiguation within an End-User Applications. In: Proc. 7th Int'l SemEval Workshop, The 2nd Joint Conf. on Lexical and Comp. Semantics (2013)
2. Miller, G., Chadorow, M., Landes, S., Leacock, C., Thomas, R.: Using a semantic concordance for sense identification. In: Proceedings of the ARPA Human Language Technology Workshop, pp. 240–243 (1994)
3. Mihalcea, R., Moldovan, D.: Automatic Generation of a Coarse Grained WordNet. In: Proc. of NAACL Workshop on WordNet and Other Lexical Resources (2001)
4. Navigli, R.: Word sense disambiguation: A survey. ACM Computing Surveys 41(2) (2009)
5. Schutze, H.: Automatic word sense discrimination. Computational Linguistics - Special Issue on Word Sense Disambiguation 24(1) (1998)
6. Pedersen, T., Bruce, R.: Knowledge lean word sense disambiguation. In: Proceedings of the 15th National Conference on Artificial Intelligence (1998)
7. Landauer, T., Dumais, S.: A solution to Platos problem: The Latent Semantic Analysis theory of the acquisition, induction, and representation of knowledge. Psychology Review (1997)
8. Pantel, P., Lin, D.: Discovering word senses from text. In: Proceedings of the 8th International Conference on Knowledge Discovery and Data Mining (2002)
9. Brody, S., Lapata, M.: Bayesian word sense induction. In: Proceedings of EACL 2009 (2009)
10. Veronis, J.: Hyperlex: lexical cartography for information retrieval. Computer Speech and Language (2004)

11. Agirre, E., Soroa, A.: Ubc-as: A graph based unsupervised system for induction and classification. In: Proc. 4th Int'l Workshop on Semantic Evaluations (2007)
12. Dorow, B., Widdows, D.: Discovering corpus-specific word senses. In: Proceedings of the 10th Conference of the European Chapter of the ACL (2003)
13. Maedche, A., Staab, S.: Discovering conceptual relations from text. In: Proceedings of the 14th European Conference on Artificial Intelligence (2000)
14. Rybiński, H., Kryszkiewicz, M., Protaziuk, G., Kontkiewicz, A., Marcinkowska, K., Delteil, A.: Discovering word meanings based on frequent termsets. In: Raś, Z.W., Tsumoto, S., Zighed, D.A. (eds.) MCD 2007. LNCS (LNAI), vol. 4944, pp. 82–92. Springer, Heidelberg (2008)
15. Nykiel, T., Rybinski, H.: Word Sense Discovery for Web Information Retrieval. In: Proc. 4th Int'l Workshop on Mining Complex Data (2008)
16. Di Marco, A., Navigli, R.: Clustering and Diversifying Web Search Results with Graph-Based Word Sense Induction. Computational Linguistics (2013)
17. Protaziuk, G., Kryszkiewicz, M., Rybiński, H., Delteil, A.: Discovering compound and proper nouns. In: Kryszkiewicz, M., Peters, J.F., Rybiński, H., Skowron, A. (eds.) RSEISP 2007. LNCS (LNAI), vol. 4585, pp. 505–515. Springer, Heidelberg (2007)
18. Zaki, M., Hsiao, C.: CHARM: An efficient algorithm for closed itemset mining. In: Proceedings 2002 SIAM Int. Conf. Data Mining (2002)
19. Zaki, M.: Closed itemset mining and non-redundant association rule mining. In: Encyclopedia of Database Systems (2009)

Review on Context Classification in Robotics

Fábio Miranda[1], Tiago Cabral Ferreira[2], João Paulo Pimentão[1], and Pedro Sousa[1]

[1] Faculdade de Ciências e Tecnologia – UNL Caparica, Portugal
f.miranda@campus.fct.unl.pt, {pim,pas}@fct.unl.pt
[2] Holos SA, Caparica, Portugal
tiago.ferreira@holos.pt

Abstract. In this paper a review on context and environment classification is presented with focus on autonomous service robots. A comprehensive research was made in order to classify the most relevant techniques, models and frameworks in use today, as well as to present possible future applications for these previous works. Most of the work done in this area has been focused in a general classification. In this sense a new possible application scenario is described and the corresponding supporting architecture.

Keywords: context-aware, reliability, environment classification, artificial intelligence.

1 Introduction

Aside from our language and the common understanding of the world and its operation, what contributes to a much more efficient exchange of knowledge between human beings, is the notion of context [1]. Humans are able to add and take into account situational information (context) in a conversation to better understand the subject, retain more information in a more efficient way and to respond appropriately to it. In other words, to increase conversational bandwidth as said in [1].

1.1 What Is Context-Awareness?

The term 'context-aware' was first introduced in [2] and they proposed "context" as software that "adapts according to its location of use, the collection of nearby people and objects, as well as changes to those objects over time.". Context-aware applications need to sense and interpret the environment around, in order to be adaptive, reactive, responsive, situated, context-sensitive and environment-directed. The main questions that context-aware applications must take into account are: who, where, when and what the user is doing, to understand and classify his situation [1].

2 State of the Art

Context and context-awareness has been a strongly addressed field for many years, but only recently was applied in real world applications like service robots. In this

M. Kryszkiewicz et al. (Eds.): RSEISP 2014, LNAI 8537, pp. 269–276, 2014.
© Springer International Publishing Switzerland 2014

sense it will be presented how data is acquired from the context-sources, what are the main techniques used in the context classification system, which are the main existing frameworks and architectures for building context applications and finally an overview on how these techniques and concepts are used in robotics.

2.1   Context Data Acquisition Methods

There are different architecture approaches to acquire context information according to [3–5]. The chosen method depends on the required application and it's relevant for the design of the whole system's architecture. Other authors refer a different classification but it seems to lack some focus thus the following enumeration is presented:

- **Direct sensor access** – drivers from the sensor directly into application. It doesn't have processing capability or support for multiple concurrent sensing.
- **Sentient objects** - may be a smart sensor with actuator, which interact with the environment and can interact with each other. They can make use of a simple rule engine and perform simple actions like, for example: a light sensor that is connected to a headlight and turns on the headlight when it's dark [6].
- **Middleware infrastructure** – implemented with a layered architecture that hides low level data and allows separation between several structure interfaces.
  − Distributed middleware – Inference results can be shared in a distributed system for various applications.
  − Centralized middleware – Where context sources, inference system and applications are running only locally and in the same system.

2.2   Context Models

There are several approaches to develop the rules which will settle the context classification system. These rules/deductions/methods establish the relation between the data sensed and the overall context situation.

Some approaches refer key-value and markup models for context applications. Key-value models use simple key-value pairs and define a list of attributes and their values[7]. Markup models use markup languages like XML, tagging data in a hierarchical way. There are some hybrid models than can be both, since they can store key-value pairs under appropriate tags [7]. These models have known limitations about reasoning support, timelines and more complex relations and dependencies[8].

There are also graphical models that use visual implementation such as UML (Unified Modeling Language) like in [9]. In extension of graphical models there is object-based modeling (like Object-Role Modeling - ORM) techniques that supports better processing and reasoning to satisfy the more demanding requirements; CML (Context Model Language) is an example. CML and ORM emerged from conceptual modeling databases[7].

There are logic-based models in which context is defined based on a set of rules, facts and statements. Logic defines the conditions needed to reach a fact or expression derived from the facts and rules[10].

Finally there are the ontology-based models. Between knowledge representation languages, DL (description logic) has emerged with a good tradeoff between expressiveness and complexity[7]. It gives us a description about the world under what is known in terms of concepts, roles and individuals [11].

In[10] the authors claim that ontology based models are the more expressive for their requirements. Since ontologies consist in descriptions of concepts and their relationships, they are a powerful instrument for modeling contextual information [5]. A formalism called OWL (Web Ontology Language) is used to represent context information distinguished by categories of objects and objects interrelation [12].

There is also hybrid models that combine different formalisms and techniques [7]. In [7] the authors believe this is a promising direction since different models and reasoning tools need to be integrated with each other towards a better tradeoff between expressiveness, complexity and support for uncertainty.

2.3   Context Frameworks

When choosing the design of a context model it should be analyzed what features are best suited to the required application, regarding the architecture, inference system and how the data is processed and sensed.

According to [5] the most common architectures use layers, are hierarchical and have one or many centralized components.

In the Context-Awareness Sub-Structure (**CASS**) architecture, the middleware contains an interpreter, a context retriever, a rule engine and a sensor listener. The sensor listener senses for data updates from sensor nodes and stores that information in the database. The context retriever retrieves the stored data from the database. That data passes through a rule engine, where it verifies some required conditions and can also use the interpreter to consider what context type is checked[13].

The **Hydrogen** project consists in another layered architecture focused on mobile devices. In this system instead of a centralized component, they have a distributed system. Hydrogen is capable of exchanging (sharing) context information with a remote client [5]. There are three layers: Application, Management and Adaptor layer. The adaptor layer is directly connected to the sensors. The management layer is where the context-server is located, which stores the data from the sensors and provide contexts to the application layer. At last, the application layer is where the actions will be implemented according to context changes[5] [14].

The **CORTEX** project is more oriented to research, cooperation and interaction between sentient objects (defined in section 2.1) and its environment[6].

CoBrA systems were design to support context-aware applications in intelligent spaces (rooms with intelligent systems)[5]. CoBrA has a centralized context broker that manages and share information between agents in a particular environment. In large smart spaces they can group multiple brokers and exchange knowledge so in table 1 it's classified as a distributed middleware [4].

The **Gaia** project consists in a layered architecture with operating system concepts. Gaia systems are more focused on the interaction between users and smart/intelligent spaces. For the Gaia systems, each space is self-contained, although they may interact between spaces. The Gaia system implements ontologies to describe context predicates[15].

Table 1. Overview of the main features of the existing context frameworks[5][16][6] ("Ont" refers to ontology-based formalism, "Obj" refers to object-based formalism, "logic" to a logic based model formalism and "KNW" to a knowledge-based approach)

System	Context Model	Context History	Data Acquisition Methods	Learning ability
CASS [13]	Logic + KNW	Yes	Centralized middleware	No
CoBrA	Ont. (OWL)	No	Distributed Middleware	No
SOCAM[17]	Ont.(OWL)	No	Distributed Middleware	No
GAIA	Ont. (DAML+OIL)[15]	Yes	Distributed Middleware	Yes [15]
Hydrogen[14]	Obj.	No	Distributed Middleware	No
CORTEX	Logic	Yes	Sentient Object	Yes[6]
COMANTO	Ont.	No	-	-

SOCAM is a service oriented context aware middleware architecture for mobile services[17]. In the SOCAM architecture, although it is a general ontology-based model, it's difficult to relate data with different granularity and constraining important data to specific contexts[16].

COMANTO (COntext MAnagement oNTOlogy) is a semantic vocabulary based on ontologies that can describe generic context types[18]. According to [16] it lacks the possibility to discard useless contexts, but it's a very expressive formal model.

In Table 1 there is a summary about various context modeling systems and its capabilities and can be seen that there is a clear trend for using ontologies connected with a distributed data acquisition method.

2.4   Context in Robotics

Systems like social, cooperative or service robots have to perceive and interpret its surroundings and take into account many variables to make correct decisions. Robots must perceive the same things that humans find to be relevant and interact with the environment similarly to living beings [19].

In [20] the authors introduce CAMUS (Context Middleware for URC Systems) which is a framework to support context-aware services for network-based robots. This article is inserted in a relatively new concept named URC (Ubiquitous Robotic Companion) which are ubiquitous service robots that provide services to users, anytime and anywhere in ubiquitous computing environments[21]. Returning to [20], this system uses UDM (Universal Data Model) to represent context information (relations between nodes) using OWL.

In [22], the authors use context-awareness to improve vehicle-to-vehicle applications in terms of its driving safety and efficiency. The techniques used for this approach, for modeling context and situations, are Context Spaces combined with the Dempster-Shafer rule of combination, for situation reasoning.

In [23] the authors presented a machine-understandable representation of objects (their shapes, functions and usages). They analyze certain combinations or sets of objects and its features in order to context/scene understanding and to deduce corresponding possible activities in those scenes. Both reasoning about object and scene recognition use an ontological representation (OWL) and relational databases.

Still in object recognition, in [24] the authors present an indoor furniture and room recognition combined with online sources. It is used a Markov Random Field (MRF) as a final probabilistic classifier to model object-object and object-scene context.

There is also an application of context in robots localization/navigation. For example, in [25] the authors use a semantic representation and a Bayesian model for robot localization measuring the distance relations between known objects in the environment. The authors named it a topological-semantic distance map.

The authors in [26] describe the development of an ontology for robots in the field of urban search and rescue, based on OWL.

There is also an attempt to use ontologies for autonomous vehicle navigation planning in [27] in order to increase performance of route planning and improve capabilities of autonomous vehicles.

In [28] the authors propose an architecture for context-aware applications and ubiquitous robotics, where robots navigate in smart environments. The proposed architecture integrates ontologies with logic approach.

Table 2. Overview of context models in robotic systems their application purpose

Application Purpose	Context Model/Reasoning Technique
Object Recognition	Markov Random Field (probabilistic classifier)[24], Ontologies[26]
Safety	Dempster-Shafer[25], Ontologies[26]
Human-machine interface	Polynomial classifiers[31], Hidden Markov Model[32]
Localization	Ontologies + Bayesian model[25]
Navigation	Ontologies[27]
Smart Environments	Ontologies + Logic[28], Ontologies[20]

The authors in [32] develop CAIROW (Context-aware Assisted Interactive RObotic Walker) for Parkinson disease patients. They use a Hidden Markov Model (HMM) to analyze the gait of the patient. Both patient and road conditions contexts are considered. The robot should adjust their speed or direction according to user gait and road conditions.

3   Implementation

3.1   Problem Approach

As it was stated before there are several applications for context in the robotics field but context can be integrated in others. It is believed that the reliability field hasn't explored context in complete way.

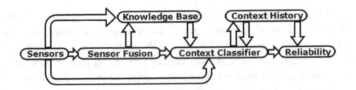

Fig. 1. Reliability context integration architecture

In Fig. 1 the reliability architecture is presented. It will first receive the sensor inputs and process its signals in order to reduce any noise and outlier values as well as necessary sensor fusion(SF). In SF, since the information can be uncertain and inconsistent, the application of rough sets is useful as rough set theory, introduced by Pawlak[33], is a technique for dealing with this kind of issues. In the knowledge base the information from sensors and sensor fusion are stored. It will be created a context classifier that will provide contextual information to the reliability calculation. Those classifications will be also stored in a context history module. This history can be re-fed to the classification module to improve performance.

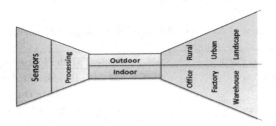

Fig. 2. Context classification hierarchy

In the proposed architecture the objective of the context integration is to classify the data output after processing in outdoor/indoor environment. Done that, there will be another model created to sub-classify the type of environment present.

Environmental conditions might influence a great deal the reliability calculation. The key elements present in the surroundings may be crucial for optimizing reliability assessment. If a mobile robot is working under a dusty environment it may be more prone to have mechanical failures thus reducing its reliability.

4   Conclusions

Independently of the application, whether it is social, entertain, service, educational, security, etc. robotic systems need to have the notion of its context in order to behave appropriately and provide useful information to users and applications that depend on that information. There are several context models, techniques and frameworks and the ones based on ontologies seem to be the ones with most widespread use.

In context classification there is uncertainty associated with the environment sensing that can be caused by inaccuracy or incompleteness of sensed information thus creating a need for models that manage and take it into account.

In the robotics field there are already some applications but it seems that context classification can be an added value to others. In this sense a new application scenario is presented as well as the architecture to support it in the reliability field.

References

1. Dey, A.K., Abowd, G.D.: Towards a Better Understanding of Context and Context-Awareness. Handheld Ubiquitous Comput. 304–307 (1999)
2. Schilit, B.N., Theimer, M.M.: Disseminating active map information to mobile hosts. Network, Dept. of Comput. Sci., Columbia Univ., New York, NY, U, vol. 8, pp. 22–32. IEEE (1994)
3. Chen, H.: An intelligent broker architecture for pervasive context-aware systems (2004)
4. Chen, H., Finin, T., Joshi, A.: An Ontology for Context-Aware Pervasive Computing Environments. Knowl. Eng. Rev. 18, 197–207 (2003)
5. Baldauf, M., Dustdar, S., Rosenberg, F.: A survey on context-aware systems. J. Int. J. Ad Hoc Ubiquitous Comput. 2, 263–277 (2007)
6. Barron, P., Biegel, G., Cahill, V., Casimiro, A., Clarke, S., Cunningham, R., Fitzpatrick, A., Gaertner, G., Hughes, B.: Preliminary Definition of CORTEX Programming Model (2003)
7. Bettini, C., Brdiczka, O., Henricksen, K., Indulska, J., Nicklas, D., Ranganathan, A., Riboni, D.: A survey of context modelling and reasoning techniques. Pervasive Mob. Comput. 6, 161–180 (2010)
8. Indulska, J., Robinson, R.: Experiences in using cc/pp in context-aware systems. Mob. Data Manag. 2574, 247–261 (2003)
9. Sheng, Q.Z., Benatallah, B.: ContextUML: a UML-based modeling language for model-driven development of context-aware web services. Mob. Bus., 206–212 (2005)
10. Strang, T., Linnhoff-Popien, C.: A context modeling survey. UbiComp 1st Int. Work. Adv. Context Model. Reason. Manag., 31–41 (2004)
11. Straccia, U.: Reasoning within fuzzy description logics. J. Artif. Intell. Res. 14, 137–166 (2011)
12. Horrocks, I., Patel-Schneider, P.F., van Harmelen, F.: From SHIQ and RDF to OWL: the making of a Web Ontology Language. Web Semant. Sci. Serv. Agents World Wide Web. 1, 7–26 (2003)
13. Fahy, P., Clarke, S.: CASS–a middleware for mobile context-aware applications. Work. Context Awareness, MobiSys (2004)
14. Hofer, T., Schwinger, W.: Context-awareness on mobile devices-the hydrogen approach. Syst. Sci. 43 (2003)
15. Ranganathan, A., Campbell, R.H., Muhtadi, A.: Reasoning about Uncertain Contexts in Pervasive Computing Environments. IEEE Pervasive Comput. 3, 62–70 (2004)
16. Bolchini, C., Curino, C.A., Quintarelli, E., Schreiber, F.A., Tanca, L., Politecnico, I., Leonardo, P.: A Data-oriented Survey of Context Models. SIGMOD Rec. 36, 19–26 (2007)
17. Gu, T., Pung, H.K., Zhang, D.Q.: A middleware for building context-aware mobile services. In: Proc. IEEE Veh. Technol. Conf., vol. 5, pp. 2656–2660 (2004)
18. Strimpakou, M.A., Roussaki, I.G., Anagnostou, M.E.: A context ontology for pervasive service provision. Adv. Inf. Netw. Appl. 2 (2006)
19. Zlatev, J.: The Epigenesis of Meaning in Human Beings, and Possibly in Robots. Minds Mach. 11, 155–195 (2001)

20. Kim, H., Kim, M., Lee, K., Suh, Y., Cho, J., Cho, Y.: Context-Aware Server Framework for Network-based Service Robots. In: 2006 SICE-ICASE Int. Jt. Conf., pp. 2084–2089 (2006)
21. Ha, Y.-G., Sohn, J.-C., Cho, Y.-J., Yoon, H.: Towards Ubiquitous Robotic Companion: Design and Implementation of Ubiquitous Robotic Service Framework. ETRI J 27, 666–676 (2005)
22. Wibisono, W., Zaslavsky, A., Ling, S.: Improving situation awareness for intelligent on-board vehicle management system using context middleware. In: IEEE Intell. Veh. Symp., pp. 1109–1114 (2009)
23. Wang, E., Kim, Y.S., Kim, H.S., Son, J.H., Lee, S.H., Suh, I.H.: Ontology Modeling and Storage System for Robot Context Understanding. In: Khosla, R., Howlett, R.J., Jain, L.C. (eds.) KES 2005. LNCS (LNAI), vol. 3683, pp. 922–929. Springer, Heidelberg (2005)
24. Varvadoukas, T., Giannakidou, E., Gomez, J.V., Mavridis, N.: Indoor Furniture and Room Recognition for a Robot Using Internet-Derived Models and Object Context. Front. Inf. Technol. 122–128 (2012)
25. Yi, C., Suh, I.H., Lim, G.H., Choi, B.-U.: Bayesian robot localization using spatial object contexts. In: 2009 IEEE/RSJ Int. Conf. Intell. Robot. Syst., pp. 3467–3473 (2009)
26. Schlenoff, C., Messina, E.: A robot ontology for urban search and rescue. In: Proc. 2005 ACM Work. Res. Knowl. Represent. Auton. Syst. - KRAS 2005, pp. 27–34 (2005)
27. Provine, R., Uschold, M., Smith, S., Stephen, B., Schlenoff, C.: Observations on the use of ontologies for autonomous vehicle navigation planning. Rob. Auton. Syst. (2004)
28. Mastrogiovanni, F., Sgorbissa, A., Zaccaria, R.: Context assessment strategies for Ubiquitous Robots. In: 2009 IEEE Int. Conf. Robot., pp. 2717–2722 (2009)
29. Choi, J., Park, Y., Lim, G., Lee, S.: Ontology-Based Semantic Context Modeling for Object Recognition of Intelligent Mobile Robots. Recent Prog. Robot. Viable Robot. Serv. to Hum. 370, 399–408 (2008)
30. Nienhiiser, D., Gumpp, T., Zollner, J.M.: A Situation Context Aware Dempster-Shafer Fusion of Digital Maps and a Road Sign Recognition System. In: 2009 IEEE Intell. Veh. Symp., pp. 1401–1406 (2009)
31. Trovato, G., Zecca, M., Kishi, T., Endo, N., Hashimoto, K., Takanishi, A.: Generation of Humanoid Robot's Facial Expressions for Context-Aware Communication. Int. J. Humanoid Robot. 10, 23 (2013)
32. Mou, W., Chang, M., Liao, C.: Context-aware assisted interactive robotic walker for Parkinson's disease patients. Intell. Robot. Syst. 329 – 334 (2012)
33. Pawlak, Z., Skowron, A.: Rudiments of rough sets. Inf. Sci (Ny) 177, 3–27 (2007)

Interestingness Measures for Actionable Patterns

Li-Shiang Tsay

North Carolina A&T State Univ., School of Tech., Greensboro, NC 27411
ltsay@ncat.edu

Abstract. The ability to make mined patterns actionable is becoming increasingly important in today's competitive world. Standard data mining focuses on patterns that summarize data and these patterns are required to be further processed in order to determine opportunities for action. To address this problem, it is essential to extract patterns by comparing the profiles of two sets of relevant objects to obtain useful, understandable, and workable strategies. In this paper, we present the definition of actionable rules by integrating action rules and reclassification rules to build a framework for analyzing big data. In addition, three new interestingness measures, *coverage*, *leverage*, and *lift*, are proposed to address the limitations of minimum left support, right support and confidence thresholds for gauging the importance of discovered actionable rules.

Keywords: Action Rule, Reclassification Model, Interestingness Measures, actionability.

1 Introduction

The ability to discover useful knowledge hidden in large volumes of data and to act on that knowledge is becoming increasingly important in today's competitive world. The knowledge extracted from data can provide a competitive advantage to support decision-making. In the last decade, the capabilities for generating and collecting data have grown explosively. Every day 2.5 quintillion (seventeen zero) bytes of data are generated from everywhere: climate sensors, emails, cell phone GPS signals, social media sites, digital pictures and videos, and purchase transaction records to name a few [1]. The 2011 IDC's annual Digital Universe Study (sponsored by EMC^2 Corporation) predicts that the amount of data in the world has doubled every two years since 2010. It also forecasted that the growth of data overall will increase by 50 times to 35 zettabytes by 2020 [2]. In 2012, the retail giant Wal-Mart was estimated to have collected more than 2.5 petabytes of data every hour from its sales transactions [3]. The amount of data in the world seems to be ever increasing and we are overwhelmed by it. Recent developments in technology and science make it too easy to save things. People gather and store so much data because they think some valuable assets are implicitly coded within it, but raw data is rarely of direct benefit. Its true value depends on the ability to extract knowledge that is useful for supporting decisions. Hence, in this data rich world, actionable knowledge discovery research has become an important research topic.

M. Kryszkiewicz et al. (Eds.): RSEISP 2014, LNAI 8537, pp. 277–284, 2014.

Actionable rule mining presents an approach to automatically construct relevantly useful and understandable strategies by comparing the profiles of two sets of targeted objects – those that are desirable and those that are undesirable. It aims to provide insight into how relationships should be managed so that objects of low performance can be improved. The notion of actionable rule was first proposed in [10] and extended in [11] [12] [13]. The expression of action rule [10] and reclassification rules [11] is similar; however, they differ in the following ways. For reclassification rule mining (RRM), the target of discovery is not pre-determined, while for action rule mining (ARM) there is one, and only one, predetermined target. RRM aims to discover all the patterns of W from a given dataset, while ARM discovers only one specific subset of W. In addition, attributes are not required to be classified as stable, flexible, or decision attributes in reclassification rule mining. Finally, action rules are built on the basis of previously discovered classification rules, and reclassification rules can be constructed directly from data. The available algorithms for mining both action rules and reclassification rules use the level-wise/breadth-first search to construct candidates. They work well in practice on typical datasets, but they are not suitable for truly big data. In this paper, we present the definition of Actionable Rules by integrating Action rule [10] and Reclassification rule [11] to pave the foundation for in-depth analysis in big data.

The existing approach formulated the mining actionable rules in a support-confidence framework. The rule is interesting if its support and confidence meet user-specified minimum left support, right support, and confidence thresholds. The anti-monotonicity property of support makes this approach quite attractive for mining actionable rules in large datasets. When the thresholds on supports are high, it tends to generate large number of rules and neglects ones with small support. Sometime, even the supports of a rule are not as frequent as defined by the thresholds; the rule is actually interesting and novel and it has a high confidence rating. When lowering these thresholds, it would usually produce large numbers of rules and would choke the overall rule mining process. In short, measures other than support and confidence are needed to ensure an efficient and useful mining project. This paper presents new interestingness measures *Coverage*, *Leverage* and *Lift* for actionable rules by adopting the popular measurement techniques from classical association rule and classification rule mining.

This paper is organized as follows: Section 2 discusses research trends in making sense from numbers. Section 3 defines actionable rules and Section 4 presents their interesting measures. This study is concluded in Section 5 along with future work.

2   Research Trends on Knowledge Discovery

The traditional way of getting information from data is via statistics, but it is still difficult for them to analyze datasets with 100 to 10^4 objects (entities/observations). The size of 10^4 is typical for a small sub-set of an encountered dataset in practice today. Tools such as SAS, R, and Matlab can support relatively sophisticated analysis, but they are not designed to support big data. One way to reduce the data size is by using either aggregation and/or sampling methods before analyzing data. However, the reduced data will not be able to provide insights for events that fall outside the norm [4]. Basically, data is too big and too hard to be managed and analyzed through traditional means [5]. The other fact is that statistical methods cannot be used to analyze data outside of its originally collected purpose. Emerging in the field of data mining (DM) is a solution for automatically processing large existing data to extract patterns and relationships based on open-ended user questions.

Most of the data mining algorithms generate predictions and describe behaviors in the form of association rules, classification rules, clusters, and other representations. Most of the time, the discovered pattern is formed as a term ($X \rightarrow Y$) to represent an association between X and Y, where X is a conjunction of attribute-value pairs and Y is a class attribute-value. These rules describe relationships between items in datasets. Rules can be generated in a given dataset if they satisfy the user-specified thresholds. They can help unearth patterns and relationships, but they fail to tell the user the value or significance of these patterns. In addition, they can identify connections between behaviors and/or variables but they fail to identify cause-effect relationships. For example, a marketing analyst may want to analyze if and how the cross-selling effect differs across several bundled packages in a database of customer transactions. Such information can lead to increased sales by helping retailers do careful marketing. Association rule mining may be used to identify related products from existing products, but not the cross-selling effect. Can this problem be solved by another standard learning method, such as classification algorithms to model actions and effects? Unfortunately, the standard methods treat populations separately and tend to fragment the knowledge space. Each group is seldom interesting by itself without meaningful context comparisons with others. In general, these types of rules are not directly useful for the business sector. Practitioners report that applying the data mining algorithms comprises no more than 20% of the knowledge discovery process and the remaining 80% relies on human experts to post-analyze the discovered patterns manually. The gap between the discovered patterns and the formulated solutions is filled manually or by semi-automatic analysis [6], which is shown to be time consuming and biased and limits the efficiency of overall knowledge discovery processes and capabilities. Furthermore, many DM algorithms do not scale beyond datasets of a few million entries.

A relatively new data analytical technology, called actionable rule mining, moves further than the classical data mining methods [7] [8] [9], by not only summarizing the data but also comparing the context of segmentations. The structure of an actionable rule is different from the classical rule representation. Given a historical database, where each record is described by a set of values, it is represented as ($A \wedge \Delta B \rightarrow \Delta D$), where A is a conjunction of constant constraints shared by both groups, ΔB represents the proposed changes in one of the groups and ΔD represents the desired effect of the action in that group. The basic principle of actionable rule mining is a process of learning a function that maps a possible transition of objects from one state to another state with respect to a target attribute. For example, actionable rule mining can provide a bank manager with actionable plans to improve his/her understanding of customers, and then improve service to make the customers more satisfied, so that they continue to do the business with the bank. It has been applied in many domains, such as medical diagnoses, homeland security, banks, etc.

The representation of actionable rules is in a symbolic notation, which is very amenable to inspection and interpretation. This characteristic allows business end users and analysts to understand the underlying action-effect in data and take actions based upon them. Based on construction methods, the actionable patterns can be further divided into two types: rule-based and object-based. Rule-based actionable rules [10] [12] are built on the foundations of pre-existing rules. The rule-based approach generally utilizes a standard data mining algorithm first and then on the basis of these mined results the actionable patterns are determined. These may work well for various applications, but it is no guarantee that the extracted patterns in the first step will lead to actionability. Recent research [13] takes a different approach, object-based analyzing, to generate actionable patterns directly from a data set without using pre-existing rules. Such approach may work well for a dataset with a predetermined discovery target, but it may be insufficient to know the space of all actionable patterns. Later on, *Reclassification Rules*, the action-effect model, was proposed in [11] to provide a way to identify the space of all actionable patterns existing in the database that satisfy some predefined thresholds.

3   Actionable Rules

An information system is used for representing knowledge and is defined as a pair $S = (U, A)$, where U is a nonempty and finite set of objects, A is a nonempty and finite set of attributes, i.e. $a : U \rightarrow V_a$ is a function for any $a \in A$, where V_a is called the domain of a [14]. The basic principle of reclassification is a process of learning a function that maps one class of objects into another class by changing the values of some conditional attributes describing them. The conditional attributes are divided into stable and flexible. Presently, we only considered a special type of information system called a decision table. It is any information system of the form $S = (U, A_{St} \cup A_{Fl} \cup \{d\})$, where $d \notin (A_{St} \cup A_{Fl})$ is a distinguished attribute called a decision [10]. The A_{St} is a set of **stable** attributes, whose values cannot be changed or influenced by the user. For example, in a bank customer table an attribute *nationality* is a stable attribute to a banker. On the other hand, A_{Fl} is a set of **flexible** attributes, whose values can be changed or influenced by the user. Referring back to the bank customer example, an attribute *interest rate* is a flexible attribute to a banker. The number of elements in $d(U) = \{ k: (\exists x \in U)[d(x)=k]\}$ is called the rank of d and it is denoted by $rank(d)$. Let us observe that the decision d determines the partition $Part_S(d) = \{x_1, x_2,..., x_{rank(d)}\}$ of the universe U, where $x_k = d^{-1}(\{k\})$ for $1 \leq k \leq rank(d)$. $Part_S(d)$ is called the classification of objects in S with respect to the decision d. Additionally, objects in $d^{-1}(\{k_2\})$ are assumed to be more preferable than objects in $d^{-1}(\{k_1\})$, for any $k_1 < k_2$ [10]. Clearly the mining goal is to shift some objects from a group $d^{-1}(\{k_1\})$ to $d^{-1}(\{k_2\})$. For example, a decision table with 7 objects $\{y_1, y_2, y_3, y_4, y_5, y_6, y_7\}$ can be described by one stable attribute $\{a\}$, two flexible attributes $\{b, c\}$ and a decision attribute $\{d\}$ in Fig. 1. Also, we assume that the mining goal is to change the objects from a group of a lower profit ranking L to a group of a higher profit ranking H, where L and H are values of the decision attribute d.

The goal of the learning process is to create an actionable model, for objects in a decision system, which suggests possible changes that can be made within values of some flexible attributes to reclassify these objects the way a user wants. In other words, reclassification is the process of showing what changes in values for some of the flexible attributes for a given class of objects are needed in order to shift them from one decision class into another more desired one. A decision system S classifies a set of objects so that for each object there exists a class label assigned to it. An actionable rule in S is defined as:

$$r = [[(A_{st_1}, \omega_1 \to \omega_1) \wedge (A_{st_2}, \omega_2 \to \omega_2) \ldots \wedge (A_{st_p}, \omega_p \to \omega_p) \wedge (A_{fl_1}, \alpha_1 \to \beta_1)$$
$$(A_{fl_2}, \alpha_2 \to \beta_2) \wedge \ldots \wedge (A_{fl_q}, \alpha_q \to \beta_q)] \Rightarrow [(d, \phi \to \varphi)]$$

where, for $i = 1, 2, \ldots, p$ and $j = 1, 2, \ldots, q$, $A_{st_i}, A_{fl_j} \in A$. $(A_{fl_j}, \alpha_j \to \beta_j)$ is called the action (antecedent/premise) and $(d, \phi \to \varphi)$ is called the effect (consequent) of the rule. A_{st_i} is called the constraints to filter out the objects that cannot truly support the actionable rule. We say that objects $o_1, o_2 \in U$ support the actionable rule r in S, if and only if:

— $(\forall i \leq p) [[A_{st_i}(o_1) = \omega_i] \wedge [A_{st_i}(o_2) = \omega_i]]$,
— $(\forall j \leq q) [[A_{fl_j}(o_1) = \alpha_j] \wedge [A_{fl_j}(o_2) = \beta_j]]$,
— $d(o_1) = \phi$ and $d(o_2) = \varphi$, and
— $(\forall i \leq p) (\forall j \leq q)[(A_{st_i} \cap A_{fl_j}) \cap d = \varnothing]$

The objects o_2 can be seen as the outcome of rule r applied on objects o_1. An actionable rule is meaningful and workable only if it contains at least one flexible condition. If we apply an actionable rule to object o_1, then the rule basically says: the values ω_i of stable attributes A_{st_i} have to remain unchanged in o_1 and then if we change the value of attribute A_{fl_j} in o_1 from α_i to β_i, then the object o_1 which is in the class ϕ is expected to move to another class φ.

4   Interestingness Measures for Actionable Rules

Since actionable plans are constructed by comparing the profiles of two sets of targeted objects, we can assume that there are two patterns associated with each actionable rule, a left hand side pattern and a right hand side pattern. The left hand side pattern of the rule r, as in the previous section, is defined as $P_L(r) = V_L \cup \{\phi\}$, where $V_L = \{ \omega_1, \omega_2, \ldots, \omega_p, \alpha_1, \alpha_2, \ldots, \alpha_q \}$. The right hand pattern of an actionable rule r is defined as $P_R(r) = V_R \cup \{\varphi\}$, where $V_R = \{\omega_1, \omega_2, \ldots, \omega_p, \beta_1, \beta_2, \ldots, \beta_q\}$. $card[P_L(r)]$ is the number of objects in S that have the property of $P_L(r)$. $card[P_R(r)]$ is the number of objects that have the property of $P_R(r)$, and $card[U]$ is the total number of objects in the decision system.

The original objective measures of interestingness for actionable rules include *Left Support*, *Right Support*, and *Confidence*. The *left support* measure defines the range of the rule, i.e., the proportion of objects that are applicable. The larger its value is, the more interesting the rule will be for a user. The *left support* of an actionable rule r is defined as:

$$supL(r) = \frac{card[(P_L(r)]}{card[U]}, \text{ where } 0 \leq supL(r) \leq 1.$$

The *right support* measure defines the feasibility of the rule, i.e., the proportion of objects that are role models for others. The higher its value is, the stronger the reclassification effect will be. The *right support* of actionable rule r is defined as:

$$supR(r) = \frac{card[(P_R(r))]}{card[U]}, \text{ where } 0 \leq supR(r) \leq 1.$$

Antecedent left support of a rule r is denoted as $supL_a(r)$ to represent the proportion of objects in S which have property $\{V_L\}$. Antecedent right support of a rule r is denoted as $supR_a(r)$ to represent the rate of objects in S that have property $\{V_R\}$. In reclassification problems, *confidence* measure is defined as the ratio of number of objects in a group of lower performance that can be correctly moved to a higher performance group. The *confidence* of the actionable rule r in S is defined as:

$$conf(r) = \frac{supL(r) \times supR(r)}{supL_a(r) \times supR_a(r)}, \text{ where } 0 \leq conf(r) \leq 1.$$

Other alternative names for *confidence* measure are *precision*, *accuracy*, *strength* and *consistency*. Sometimes, *confidence* alone is not enough for evaluating the interestingness of a rule because rules with high confidence may occur by chance. For rules with the same confidence, the one with the highest left support/right support is preferred as they are more reliable.

Three additional interestingness measurements for an actionable rule r are proposed: *coverage(r)*, *leverage(r)*, and *lift(r)*. The consequent left support of a rule r is denoted as $supL_c(r)$ to indicate the proportion of objects in S which have property $\{\phi\}$. The consequent right support of a rule r is denoted as $supR_c(r)$ for the ratio of objects in S that have property $\{\psi\}$. These new interestingness measurements are defined as follows:

$$coverage(r) = \frac{[supL(r) \times supR(r)]}{[supL_c(r) \times supR_c(r)]}, \text{ where } 0 \leq coverage(r) \leq 1.$$

$$leverage(r) = [supL(r) \times supR(r)] - \{[supL_a(r) \times supR_a(r)] \times [supL_c(r) \times supR_c(r)]\},$$
$$\text{where } -1 \leq leverage(r) \leq 1.$$

$$lift(r) = \frac{[supL(r) \times supR(r)]}{\{[supL_a(r) \times supR_a(r)] \times [supL_c(r) \times supR_c(r)]\}}, \text{ where } lift(r) \geq 1.$$

To avoid the rare object problem, the *coverage* is proposed to measure the generality of a rule. The *coverage* can also be called *completeness*, and *sensitivity*. It defines the fraction of objects covered by the body of a rule. The value range of the *coverage* is from 0 to 1. A coverage value close to 1 is expected for an important actionable rule. When the coverage is small, the rule is pretty weak. In general, coverage threshold is set reasonably high to ensure the applicability of the rule. Both *confidence* and *coverage* are important indicators of the reliability of an actionable rule. They are not independent of each other. A rule with a higher coverage may have a lower confidence, while a rule with a higher confidence may have a lower coverage.

The *leverage* is also called *novelty,* and is used to measure how much more counting is obtained from the co-occurrence of the antecedent and consequent, than the independence of them in a rule. The value range of the *leverage* is between -1 to 1. A leverage value equal to, or below, zero indicates a strong independence between the action and effect of an actionable rule. On the other hand, when it is close to 1, the rule is expected to be an important one.

To detect spurious rules, the *lift* of an actionable rule is proposed to measure how many times more often its antecedent and consequent are together, than expected, if they were statistically independent. If the lift is greater than 1, the rule is considered to be "interesting". The measure with *lift* = 4.08 in Fig. 1 means that objects are 4.08 times more likely to be transformed from lower performance groups to higher ones, other than objects in S.

Decision Table					Actionable Rule	supL	supR	conf	lift	coverage	leverage
Objects	a	b	c	d							
y_1	0	2	2	H	Sample rule r_1:	$\frac{3}{7}$	$\frac{2}{7}$	*100%*	4.08	*50%*	0.0925
y_2	0	2	2	H	$((a,0 \rightarrow 0) \wedge (b, 1 \rightarrow 2)) \Rightarrow (d, L \rightarrow H)$						
y_3	0	1	2	L	$supL(r_1) = \frac{Card[P_L(r_1)]}{crad[U]} = \frac{Occurence\ of\ a=0, b=1, and\ d=L}{Total\ objects} = \frac{3}{7}$						
y_4	2	3	1	H	$supR(r_1) = \frac{Card[P_R(r_1)]}{crad[U]} = \frac{Occurence\ of\ a=0, b=2, and\ d=H}{Total\ objects} = \frac{2}{7}$						
y_5	0	1	2	L	$supL_a(r_1) = \frac{Occurence\ of\ a=0, and\ b=1}{Total\ objects} = \frac{3}{7};\ supR_a(r_1) = \frac{Occurences\ of\ a=0\ and\ b=2}{Total\ objects} = \frac{2}{7}$						
y_6	0	1	1	L	$supL_c(r_1) = \frac{Occurence\ of\ d=L}{Total\ objects} = \frac{3}{7};$						
y_7	2	1	1	H	$supR_c(r_1) = \frac{Occurence\ of\ d=H}{Total\ objects} = \frac{4}{7};$						

Fig. 1. Example of interestingness measures

From the viewpoint of reclassification, we are not targeting all possible cases on the decisional part of reclassification. Since some states are more preferable than other states, we should basically ask users to specify in what direction they prefer to see the changes. On the conditional part of actionable rules, we have no information to verify if the rule is applicable. If the domain expert can supply prior knowledge of a given domain then some of the rules cannot be applied. For example, the size of a tumor should not increase when the status of a patient is changing from sick to becoming cured. Therefore, some combinations can be ruled out automatically just by having an expert who is involved in the application domain.

5   Conclusion

The goal of this study is to lay foundations for discovery actionable knowledge in big data. The level-wise breadth first based approach is typically used for extracting action rules or reclassification rules. It works well on typical datasets, but it is not feasible for truly big data due to the memory requirements. To address this issue, this paper presents actionable rules by integrating action rule and reclassification rule. In addition, we introduced new interestingness measures to guide the extraction of actionable patterns. In the near future, we plan to integrate parallel computing architecture and depth-first approaches to mine actionable rules in big data.

References

1. Fem, H.: Four Vendor Views on Big Data and Big Data Analytics: IBM. IBM (2012)
2. Gantz, J., Reinsel, D.: The 2011 IDC Universe Digital study... Extracting Values from Chaos. IDC (2011)
3. Troester, M.: Big Data Meets Big Data analytics. SAS (2012)
4. Dumbill, E.: Volume, Velocity, Variety: What You Need to Know About Big Data. forbes.com [Online] (2012)
5. Madden, S.: From Database to Big Data. IEEE Internet Computing 16(3), 4–6 (2012)
6. Domingos, P.: Toward knowledge-rich data mining. Data Mining Knowledge Discovery 15(1), 21–28 (2007)
7. Agrawal, R., Imielinski, T., Swami, A.: Mining association rules between sets of items in large databases. In: The ACM SIGMOD International Conference on the Management of Data, pp. 207–216 (1993)
8. Grzymala-Busse, J.: A new version of the rule induction system LERS. Fundamenta Informaticae 31(1), 27–39 (1997)
9. Quinlan, J.R.: C4.5: program for machine learning. Morgan Kaufmann (1992)
10. Raś, Z.W., Wieczorkowska, A.: Action-rules: How to increase profit of a company. In: Zighed, D.A., Komorowski, J., Żytkow, J.M. (eds.) PKDD 2000. LNCS (LNAI), vol. 1910, pp. 587–592. Springer, Heidelberg (2000)
11. Tsay, L.-S., Raś, Z.W., Im, S.: Reclassification Rules. In: ICDM Workshops Proceedings, Pisa, Italy, pp. 619–627. IEEE Computer Society (2008)
12. Raś, Z.W., Tsay, L.-S.: Discovering extended action-rules (System DEAR). In: Kłopotek, M.A., Wierzchoń, S.T., Trojanowski, K. (eds.) Proceedings of the Intelligent Information Systems Symposium. ASC, vol. 22, pp. 293–300. Springer, Heidelberg (2003)
13. Tsay, L.-S., Raś, Z.W.: Discovering the concise set of actionable patterns. In: An, A., Matwin, S., Raś, Z.W., Ślęzak, D. (eds.) Foundations of Intelligent Systems. LNCS (LNAI), vol. 4994, pp. 169–178. Springer, Heidelberg (2008)
14. Pawlak, Z.: Information systems - theoretical foundations. Information Systems Journal 6, 205–218 (1981)

Meta-learning: Can It Be Suitable to Automatise the KDD Process for the Educational Domain?

Marta Zorrilla and Diego García-Saiz

Department of Computer Science and Electronics, University of Cantabria
Avenida de los Castros s/n, 39005, Santander, Spain
{marta.zorrilla,diego.garcias}@unican.es

Abstract. The use of e-learning platforms is practically generalised in all educational levels. Even more, virtual teaching is currently acquiring a great relevance never seen before. The information that these systems record is a wealthy source of information that once it is suitably analised, allows both, instructors and academic authorities to make more informed decisions. But, these individuals are not expert in data mining techniques, therefore they require tools which automatise the KDD process and, the same time, hide its complexity. In this paper, we show how meta-learning can be a suitable alternative for selecting the algorithm to be used in the KDD process, which will later be wrapped and deployed as a web service, making it easily accessible to the educational community. Our case study focuses on the student performance prediction from the activity performed by the students in courses hosted in Moodle platform.

Keywords: Meta-learning, classification, predicting student performance.

1 Introduction

Educational data mining is a recent field of research which rose as a result of the appearance of the current computer-supported interactive learning methods and tools (e.g. e-learning platforms, tutoring systems, games,etc.). These have given the opportunity to collect and analyze student data, to discover patterns and trends in this data, and to make new discoveries and test hypotheses about how students learn [12].

Although the contributions in this field are numerous, there is still a lot of work left to be done. One of these contributions is, what we name, the democratization of the use of data mining in the educational arena. That means that people involved in this field and, in particular, instructors, can gain richer insights into the increasingly amount of available data and take advantage of its analysis. However, non-expert users may find it complex to apply data mining techniques to obtain useful results, due to the fact that it is an intrinsically complex process [5].

In order to advance towards our end goal, in this paper, we still deal with the search of a mechanism which recommends us the predictive algorithm that better works in a certain problem at hand. We concretely address one of the oldest and

M. Kryszkiewicz et al. (Eds.): RSEISP 2014, LNAI 8537, pp. 285–292, 2014.

best-known problems in educational data mining [15] that is predicting students performance. Although, a wide range of algorithms have been applied to predict academic success, choosing the most adequate algorithm for a new dataset is a difficult task, due to the fact that there is no single classifier that best performs on all datasets.

Different approaches can be followed to achieve this goal: a) a traditional approach based on a costly trial-and-error procedure; b) an approach based on the advice of a data miner expert in the domain, which is not always straightforward to acquire; or, c) an approach based on meta-learning, able to automatically provide guidance on the best alternative from a set of meta-features.

We consider that the latter approach is the most suitable for our goal, since it allows us to automate the KDD process and deploy it as a service (or plug-in) for e-learning platforms. In fact, this functionality is offered in our E-learning Web Miner tool [4] but the algorithm is currently pre-set. Therefore, this work will help us to substitute the pre-set algorithm, in this case J48, for a recommender that will be built by an educational data mining expert from the experimental database designed and fed. Thus, the main aim of this paper is to propose a set of measurable features on educational data sets and assess to what extent a recommender based on these meta-features is suitable for our aim.

This paper is organised as follows: Section 2 relates works on meta-learning. Section 3 describes the methodology used in our case study and the setting of our experiment. Section 4 presents and discusses the results obtained. Finally, conclusions and future works are outlined in Section 5.

2   Related Work

Meta-learning is a subfield of machine learning that aims to apply learning algorithms on meta-data extracted from machine learning experiments in order to better understand how these algorithms can become flexible in solving different kinds of learning problems, hence to improve the performance of existing learning algorithms [17] or to assist the user to determine the most suitable learning algorithm(s) for a problem at hand [10], among others. The field of application is wide although, generally, the meta-learners are implemented for each specific domain (e.g. time series forecasting models [9] or bio-informatics [3]). After an intense search, we only found two works focused on the educational field [13,2].

There are different approaches about what features can be used as meta-data. In most cases measurable properties of data sets and algorithms are chosen. For instance, some authors [17][16] utilize general, statistical and information-theoretical measures extracted from data sets whereas others use landmarkers [1]. On the other hand, there are works which use model properties as meta-data, such as the average ratio of bias, or their sensitivity to noise [7] or structural shape and size of the model as in [14]. Recently, a new approach called meta-learning template [11] has arisen with the aim of recommending a hierarchical combination of algorithms, instead of only one.

3   Methodology and Experiment Design

The methodology followed in this experiment comprises two stages: 1) one, focused on the task of feeding our meta-database and 2) the second one, oriented to build the recommender and check its performance. The first stage, in turn, consists of several steps: the selection of courses and activity data extraction from e-learning platforms, the building of classification models for each data set and the extraction of the meta-features of each data set. The second stage comprises the selection of the technique which has best performed for each data set in terms of accuracy and TP rate (fail rate) by means of a statistical test, the creation of the corresponding data sets with their meta-features establishing the algorithm previously chosen as class attribute and finally, the building of the recommender and its evaluation in terms of the number of times that its answer matches the algorithm that better classifies the data set.

Regarding data sets, we selected 32 courses from two e-learning platforms, Moodle and Blackboard. The profile of the courses is diverse. There are cross-curricular courses which any student can enroll in and others which belong to a specific degree. There are also courses that are completely virtual and others that follow a blending-learning approach. As input attributes, these data sets contain a variety of information about the interaction of students in the e-learning platform and the class to be predicted is the final mark (pass/fail) obtained by students in the courses. We created 64 data sets from these 32 courses, two for each course. One, where all the attributes are numeric except the class, and the other, where the attributes were discretized by means of PKIDiscretize from Weka. Next, we applied a feature selection technique on all data sets, in particular CfsSubsetEval offered in Weka, in order to remove those attributes which are not relevant for the classification task. Thus, the data sets used in our experimentation present distinct features as can be observed in Table 1. From the 64 data sets, we selected 60 to build the recommender and the remaining ones were used to test the performance of the latter.

Table 1. Data sets description

# Instances	range from 13 to 504
# Attributes	range from 5 to 28
# cross-curricular courses	36 out of 64 datasets
# online courses	32 out of 64 datasets

Models generation was performed by applying 4 classification algorithms, which follow a different paradigm, on 64 data sets. The algorithms chosen were: NaiveBayes, NearestNeighbours, Jrip and J48. All were run with default parameters using their implementation in Weka. These classifiers were selected among those frequently used for the prediction of students' performance.

The meta-features used in this work are divided in the following groups:

- simple/general features such as the number of attributes, the number of instances and the type of attributes (numerical, categorical or mixed)
- complexity features that characterize the apparent complexity of data sets for supervised learning [8] provided by DCoL (data complexity library), such as the maximum Fisher's discriminant ratio (F1), the overlap of the per-class bounding boxes (F2), the maximum (individual) feature efficiency (F3), the collective feature efficiency (sum of each feature efficiency)(F4), the fraction of points on the class boundary (N1), the ratio of average intra/inter class nearest neighbor distance (N2), the training error of a linear classifier (N3), the fraction of maximum covering spheres (T1), and the average number of points per dimension (ratio of the number of examples in the data set to the number of attributes)(T2).
- domain features such as if the course is wholly virtual (online or blending) and the target audience (cross-curricular or specific).

Finally, we must mention that we have not considered statistical features such as the skewness or the kurtosis of each attribute [17] since the attributes which characterise each course are different as a consequence of the fact that they are hosted in different platforms and the resources which they manage are also different.

4   Results and Discussion

This section shows the results achieved in each relevant step of our methodology. Once the data sets were created, we proceeded to build their corresponding classification models with each one of the four algorithms chosen. In order to evaluate their performance, we stored their accuracy as well as their sensitivity (the percentage of correctly classified students which failed the course) as part of the meta-database. A sample of these figures are reported in Table 2. The algorithm which achieved the best performance for each data set is marked in bold. The selection of the best algorithm was performed by means of a t-test at a significance level of alpha=0.05 taking accuracy and sensitivity measures into account. The algorithm which worst performed was chosen as the base algorithm. In most cases, the classifier with higher accuracy was selected. The same happened when sensitivity was assessed.

Subsequently, we created our meta-datasets, i.e., we extracted the meta-features, mentioned in the previous section, of each data set and built a new data set in which each instance gathered the meta-features of each training data set and, as target attribute, had the name of the algorithm which better performed. We generated two meta-datasets, one which takes the accuracy into account, named m1, and the other, m2, according to the sensitivity.

Table 3 shows the meta-features of our four test data sets along with the algorithm which achieved the best accuracy (BestAlgAcc) and the best sensitivity (BestAlgSens). It must be observed that, regarding accuracy, each data set was better classified with a different technique, whereas J48 and NaiveBayes were the algorithms which got the best models regarding sensitivity.

Table 2. Accuracy and sensitivity achieved by the four classifiers chosen in 4 out of 64 data sets

	J48		JRip		NNge		NaiveBayes	
	Acc.	Sens.	Acc.	Sens.	Acc.	Sens.	Acc.	Sens.
dataset1	60.50	50.00	59.50	37.00	**69.00**	46.00	61.00	**65.00**
dataset2	**91.15**	60.00	90.40	**65.00**	87.84	63.00	89.90	39.00
dataset3	72.38	**84.00**	67.63	82.00	72.13	81.00	**73.25**	73.00
dataset4	**68.00**	50.00	59.50	37.00	65.00	46.00	62.50	**65.00**
...	...	...	...	...	...	...	...	...

Table 3. Meta-features of our test datasets

	N#ins.	N#att.	Course	Teaching	Att.type	BestAlgAcc	BestAlgSens
test1	17	11	crosscurricular	online	numeric	JRip	J48
test2	65	12	crosscurricular	online	nominal	NaiveBayes	NaiveBayes
test3	504	17	specific	blending	numeric	J48	J48
test4	80	14	specific	blending	nominal	NNge	NaiveBayes

	F1	F2	F3	F4	T1	T2	N1	N2	N3
test1	1.16	0	0.53	1	0.88	1.7	0.65	0.84	0.35
test2	4.06	0.01	0.54	0.97	1	5.91	0.31	0.67	0.19
test3	1.83	0	0.33	0.60	1.00	31.50	0.17	0.42	0.112
test4	2.99	0.4	0.05	0.06	1	6.15	0.51	0.86	0.36

Next, we built two recommenders using the meta-datasets m1 and m2. The learning algorithm utilised was J48. These recommenders are displayed in Figures 1 and 2 respectively. As can be observed, the most relevant attribute in both recommenders is the type of the course (cross-curricular or specific), but there are other important measures as F1, F2 or the number of instances for the first recommender and the number of attributes for the second one.

Then, we passed the test meta-datasets through both recommenders and the algorithms which were offered as a result are shown in Table 4. It can be observed, that 3 out of 4 datasets were correctly classified. The first recommender fails to determine the best algorithm for test4, nevertheless it offers the second best. The same occurs with the test1 when the sensitivity-based recommender is used, which suggests the use of NaiveBayes instead of J48, but once again, it is the second best option. Thus, we can draw that the recommenders built from the meta-features chosen work properly.

Table 4. Algorithm recommended for each test data set

	BestAlgAcc.	RecomAlgAcc.	RecomAcc	BestAlgSens.	RecomAlgSens.	RecomSens
test1	JRip	JRip	68.63	J48	NaiveBayes	75.00
test2	NaiveBayes	NaiveBayes	89.23	NaiveBayes	NaiveBayes	84.09
test3	J48	J48	90.18	J48	J48	57.75
test4	NNge	NaiveBayes	68.75	NaiveBayes	NaiveBayes	70.40

```
CourseType = crosscurricular
|   F1 <= 3.07
|   |   attributesType = nominal
|   |   |   T2 <= 2.26: NaiveBayes
|   |   |   T2 > 2.26: J48
|   |   attributesType = numeric
|   |   |   F2 <= 0.000012
|   |   |   |   T1 <= 0.98
|   |   |   |   |   T1 <= 0.96: JRip
|   |   |   |   |   T1 > 0.96: NNge
|   |   |   |   T1 > 0.98: J48
|   |   |   F2 > 0.000012: J48
|   F1 > 3.07: NaiveBayes
CourseType = specific
|   T2 <= 2.83: JRip
|   T2 > 2.83
|   |   Ninstances <= 115
|   |   |   F2 <= 0.03: J48
|   |   |   F2 > 0.03: NNge
|   |   Ninstances > 115
|   |   |   F2 <= 0.000006: J48
|   |   |   F2 > 0.000006: NaiveBayes
```

Fig. 1. J48 Recommender for md1 meta-dataset (Best Accuracy)

```
CourseType = crosscurricular
|   Natt <= 18
|   |   F4 <= 0.58
|   |   |   F1 <= 0.75: J48
|   |   |   F1 > 0.75: NaiveBayes
|   |   F4 > 0.58: J48
|   Natt > 18
|   |   F2 <= 0.000068: NaiveBayes
|   |   F2 > 0.000068: JRip
CourseType = specific
|   Natt <= 19
|   |   Ninstances <= 66: J48
|   |   Ninstances > 66
|   |   |   T2 <= 11.46
|   |   |   |   Natt <= 13
|   |   |   |   |   F2 <= 0.000068: NaiveBayes
|   |   |   |   |   F2 > 0.000068: J48
|   |   |   |   Natt > 13: NaiveBayes
|   |   |   T2 > 11.46: J48
|   Natt > 19: J48
```

Fig. 2. J48 Recommender for md2 meta-dataset (Best Sensitivity)

Before the end of this section, we would like to show the performance of the classifiers built for each test data set when we, as expert data miners, managed to generate the best model for each one. Table 5 reports that in most cases the difference between accuracy and sensitivity achieved is lower than 1%. This difference was achieved as a result of performing a context-aware outlier detection task [6] and changing the parameter setting.

It is important to say that data sets from e-learning platforms present certain features which make them different from other data sets. On the one hand, its quality is usually very high, that means that the attribute values are correct and do not present missing-values although, as in any data sample, can present outliers which in this case are a consequence of how students work in online courses [6]. On the other hand, the experimentation phase is more difficult because the data is very dynamic and can vary a lot among samples (different course design, students with different skills, different methods of assessment, different resources used, etc.). Because of that, the amount of data available to mine is reduced.

Table 5. Performance difference between the models built from the recommendation and those built by the expert

	Recom.Acc.	ExpertAcc.	Difference	Recom.Sens.	ExpertSens.	Difference
test1	68.63	69.27	0.64	75.00	77.50	2.50
test2	89.23	91.43	2.20	84.09	85.00	0.81
test3	90.18	91.06	0.88	57.75	58.25	0.50
test4	68.75	71.95	3.20	70.40	72.20	1.80

In view of these results, we can state that the use of recommenders based on meta-learning are suitable for automatising our E-learning Web Miner tool, in such a way that it would rely on the recommender instead of using the pre-set algorithm and therefore, enabling it to build more accurate classification models.

5 Conclusions

In this paper, we show the feasibility of using an algorithm recommender in order to automatise the KDD process, in such a way that this process can be deployed as a service. This is convenient and necessary in contexts such as the educational one, where people involved are not experts in data mining techniques but, require to extract knowledge from the trail left by the students during their online learning in order to improve the learning-teaching process. We rely on meta-learning since we have checked that choosing a suitable set of meta-features, it is possible to build a recommender which selects one of the best algorithms for a problem at hand.

In a near future, we will extend our experimental database and include more algorithms in our recommenders. Likewise, we will add the parameter setting to our meta-database so that our recommenders can provide the best configuration for the best technique.

References

1. Abdelmessih, S.D., Shafait, F., Reif, M., Goldstein, M.: Landmarking for meta-learning using rapidminer. In: RapidMiner Community Meeting and Conference (2010), http://madm.dfki.de/publication&pubid=4948
2. Romero, C., Olmo, J.L., Ventura, S.: A meta-learning approach for recommending a subset of white-box classification algorithms for Moodle datasets. In: Proc. 6th Int. Conference on Educational Data Mining, pp. 268–271 (2013)
3. de Souza, B.F., de Carvalho, A.C.P.L.F., Soares, C.: Empirical evaluation of ranking prediction methods for gene expression data classification. In: Kuri-Morales, A., Simari, G.R. (eds.) IBERAMIA 2010. LNCS, vol. 6433, pp. 194–203. Springer, Heidelberg (2010)
4. García-Saiz, D., Palazuelos, C., Zorrilla, M.: Data Mining and Social Network Analysis in the Educational Field: An Application for Non-expert Users. In: Peña-Ayala, A. (ed.) Educational Data Mining. SCI, vol. 524, pp. 411–439. Springer, Heidelberg (2014)
5. Fayyad, U., Piatetsky-Shapiro, G., Smyth, P.: The kdd process for extracting useful knowledge from volumes of data. Commun. ACM 39(11), 27–34 (1996)
6. García-Saiz, D., Zorrilla, M.: A promising classification method for predicting distance students' performance. In: Yacef, K., Zaïane, O.R., Hershkovitz, A., Yudelson, M., Stamper, J.C. (eds.) EDM, pp. 206–207 (2012), www.educationaldatamining.org
7. Hilario, M., Kalousis, A.: Building algorithm profiles for prior model selection in knowledge discovery systems. Engineering Intelligent Systems 8, 956–961 (2002)
8. Ho, T.K.: Geometrical complexity of classification problems. CoRR cs.CV/0402020 (2004)
9. Jankowski, N., Duch, W., Grąbczewski, K. (eds.): Meta-Learning in Computational Intelligence. SCI, vol. 358. Springer, Heidelberg (2011)
10. Kalousis, A., Hilario, M.: Model selection via meta-learning: a comparative study. In: Proc. 12th IEEE International Conference on Tools with Artificial Intelligence, pp. 406–413 (2000)
11. Kordík, P., Cerný, J.: On performance of meta-learning templates on different datasets. In: IJCNN, pp. 1–7. IEEE (2012)
12. Feng, M., Bienkowski, M., Means, B.: Enhancing teaching and learning through educational data mining and learning analytics: An issue brief. Tech. rep., U. S. Department of Education (2012)
13. Molina, M.M., Luna, J.M., Romero, C., Ventura, S.: Meta-learning approach for automatic parameter tuning: A case study with educational datasets. In: Proc. 5th International Conference on Educational Data Mining, pp. 180–183 (2012)
14. Peng, Y.H., Flach, P.A., Soares, C., Brazdil, P.B.: Improved dataset characterisation for meta-learning. In: Lange, S., Satoh, K., Smith, C.H. (eds.) DS 2002. LNCS, vol. 2534, pp. 141–152. Springer, Heidelberg (2002)
15. Romero, C., Ventura, S.: Data mining in education. Wiley Interdisciplinary Reviews: Data Mining and Knowledge Discovery 3(1), 12–27 (2013)
16. Segrera, S., Pinho, J., Moreno, M.N.: Information-theoretic measures for meta-learning. In: Corchado, E., Abraham, A., Pedrycz, W. (eds.) HAIS 2008. LNCS (LNAI), vol. 5271, pp. 458–465. Springer, Heidelberg (2008)
17. Vilalta, R., Drissi, Y.: A perspective view and survey of meta-learning. Artificial Intelligence Review 18, 77–95 (2002)

Discovering Collocation Rules and Spatial Association Rules in Spatial Data with Extended Objects Using Delaunay Diagrams

Robert Bembenik[1], Aneta Ruszczyk[2], and Grzegorz Protaziuk[1]

[1] Institute of Computer Science, Warsaw University of Technology,
Nowowiejska 15/19, 00-665 Warszawa, Poland
{R.Bembenik,G.Protaziuk}@ii.pw.edu.pl
[2] ATOS, Woloska 5, 02-675 Warszawa, Poland
annie.r86@gmail.com

Abstract. The paper illustrates issues related to mining spatial association rules and collocations. In particular it presents a new method of mining spatial association rules and collocations in spatial data with extended objects using Delaunay diagrams. The method does not require previous knowledge of analyzed data nor specifying any space-related input parameters and is efficient in terms of execution times.

Keywords: extended objects, spatial collocations, Delaunay diagram, spatial data mining.

1 Introduction

Spatial collocation [1] techniques aim to discover groups of spatial features frequently located together. These features are fuzzy or Boolean which means that they occur or not in the analyzed space. Discovering collocation patterns is in some ways similar to discovering association rules in non-spatial relational datasets. It is however a more challenging task as no transactions are defined in spatial databases. Instead, explorers have to deal with distribution of Boolean spatial features set in the continuous space. Some spatial or geometric information can be described by means of alphanumeric data, but it is more complicated when it comes to describing their structures or relations due to the fact that space is continuous rather than discreet. For that purpose topological relations [2] are utilized.

Spatial rules discovery may reveal many aspects of spatial objects. Co-occurrences of objects can be discovered by means of collocation rules [4] aiming to find spatial collocation patterns [1] – subsets of Boolean spatial features frequently located in close proximity. An instance of collocation pattern is a set of spatial objects which satisfy feature and neighborhood constraints. Collocation rule is a rule of the form that collocation pattern implies another collocation pattern to occur. It refers to patterns of features that do not intersect.

Spatial association rules [4] discovery is a similar problem, but focuses on finding patterns related to defined reference features. Hence, spatial association pattern specifies

M. Kryszkiewicz et al. (Eds.): RSEISP 2014, LNAI 8537, pp. 293–300, 2014.
© Springer International Publishing Switzerland 2014

reference features and a set of features which tend to occur in their neighborhood. This does not suggest whether they are neighbors to each other as well, but provides information on frequently occurring features in the proximity of these specified as a reference.

Extended objects examples are rivers, roads, streets. Their shapes are widely represented in GIS by means of polygons, arcs and lines. Regular spatial objects examples are houses, schools, shops, etc., and they can be represented as points (for the purpose of exploratory analyses). To the best of our knowledge the only approach to discovering collocation patterns in data sets with extended spatial objects was given in [1]. The approach uses a buffer-based model, quite computationally expensive. The authors define notions of Euclidean neighborhood for a feature and Euclidean neighborhood for a feature set, which are then used to compute coverage ratio serving as a prevalence measure in the buffer-based model. The coverage ratio is computed by division of the Euclidean neighborhood for a given feature set by the total area of the plane. If the coverage ratio for a given feature set is greater than the user-specified threshold then the feature set is a collocation pattern.

Spatial collocation rules and spatial association rules concerning extended objects may be utilized in many areas. One example can be an electronic service providing information on local services such as good pizza or cheap gas stations; spatial rules could be used here to adapt the offers to the needs of clients in specific locations or travelers on a given route. As another example one could be interested in finding all highways adjacent to a particular one. Spatial rules could also be utilized in areas of decision-making, environmental management, public health, tourism, etc.

The rest of the paper is organized as follows. Section 2 introduces a method for mining extended spatial objects using spatial tessellations. Section 3 presents experimental evaluation of the introduced method. Section 4 concludes the paper.

2   Mining Extended Spatial Objects Using Spatial Tessellations

In this section we introduce a new method of mining extended spatial objects using spatial tessellations called DEOSP (Discovery of Spatial Patterns with Extended Objects).

2.1   Spatial Tessellations

Spatial tessellations utilized to mining collocations including extended spatial objects in our approach are Voronoi diagram, Delaunay triangulation, as defined in [5], constrained Delaunay triangulation and conforming Delaunay triangulation, whose definitions are given below.

Definition 1. [5] Constrained Delaunay Triangulation
For a given planar straight-line graph $G(P_g, L_g)$ representing obstacles and a set Q of points, the constrained Delaunay triangulation is a triangulation spanning $P=P_g \cup Q$ satisfying the condition that the circumcircle of each triangle does not contain in its interior any other vertex which is visible from the vertices of the triangle.

Constrained triangulation is generally different from a Delaunay triangulation. But if we add a set S of points on L_g, then the constrained Delaunay triangulation spanning $P \cup S$ may coincide with the ordinary Delaunay triangulation spanning $P \cup S$. Such a special Delaunay triangulation is called a conforming Delaunay triangulation.

Definition 2. [5] Conforming Delaunay Triangulation
For a given planar straight-line graph $G(P_g, L_g)$ representing obstacles and a set Q of points, we consider a set S of additional points and construct the ordinary Delaunay triangulation $\mathcal{D}(P_g \cup Q \cup S)$ spanning $(P_g \cup Q \cup S)$. If all line segments in L_g are the union of the edges of $\mathcal{D}(P_g \cup Q \cup S)$, we call $\mathcal{D}(P_g \cup Q \cup S)$ the *conforming Delaunay triangulation*.

2.2   Mining Extended Spatial Objects – Method Outline

In order to accomplish the process of mining collocations with extended spatial objects, we make the following assumptions:

- line segments are represented as end points without intermediate points on the line segment;
- additional points are selected in such a way, that there exist edges between endpoints of the line segments or the existing edges build a line segment that links them;
- constrained Delaunay triangulation corresponding to the topology of the triangulation for line segments is utilized;
- only nearby objects are taken into consideration;

To generate candidates, spatial association rules and collocations *Apriori*-like operation is employed.

Taking into consideration the aforementioned spatial tessellations as well as the assumptions, the mining process for extended spatial objects can be realized in the following steps: (1) Construction of a constrained Delaunay triangulation, (2) Construction of a conforming Delaunay triangulation, (3) Selection of triangles for processing, (4) Removal of relations among objects that do not belong to the same groups, (5) Finding collocation instances, (6) Discovering collocation rules and spatial association rules in spatial data.

2.3   Detailed Description of Method Steps

Construction of a Constrained Delaunay Triangulation. Based on the input data being coordinates of point objects and line segment endings a constrained Delaunay triangulation is created between endings of the imposed line segments. In the process of triangulation construction each node stemming from point object is labeled with its type and instance. Line segments appearing in line objects are constraints defined by the beginning and ending of the line segment. Endings of the imposed segments have the same type and instance. Only the first segment ending receives a label. Sample input data and constrained Delaunay triangulation is shown in Fig. 1.

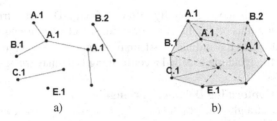

Fig. 1. a) Sample input data; b) constrained Delaunay triangulation with labeled vertices

Construction of a Conforming Delaunay Triangulation. Based on the triangulation created in the First step of the method, a conforming Delaunay triangulation is created, that is a triangulation having all edges being Delaunay edges. The process is based on the following property: each constrained triangulation can be transformed to a conforming Delaunay triangulation [6]. In most cases it is necessary to introduce additional points being vertices to the triangulation. They are labeled with type and instance. This is achieved in such a way, that all vertices of the imposed edges receive label of type and instance of the first ending of the edge. Because of possible intersections of objects having a common ending point, as shown in Fig. 2a, for segments S.2 and S.3, each point can have many labels.

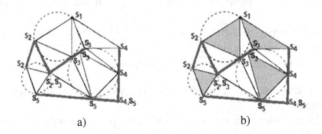

Fig. 2. a) Constrained Delaunay triangulation consistent with Delaunay triangulation. The blue line denotes imposed edges. Labels denote object instances represented by vertices. b) Selection of triangles with non-redundant information. The triangles kept are marked green.

In this example the common vertex has two labels: S.2, S.3. Adding a new point to the triangulation is yet dependent on the imposed edges of one segment, so adding labels refers only to the label of the edge ending related to this segment, ignoring the remaining edges. It is visible for added points labeled S.3 that do not constitute the ending of this segment. The outcome of this step is a constrained Delaunay triangulation consistent with Delaunay triangulation for points with labels representing object instances, which is shown in Fig. 2a.

Selection of Triangles for Processing. The purpose of *Selection of triangles for processing* step is elimination of triangles that bring in redundant information concerning topology of objects. It is noteworthy, that occurrence of a combination, such as S.2, S.2, S.5 does not indicate the existence of an imposed edge in the triangle. Such is the case, e.g., for the triangle in Fig. 2b with such a combination. The triangle can be rejected when there is no other correct combination for it. It is then not consistent with the assumptions of a Voronoi diagram where all generators are different. One should however expect that the relation between two objects is included in a different triangle of the diagram. The outcome of this step consists in selected triangles of the constrained Delaunay triangulation for the labeled points representing instances of different objects. Rejected triangles are not considered in further processing.

Removal of Relations among Objects That Do Not Belong to the Same Groups. Removal of relations among objects belonging to various groups follows the method used in NSCABDT algorithm [7]. For all retained triangles, further referred to as correct, the mean length of edges (*Global_mean*) and global standard deviation (*Global_stddev*) are calculated. To reject or retain an edge between two vertices it is necessary to compare the length of edges to the value of a discriminating function given by the following formula [7]:

$$F(v) = Global\_mean + Global\_stddev \cdot \frac{Global\_mean}{Local\_mean(v)}$$

where *Local_mean(v)* denotes the mean length of the incident edges of vertex v, computed with the following formula $Local_{mean(v)} = \frac{\sum_{k=1}^{K} Len\,(e_k)}{K}$.

Edges, for which the relation $Len(e_k) > F(v)$ holds are removed. The outcome of the step are edges connecting objects belonging to the same clusters. They can be regarded as edges of triangles with missing edges.

Finding Collocation Instances. The purpose of *Finding collocation instances* step is to determine which objects build up cliques. Having information on cliques it is feasible to search for spatial association rules as well as collocation rules. It results from the fact that each clique is equivalent to all association rules with centers corresponding to clique elements. For example, in a 3-element clique pattern (A, B, C) it is possible to find three star patterns: <A, {B, C}>, <B, {A, C}>, <C, {A, B}>. For triangles from which no edges were rejected the cliques build up object clusters described by their correct label combinations. If one or two edges in the triangle were rejected, each retained edge is two-element clique for the combination of labels of different types and instances. So, for a sample triangle with edges (1, 1) – A.1, (2, 2) – B.1, (3, 3) – A.2, C.2, after rejecting edges (1, 1) – (2, 2) we will get instances of a clique for the retained:

A.1, C.2 – the correct combination of edges (1, 1) – (3, 3),

B.1, A.2 and B.1, C.2 – instances of a clique for the edges (2, 2) – (3, 3).

Discovering Collocation Rules and Spatial Association Rules in Spatial Data. The last step of the method is realized in a similar fashion to the *Apriori* algorithm. Candidates are generated using T-tree structure [3]. The structure of the tree depends on the type of a pattern, as we consider collocation patterns and spatial association rules. To compute spatial association rules and collocation rules we need to use the notions of participation ratio, participation index for a group of types, participation index for an association pattern as well as the notion of a spatial association rule together with the confidence of a spatial association rule and a collocation rule as well as the confidence of a collocation rule. We used those concepts as they are defined in [3].

A detailed description of the T-tree together with the way of computing prevalences using this structure can be found in [3]. A modification of the T-tree for computing frequent itemsets for spatial association rules as well as a description of the way it computes the necessary values is also given in [3]. Prevalence computed with the T-tree is expressed in terms of participation index. At a given value of minimum prevalence patterns with higher values of prevalence are frequent. From them candidate spatial rules are generated. For each candidate (collocation or association) spatial rule appropriate confidence is computed based on definitions of confidence of a spatial association rule and confidence of a collocation rule given in [3]. Again, if the confidence is above a given confidence threshold, such rules are treated as valid.

3   Experimental Evaluation

To evaluate the proposed method we implemented it and conducted tests on real datasets. All tests were performed using a PC with an Intel Core 2 Duo 2.27GHz processor, 3.00 GB of RAM running Windows 7 Professional. We used the following datasets from MetroGis website (www.datafinder.org) in shapefile format from Twin Cities area: main roads, bus garages, largest shopping centers, and geographic and cultural objects of Twin Cities. The database of main roads contains 199 linear objects belonging to 4 road types: county road, interstate, state highway and US highway. Each objects is composed of many segments. Bus garages database contains 16 objects of 2 types: Metro Transit MT Garages and Regional Provider RP Garages. Shopping centers database contains 330 objects of 6 types: community center, downtown center, mega center, neighborhood center, regional center and sub-regional center. Database of geographic and cultural objects contains 2974 objects of 34 types, including among others: airports, schools, churches, bridges, hospitals. A sample of the test data is given in Fig. 3.

Fig. 3. Visualization of shopping centers and roads

We divided the tests based on the data we used together with the roads dataset. In the first test we applied the garages dataset to generate spatial association and collocation rules. The generated association rules had one element in the rule antecedent and many elements in the rule consequent. There was no such limitation for collocation rules.

Table 1. Execution times and number of spatial association and collocation rules for the three tests

minPrev, minConf	Collocation rules						Spatial association rules					
	test 1		test 2		test 3		test 1		test 2		test 3	
	t [s]	nb	t [s]	nb	t [s]	nb	t [s]	nb	t [s]	nb	t [s]	nb
0.1, 0.8	0.7	8	2	8	31	8	0.4	12	2	52	191	840
0.05, 0.1	0.7	53	2.5	110	35	286	0.4	100	2.6	650	315	76439
0, 0	0.73	86	3	354	70	12077	0.48	186	14	5110	419	247761

In the second test we used shopping center data together with the roads dataset. In the third test we took into consideration geographic and cultural objects of Twin Cities together with the roads dataset. The summary of computation times and the number of generated spatial association and collocation rules for the conducted tests are given in Table 1.

We generated spatial association rules and collocations. Sample spatial association rules that were discovered are:

```
State_Highway-->Neighborhood_Center(0.5792079,0.9375)
State_Highway-->
Interstate, Neighborhood_Center(0.5792079,0.8125)
Interstate -->
Community_Center Neighborhood_Center (0.5990099,1)
```

The rule `Interstate --> Community_Center Neighborhood_Center` (0.5990099,1) means that in the neighborhood of objects of type `Interstate` we will certainly find objects of type `Community_Center` and `Neighborhood_Center`. Minimum probability, that an object belonging to one of the types `Interstate`, `Community Center` or `Neighborhood Center` observed on the map belongs to a spatial association rule is equal to 0,5990099.

Sample spatial collocation rules generated from the dataset discussed in the paper are:

```
Mega_Center-->Interstate (0.1428571,1)
Interstate, US_Highway -->
Neighborhood_Center (0.1287129,0.7777778)
Interstate, County_Road-->
Neighborhood_Center (0.06435644,0.8571429)
```

The last example spatial collocation rule means that when on a map interstate and county roads are spotted close to each other then with a probability of 85% a neighborhood center will also be nearby.

Based on the conducted experiments we can tell that DEOSP, the new method proposed in this paper, is useful in mining spatial association rules and collocations. It successfully and quickly leads to the generation of rules and specifies their prevalence and confidence. Their usefulness can be justified by potential usage in tasks related to infrastructure of transit, information systems, delivery of goods, commuting, communication between organizational units, etc.

In relation to the number of generated rules the algorithm of mining spatial association rules is faster than the algorithm of mining collocations. One has to remember that because of the monotonicity property the value of prevalence influences the speed of rules generation.

4   Conclusions

The significance of knowledge derived by the means of collocation rules is increasing in various areas. Not all traditional data mining methods can be used in spatial analysis. The most serious problem with the definition of the transaction and the neighborhood in space can be overcome by using geometric methods utilizing properties of Delaunay diagrams. DEOSP, a method presented in this paper, allows for efficient discovery of dependencies among spatial objects in the form of spatial association and collocation rules. It does not require prior knowledge of the analyzed data (e.g. to define buffer size) – it is based on the geometric configurations and data characteristics. Geometric arrangements created by objects are discovered automatically using the properties of Voronoi diagrams and Delaunay triangulations: constrained Delaunay triangulation and conforming Delaunay triangulation are used to determine relations among objects. The method enables to generate spatial rules for which parameters of prevalence and confidence are computed.

References

1. Xiong, H., Shekhar, S., Huang, Y., Kumar, V., Ma, X., Yoo Soung, J.: A Framework for discovering co-location patterns in datasets with extended spatial objects. In: SDM (2004)
2. Egenhofer, M.J.: J Herring, Categorizing Binary Topological Relations Between Regions, Lines, and Points in Geographic Databases, Technical Report, Department of Surveying Engineering, University of Maine, Orono (1991)
3. Bembenik, R., Rybiński, H.: FARICS: a method of mining spatial association rules and collocations using clustering and Delaunay diagrams. Springer, Netherlands (2009)
4. Shekhar, S., Xiong, H.: Encyclopedia of GIS. Springer (2008)
5. Okabe, A., et al.: Spatial tessellations: concepts and applications of Voronoi diagrams. Wiley (2000)
6. Rineau, L.: 2D conforming triangulations and meshes. CGAL User and Reference Manual, http://www.cgal.org/Manual/latest/doc_html/cgal_manual/ Mesh_2/Chapter_main.html
7. Yang, X., Cui, W.: A novel spatial clustering algorithm based on Delaunay triangulation. Journal of Software Engineering and Applications (JSEA) (2010)

Potential Application of the Rough Set Theory in Indoor Navigation

Dariusz Gotlib and Jacek Marciniak

Faculty of Geodesy and Cartography, Warsaw University of Technology, Warsaw, Poland
{d.gotlib,j.marciniak}@gik.pw.edu.pl

Abstract. The paper presents concepts of using the Rough Set Theory in indoor navigation. In particular, attention was drawn to potential verification of a position received from the positioning system and generation better quality navigation guidelines. The authors proposed the use of expert systems, spatial data for buildings and spatial analysis techniques typical for Geographic Information Systems (GIS). The presented analysis lies within the scope of the research conducted at the Laboratory of Mobile Cartography at Warsaw University of Technology.

Keywords: indoor navigation, routing algorithms, building models, rough sets.

1 Introduction

The research on indoor navigation has been intensified in recent years. After marine, aeronautical and car navigation we observe a raise of interest in navigation systems for pedestrians in buildings where GNSS signal is weak.

Most of the research focuses on the development of positioning systems, specific route calculations algorithms and modelling of spatial data indoors. The authors have published a number of articles presenting conclusions from the research on modelling of building interiors and navigation directions [1,2,3].

This paper introduces two experiments related to indoor navigation dealing with the problems of tracking a user and generating intuitive directions. The expert systems based on the Rough Set Theory are used for supporting the decision processes in a navigation system.

2 The Issues of Indoor Navigation

Implementation of an indoor navigation system involves common tasks related to: (1) location methods and technology, (2) modelling of the interior of buildings, (3) visualization of the interior of buildings, and (4) routing and generating of voice navigation commands.

The user of a navigation system calculates the route in order to get information how to reach a certain place. A navigation graph is used to calculate a route optimal in terms of a cost function, i.e. the length and the route is visualized to the user to

M. Kryszkiewicz et al. (Eds.): RSEISP 2014, LNAI 8537, pp. 301–308, 2014.

imagine the suggested trajectory, and, subsequently, to move along it. The process of tracking of the user is usually supported by navigation directions in the form of graphical signs on a map or voice commands. The tracking module of the system should react on the event of leaving the calculated route by the user and propose a new route.

The way how routes are calculated and graphical and voice messages are generated has a significant impact on the usability of navigation applications. A number of aspects reflecting specific indoor conditions should be taken into consideration when route directions are provided for the user [2].

Generating intuitive directions is a complex problem due to the fact that the geometry of edges in navigation graph provides generalized information about real space [4]. The algorithms must determine the decision points – when the user needs guidelines what usually happens only in selected nodes, and then provide the adequate direction. This process requires the analysis of the geometry of the neighbor edges in the context of the sequence of commands [5]. In more advanced systems also topographical data can be analyzed to use information about landmarks, which helps in better understanding of a spatial message.

3   Optimization of Route Recalculation

3.1   Definition of the Problem

The main aim of the research is to find an algorithm that will correct the position calculated by the localization system. For this purpose, it is assumed that the algorithm use (1) information about the topography in the form of the building model, (2) methods of Geographical Information Systems (GIS) analysis and (3) an expert system using Rough Set Theory. An adjustment of the position is desirable when, due to the inaccuracy of the location system, the position is determined in a different room than the actual user location. Moreover, in the case of indoor navigation systems a distance of 1 meter is essential – rooms are separated by a wall of a few inches in width. The position error of 2 meters may indicate that the user is "located even outside" the building, although in fact they are inside. When the navigation system provides information about the topography of the building, and we know that the user is in a room on the 3rd floor, it can be concluded that if they did not go out of the hall and then stairs or an elevator, they cannot be outside (except in special, extremely rare situations).

This is particularly important when starting route calculation and its recalculation during tracking of the user. An incorrect position is likely to create a significant deviation from the planned route and, in consequence, the recalculation will be needed using an actual position of the user. It can cause too frequent and erroneous route recalculations, what in turn leads to misleading navigation guidance.

3.2   Assumptions

A typical situation of tracking of the user is presented in Fig. 1.

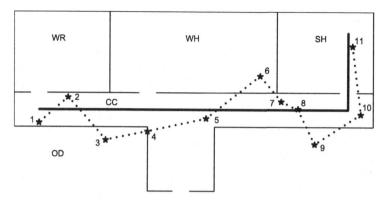

Fig. 1. The example of a planned route inside the building (solid line) and the movement of the user which is pointed by the localization system (dotted line). OD – area outside of the building, CC – corridor, WR – working room, WH, warehouse, SH – shop.

The first step is to perform a typical spatial analysis that allows determining the type of the room in which the user is located, e.g. hallway, room, warehouse, technical room, elevators, and conference room. This analysis is based on a comparison of the coordinates obtained from the positioning system (which always carries a certain error) with the topography of the building (GIS model). To increase the probability of correct classification of the position biased with an error, it is proposed to analyze historical locations, speed, type of the user (customer of a mall, security guard, an employee of technical services), navigation destination (shopping, rescue action) and the distance from the nearest door.

The application of an expert system is to minimalize errors in determining the position and in consequence reduce number of redundant route recalculations during the tracking of the user.

3.3   Examples

The information used by the expert system based on the proposed algorithm may look like the one presented in Table 1 and Table 2.

Table 1. Attributes in the expert system

Att.	Meaning	Domain
V	a velocity of the user	no move (N), slow (S), medium (M), fast (F)
UCAT	a category of the user	a customer (C), a security officer (S), etc.
DEST	a category of a destination point	shopping (SP), a rescue action (RS), etc.
DOOR	a distance to the nearest door	far (F), near (N)
LOC	a current position from the localization system	a corridor (CC), a working room (WR), a warehouse (WH), etc.
C2, C1	classifications in 2 previous steps	a corridor with a level of confidentiality: proba-
CP	a decision attribute, classification of the current location	bly (CC1), very likely (CC2), certainly (CC3); a working room (WR1), (WR2), (WR3), etc.

Table 2. The attributes in the system for the trajectory presented in Fig. 1

ID	UCAT	DEST	V	DOOR	LOC	C2	C1	CP
1	C	SP	S	F	CC	none	none	CC1
2	C	SP	S	N	CC	none	CC1	CC3
3	C	SP	S	F	OD	CC1	CC3	CC2
4	C	SP	S	F	CC	CC3	CC2	CC3
5	C	SP	F	F	CC	CC2	CC3	CC3
6	C	SP	F	F	WH	CC3	CC3	CC3
7	C	SP	F	F	CC	CC3	CC3	CC3
8	C	SP	S	F	CC	CC3	CC3	CC3
9	C	SP	S	F	OD	CC3	CC3	CC2
10	C	SP	S	F	CC	CC3	CC2	CC3
11	C	SP	N	N	SH	CC2	CC3	SH1

It is worth paying attention to the position 6 in the example (see Table 2). Although the localization system indicates that the user is in a warehouse (WH), the expert system corrects the information and provides the classification of this position as "within the corridor" (CC3). This is because the previous user locations indicated a high probability that the user is moving along the corridor, is a customer of the mall, moves quickly, the destination is near the shop, and there is no door in the neighborhood. The probability that this user under these conditions is in the warehouse is very low.

4   Supporting Automatic Generation of Navigation Directions

4.1   Definition of the Problem

The process of generating navigation directions takes the calculated route and the navigation graph [5], and as an output provides a list of decision points with associated commands. Decision points are usually defined for nodes, so we can consider the problem of classification of all nodes in the route to the classes representing a predefined set of all possible commands. The process is not straightforward and the decision what direction (if any) should be given in a certain point may depend on many conditions that are not easily identifiable. The issue is even more complex indoors as the movement of pedestrians is not well organized [6].

This problem is addressed in research. For example, a semantic graph (related to the navigation graph) introducing ontology to the system and may be used. Depending on the concrete implementation, various information may be deduced. Reasoning rules use relationships between rooms, walls, furniture, landmarks and other objects to provide the implicit meaning of the areas and select the most adequate landmarks for building navigation directions [7]. Another interesting example of hierarchical approach distinguishing 4 levels of ontology is introduced in [8] – upper ontology with high-level concepts, domain ontologies (artefacts of indoor and outdoor space), navigation task ontology, and application ontologies.

The ontology provided by a semantic graph may be specifically used for the determination of adequate guidelines during navigation. This approach allows for creation of intuitive directions, however, it requires a huge effort to model the surroundings of each individual decision point, store considerably large data and implement algorithms for efficient information extraction.

As presented in this paper, an expert system based on the Rough Set Theory may be built to support the automatic generation of directions using only a navigation graph and the geometry of its edges.

4.2   Simplification of the Problem

In order to build an expert system the problem needs to be simplified and expressed in the Rough Set Theory terms. Firstly, we may divide the task into two separate ones: (1) the determination of the directions for all decision points, in our case all nodes of the calculated route, and (2) the decision if the direction should be presented to the user. The second task may be done by selecting the most accurate points from the complete list taking into consideration not only the spatial information but also the context of preceding and succeeding directions to aggregate a few of them together or omit less important information. This research focuses only on the first task.

The system is expected to provide 'turn left/turn right' commands. The distinction between left and right direction as well as description how steep the turn is may be deduced from the geometry of the edges in the graph. Hence, the decision focuses only on determination whether the situation represents go 'straight' or take a 'turn' for each node.

4.3   Examples

The analysis of a number of manoeuvers allows defining conditional attributes for classification. This section discusses representative cases and concludes the selection.

A situation shown in Fig. 2 illustrates the importance of attributes αL, αR – the angles between the current edge and edges outgoing from the decision node from the left and the right side. The angle between route segments E1 and E2 is the same in 2a) and 2b) but the existence of the outgoing edge E4 changes the expected direction.

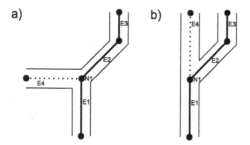

Fig. 2. Impact of the additional edge on the expected direction: a) manoeuver M1 – expected 'go straight'; b) manoeuver M2 – expected 'turn right'

Looking at the graph only in Fig. 3a) we may conclude that the 'turn right' direction is a proper decision in the node N1. However, the graph may represent the corridor system as in 3b), and the turn in node N1 may come from just an imperfection of the model while the expected direction is 'go straight'. The analysis leads to the conclusion that we should add to the system attributes describing this situation: L, L1, L2 (lengths of the current and the 2 next edges) and α1 (the next angle of the route in the next decision point.

The next observation illustrates Fig. 4. It shows how the lengths of the edges outgoing from the decision node change the perception of the situation.

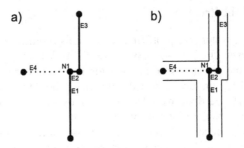

Fig. 3. Imperfection of the model may lead to unambiguity: a) a bare graph does not allow concluding a proper direction; b) manoeuver M3 – expected 'go straight'

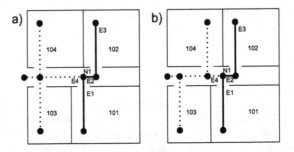

Fig. 4. The perception of the manoeuver may depend on the length of an additional edge outgoing from the current node: a) manoeuver M4: 'go straight', b) manoeuver M5: 'turn right'

Table 3 summarizes the attributes chosen for the expert system. Table 4 presents the decisions of the expert system for the cases from figures 2-4.

Table 3. Attributes describing maneuvers

Att.	Meaning	Domain
α , α1	the current and the next angles on the route	a value in degrees rounded
αL, αR	angles between the next edge and the left or right side edge in the decision node	to multiples of 45
L, L1, L2	lengths of the current and the 2 next edges	short (S), medium (M),
LL, LR	lengths of the left side and right side edges	long (L), very long (VL)
Dir	a decision attribute	straight (ST), turn (TR)

Table 4. The values of the attributes for the considered maneuvers (M1-M5)

M	A	α1	αL	αR	L	L1	L2	LL	LR	Dir
M1	45	-45	135	none	M	M	S	M	M	ST
M2	45	-45	45	none	M	M	S	M	none	TR
M3	90	-90	180	none	M	S	M	M	none	ST
M4	90	-90	180	none	M	S	M	M	none	ST
M5	90	-90	180	none	M	S	M	S	none	TR

5   Expert Systems in Indoor Navigation

In previous sections of the paper two expert systems were proposed for supporting the decisions during indoor navigation. In both cases we deal with typical problems that can be addressed by Rough Set Theory.

Building expert systems starts with the analysis of concrete cases and the anticipation of possible attributes that may be significant in the decision process. At this stage, the designer of the system does not have to decide if some attributes are more important than others. The choice can be done excessively in order to ensure that all conditions having impact to decisions are taken into consideration. The Rough Set Theory allows replacing the attribute set with its reduct [9] to keep the same decisions. As the consequence of this process it will be determined which attributes are really important and only they will be implemented in the system.

The expert systems are trained with examples representing real life cases. To create a learning database, it is necessary to carry out tests with the users of the system, the best in representative buildings. During this process testers provide expected classifications – values of decisions for conditional attributes calculated by the system. Having a large number of classified cases Rough Set Theory allows to generate a decision rule set of a reasonable size involving trade-off between the size of data, the time of processing and the adequacy of decisions. In the further research the authors plan to compare selected algorithms for optimization of a number of rules [10].

The expert systems will use the preloaded decision rules that will work also for new situations not included in the training set. Further tweaking of the navigation system in order to improve its behavior will be limited in many cases to upgrading only the decision rule set.

6   Conclusions

Similarly to other systems employing human-computer interaction, algorithms in Indoor Navigation make use of Artificial Intelligence methods for the adaptation of their behavior to human expectations. The majority of them lead to typical classification problems without full knowledge about decision domain. The system is expected to take various decisions such as displaying the user's position or playing adequate navigation directions in accordance with human intuition, while it is not clear what events or specific conditions have impact on the decision.

Such problems may be addressed by methods of the Rough Set Theory. The reduction of attributes in an expert system and the creation of reasoning rules allow to extract the knowledge from predefined test cases and in consequence build the components of the system that may adapt to changing conditions of navigation.

References

1. Gotlib, D., Marciniak, J.: Cartographical aspects in the design of indoor navigation systems. In: Annual of Navigation no 19/2012, Polish Navigation Forum (2012)
2. Gotlib, D., Gnat, M., Marciniak, J.: The Research on Cartographical Indoor Presentation and Indoor Route Modeling for Navigation Applications. In: IEEE Proceedings 2012 International Conference on Indoor Positioning and Indoor Navigation (2012)
3. Gotlib, D., Gnat, M.: Spatial Database Modeling For Indoor Navigation Systems. Reports on Geodesy and Geoinformatics 95, 49–63 (2013), doi:10.2478/rgg-2013-0012
4. Lorenz, B., Ohlbach, H.J., Stoffel, E.-P.: A hybrid spatial model for representing indoor environments. In: Carswell, J.D., Tezuka, T. (eds.) W2GIS 2006. LNCS, vol. 4295, pp. 102–112. Springer, Heidelberg (2006)
5. Raubal, M., Worboys, M.F.: A Formal Model of the Process of Wayfinding in Built Environments. In: Freksa, C., Mark, D.M. (eds.) COSIT 1999. LNCS, vol. 1661, pp. 102–112. Springer, Heidelberg (1999)
6. Gudmundsson, J., Laube, P., Wolle, T.: Computational movement analysis. In: Kresse, W., Danko, D. (eds.) Springer Handbook of Geographic Information, pp. 725–741. Springer, Berlin (2012)
7. Mast, V., Wolter, D.: A Probabilistic Framework for Object Descriptions in Indoor Route Instructions. In: Tenbrink, T., Stell, J., Galton, A., Wood, Z. (eds.) COSIT 2013. LNCS, vol. 8116, pp. 185–204. Springer, Heidelberg (2013)
8. Yang, L., Worboys, M.: A navigation ontology for outdoor-indoor space. In: Proceedings of the 3rd ACM SIGSpatial International Workshop on Indoor Spatial Awareness. ACM (2011)
9. Pawlak, Z., Skowron, A.: Rudiments of rough sets. Information Sciences, Informatics and Computer Science Intelligent System Applications 177, 3–27 (2007)
10. Zhou, J., Shu-You, L., Wen-Shi, C.: An approach for optimization of certainty decision rule sets. In: 2008 International Conference on Machine Learning and Cybernetics. IEEE (2008)

Mobile Indicators in GIS and GPS Positioning Accuracy in Cities

Artur Janowski[1], Aleksander Nowak[2], Marek Przyborski[2], and Jakub Szulwic[2]

[1] University of Warmia and Mazury, Faculty of Geodesy and Land Management,
Institute of Geodesy, Olsztyn, Poland
artur.janowski@geodezja.pl
[2] Gdansk University of Technology, Faculty of Civil and Environmental Engineering,
Department of Geodesy, Gdansk, Poland
{aleksander.nowak,marek.przyborski,jakub.szulwic}@pg.gda.pl

Abstract. The publication describes the possible use of tele-geoprocessing as a synergy of modern IT solutions, telecommunications and GIS algorithms. The paper presents a possibility of urban traffic monitoring with the use of mobile GIS indicators of dedicated monitoring system designed for taxi corporation. The system is based on a stationary and mobile software package. The optimal and minimal assumptions for the monitoring of urban traffic are described. They can be implemented as a verification or supplementary tool for complex and high cost transportation management systems or throughput of city streets monitoring systems. The authors show limitations of standard monitoring and GNSS positioning in urban area. They indicate the possible improvement in the functionality of the application to the calculation of supplementary vector data of possible trajectories and based on it, the correction of the data received from the satellite positioning.

Keywords: mobile GIS indicators, urban traffic monitoring, limitation of GNSS in traffic monitoring.

1 Introduction

Contemporary technological availability of GNSS positioning (Global Navigation Satellite System), in particular GPS-NAVSTAR (Global Positioning System - NAVigation Signal Timing And Ranging), the ability to use a wide range of solutions and algorithms in the field of information systems (GIS, Geographic Information System) and still permanently increasing and required fulfilling expectations of the information society implies a significant development of all services dedicated to the general urban population. The consequence of this is also the noticeable and continuous development of the IT service of taxi corporations. Increasingly important is also the reliability and availability of GNSS positioning in compact urban development. The problem is connected with the constantly moving GNSS receiver and the variability of this movement (which is the consequence of this movement) and impact of field screens. It requires the use of appropriate corrective algorithms,

M. Kryszkiewicz et al. (Eds.): RSEISP 2014, LNAI 8537, pp. 309–318, 2014.

evaluating the accuracy of the determined position. This is the starting point for the other algorithms of movement evaluation in urban agglomerations. Issues of determining the optimal route and traffic monitoring are often undertaken in the world, as an example we can cite [1, 2, 3, 4]. Taking the issue of satellite positioning use in relation to GIS algorithms is also associated with noticeable technological progress in the segment of individual transport in cities. With the computerization of taxi corporations and the increasing formal extent (taxi companies conglomerates operating along the whole country), as well as the limited area supported strictly by a single corporation, it is possible to reflect the positioning of taxi vehicles by the analyses of streams and transport networks in specific cities. The authors notice the problem of ambiguous polarity of taxi transport in some cities, where taxis can use bus-lanes. Then, taxi traffic analysis can be fraught with unfavorable for analysis indicator (if the expected outcome is to determine the operability of non-privileged vehicles) to facilitate their passage. However, when this parameter will be included in the modeling of traffic (adoption of vector reference and the introduction of the estimated passability correlation parameters for favored lines and others), so far its ambiguous status can be an asset expanding the possibility of using taxis traffic analysis in the study and optimization of transport network as an individual, as well as collective. Hence the choice of indicators of urban traffic in the form of taxi vehicles is not accidental.

2   Analysis of the Problem and Preliminary Assumptions to the Test System

In the studies related to the modeling of transport in the cities, the analysis of public transport in urban areas has particular social and economic meaning [5, 6]. It is indicated by the authorities as the most important and this is where the possibility of solving people within the agglomeration is seen. As a basic, and at the same time, the most important parameter affecting the travel modeling, the speed (average) is indicated. This parameter is indeed critical in the context of planning and capacity assessment and modeling of traffic flows for area solutions. However, the purpose of modeling and analysis of the journey is the final positive verification of travel carried out by a standardized model individual user of the transport system.

There are essential parameters for assessing the quality of the transport system. The users of urban space, using the available means of transport, evaluate the travel using parameters as: price, time and effort [7], also in functional relation to the time, day of the week, special days on the calendar and terrain topography. Sometimes, the "effort" is replaced with the parameter of "quality of transport services", which is identified generally as the traveling comfort. In the presented analysis, this has been omitted because of the difficulty of standardization and its dependence on the personal assessment of each user of service. Each mass user [8] (passenger) defines individually the limits of the maximum prices and the longest acceptable travel time, and also the following which is hard to be mathematically analyzed: efforts to drive / ride comfort. For a more complete assessment of the transport network, it is also significant the local nature of the terrain and transport infrastructure resulting from the location of functional areas of the city (industrial sites, services, housing, etc.).

Nowadays taxi corporations are supported by dedicated systems, created by IT companies (based on solutions and AVL, less on AVL-LBS; AVL: Automatic Vehicle Location, LBS: Location-Based Services [9, 10]), designated by the authors as AVL-TAXI and AVL-LBS-TAXI. Methods of the transport network analysis based on mapping portals typically use feedback information from navigation in vehicles. The principles used in the algorithms applied in these systems are usually not explicit, and thus can not be subjected to explicit verification and standardization with other similar compilation. At the same time, they are directly related to the speed parameter and they use it as a primary, or even as sole parameter of analysis. In the urban space, permanent places for which the mapping portals indicate a lot of traffic can be identified, but in reality it is only a local slow of traffic resulting from road parameters. This problem indicates that it is possible to create system which uses modern methods of positioning, but using too many general algorithms for the process of road network analysis does not correspond to the level of detail which is expected by the mass user. Solutions based on portals and mapping systems using the universal speed analysis has been criticized by the authors of this article. This is also because the navigation user does not subject to calibration. For individual drivers there is no weight affecting the feedback result which has an influence on the general situation on the road.

Modeling the throughput may be based on the examination of the road network by means of speed parameter, but taking into consideration the factors limiting the speed of travel. We find analyzes based on the measurement of traffic flow, busy condition of the route section, the speed of vehicles in the function of the vehicle length [5] or simplified analyses making the road section resistance dependent on variables, such as traffic flow, throughput and speed of the vehicle in free motion [11]. In city traffic, it is important to properly assess in its model the congestion effect of the road network or the congestion effect of selected section. In such situation, there is no place for network analyzes. It has point or linear nature or it is characterized by the discontinuity of movement.

The experimental results show that these models should be verified and adjusted on the basis of measurements based on fixed and permanent measuring points (inductive loop systems, photo registration systems). However, the correct approximation of the situations on roads in the road network model is possible to be determined by input data from numerous test or calibrating / verifying transits. This type of data is directly accessible by means of position AVL (taking into account the availability of positioning [12]), and services based on LBS along with the assigned and unique in time and space ID and time mark. At that time, the *mobile indicators* (MI) are generated in the model [14], where each of them is available for the selected space (in terms of geo-reference) and for preset time and event circumstances (occurring in the data coming from the past observation epochs). Field infrastructure (not only for vehicular traffic, but also for rail transport in urban areas) can be obtained from the cartographic materials (e.g.: official resource of geoportal.gov.pl, or alternatively from mapping portals such as Open Street Map) or with reference to the measurements using precise satellite positioning [14].

3   Problems with the Accuracy of GNSS Positioning in Urban Areas

Nowadays, the accuracy of positioning achieved even by low-cost satellite navigation receivers without the differential support varies at the level of meters in open space, so it is sufficient for most navigation and localization tasks. The situation changes dramatically when the vehicle moves between high buildings. Shadow effect (associated with signal blocking) and multi-path effect (caused by reflection of signals) make that the fixed coordinates may be burdened by significant errors (even a few hundred meters). These are not errors related to the faulty work of the system, but to the conditions of measurement.

The problem of outliers in pseudo-range measurements when shadow and multi-path effects are present is shown in the Fig. 1.

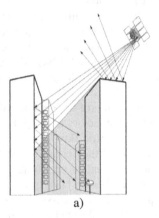
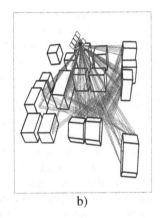

a)   b)

Fig. 1. Errors of pseudorange measurement caused by shadow and multipath effects [12]

Exemplary GNSS fix errors recorded in urban areas during tests are shown in Fig.2. The accuracy of GNSS fixes is estimated by using the following dependencies:

$$MRE = RMS_\rho \cdot DOP \tag{1}$$

where:

MRE – Mean Radial Error of GNSS fixes,
RMS_ρ – Root Mean Square error of pseudorange measurements,
DOP – Dilution Of Precision.

DOP (Dilution of Precision) is the factor determining the impact of the deployment of the satellites on the fixes accuracy. It is dependent on the number of visible satellites and their geometric layout. Therefore, blocking the signals by buildings in the city will significantly affect the value of the DOP. On the one hand, buildings will reduce the number of observed satellites, on the other hand, only satellites with high elevation (greater than buildings) will be seen. Even if it is assumed that the buildings will not cause an increase in RMS_ρ, in accordance with (1) the same increase in DOP will entail greater fix errors.

Fig. 2. Exemplary GNSS fix errors recorded in urban areas [own researches]

Using the author's software it was possible to carry out an experiment for which the objective was to assess the impact of screen as a form of buildings on the accuracy of satellite positioning. The 24 hour static measurement session was simulated, in which the virtual receiver was placed between the buildings modeled as shown in Fig. 3.

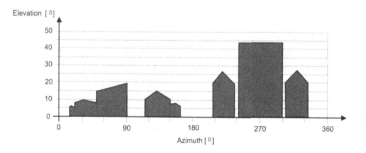

Fig. 3. Simulated urban area

The software allowed the simultaneous registration of the number of visible satellites and values of DOP factors in simulated conditions (with clear horizon and when the obstacles are present). On the basis of (1) the estimated values of GNSS fix errors were computed, assuming that the $RMS_\rho = 5$ m. Simulations were carried out for the average values of latitude and longitude, determined for the area of Poland.

According to the assumptions, the introduction of the buildings caused decline in the number of visible satellites, and this resulted in an increase in the values of DOP factors and consequently an increase in fix errors [15]. Comparison of the values are presented in Fig. 4.

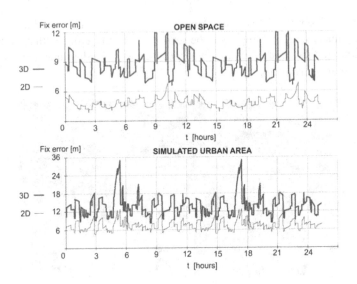

Fig. 4. GNSS fix errors as a function of time. In the open space – top chart; where there are obstacles – the lower chart.

4   Implementation of the AVL-LBS-TAXI Authors System

The basic structure of the AVL-LBS-TAXI system is based on three modules: managing module in the control room, a terminal module in a taxi and integrating module (Fig. 5).

The management module's task is to provide the taxi corporate headquarters with the logistically useful information (service taxi orders and identification of calls from customers, geolocation of taxi vehicles along with zoning, status, and optional information about the road event, geocoding and geoprocessing of transport orders for taxis).

Terminal module (in taxis) ensures communication with other system modules via Internet, the spatial localization of a vehicle, determination of the vehicle status (busy, free, on the road to the passenger, etc.), maintenance of road accidents and emergency messages and orders receiving.

The integrating module operates as the verifier and intermediary in communication between two other modules. It provides: parameters to support risk assessment when taking a new order (order verification based on the history of orders from a registered client, phone number or address), monitoring the realization of current and pending orders, including dedicated and customized ones, navigational support in spatially ambiguous situations (GNSS positioning problems), statistics and a statistical analysis of orders, parameterized record of the instantaneous location of mobile indicators: time (UTC, Coordinated Universal Time) of the server module, φ, λ (latitude, longitude) in WGS84 (World Geodetic System '84), velocity, move-ment azimuth - the parameters derived from the message of NMEA (National Marine Electronics Association) from GPS receivers mounted in MI (*mobile indicators*).

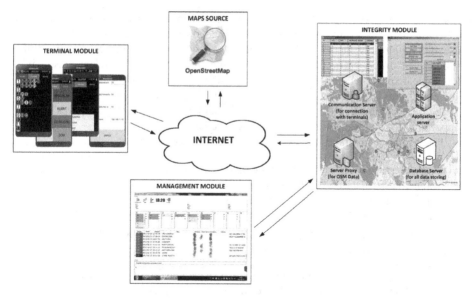

Fig. 5. Functional diagram of presented AVL-LBS-TAXI system

The last indicated possibility may be less important from the point of view of the corporation, but actually, it is the basis for the described in the article possibilities related to the analytical predictions of vehicular transport network of the city.

Additionally, on the basis of the synchronous communication with terminal module, forced by the necessary stability disconnects automatically the vehicles that do not maintain the cooperation with the system (no transfer of spatial position for a set period of time or lack of response upon sent order).

The module is responsible for re-allocation to regions and mutual queuing of orders and vehicles. Despite the use of filters and algorithms for the improvement of positioning [16], the authors of the AVL-LBS-TAXI system take into account the limited precision of satellite positioning receivers used in the system (receivers integrated with mobile devices including spartphone, phablet, tablet). Therefore, the associated possible oscillation of position between areas has been limited by establishing verifying procedures for the correctness of assigned allocation by sending a query to the driver.

This functionality is active only in the border areas of parameterized regions as a buffer zone with a width set by the administrator. This standard GIS functionality is one of many spatial analysis algorithms used in the system.

A preliminary model of the transport network used in the system is constructed in Euclidean space, but becomes closer to reality only through support on analyzes in a topological space. The idea is to apply the Voronoi tessellation, in which the centers are defined as points of maximized convergence of vehicle streams (intersections, shopping malls, train stations, etc.). Optimizing the number of centers can be developed on the basis of arbitrary acceptance of traffic limits. As a target, the model of the transport network can be based on the topological model supported by directed graphs.

Principles of creating topological model, useful in AVL-LBS-TAXI system, and based on areas (areas into which the city is divided by the taxi corporation) permit to use the rough sets theory [17]. This approach, however, is not necessary in the situation of urban traffic analysis. Each region may be an independent cluster in a topological space, which, together with the other clusters exists in single city system and constitutes compact topological space and this can be a basis for analyzes. Access to the archive and test drives observation, guaranteed by the system gives you an opportunity of an efficient estimation of expected events.

The primary advantages of AVL-LBS-TAXI system include:

- Current positioning of taxi cab on the map,
- Automatic notification for the passenger about waiting taxi,
- Automatic assignment of taxis to the regions and zones,
- Shortening the time for receiving orders to only a few seconds,
- Changes in the number and range of areas because of the time of day, day of week, time of the year.

The use of solutions based on a topological space may allow to calculate the fare at the time of accepting the order and determining travel routes.

During the empirical process of the system construction (in 2001-2013, i.e. within the period of access to the corporate taxi and construction of dedicated IT systems for transport services for third parties [18, 19]) a numerous mechanisms conditioning the resistance to personal factors were implemented. Logistics analysis was conducted and functional and technological characteristics of the system were presented [13]. Important aspect for the analysis seems to be a standardization of taxi transport logistics systems according to the assumptions mentioned in the article and the principles of decision-making in transport [20], so that they could be considered as effective in monitoring and optimization of the city transport situation. Non-standardized systems may also be approved, but additional filter analysis for unwanted events or for sole interest of one of the participants in the system (i.e. a corporation, a taxi driver or passenger) is particularly important.

5   Conclusions

In the context of the development of satellite positioning available for urban mass market, it seems a natural approach to use the mobile indicators for approximation and traffic modeling in cities. The specificity of taxi transport is its commitment to urban spaces, hence it gives a significant opportunity to use them as mobile indicators.

Further expansion of MI (*mobile indicators*) - including improved determination of their position - allows you to draw a thesis about the possibility of high precision, quality, accuracy, the value of the forecast results compared to observed facts. The authors draw attention to the implementation of solutions based on growth opportunities and expanding the use of existing systems and solutions without any noticeable increase in IT costs and the burden on the company's IT infrastructure. Recording and storage of location data has particular meaning here. There is an opportunity to extend the functionality of AVL-LBS-TAXI system with temporal logic. Referring to the topological space supported by GIS algorithms, taking into account the satellite positioning analysis of urban space communication may be carried out for the individual characteristics of a particular city. Location of mobile indicators in GIS allows you to create models which enable the preparation of forward-looking (hypothetical) trip.

The presented concept of the system is a modern and very promising direction for the development of such systems. The current version of the AVL-LBS-TAXI system is still being developed and can be used to integrate data from GIS mobile indicators as a support for the analysis and road network management. It is obvious, that the system itself can serve to improve the quality of passenger service within a particular taxi company. By using a sufficiently large number of mobile indicators, reducing the difficulties and increasing the accuracy of the classic mobile GNSS positioning in urban areas, it is possible to carry out the analysis of the city road network. The proposed solution can also be implemented in order to monitor traffic on inland waters and coastal areas. However, the inclusion of data obtained from mobile indicators to GIS and advanced traffic management systems may be an independent element of verification and support to improve transport logistics in isolated areas. Experiments conducted by the authors of the article, based on the introduction of mobile indicators supported by GIS analysis in a topological space provide basis to indicate the use of this solution in the analysis of the road network. Obtained results are a starting point for further research in which the matter of GNSS positioning is significant in connection with the localization technology which uses mobile phones network and radio-navigation methods in areas inaccessible for GPS / GLONASS (Russian: ГЛОНАСС, IPA: [glɐˈnas]; Глобальная навигационная спутниковая система), acronym for GLObalnaya NAvigatsionnaya Sputnikovaya Sistema).

References

1. Romaniuk, R.: Selection of GRNN network parameters for the needs of state vector estimation of manoeuvring target in ARPA devices. In: Romaniuk, R.S. (ed.) Photonics Applications in Astronomy, Communications, Industry, and High-Energy Physics Experiments IV. Proceedings of the Society of Photo-Optical Instrumentation Engineers (SPIE), vol. 6159, pp. F1591–F1591. Wilga (2006)
2. Stateczny, A.: Artificial neural networks for comparative navigation. In: Rutkowski, L., Siekmann, J.H., Tadeusiewicz, R., Zadeh, L.A. (eds.) ICAISC 2004. LNCS (LNAI), vol. 3070, pp. 1187–1192. Springer, Heidelberg (2004)
3. Stateczny, A.: Methods of comparative plotting of the ship's position. In: Brebbia, C.A., Sciutto, G. (eds.) Maritime Engineering & Ports III. Water Studies Series, vol. 12, pp. 61–68. Rhodes (2002)
4. Stateczny, A.: Neural manoeuvre detection of the tracked target in ARPA systems. In: Katebi, R. (ed.) Control Applications in Marine Systems 2001 (CAMS 2001). IFAC Proceedings Series, pp. 209–214. Glasgow (2002)
5. Birr, K., Jamroz, K., Kustra, W.: Analysis of factors that affects the speed of public transport vehicles on the example of Gdańsk city, The Scientific Work Warsaw University of Technology - Transport: 96 (2013)

6. Karoń, G., Janecki, R., Sobota, A.: Traffic modeling in Upper Silesian conurbation - public transport network. Scientific Sheets of Silesian University of Technology, Gliwice: Series Transport: 35 (2010)
7. Geurs, K.T., Ritsema van Eck, J.R.: Accessibility Measures: Review and Applications. Evaluation of Accessibility Impacts of Land-Use Transport Scenarios, and Related Social and Economic Impacts. RIVM report 408505 006, National Institute of Public Health and the Environment, Bilthoven (2001)
8. Janowski, A.: Selection of optimal IT tools when constructing SIP applications aimed to admass. University of Warmia and Mazury, dissertation (2003)
9. Balqies, S., Al-Bayari, O.: LBS and GIS technology combination and applications. In: IEEE/ACS International Conference on Computer Systems and Applications, AICCSA 2007. IEEE (2007)
10. Sotelo, M.A., Van Lint, J.W.C., Nunes, U., Vlacic, L.B., Chowdhury, M.: Introduction to the special issue on emergent cooperative technologies in intelligent transportation systems. IEEE Transactions on Intelligent Transportation Systems 13(1), 1–5 (2012)
11. Dailey, D.J.: A statistical algorithm for estimating speed from single loop volume and occupancy measurements. Transportation Research Part B: Methodological 33(5), 313–322 (1999)
12. Nowak, A.: Problems of satelite positioning in urban agglomerations. Logistics / Logistyka (2011)
13. Janowski, A., Szulwic, J.: GIS mobile indicators in the urban traffic analysis. Logistics / Logistyka (2014)
14. Specht, C., Koc, W., Nowak, A., Szulwic, J., Szmagliński, J., Skóra, M., Specht, M., Czapnik, M.: Availability of phase solutions of GPS / GLONASS during the geodesic inventory of railways – on the example of tram lines in Gdańsk, Rail Transport Technique TTS, vol. 01/2012, pp. 3441–3451 (2012)
15. Nowak, A., Specht, C.: Snapshot RAIM algorithms availability in urban areas. Annual of Navigation, No. 11/2006 (2006)
16. Stateczny, A., Kazimierski, W.: A comparison of the target tracking in marine navigational radars by means of GRNN filter and numerical filter. In: 2008 IEEE Radar Conference, Rome, pp. 1994–1997 (2008)
17. Ge, X., Bai, X., Yun, Z.: Topological characterizations of covering for special covering-based upper approximation operators. Information Sciences 204, 70–81 (2012)
18. Janowski, A., Szulwic, J.: Present development tools at the service of photogrammetry and GIS. Archives of Photogrammetry, Cartography and Remote Sensing 14 (2004)
19. Bednarczyk, M., Janowski, A.: Mobile Application Technology in Levelling. Acta Geodynamica et Geomaterialia 11(2(174), xx1–xx5 (2014)
20. Zak, J.: Decision support systems in transportation. In: Handbook on Decision Making, pp. 249–294. Springer, Heidelberg (2010)

Analysis of the Possibility
of Using Radar Tracking Method
Based on GRNN for Processing Sonar Spatial Data

Witold Kazimierski and Grzegorz Zaniewicz

Institute of Geoinformatics, Maritime University of Szczecin, Poland
{g.zaniewicz,w.kazimierski}@am.szczecin.pl

Abstract. This paper presents the approach of applying radar tracking methods for tracking underwater objects using stationary sonar. Authors introduce existing in navigation methods of target tracking with particular attention to methods based on neural filters. Their specific implementation for sonar spatial data is also described. The results of conducted experiments with the use of real sonograms are presented.

Keywords: sonar, target tracking, geo-data, spatial data processing.

1   Introduction

Sonar is one of the basic equipment used in hydrographic work. Based on sonar acoustic images of a bottom, detection and initial assessment of underwater objects is possible. Thus, the basic way to use sonar is to conduct a survey of the bottom, or a search of the bottom. However with the development of sonar technology, the scope of sonar works has increased. Nowadays more and more often sonar is used to monitor underwater moving objects such as ROVs (Remotely Operated Vehicle) or AUVs (Autonomous Underwater Vehicle) based on sonogram. The sonogram itself is treated as an example of raster spatial data, mainly requiring calibration, and then processed to obtain the necessary information. In case of tracking, the required information are location and object motion parameters used to allow prediction of targets movement.

The article presents issues related to tracking objects based on sonograms. In the first part the motivation and essence of the problem is given. Based on radar tracking experiences authors suggests to use neural filters based on polar coordinates instead of the commonly used Cartesian system and to adjust radar tracking methods for sonar. Article is completed by the summary and conclusions.

2   The Problem of Underwater Vehicle Location and Tracking in Hydrographic Practice

Hydrographic works are performed to obtain spatial information about objects under water. The importance of these data for navigational safety is increasing while including

M. Kryszkiewicz et al. (Eds.): RSEISP 2014, LNAI 8537, pp. 319–326, 2014.

them in navigational systems and decision support systems like [1]. Information about the location of acquired information is here crucial, especially while working with autonomous instruments, like ROV or AUV. The influence of position uncertainty on quality of data was discussed for example in [2, 3]. In practice, the most popular are three approaches: inertial systems on underwater vehicle, hydroacoustic positioning systems (such as LBL, USBL), positioning and tracking based on geodata sonar. Common drawback of inertial systems is increase of positioning inaccuracies in time due to the summation of the subsequent measurement uncertainty and thus the necessity for developing course stabilization algorithms like in [4]. Hydroacoustic systems, commonly used on large ships, are accurate, but their main disadvantage is their price, which makes them essentially unreachable for a large part of the hydrographers [5]. Thus positioning and tracking based on sonar data seems to be an interesting alternative.

There are two basic methods of working with sonar: side scanning and stationary scanning. From the point of view of objects tracking, it is easier and more accurate to use stationary sonar, where sonar head is hanged on a tripod on the seabed and polar scanning by rotated transducer is performed. The measurement errors resulting from the movement of vessels (instability of the head) are then omitted. Processing and mosaicking of such images in order to obtain a consistent sonogram is also much easier.

In [6] the summary of literature discussion is given, identifying the basic steps in the process of tracking, including moving target extraction, target association, tracking and movement prediction. There are many methods of detecting motion in the sonar image, taken from the field of computer science called computer vision. Examples of such methods in the classical recursive approach are given in [7]. After identifying of the objects, their association in the consecutive scans is needed, usually with the Kalman filter, but other methods are also possible [8]. After association suitable tracking algorithm can be applied.

3   Overview of Tracking Algorithms

Tracking at this stage means in fact estimation of object's movement parameters or speaking more precisely estimation of state vector. Usually classical system description with system of two equations – state equation (1) and measurement equation (2) is used:

$$x_{i+1} = f(x_i, U_i, i) + W_i \tag{1}$$

$$z_{i+1} = g(x_{i+1}) + V_{i+1} \tag{2}$$

where:

x_i – *n – dimensional state vector; z_{i+1} – measurement vector,*
U_i – *p – dimensional control (known external input) vector,*
W_i – *state noise vector, V_i – measurement noise vector, i –   discrete time.*

Exact definition of the elements of this model varies in different tracking algorithm. Commonly some kind of numerical filter, being mostly modification of Kalman Filter is used. Interesting cases of using numerical filters (like EKF or UKF) for sonar target tracking are presented for example in [6] or [9]. Results given in those works confirms the possibility of using Kalman filtering for tracking UAV based on sonar measurements.

Other possible approaches often presented in literature are particle filters. The examples can be found in [10, 11]. Sometimes also some other concepts of sonar target tracking can be found including for example tree algorithm [12].

4   Neural Alternative

Common issue in all presented approaches is noting sonograms as a graphical picture, which causes the need for defining state vector and measurement vector in Cartesian coordinate system. In case of scanning sonar, used in this research, it is more convenient to use polar coordinates. In this approach, commonly used also in radars, state vector can be defined as movement vector (3) and measurement vector as the difference between the measured position and position estimated in previous step of estimation (4).

$$x_{i+1} = \begin{bmatrix} d_x \\ d_y \end{bmatrix} = t \cdot \begin{bmatrix} V_x \\ V_y \end{bmatrix} = t \cdot \begin{bmatrix} V \cdot \cos(KR) \\ V \cdot \sin(KR) \end{bmatrix} \tag{3}$$

$$z_{i+1} = \begin{bmatrix} BE_o - BE_e \\ d_o - d_e \end{bmatrix} \tag{4}$$

where:
   BE_o, d_o - *measured bearing and distance;*
   BE_e, d_e - *estimated bearing and distance.*

With this assumption it is possible to define numerical filter based on Kalman Filter. An alternative for numerical filters, might be however neural filter, presented for example in [13]. The idea was to implement the structures of neural networks, which are non-linear in its nature. Thus the filter dedicated for radar target tracking based on General Regression Neural Network (GRNN) was constructed. The usability of neural networks in processing navigational data was proved many times for example in [14, 15] for vector estimation or in [16, 17, 18] for bathymetric modelling. The problem of reliability of navigational data in view of artificial intelligence was also discussed in [19]. The neural filter for target tracking itself was presented and discussed in [20, 21, 22].

GRNN is a special purpose case of probabilistic neural network (PNN). It is in fact neural implementation of kernel regression algorithm. This is a convenient method of non-parametric estimation with the use of so called kernel functions. The only parameter to be set in this network is smoothing factor, which represents width of Gaussian function. The network described above performs regression according the following equation [21]:

$$
\begin{bmatrix} Vxe_i \\ Vye_i \end{bmatrix} = \begin{bmatrix} \sum_{i=1}^{n} Vxo_i * e^{-\left(\frac{\|t-t_i\|}{2\sigma}\right)^2} / \sum_{i=1}^{n} e^{-\left(\frac{\|t-t_i\|}{2\sigma}\right)^2} \\ \sum_{i=1}^{n} Vyo_i * e^{-\left(\frac{\|t-t_i\|}{2\sigma}\right)^2} / \sum_{i=1}^{n} e^{-\left(\frac{\|t-t_i\|}{2\sigma}\right)^2} \end{bmatrix} \tag{5}
$$

where:
Vxe, Vye – estimated speed along X and Y axis;
Vxo, Vyo – observed speed along X and Y axis;
σ – smoothing factor for Gaussian kernel function;
t – current time moment, given as estimation step number;
t_i – previous time moments, given as previous estimation step numbers.

Teaching process in this network means setting the weights between radial and summing layer as the values of observed speed in previous steps. Thus teaching process does not need any teacher and it means in fact single copying of teaching values into network structure.

5   Research Experiment

The research presented in this paper was performed using real measurements of sonar. Data recorded in the form of NMEA messages (TTM) were saved in txt format (together with other NMEA data) and replayed off-line with the use of different tracking filters in dedicated software developed in VB.Net. In the research, Kongsberg MS1000 scanning sonar and remotely operated underwater vehicle ROV VideoRay Explorer, were used. Main parameters of this sonar are: high frequency acoustic waves - 675kHz and good angle parameters of acoustic beam - 0.9x30°. Acquisition of image data is carried out using PC and the signal interface connected to the sonar's head. The research focused on target tracking stage. Echo extraction process itself, and therefore the image processing was made by software of Kongsberg sonar.

5.1   Research Scenarios

The research covered two test scenarios. The idea was to apply tracking technique both in case of uniform motion and course alteration. In the first scenario object was initially moving towards north, then turned around and headed south, and in the second scenario target was moving north at first, and then made a return to the east (Fig. 1).

In both cases, the vehicle was moving at a slow speed and was controlled manually by an operator. Both situations have occurred at short distances. In the first situation 75 consecutive measurements were made and in second 42 measurements. Note periodical signal fading - sonar (software) sometimes has lost tracking object.

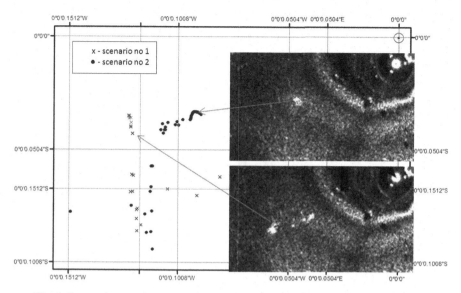

Fig. 1. Research scenarios (coordinates are determined in relation to the sonar head)

5.2   Research Results

The research included estimation with Kalman Filter and with GRNN filter. Two variants of neural filter were analyzed with different parameters. In the first one shorter teaching sequence (3 radial neurons) was used, while in the second one longer (10 radial neurons). Bigger smoothing factor was also used in the second case. Based on the analysis of the literature and earlier research on using GRNN for radar tracking, second filtration stage was omitted to avoid large delays. The covariance matrix for the measurement was set to $R=diag[0,9\ 1]$, based on sonar technical parameters.

Figure 2 presents the graph of target's course estimated with various tracking filters in first test scenario. The graph includes four lines – measurement (without filtration), Kalman Filter (numerical filter), GRNN (neural filter in 1st variant), GRNN2 (neural filter in 2nd variant).

It can be said, based on the presented graph, that course calculated directly from the measured positions is much variable und in fact useless for further processing (for example for predicting collisions with other targets). This confirms the need for using filtration while tracking sonar targets. Analyzing the results given by various filters, it can be said in general that all of them smooth the graph in some degree. However differences between filters (especially at the later phase, when target was moving slower) are significant. The practice shows that values after filtration are more reliable.

The smoothest graph is the one given by GRNN in second variant. This could have been easily foreseen as this network has longer teaching sequence and bigger smoothing factor. Larger number of neurons causes bigger averaging of values. Simultaneously it means slower reaction for the maneuvers, which are initially treated as noises or measurement errors. It shall be noticed that the speed of moving target was very small (comparing to the speed of radar targets) so the filters had some problems with tracking. In general it can be pointed out that when targets are moving slower,

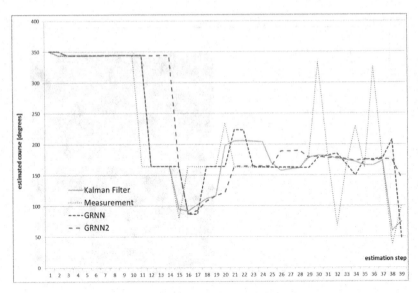

Fig. 2. Estimated course - scenario 1

the errors and noises have more impact on estimation, resulting in less stable estimated vectors. The research showed also that, while analyzing estimated track of target in Cartesian coordinates, based on estimated course and speed, the effect of summing differences between measurement and estimate can be noticed.

The results achieved in the second test scenario confirmed above conclusions and thus graphs will not be presented due to lack of space.

6   Conclusions

The article presents an attempt of using radar tracking methods for tracking underwater objects using sonar imaging. Complex analysis of the results presented in the article allows stating the following findings:

- The use of tracking filter is necessary, if estimated motion parameters are to be useful from navigating point of view.
- Selection of the parameters of filters has major impact on the results - longer teaching sequence and higher smoothing coefficients allow better smoothing of motion vector, but cause slower response to the maneuver.
- In the analyzed cases the object has been moving close to sonar head and very slowly, which also made it difficult to track.
- Introduction of object position information into the state vector might be useful.

It should also be noted that crucial for the quality of the tracking is the process of extracting echoes of the image. It is necessary to apply advanced image processing techniques to identify echoes and motion detection.

It has to be pointed out as a final conclusion that tracking target with sonar has its own characteristic that varies from radar. Targets are usually closer, measurement errors are different, noise is bigger, movement is slower and interruption of tracking is more likely. This leads to a conclusion that there is a need of developing tracking a tracking method dedicated for sonar, although some "radar experiences" may be useful. Final results achieved with such a new method might be verified with independent positioning system like LBL or direct measurement with sensors mounted on ROV. It would be then possible to calculate errors to show performance of course estimation.

References

1. Pietrzykowski, Z., Borkowski, P., Wołejsza, P.: Marine integrated navigational decision support system. In: Mikulski, J. (ed.) TST 2012. CCIS, vol. 329, pp. 284–292. Springer, Heidelberg (2012)
2. Maleika, W.: The influence of track configuration and multibeam echosounder parameters on the accuracy of seabed DTMs obtained in shallow water. Earth Science Informatics 6(2), 47–69 (2013)
3. Maleika, W., Palczynski, M., Frejlichowski, D.: Interpolation Methods and the Accuracy of Bathymetric Seabed Models Based on Multibeam Echosounder Data. In: Pan, J.-S., Chen, S.-M., Nguyen, N.T. (eds.) ACIIDS 2012, Part III. LNCS (LNAI), vol. 7198, pp. 466–475. Springer, Heidelberg (2012)
4. Borkowski, P.: Ship course stabilization by feedback linearization with adaptive object model. Polish Maritime Research 211(81), 14–19 (2014)
5. Lekkerkerk, H.-J., Theijs, M.J.: Handbook of offshore surveying. Skilltrade (2011)
6. Modalavalasa, N., SasiBhushana Rao, G., Satya Prasad, K.: An Efficient Implementation of Tracking Using Kalman Filter for Underwater Robot Application. IJCSIT 2(2) (2012)
7. Velastin, S.A., Remagnino, P. (eds.): Intelligent distributed video surveillance systems. IET, London (2006)
8. Bar Shalom, Y., Li, X.R.: Estimation with Applications to Tracking and Navigation: Theory Algorithms and Software. John Wiley & Sons, Inc., NY (2001)
9. Wood, T.: Mathematical Modelling Of Single Target SONAR and RADAR Contact Tracking, PhD thesis. University of Oxford, Michaelmas Term (2008)
10. Hue, C., Le Cadre, J.P., Perez, P.: Tracking multiple objects with particle filtering. IEEE Trans. on Aerospace and Electronic Systems 38(3), 791–812 (2002)
11. Clark, D.E., Bell, J., de Saint-Pern, Y., Petillot, Y.: PHD filter multi-target tracking in 3D sonar. In: Oceans 2005 – Europe, vol. 1 (2005)
12. Lane, D.M., Chantler, M.J., Dai, D.: Robust Tracking of Multiple Objects in Sector-Scan Sonar Image Sequences Using Optical Flow Motion Estimation. IEEE Journal of Oceanic Engineering 23 (1998)
13. Stateczny, A., Kazimierski, W.: A comparison of the target tracking in marine navigational radars by means of GRNN filter and numerical filter. In: 2008 IEEE Radar Conference, Rome. IEEE Radar Conference, vol. 1-4, pp. 1994–1997 (2008)
14. Stateczny, A.: Artificial neural networks for comparative navigation. In: Rutkowski, L., Siekmann, J.H., Tadeusiewicz, R., Zadeh, L.A. (eds.) ICAISC 2004. LNCS (LNAI), vol. 3070, pp. 1187–1192. Springer, Heidelberg (2004)
15. Stateczny, A.: Neural manoeuvre detection of the tracked target in ARPA systems. In: Katebi, R. (ed.) Control Applications in Marine Systems 2001 (CAMS 2001), Glasgow. IFAC Proceedings Series, pp. 209–214 (2002)

16. Stateczny, A.: The neural method of sea bottom shape modelling for the spatial maritime information system. In: Brebbia, C., Olivella, J. (eds.) Maritime Engineering and Ports II, Barcelona. Water Studies Series, vol. 9, pp. 251–259 (2000)
17. Lubczonek, J.: Hybrid neural model of the sea bottom surface. In: Rutkowski, L., Siekmann, J.H., Tadeusiewicz, R., Zadeh, L.A. (eds.) ICAISC 2004. LNCS (LNAI), vol. 3070, pp. 1154–1160. Springer, Heidelberg (2004)
18. Lubczonek, J., Stateczny, A.: Concept of neural model of the sea bottom surface. In: Rutkowski, L., Kacprzyk, J. (eds.) Neural Networks and Soft Computing. AISC, vol. 19, pp. 861–866. Springer, Heidelberg (2003)
19. Przyborski, M., Pyrchla, J.: Reliability of the navigational data. In: Klopotek, M.A., Wierzchon, S.T. (eds.) IIS: IIPWM 2003. AISC, vol. 22, pp. 541–545. Springer, Heidelberg (2003)
20. Stateczny, A., Kazimierski, W.: Selection of GRNN network parameters for the needs of state vector estimation of manoeuvring target in ARPA devices. In: Romaniuk, R.S. (ed.) Photonics Applications in Astronomy, Communications, Industry, and High-Energy Physics Experiments IV. Proceedings of (SPIE), vol. 6159, p. F1591. Wilga (2006)
21. Kazimierski, W.: Two – stage General Regression Neural Network for radar target tracking. Polish Journal of Environmental Studies 17(3B) (2008)
22. Stateczny, A., Kazimierski, W.: Determining Manoeuvre Detection Threshold of GRNN Filter in the Process of Tracking in Marine Navigational Radars. In: Kawalec, A., Kaniewski, P. (eds.) Proceedings of IRS 2008, Wrocław, pp. 242–245 (2008)

Supporting the Process of Monument Classification Based on Reducts, Decision Rules and Neural Networks

Robert Olszewski and Anna Fiedukowicz

Warsaw University of Technology, Faculty of Geodesy and Cartography,
Department of Cartography, Warsaw, Poland
{r.olszewski,a.fiedukowicz}@gik.pw.edu.pl

Abstract. The present article attempts to support the process of classification of multi-characteristic spatial data in order to develop the correct cartographic visualisation of complex geographical information in the thematic geoportal. Rough sets, decision rules and artificial neural networks were selected as relevant methods of spatially distributed monument classification. Basing on the obtained results it was determined that the attributes reflecting the spatial relations between specific objects play an extremely significant role in the process of classification, reducts allow to select exclusively essential attributes of objects and neural networks and decision rules are highly useful for the purposes of classification of multi-characteristic spatial data.

Keywords: rough sets, neural networks, spatial data mining, multi-characteristic spatial data, classification.

1   Introduction

Currently created databases (including those containing spatial data) are characterised both by a complex structure and by numerous attributes describing individual objects. The essence of the cartographic message – e.g. of a digital map presented on an Internet geoportal – is, however, the unambigouty and graphic simplicity that allows the user to unambigously interpret the image created as a result of visualisation of the spatial data base content [1,2].

Thus, it is necessary to develop a methodology for reliable and universal classification of multi-attribute spatial data that will allow for a correct cartographic visualisation of complex geographical information, characterised by numerous, diverse descriptive and spatial attributes [3,4,5,6].

2   Inspiration

The present study was inspired by the research and development (R&D) works, conducted for the Faculty of Geodesy and Cartography and the Faculty of Architecture of the Warsaw University of Technology for the National Heritage Board (NHB) of Poland. These works are related to the preparation of a visualisation of the NHB

M. Kryszkiewicz et al. (Eds.): RSEISP 2014, LNAI 8537, pp. 327–334, 2014.

database for a specialised geoportal. The requirement to create such geoportal results from the realisation of the EU INSPIRE Directive by Poland [7].

Among the challenges faced during the works on the creation of the NHB geoportal one of the major issues was the adjustment of the number of displayed objects to the current zoom of the map view. It was important to address both issues: how many objects of the total group of nearly 90 thousand should be shown at the given scale, how to display them (e.g. whether objects should be grouped) and which of these objects to select.

During the implementation works carried out on the initial stage of creation of the NHB portal it was decided to apply the method consisting in grouping the objects. However, the authors of this article would like to present an alternative methodology, which would allow for the display of selected, important objects at low zoom and to increase their number and show the less important ones as the map is zoomed in.

3   Problem of Classification of Real-Property Monuments

In order to solve the problem formulated in such way it is required to conduct a classification of multi-attribute objects, i.e. real property monuments gathered in the database of the NHB. Such classification may be of a supervised or unsupervised.

An example of the supervised - expert classification is the classification of monuments used in Poland in the years 1961-1973, aimed at the determination of their artistic, historical and scientific value and the degree of protection. It consisted of 5 degrees, where "0" meant the "most valuable" monuments and "4" – the "least valuable" ones. This classification was abandoned due to the fact that it led to the destruction of numerous class 4 monuments, which were insufficiently protected by law.

However, the application of such or similar classification for practical purposes, i.e. for the selection of objects during the visualisation of spatial data, might be quite useful. It would not be aimed at the differentiation of the degree of protection of monuments (which, as previous experience shows, may lead to disastrous results).The aim of the classification defined in such way would be only to decide which of the objects should be visible on the map on the national, voivodeship, district and finally local level.

4   Test Data

For the purposes of the conducted experiment a test set was prepared, consisting of 100 objects. Such solution enables to carry out experiments on a well-defined sample of multi-attribute spatial data without detriment to the generality of reasoning. Each of the objects was described by 10 attributes, which included both descriptive attributes originating from the NHB database (first 5) and geometric attributes determined with use of GIS techniques, resulting from the spatial location of objects (next 5). The attributes and their values are described in detail in Table 1. The location of the objects and spatial relations, as well as the classification according to the "expert" method are presented in Figure 1.

Table 1. List of attributes and their values

Nr	attribute	description	possible values
1	type	object type	apartment house, chateau, church, monastery, windmill
2	material	of which the object is built	brick/stone, wooden
3	condition	the condition of the object	perfect, good, ruin
4	style	dominant architectural style of the object	roman, gothic, renaissance, baroque, classicist
5	date	year of establishment	year of establishment
6	location	degree of urbanization of the area where the object is located	center of a big city, big city, small town, rural area
7	NS1	distance to the nearest monument	in km (with 10m precision)
8	NS5	averaged distance to the 5 closest monuments	in km (with 10m precision)
9	NSDT1	distance to the nearest monument of the same type	in km (with 10m precision)
10	NSDT5	averaged distance to the 5 closest monuments of the same type	in km (with 10m precision)

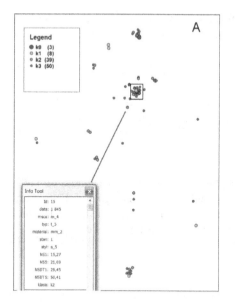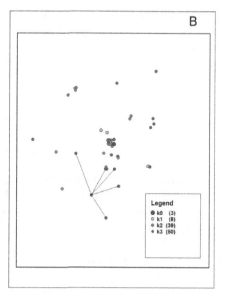

Fig. 1. Expert's object classes (method 2): A — The general view with the characteristic of chosen object, B — Fragment selected in Fig. 1a with the distances to 5 closest objects of the same type marked

The last attribute (the decisive attribute) is the suggested class of the object (adopting values from 0 to 3). This classification was generated with use of several different methods; obtaining de facto 5 independent test sets (the same attributes describing monuments but different class):

1. random method, i.e. random assignment of classes basing on the following assumed distribution among classes: class 0 - 3% objects, class 1 - 12% objects, class 2 - 35 % objects, class 3 - 50% objects;

2. "expert" method, i.e. assignment of classes basing on the knowledge of the authors of the experiment (inspired by expert publications), proportions similar to those in random method;

3. "multi attribute" method – determination of subsequent classes taking into account the maximum of one-two attributes at the same time, proportions as above;

4. method "oldest first" – this method consisted in arranging the objects according to the date of creation and assignment of classes basing on the assumption that the oldest ones are the most valuable; proportions between classes – as above;

5. method using unsupervised classification realised by a self-organising Kohonen neural network; in this approach the numbers of objects in individual classes are similar, but the objects may be distinguished according to class, although there is no hierarchy of classes.

5   Proposed Methodology

The authors attempted to develop a classification of multi-attribute model data both with use of all 10 attributes (5 descriptive and 5 geometrical ones) and through the determination of a representative sub-set of input variables. In order to determine which of the attributes would enable us to make a decision about the class of the given "monument" object, three methods were applied:

1. reduct method based on the theory of rough sets developed by Professor Zdzisław Pawlak [8,9] - the exhaustive algorithm was used,

2. method of decision rules [10,11],

3. artificial neural networks [12,13,14].

Methods mentioned above are rather simple ones as authors concentrate on inclusion of the spatial aspect of the data which is usually omitted. The goal of the research is to combine GIS solutions (and solve cartographic problem) with the methods which are not commonly known in the society of cartographers, geographers etc.

6   Obtained Results

6.1   Reducts

Regardless of the adopted test set (the manner of determination of monument class), the determination of reducts allowed us to reduce the number of attributes necessary for taking the decision. These were usually two-element reducts (already two attributes allow to take as good a decision about the class of the object as it is with use of all 10 attributes). Less frequently these were 3-element reducts, and in one case the reduct was composed of only one attribute.

Table 2. Examples of reducts for different data sets

Nr	Set	Reduct	Nr	Set	Reduct
1	Random	{Date, NS5}	6	Multi-attribute	{NS5, NSDT5}
2	Random	{NS1, NSDT5}	7	Date	{Date}
3	Expert	{Object type, NSDT5}	8	Date	{Architectural style, NS1}
4	Expert	{Material, NS1, NS5}	9	Kohonen networks	{Place, State, NS1}
5	Multi-attribute	{Architectural style, NSDT1}	10	Kohonen networks	{NS1, NS5}

The last case can be easily explained, as it appeared when method 4 was applied and it contained the date of creation of the object, which, in this case, was the only factor deciding about the class of the object. What is interesting is the fact that even in this set (where the only classification criterion was the date) also 7 other reducts were determined, of the length of 2 (examples are listed in Table 2). These included the architectural style of the object connected with one of the attributes related to neighbourhood (attributes 7, 8, 10) and various combinations of attributes related to neighbourhood. The presence of style is not surprising, as it is strongly correlated with the date of creation of the object, although it does not reflect the date unambigouty (during the creation of sets we adopted the assumption that individual styles overlap as far as the chronology of creation of objects is concerned).

However, the strong representation of attributes related to the geometrical neighbourhood of monuments among reducts seems important. They occur in form of independent reducts or in connection with descriptive attributes (date, style, material etc.) from the database of the NHB, in all reducts of sets 1, 2, 4 and in nearly all reducts in classes obtained with use of methods 3 and 5. This proves the high potential of spatial information hidden in the data, which is usually not explicitly disclosed in form of attributes assigned to objects.

6.2   Rules

For each of the analysed sets over 500 association rules deciding about the class of the object were obtained, although in a majority of the cases these rules had very low support. Sample rules obtained:

> If the object is located in a large city and it is an apartment house, its class is 3
> If the object is located in a large city and it is an apartment house, its class is 3
> If the object is from the 16th century, its class is 2

The lowest support was characteristic for set 1 in which the classes assigned randomly had no connection whatsoever with the attributes of objects (the highest support of the rule was only 5). The best adjusted rules were found in the set where the class of object was assigned as a result of activity of an unsupervised artificial neural Kohonen network (support for some of the "best" rules even exceeded 30). It is also worth noting that this was the only set where all 4 classes were results among rules with high support. Quite high support for several initial rules was also noted for the set classified according to the date of creation of objects. The expert set was characterised by lower support for rules, which proves the non-trivial nature of decisions taken in this way.

It is worth noting that the rules with the highest support included mainly those which allowed to assign the monument to classes 2 or 3 (i.e. to one of the two lowest classes). Furthermore, most of the cases (except set 5) the proportions of classes reflected the actual state, i.e. the fact that the most valuable monuments (class "0") are the least frequent. This is main reason why the support for rules among these "high" classes is quite low – not many objects of this type exist in the database. Another reason might be the complex process of taking the decision to assign the object to the most valuable group, based not only on attributes available in the database but also on the historical context of a given monument, its uniqueness, the reputation of the architect and other attributes known to experts and usually unavailable even in the most sophisticated databases.

6.3   Artificial Neural Networks

In the conducted experiment RBF networks and MLP type artificial neural networks (with one or two hidden layers) were used. Test sets were randomly divided: 70% objects were used as the teaching set, while 30% as the validation set for the determination of the correctness of results. Individual neural networks were taught in the suppervised mode, with use of the methods of reverse error propagation, Quasi-Newton, Levenberg-Marquardt and Delta-Bar. The analysis of the obtained results was conducted basing on the quality (correctness) of the classification, the value of error for the validation and teaching sets and the complexity of the network – the number of attributes taken into account and the number of neurones in hidden layers [15].

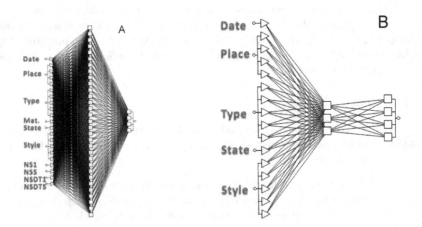

Fig. 2. Schemas of two of obtained neural networks, with the input attributes needed for decision: A — MLP type network for "random" classification (method1), B — MLP type network for "expert" classification (method 2)

The obtained results are presented in Table 2 and Figure 2. The complexity of the neural networks and the type of input attributes used are shown in Figure 2. It should be emphasised that in most of the applied classification methods we managed to successfully reduce the number of necessary functional attributes, at the same time maintaining a high accuracy of typology of objects. An exception is the random classification, where, in spite of application of 10 input attributes, the level of classification is very low (51%). For methods: attribute-based, according to date and the method using unsupervised Kohonen classification it was also necessary to use geometrical attributes (e.g. average distance between the object and 5 nearest neighbours of the same type), extracted from the spatial data base. Also the ratio of serious errors (assigning not neighboring class) was calculated (Tab. 3) which is (besides random classification) not higher than 3%. Obtained results give much better results than simple clustering methods like k-means or EM (Maximization Expectation) available in STATISTICA software. Also comparing with c-means method based on fuzzy clustering our results seem to be promising.

Table 3. Examples of reducts for different data sets

Classification method	Quality of classification	Serious errors	Number and type of attributes used
Random	51%	14%	10 – all functional attributes
Expert	91%	3%	Date, Place, Type, State, Style
Multi-attribute	94%	3%	Date, Place, Type, State, Style, NS5
According to date	81%	0%	Date, Place, Type, Style, NSDT1, NSDT5
Kohonen	100%	0%	Place, Type, Material, NSDT5

7   Conclusions

The conducted studies show that the attributes reflecting the spatial relations between specific objects play an extremely significant role. This is interesting, as the information related, for example, to the neighboring monument objects is usually not explicitly disclosed in databases. However, its important role in the decision on the class of objects proves, that such information should be added. It constitutes a specific example of knowledge "hidden" in the spatial data base (which contains information about the location of individual objects, but not about the relations between them).This type of knowledge is probably taken into account intuitively in the process of expert, subjective classification of objects. Disclosing such information in the database has another advantage, consisting in the fact that it can be done automatically – it does not require the work of an expert, only simple geometrical calculations based on the geometry of objects in the database. In this way, underestimated so far, the potential of spatial data can be used for decision-making purposes.

The experiments described above also prove the complexity of the issue of monument classification, or, more generally – the classification of multi-attribute objects. It is doubtlessly impossible to fully automatize this process, although the tests described herein seem to prove that it is possible to apply such methods as reducts, association rules or artificial neural networks to support the decision-making process. Classes of objects determined in this way can be treated as the first approximation of the final classification. Of course there is number of other data mining methods which might be implemented and compared. However, for "the highest" classes it is recommended to retain the key role of the expert, who can take the decision basing on a series of information (e.g. connected with the cultural or historical role of the monument) that cannot be fully expressed with use of the database structure.

In the analysed case obtained automatic classification may constitute a type of weight facilitating the selection of objects in the case of excessive density on the electronic map. However, it may not be recommended to provide explicit information of a weight calculated in such way, due to reasons listed in the introduction. Generalising the conducted reasoning, it is possible to show potential applications of this methodology for the purposes of classification of spatial objects characterised by numerous complex descriptive and/or spatial attributes

References

1. Longley, P.A., Goodchild, M.F., Maguire, D.J., Rhind, D.W.: GIS, Geographic Information Systems and Science, 3rd edn. Wiley (2011)
2. Robinson, A.H.: Elements of Cartography. John Wiley&Sons (1995)
3. Olszewski, R.: Kartograficzne modelowanie rzeźby terenu metodami inteligencji obliczeniowej. Prace Naukowe - Geodezja, z. 46. Oficyna Wydawnicza Politechniki Warszawskiej, Warszawa (2009)
4. Weibel, R.: Amplified intelligence and rule-base systems. In: Buttenfield, B., McMaster, R. (eds.) Map generalization: Making rules for knowledge representation, Red, Longman, London (1991)
5. Makowski, A.: Cartography versus spatial information systems. Geodezja i Kartografia, t, XLVI, z. 3,Warszawa (1997)
6. Mackaness, W.: Understanding geographic space. In: Mackaness, W., Ruas, A., Sarjakoski, T. (eds.) Generalisation of Geographic Information: Cartographic Modelling and Application. Elsevier (2007)
7. Gotlib D., Olszewski R., Rola bazy danych obiektów topograficznych w tworzeniu infrastruktury informacji przestrzennej w Polsce. Główny Urząd Geodezji i Kartografii, Warszawa (2013)
8. Pawlak, Z.: Rough sets. International Journal of Parallel Programming 11(5), 341–356 (1982)
9. Pawlak, Z.: Rough Sets: Theoretical Aspects of Reasoning About Data. Kluwer Academic Publishing, Dordrecht (1991)
10. McGraw, K.L.: Harbison-Briggs, Knowledge Acquisition: Principlesand Guidelines. Prentice Hall, Englewood Cliffs (1989)
11. Miller, H.J., Han, J.: Geographic data mining and knowledge discovery. Taylor&Francis, London (2001)
12. Fausett, L.: Fundamentals of Neural Networks. Prentice-Hall, New York (1994)
13. Patterson, D.: Artificial Neural Networks. Prentice-Hall, Singapore (1996)
14. Winston, P.: Artificial intelligence, 3rd edn. Addison Wesley (1992)
15. Olszewski, R.: Utilisation of artificial intelligence methods and neurofuzzy algorithms in the process of digital terrain model generalisation. In: ICA Conference, La Coruna (2005)

Self-organizing Artificial Neural Networks into Hydrographic Big Data Reduction Process

Andrzej Stateczny[1] and Marta Wlodarczyk-Sielicka[2]

[1] Marine Technology Ltd., Szczecin, Poland
a.stateczny@marinetechnology.pl
[2] Maritime University, Szczecin, Poland
m.wlodarczyk@am.szczecin.pl

Abstract. The article presents the reduction problems of hydrographic big data for the needs of gathering sound information for Navigation Electronic Chart (ENC) production. For the article purposes, data was used from an interferometric sonar, which is a modification of a multi-beam sonar. Data reduction is a procedure meant to reduce the size of the data set, in order to make them easier and more effective for the purposes of the analysis. The authors` aim is to examine whether artificial neural networks can be used for clustering data in the resultant algorithm. Proposed solution based on Kohonen network is tested and described. Experimental results of investigation of optimal network configuration are presented.

Keywords: Kohonen network, big data, hydrography.

1 Introduction

The main component that contributes significantly to the safety of navigation is the information on the depth of the area. A navigational chart is the primary source of information for the navigator. The information on depth is the one of the most important layer of Electronic Navigational Chart (ENC) according to International Hydrographic Organization (IHO) standards. Although in Inland ENC there is no obligation to provide bathymetric data but it is often provided by electronic chart producers. Some aspects of electronic charts for inland and maritime navigation, chart production planning and navigational data evaluation were presented in [1,2,3,4,5].

Usually bathymetric data are gathered by multibeam echosounder (MBES), which uses acoustic waves is a device for bathymetric measurements and it measures the vertical distance between the head and the bottom or an object located at the bottom. For the purposes of the article, data was used from an interferometric sonar, which is a modification of a multi-beam echo sounder. During survey process large amount of data was collected. Data reduction is a procedure meant to reduce the size of the data set, in order to make them easier and more effective for the purposes of the analysis. There are several ways to perform data reduction. One of them is to transform a large quantity of variables into a single, common value. Another way to reduce data is to

M. Kryszkiewicz et al. (Eds.): RSEISP 2014, LNAI 8537, pp. 335–342, 2014.
© Springer International Publishing Switzerland 2014

use advanced statistical methods, that will make it possible to decrease the size of a data pack by breaking it down into basic factors, dimensions or concentrations, pinpointing the basic relations between the analyzed instances and variables. Another method is to deduct given number of instances from a large array, while maintaining its overall suitability for the analyzed population. The problem of bathymetric big data processing has been taken up by several authors [6,7,8,9,10,11,12].

Modern idea to solve computational problems with large amount of data processing is to use artificial intelligence methods, especially artificial neural networks (ANN). Some aspects of ANN solutions for navigation purpose are described in [13,14,15,16,17]. Very interesting idea is to process big data for compression or reduction purpose by using self-organized ANN. Especially useful to data reduction are self-organized Kohonen network [18,19,20,21,22,23].

Kohonen network idea was described in many articles and books [24,25]. Kohonen networks are an example of ANN taught through the self-organization. The idea of their functioning consists on self-adapting state of the network to present entrance data. In the process of the network's learning there is no association between input signals and output of the network. The learning consists in the specific concept of adapting of networks passed to images for their input. In case of Kohonen networks a competition between neurons is a philosophy for the value of its weights updating. During the competition between neurons the neuron closest to the input sources in the meaning of chosen distance method calculation is a winner for input data set. Euclidean distance is usually used. The degree of adjustment depends on the distance of the neuron from the input data.

In the article Kohonen networks are examined and offered as a step in the process of reduction of hydrographic big data.

2   Hydrographic Big Data Reduction Problem in Electronic Navigation Chart Production Process

One of the issues connected to bathymetric measurements is registering an large amount of the data, as well as various types of interference. The echo sounder is a device used to measure the vertical distance between the head of the equipment and the sea bottom, or an object located on the sea bed, using a acoustic wave. The establishment the depth of water is achieved by measuring the time in which the acoustic wave needs to reach the object, as well as to return to the receiving transducer as a reflected wave. The angle measurement is carried out in various manners, depending on the type of the echo sounder. The simplest example is the single beam sonar, which operates by sending a narrow acoustic signal beam vertically downwards. The MBES emits several signal beams in multiple directions, monitoring in all these directions. This solution allows for a much wider area of measurement, when compared to a single beam sonar, by increasing the width of the scanning zone. For collecting bathymetric data authors used the interferometric sonar system GeoSwath Plus 250kHz, which is able to gather high density data. It allows to simultaneously collect vertical data (like a standard multi-beam sonar), as well as horizontal data (like a side scan sonar). The depth data is received not only based on the measurement of time in which the acoustic wave reflected off the object returns to the receiving transducer, but also based on measuring the difference between phases of the wave reaching the piezoelectric sensors installed within the head. Bathymetric data processing is realized in several stages. First off, all values of corrections influencing the accuracy of the measurements which are taken into consideration, such as: water properties, head submersion, errors in the average speed of sound in water and erroneous offset inputs in the measurement devices. Subsequently, the system operator performs the initial rough filtration, using predefined data processing filters. The next step is to process data. The first stage of processing is to convert the raw data into 'swath' files. There are four basic types of filters used to process the data: the amplitude filter, the limit filter, the across track filter and the along track filter. If properly applied, the combination of these filters guarantees the optimal size and content of the files [26]. The authors` main purpose is to create a data reduction algorithm in aspects of production electronic navigational chart. The whole process is shown in the figure below.

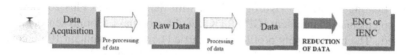

Fig. 1. Processing of bathymetric data

It is worth noting if the filtered samples are big amount of data sets. Data reduction is a procedure meant to reduce the size of the data set. Generally the hydrographic systems generate the grid with use the following methods: mean (select a mean depth value) or weighted mean (uses amplitude values to give higher weighting to data points which are higher in amplitude when calculating the mean depth value).

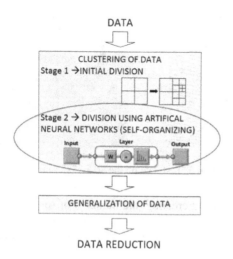

Fig. 2. Reduction algorithm of bathymetric data

The main objective of authors is compilation of reduction algorithm. The clustering of data is first part of search algorithm, as shown in figure above. The next step will be generalization of bathymetric data. The aim of article is to examine whether artificial neural networks can be used for clustering data in the resultant algorithm.

3   Experimental Results of Kohonen Network for Hydrographic Data Reduction

For researches Matlab software with the Neural Network toolbox was used. For the purpose of clustering problems, the self-organizing feature map (SOFM) was applied. SOFM has one layer with neurons organized in a grid. Self-organizing maps learn both the distribution and topology of the input vectors they are trained on. During the training, network applies the rule WTM (Winner Take Most) according to which the weight vector associated with each neuron moves to become the centre of a cluster of input vectors. In addition, neurons that are adjacent to each other in the topology are also moved close to each other in the input space.

The floating laboratory Hydrograf XXI, with the GeoSwath Plus 250 kHz sonar and supplementary equipment (GPS/RTK, satellite compass, motion sensor) installed was used . The measurement profiles were carried out due to maintain 100% coverage of the measured body of water. Test data was collected within the Szczecin Harbour, near the Babina Canal, on May 23, 2010.

Test area included very high density data , which is the main limitation of neural network usage. It is impossible to train the network from whole data set using standard computer. In order to solve this problem, authors divide original data points set into smaller subsets, which could be trained separately. Training data set is the square measuring 25 to 25 metres and it includes 28911 samples of 3 elements (X,Y,Z), as shown on figure 3.

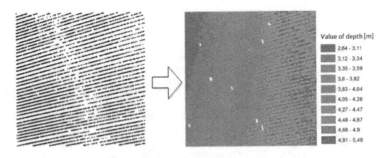

Fig. 3. Test data on the area 25x25 metres

While network was being creating, the numbers of rows and columns were specified in the grid. During tests, these values were set to 3x3, 7x7 and 10x10 which provided the number of neurons 9, 49 and 100, respectively. Authors decided to select the hexagonal topology of network, where each of the hexagons represents a neuron. Distances between neurons are calculated from their positions with a distance function. The most common link distances were selected in those researches: the number of links, or steps, that must be taken to get to the neuron under consideration. Initial neighbourhood size was set at 3 and ordered phase steps was set at 100. The training was running for the selected number of epochs: 10, 50, 100, 200, 500 and 1000.

There are many visualization tools that can be used to analyse the resulting cluster. One visualization for SOFM is the weight distance matrix presented on figure 4.

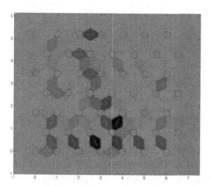

Fig. 4. Weight distance matrix for 49 neurons and 200 epochs

The grey hexagons illustrate the neurons and the red lines connect the neighbouring. The colours in the regions contain the red lines designate the distance between the received neurons: the lighter colour indicates smaller distance and darker colours indicates bigger distance.

Next type of presentation of the results is illustrating the locations of neurons in the topology and it indicates numbers of the bathymetric data associated with each of the cluster centres (Fig.5.).

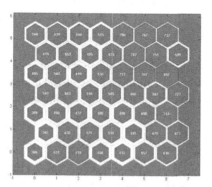

Fig. 5. Sample hits for 49 neurons and 200 epochs

Another useful visualization of the results is to show the locations of the data points and the weight vectors (Fig.6.).

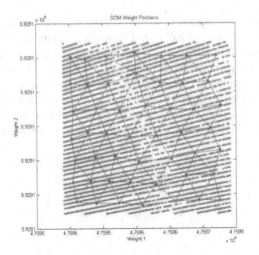

Fig. 6. Weight positions for 49 neurons and 200 epochs

It was found, that after only 200 epochs the map is well distributed through the input space. Results received at these settings are sufficient for clustering bathymetry data. The authors test sample data with the following settings: 49 neurons and 200 epochs one hundred times.

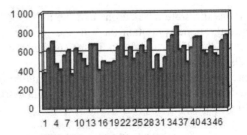

Fig. 7. Frequency distribution

Frequency distribution of mean outcomes is presented on figure above The horizontal axis represents number of neurons and the vertical axis shows bathymetric data points. The examples in a set of data are evenly distributed in each cluster.

4   Conclusion

Data reduction is a procedure meant to reduce the size of the data set. Generally the hydrographic systems generate the grid with use the following methods: mean or weighted mean. The main purpose of authors is elaboration of reduction algorithm of bathymetric data. The clustering of data is first part of search algorithm, as shown in figure above. The next step will be generalization of bathymetric data. The use of 200 epochs during training have the result in decomposition of the neurons – they are very evenly spaced. Number of neurons will be selected depend on the compilation scale and geographical coverage of electronic navigational chart. The test examples in a set of data are regularly distributed in each cluster. The main limitation of neural network usage is high density data. It is long-term to train the network from whole data set using standard computer. The results point that artificial neural networks are good methods of clustering large amount of data. Application of SOFM in Matlab involves grouping data by similarity and it can be used as one of the steps in reduction algorithm of bathymetric data.

References

1. Lubczonek, J.: Application of GIS Techniques in VTS Radar Stations Planning. In: Kawalec, A., Kaniewski, P. (eds.) 2008 International Radar Symposium, Wroclaw, pp. 277–280 (2008)
2. Weintrit, A., Kopacz, P.: Computational Algorithms Implemented in Marine Navigation Electronic Systems. In: Mikulski, J. (ed.) TST 2012. CCIS, vol. 329, pp. 148–158. Springer, Heidelberg (2012)
3. Lubczonek, J., Stateczny, A.: Aspects of spatial planning of radar sensor network for inland waterways surveillance. In: 6th European Radar Conference (EURAD 2009). European Radar Conference-EuRAD, Rome, pp. 501–504 (2009)
4. Przyborski, M., Pyrchla, J.: Reliability of the navigational data. In: Kłopotek, M.A., Wierzchoń, S.T., Trojanowski, K. (eds.) Intelligent Information Processing and Web Mining. ASC, vol. 22, pp. 541–545. Springer, Heidelberg (2003)
5. Stateczny, A.: Artificial neural networks for comparative navigation. In: Rutkowski, L., Siekmann, J.H., Tadeusiewicz, R., Zadeh, L.A. (eds.) ICAISC 2004. LNCS (LNAI), vol. 3070, pp. 1187–1192. Springer, Heidelberg (2004)
6. Przyborski, M.: Possible determinism and the real world data. Physica A-Statistical Mechanics and its Applications 309(3-4), 297–303 (2002)
7. Stateczny, A.: Methods of comparative plotting of the ship's position. In: Brebbia, C., Sciutto, G. (eds.) Maritime Engineering & Ports III, Rhodes. Water Studies Series, vol. 12, pp. 61–68 (2002)
8. Maleika, W., Palczynski, M., Frejlichowski, D.: Effect of Density of Measurement Points Collected from a Multibeam Echosounder on the Accuracy of a Digital Terrain Model. In: Pan, J.-S., Chen, S.-M., Nguyen, N.T. (eds.) ACIIDS 2012, Part III. LNCS (LNAI), vol. 7198, pp. 456–465. Springer, Heidelberg (2012)
9. Lubczonek, J., Stateczny, A.: Concept of neural model of the sea bottom surface. In: Rutkowski, L., Kacprzyk, J. (eds.) Neural Networks and Soft Computing. ASC, vol. 19, pp. 861–866. Physica, Heidelberg (2003)
10. Lubczonek, J.: Hybrid neural model of the sea bottom surface. In: Rutkowski, L., Siekmann, J.H., Tadeusiewicz, R., Zadeh, L.A. (eds.) ICAISC 2004. LNCS (LNAI), vol. 3070, pp. 1154–1160. Springer, Heidelberg (2004)
11. Stateczny, A.: The neural method of sea bottom shape modelling for the spatial maritime information system. In: Brebbia, C., Olivella, J. (eds.) Maritime Engineering and Ports II, Barcelona. Water Studies Series, vol. 9, pp. 251–259 (2000)
12. Maleika, W.: The influence of track configuration and multibeam echosounder parameters on the accuracy of seabed DTMs obtained in shallow water. Earth Science Informatics 6(2), 47–69 (2013)

13. Stateczny, A., Kazimierski, W.: Determining Manoeuvre Detection Threshold of GRNN Filter in the Process of Tracking in Marine Navigational Radars. In: Kawalec, A., Kaniewski, P. (eds.) 2008 Proceedings International Radar Symposium, Wroclaw, pp. 242–245 (2008)

14. Balicki, J., Kitowski, Z., Stateczny, A.: Extended Hopfield Model of Neural Networks for Combinatorial Multiobjective Optimization Problems. In: 2nd IEEE World Congress on Computational Intelligence, Anchorage, pp. 1646–1651 (1998)

15. Stateczny, A., Kazimierski, W.: A comparison of the target tracking in marine navigational radars by means of GRNN filter and numerical filter. In: 2008 IEEE Radar Conference, Rome, vol. 1-4, pp. 1994–1997 (2008)

16. Stateczny, A., Kazimierski, W.: Selection of GRNN network parameters for the needs of state vector estimation of manoeuvring target in ARPA devices. In: Romaniuk, R.S. (ed.) Photonics Applications in Astronomy, Communications, Industry, and High-Energy Physics Experiments IV, Wilga. Proceedings of the Society of Photo-Optical Instrumentation Engineers (SPIE), vol. 6159, pp. F1591–F1591(2006)

17. Stateczny, A.: Neural manoeuvre detection of the tracked target in ARPA systems. In: Katebi, R. (ed.) Control Applications in Marine Systems 2001 (CAMS 2001), Glasgow. IFAC Proceedings Series, pp. 209–214 (2002)

18. Chung, K., Huang, Y., Wang, J., et al.: Speedup of color palette indexing in self-organization of Kohonen feature map. Expert Systems with Applications 39(3), 2427–2432 (2012)

19. Ciampi, A., Lechevallier, Y.: Multi-level Data Sets: An Approach Based on Kohonen Self Organizing Maps. In: Zighed, D.A., Komorowski, J., Żytkow, J. (eds.) PKDD 2000. LNCS (LNAI), vol. 1910, pp. 353–358. Springer, Heidelberg (2000)

20. de Almeida, C., de Souza, R., Candelas, A.: Fuzzy Kohonen clustering networks for interval data. Neurocomputing 99, 65–75 (2013)

21. Du, Z., Yang, Y., Sun, Y., et al.: Map matching Using De-Noise Interpolation Kohonen Self-Organizing Maps. In: Conference: International Conference on Components, Packaging and Manufacturing Technology, Sanya. Key Engineering Materials, vol. 460-461, pp. 680–686 (2011)

22. Guerrero, V., Anegon, F.: Reduction of the dimension of a document space using the fuzzified output of a Kohonen network. Journal of the American Society for Information Science and Technology 52(14), 1234–1241 (2001)

23. Rasti, J., Monadjemi, A., Vafaei, A.: Color reduction using a multi-stage Kohonen Self-Organizing Map with redundant features. Expert Systems with Applications 38(10), 13188–13197 (2011)

24. Kohonen, T.: Self-Organized Formation of Topologically Correct Feature Maps. Biological Cybernetics 43(1), 59–69 (1982)

25. Kohonen, T.: The Self-Organizing Map. Proceedings of The IEEE 78(9), 1464–1480 (1990)

26. Wlodarczyk-Sielicka, M.: 3D Double Buffering method in the process of hydrographic chart production with geodata taken from interferometry multibeam echo sounder. Annals of Geomantic, issue X 7(57), 101–108 (2012)

Managing Depth Information Uncertainty in Inland Mobile Navigation Systems

Natalia Wawrzyniak[1] and Tomasz Hyla[2]

[1] Institute of Geoinformatics, Maritime University of Szczecin, Poland
n.wawrzyniak@am.szczecin.pl
[2] Marine Technology Ltd., Szczecin, Poland
t.hyla@marinetechnology.pl

Abstract. Rough sets theory allows to model uncertainty in decision support systems. Electronic Charts Display and Information Systems are based on spatial data and together with build-in analysis tools pose primary aid in navigation. Mobile applications for inland waters use the same spatial information in form of Electronic Nautical Charts. In this paper we present a new approach for designation of a safety depth contour in inland mobile navigation. In place of manual setting of a safety depth value for the need of navigation-aid algorithm, an automatic solution is proposed. The solution is based on spatial characteristics and values derived from bathymetric data and system itself. Rough sets theory is used to reduce number of conditional attributes and to build rule matrix for decision-support algorithm.

Keywords: depth uncertainty, spatial information, rough sets, inland navigation, electronic charts, decision support.

1   Introduction

Spatial information and positioning are two most important features in any navigation support system. Existing systems for inland navigation are based on the achievements of marine applications, which started being developed many years earlier. These systems are strictly dependent on existing legislation from International Maritime Organization and their standards [1]. Main spatial information systems used on waters are Electronic Charts Display Systems, which exists also in their adapted form for Inland waters [2]. Functionality of such systems provides necessary analysis, but automation in decision making is vastly reduced due to its specifics – marine navigation concerns mainly commercial vessels of significant material value. Standardization of such aid is complicated because it needs cooperation and consent of ship-owners and insurance companies. Nevertheless, there are many studies on support and automation for navigation using artificial intelligence methods [3,4]. Inland navigation has its own characteristic and therefore requires different approach [5]. The environment is quickly alternating, vessels are smaller, the area more confined and the participation of recreational units in the traffic much greater. On one hand the spatial data used to produce navigational charts is more accurate and the precise positioning more accessible.

M. Kryszkiewicz et al. (Eds.): RSEISP 2014, LNAI 8537, pp. 343–350, 2014.

But on the other - inland charts in many areas are rarely updated, not all of the area is covered and the information about water levels and navigational situation depend on local information services.

MOBINAV system is a mobile application dedicated for inland recreational units which is being developed currently by Marine Technology. It uses collections of spatial data in form of inland ENCs and navigational data provided by whatever devices and means possible on board. It allows loading and using any charts available for required area in standardized format. Bathymetric data in navigational charts used by the system can exists in up to three vector layers representing depth area, depth contours and soundings [6]. Positioning can be obtained by GPS receiver of any kind - either build-in mobile device or externally used on board. Any additional, non-spatial information, that aid navigation can be acquired from accessible online services, e.g., River Information System [7], or entered manually.

In existing ECDIS systems to help avoid navigational danger a navigator analyses provided spatial and other situational information and defines a safety depth – a minimum depth value the ship cannot cross and with which any navigational system highlight or dim areas shallower or deeper than this value [8]. Bathymetric data can be the most complex information in IENC's, depending on its resolution and representation. The IENC standard does not dictate the density of depth contours nor quality of data used for chart generation itself [2]. Also many attributes that help defining the quality and reliability of data are only optional according to this standard [9] and in practice are rarely filled out. That means that even a user with advanced knowledge about depth data characteristic has problems to assess the data precision and reliability.

In this paper we propose an automatic analysis based on available spatial and non-spatial data in place of manual definition of safety depth to define areas that are non-navigable. Taking into consideration high uncertainty of the information our approach is based on rough sets theory in creating decision rules.

The paper is organized as follows. Section 2 contains description of a problems related to processing of special data containing depth information and issues related to navigation. Section 3 describes our new methodology that enables to assess the quality of safety depth contour using rough sets. The paper ends with conclusions.

2   Background

The problem of uncertainty of spatial data in information systems is present in research papers for over 30 years [10,11]. The degree of generalization of real entities adopted at the stage of obtaining the spatial data makes it burdened with sense of uncertainty. It results from lack of assurance of acquired knowledge, limitation of measurement precision, and also doubt in their further analysis outcome [12]. The uncertainty associated with this generalization is an inseparable part of geoinformation systems. The actual complexity of the world makes its description in a digital data representation currently not achievable in freely demanded accuracy. The recipient of such data is therefore accompanied by a feeling of insecurity.

The importance of uncertainty in bathymetric data lays even deeper as its main purpose is to ensure safety in navigation. The problem has many layers and can be identify on many levels in the processing flow of bathymetric data (Fig.1) – from acquisition to the final form of presentation in electronic charts [13,14,15].

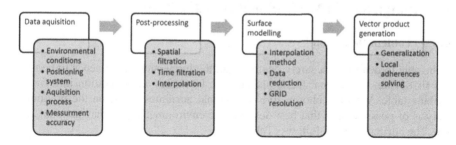

Fig. 1. Main grounds for uncertainty in bathymetric data processing flow

The acquisition of data takes place in highly changeable environment with the use of complex system contained of many devices incl. echosounders, sound velocity sensor, positioning system, motion reference unit etc. Each of the devices brings possible measurement errors. The technical characteristic of echosounder ensures particular density of measurement when the data coverage is also depended on survey profile planning. The raw data then needs to be post-processed to remove noises and potential navigational errors which include time and space oriented filtering. It is done semi-automatically with supervision of authorized hydrograph. Cleared data is then transformed in to DTM model using one of many existing interpolation algorithms [16,17,18]. The size of a grid cell and used gridding method has enormous influence on size and precision of created model. Then the final vector products are produced based on this model: depth areas, contours and soundings. Their density does not depend on the acquired bathymetric nor the standard but on a producer of final charts.

But for an inland navigator as a receiver of a final product (a chart), other factors besides spatial data itself are as much of importance. How up-to-date is given information and is it fully usable considering other navigational data? As mentioned earlier, in ECDIS systems all of this information needs to be pre-processed by the navigator himself in order to define safety depth contour. The system then will indicate the next shallower (or next deeper [1]) contour existing in the chart and will generate an alarm if the ship, within a specified time set by the mariner, is going to cross the safety contour.

In MOBINAV application we propose to automatically appoint a safety depth based on set of attributes and immersion value of the vessel. The strength of alarm and the buffer size presented on the chart could depend on the strength of the activated rule in decision rules table. The rough sets theory was adapted to model such decision support system.

Rough sets theory [19] is used to examine imprecision, vagueness and uncertainty in data analysis. Information in system based on this method is stored in tabular form. For different combinations of values of conditional arguments - different or same values of decision arguments appear. Information table becomes inconsistent and some uncertain rules appear. The advantage of such system is processing of such rules. Also using rough sets allows simplifying the rules by reducing some correlated attributes. In consequences the decision systems takes on characteristics of generalization [9].

3   Rough Sets in Navigational Decision Making

3.1   Problem Definition

To be able to determine a safety depth value, the immersion value needs to be specified first. Then all external factors need to be taken into consideration in a form of attribute table. We have chosen five conditional attributes based on the preliminary analysis of possible ones that best describe the environmental and navigational situation. The attributes are as follows (Tab. 1):

1. Region type (divided into types based on maintainability level);
2. Geographical position accuracy, i.e., accuracy of vessels GPS device;
3. Age of data, i.e., when the chart data was updated;
4. Water level information, i.e., if water level correction data was available for navigation software;
5. Density of plotted isobaths on a chart, i.e., if more contours are available for given region, then safety contour depth can be more exactly calculated.

All the continuous values of attributes had to be discretized and coded into form represented by linguistic value (V-very, A-average, N-not/poor). The decision attribute is information about danger, i.e., if the safety depth contour can be trusted or not (Y-yes, N-no).

Table 1. Conditional and decisional attributes

Region (A₁)	Coded form	GPS Position accuracy (A₂)	Coded form
Port (very good maintenance)	V	Below 1m	V
Waterway (good maintenance)	A	Between 1m and 5m	A
Other (poor maintenance)	N	Above 5m	N
Age of data (A₃)	**Coded form**	**Water level information (A₄)**	**Coded form**
Old (more than 2 years)	V	Lack of information (average	N
Average (between 1 and 2 years)	A	water level is assumed)	
Up-to-date(below 1 year)		Information available	
	N		V
Density of isobaths (A₅)	**Coded form**	**Danger** *(decisional attribute)*	**Coded form**
High	V	Depth below safety depth con-	Y
Average	A	tour	
Poor	N	Depth equal or above safety depth contour	N

3.2   Experimental Data Source

The information system contains 61 objects that were collected during last year summer months at inland waters near Szczecin, i.e., Dabie Lake, Szczecin Cargo Port and parts of the Odra River. We had different types of IENC cells and GPS devices to our disposal and fully professional high-density bathymetric system (Geoswath +) to acquire needed experimental data. The echosounder was used to determine if actual depth was below the safety depth contour or if it was greater or equal. The actual value being lower than the designated value by the navigator was considered as navigational danger. Tab. 2 contains a few exemplary objects.

Table 2. Extract from table with information system objects

Object No.	Conditional attributes $A = \{A_1, A_2, A_3, A_4, A_5\}$	Decisional attribute
1	V V V V V	Y
2	V V V V A	Y
..	..	..
60	N A A N N	N
61	A V V V A	Y

3.3   Analysis

The coded information table was analysed using our proprietary software for rough sets analysis. Let $T = (U, A)$ will be information system (information table) and let $B \subseteq A$ and $X \subseteq U$. The 61 objects can be divided into 26 equivalence classes. The cardinality of the information table $|U| = 61$ and cardinality of positive region $|\underline{B} X| = 54$. Accuracy of the lower approximation is equal:

$$\gamma_B(X) = \frac{|\underline{B}X|}{|U|} = \frac{54}{61} \approx 0.89 \tag{1}$$

Certain rules can be calculated using 89% of objects. The accuracy of the rough-set representation is as follows:

$$\alpha_B(X) = \frac{|\underline{B}X|}{|\overline{B}X|} = \frac{54}{68} \approx 0.79 \tag{2}$$

All conditional attributes where analysed to check if some attributes are superfluous. The Tab. 3 contains parameters related to reduced spaces.

Attribute sets $A'^1 = \{A_2, A_3, A_4, A_5\}$, $A'^2 = \{A_1, A_3, A_4, A_5\}$, $A'^3 = \{A_1, A_2, A_4, A_5\}$ and $A'^5 = \{A_1, A_2, A_3, A_4\}$ are not absolute reducts of set $A = \{A_1, A_2, A_3, A_4, A_5\}$. This is due to the fact that equivalence classes differ. Relative significance of attributes $A1$, $A2$, $A3$, $A5$ is greater than 0, so they cannot be removed from set A. However, relative significance of attribute A_4 equals 0 and equivalence classes of A and A'^4 are the same. In this way the attribute concerning water level information has been excluded.

Table 3. Parameters of reduced spaces

Parameter \ Attribute	A₁	A₂	A₃	A₄	A₅
Cardinality of positive region when *i* attribute is removed	49	44	44	54	50
γ_i - the accuracy of the lower approximation when *i* attribute is removed	0,80	0,72	0,72	0,89	0,82
δ_i - significance of the attribute *i:* $$\delta_i = 1 - \frac{\gamma_i}{\gamma}$$	0,093	0,185	0,185	0	0,074

The reduced information table has accuracy of the lower approximation that equals 0.89 and the accuracy of the rough-set representation is equal to 0.79.

In the next step the rules were calculated. From complete set of 29 rules 24 is useful and 5 are not. Tab. 4 presents useful rules. These rules can be further simplified, so they can be stored in more human readable form.

Table 4. Useful decision rules

Rule No.	Conditional attributes {A₁,A₂,A₃,A₅}	Decisional attributes	Quantity	Strength	Certainty
1	V V V V	Y	5	0.08	1.00
2	V V V A	Y	5	0.08	1.00
3	V A V A	Y	7	0.11	1.00
4	V A A A	N	2	0.03	1.00
5	V N V N	N	1	0.02	1.00
7	V N A N	N	2	0.03	0.67
8	A V V N	N	1	0.02	1.00
9	A V V A	Y	3	0.05	1.00
10	A A V A	Y	2	0.03	1.00
11	A A V V	Y	1	0.02	1.00
12	A A A V	Y	2	0.03	1.00
13	A N A V	N	2	0.03	1.00
14	A N A A	N	2	0.03	1.00
15	A N A N	N	1	0.02	1.00
16	A V A N	N	1	0.02	1.00
19	N A A N	N	4	0.07	1.00
20	N A N N	N	1	0.02	1.00
21	N A N A	N	3	0.05	1.00
22	N N N A	N	2	0.03	1.00
23	N N A V	N	2	0.03	1.00
24	N N V V	N	2	0.03	1.00
25	V V V N	Y	3	0.05	1.00
26	A A N A	N	1	0.02	1.00
27	N N A A	N	1	0.02	1.00

The rules of insufficient certainty (Tab. 5) were eliminated. They presented experimental cases conducted in areas with poor maintenance (e.g., lakes and river outside of cargo and commercial traffic). It shows the importance of this characteristic in estimation of navigational risks.

Table 5. Useful decision rules

6	N V V V	Y	1	0.02	0.5
17	N V V V	N	1	0.02	0.5
18	N A V V	Y	1	0.05	0.5
28	N A V V	N	1	0.05	0.5
29	V N A N	Y	1	0.03	0.33

4   Conclusions

The main purpose of our research was to find a way to determine the level of assurance of safety depth contour to enable more precise navigation. On inland waters spatial data containing depth information is one of the main factors that are used by a navigator (or automatically by navigation software) to calculate a route. Thus the proper calculation of safety depth contour is important, especially when navigating on shallow waters. However, when spatial data containing depth information is old or insufficient, the safety depth contour might not reflect a real world. In practice, a navigator always tries to remain in certain distance from a safety depth contour or he sets this value with some margin. We have used rough sets methodology which resulted in creating decision table (Tab. 4) to show on chart if displayed safety depth contour can be trusted. It means that a navigator can securely approach the isobath indicated by safety depth contour.

The decision rules were obtained using the knowledge from the inland waters in Szczecin area. Hence, they can be used for navigating specifically these waters or the waters with similar properties. In practice, the inland environment differs among regions, e.g., for fast flowing rivers which often floods, this decision rules will probably not work. This is main drawback of this technique as it requires to gather data and to process it into decision rules for every region where it might be used.

Future work will be carried out simultaneously in two directions. In the first one we will focus on finding best way to visualise information obtained from the decision table on-top of spatial data. The second direction will concern a methodology which will allow creating a rule set for any region of inland waters.

Acknowledgment. This scientific research work is supported by NCBiR of Poland (grant No LIDER/039/693/L-4/12/NCBR/2013) in 2013-2016.

References

1. IHO, S-52 standard, Specifications for chart content and display aspects of ECDIS
2. IMO, Resolution MSC.232(82), Adoption of the revised performance standards for electronic chart diplay and information systems, ECDIS (2006)

3. Stateczny, A., Kazimierski, W.: A comparison of the target tracking in marine navigational radars by means of GRNN filter and numerical filter. In: 2008 IEEE Radar Conference, Rome, vol. 1-4, pp. 1994–1997 (2008)
4. Stateczny, A., Kazimierski, W.: Determining Manoeuvre Detection Threshold of GRNN Filter in the Process of Tracking in Marine Navigational Radars. In: Kawalec, A., Kaniewski, P. (eds.) 2008 Proceedings IRS, Wrocław, pp. 242–245 (2008)
5. Kazimierski, W., Wawrzyniak, N.: Modification of ECDIS interface for the purposes of geoinformatic system for port security. Annuals of Navigation, 51–70 (2013)
6. IHO, S57 standard, Transfer Standard for Digital Hydrographic Data (2000)
7. Stateczny, A.: Rzeczny system informacyjny dla dolnej Odry. Roczniki Geomatyki (Annals of Geomatics) 5, 49–54 (2007)
8. Vetter, L., Jonas, M., Schroeder, W., Pesh, R.: Marine Geographic Information Systems. In: Kresse, W., Danko, D. (eds.) Handbook of Geographic Information, pp. 761–776. Springer (2012)
9. Slowinski, R., Greco, S., Matarazzo, B.: Rough Sets in Decision Making. In: Mayers, R. (ed.) Encyclopedia of Complexity and Systems Science, pp. 7753–7787. Springer (2009)
10. Devillers, R., Stein, A., Bédard, Y., Chrisman, N., Fisher, P., Shi, W.: Thirty years of research on spatial data quality: achievements, failures, and opportunities. Transactions in GIS 14(4), 387–400 (2010)
11. Wenzhong, S., Fisher, P., Goodchild, M.: Spatial data quality. CRC Press (2003)
12. Felcenloben, D.: Uncertainety of spatial data in GIS systems. Acta Sci. Pol. Geod. Descr. Terr. 9(3), 3–12 (2010)
13. Przyborski, M., Pyrchla, J.: Reliability of the navigational data. In: Kłopotek, M.A., Wierzchoń, S.T., Trojanowski, K. (eds.) IIS: IIPWM 03. AISC, vol. 22, pp. 541–545. Springer, Heidelberg (2003)
14. Maleika, W., Palczynski, M., Frejlichowski, D.: Effect of Density of Measurement Points Collected from a Multibeam Echosounder on the Accuracy of a Digital Terrain Model. In: Pan, J.-S., Chen, S.-M., Nguyen, N.T. (eds.) ACIIDS 2012, Part III. LNCS, vol. 7198, pp. 456–465. Springer, Heidelberg (2012)
15. Maleika, W., Palczynski, M., Frejlichowski, D.: Interpolation Methods and the Accuracy of Bathymetric Seabed Models Based on Multibeam Echosounder Data. In: Pan, J.-S., Chen, S.-M., Nguyen, N.T. (eds.) ACIIDS 2012, Part III. LNCS (LNAI), vol. 7198, pp. 466–475. Springer, Heidelberg (2012)
16. Lubczonek, J.: Hybrid neural model of the sea bottom surface. In: Rutkowski, L., Siekmann, J.H., Tadeusiewicz, R., Zadeh, L.A. (eds.) ICAISC 2004. LNCS (LNAI), vol. 3070, pp. 1154–1160. Springer, Heidelberg (2004)
17. Stateczny, A.: The neural method of sea bottom shape modelling for the spatial maritime information system. In: Brebbia, C., Olivella, J. (eds.) Maritime Engineering and Ports II, Barcelona. Water Studies Series, vol. 9, pp. 251–259 (2000)
18. Lubczonek, J., Stateczny, A.: Concept of neural model of the sea bottom surface. In: Rutkowski, L., Kacprzyk, J. (eds.) Neural Networks and Soft Computing. AISC, vol. 19, pp. 861–866. Springer, Heidelberg (2003)
19. Pawlak, Z.: Rough sets. International Journal of Computer & Information Sciences 11(5), 341–356 (1982)

Optimal Scale in a Hierarchical Segmentation Method for Satellite Images

David Fonseca-Luengo[1], Angel García-Pedrero[2], Mario Lillo-Saavedra[1],
Roberto Costumero[2], Ernestina Menasalvas[2], and Consuelo Gonzalo-Martín[2]

[1] Faculty of Agricultural Engineering, Universidad de Concepción,
Vicente Méndez 595, Chillán, Chile
`davidfonseca@udec.cl`
[2] Centro de Tecnología Biomédica at Universidad Politécnica de Madrid Campus
Montegancedo, 28223 Pozuelo de Alarcón, Madrid, Spain

Abstract. Even though images with high and very high spatial resolution exhibit higher levels of detailed features, traditional image processing algorithms based on single pixel analysis are often not capable of extracting all their information. To solve this limitation, object-based image analysis approaches (OBIA) have been proposed in recent years.

One of the most important steps in the OBIA approach is the segmentation process; whose aim is grouping neighboring pixels according to some homogeneity criteria. Different segmentations will allow extracting different information from the same image in multiples scales. Thus, the major challenge is to determine the adequate scale segmentation that allows to characterize different objects or phenomena, in a single image.

In this work, an adaptation of SLIC algorithm to perform a hierarchical segmentation of the image is proposed. An evaluation method consisting of an objective function that considers the intra-variability and inter-heterogeneity of the object is implemented to select the optimal size of each region in the image. The preliminary results show that the proposed algorithm is capable to detect objects at different scale and represent in a single image, allowing a better comprehension of the land-cover, their objects and phenomena.

Keywords: Remote sensing, hierarchical segmentation, multi-scale, high resolution images.

1 Introduction

High resolution images are an important part of the big volume of information generated in almost every domain: health, Earth observation, biology, to name a few. Two main image analysis approaches can be found in the literature. On the one hand, pixel-based approaches interpret images as a set of disconnected pixels, employing only spectral information. On the other hand, images in OBIA (Object Based Image Analysis) paradigm are interpreted as a set of homogeneous areas in the scene (segments), which are considered the minimal unit instead the pixels [1]. The main advantages of the OBIA approach are: i)

M. Kryszkiewicz et al. (Eds.): RSEISP 2014, LNAI 8537, pp. 351–358, 2014.

its capacity of extracting information such as shape, size, texture and contextual relationship from the objects and ii) reducing the computational requirements [2]. Even though in the last years more of the OBIA approaches have been developed in the area of Remote Sensing and Geographical Information Systems [3,4], applications in other areas like medicine [5] or neurosciences [6] can also be found.

The OBIA approach consists of two main processes: an initial segmentation for generating segments by grouping neighboring pixels according to some homogeneity criteria, such as color, texture, among others. And a second step, a classification process in which each segment is assigned to a class depending on its features [7]. Numerous researches have shown that the segmentation process is a critical step in the OBIA approach. This is due to the fact that the results of the classification step strongly depend on the initial segmentation [8,9].

Nowadays it is fully accepted that natural processes are scale dependent, and most of them are hierarchically structured. Therefore an appropriate understanding of these processes and interactions among them requires the determination of optimal scales [10]. Different segmentations will allow extracting different information from the same image in multiple scales [11]. Some studies have designed methodologies for estimating a single optimal scale of segmentation for the scene [12,13]. Other studies have implemented a multi-scale approach [14,15] to address this problem, but in this case the selection of the segmentation scales typically is based on an extensive knowledge of the study site [16]. In our approach, image segmentation is carried out by an adaptation of SLIC algorithm to deal with objects that are scale dependent and hierarchically related in a scene. On the other hand, in our case the selection of the optimal scale segmentation is performed by evaluating the intra-variability and inter-heterogeneity of the regions defined by the coarsest scale in contrast to the evaluation method presented in [16] where the complete scene is used for evaluation.

2   Materials and Methods

2.1   Data Set

The satellite image was collected on January 18th, 2013 by Pléiades Satellite, in the central irrigated valley of Chile (36° 31' 54" S; 72° 08' 03" O). The area of the scene has 420 hectares and corresponds to 1024 × 1024 pixels in the multispectral image. The acquired image has a spatial resolution of 2 meters and corresponds to ORTHO-BASIC products, characterized by a basic radiometric normalization (for detector's calibration) and geometric correction (WGS84/UTM 18S projection).

In this work Blue, Green, Red and Near Infrared bands were used to segmentation process. Furthermore, these 4 bands were used in the calculation of the metrics to determine the optimal scale for each parent superpixel.

2.2 Methodology

The methodology we propose, inspired by the one presented in [16], is composed of two main processing modules: i) image segmentation, and ii) segmentation evaluation. Though it differs from that one, both in the algorithm used for image segmentation and in the evaluation process followed. In what follows we detail the components of our approach.

Image Segmentation. Simple Linear Iterative Clustering (SLIC) [17] is a segmentation procedure based on the well-known *k-means* clustering algorithm to group image pixels into disjoint regions called superpixels. Superpixels are generated according to two criteria: i) color similarity and ii) spatial proximity. Moreover, a weighted distance that combines color and spatial proximity allows the control of the size and compactness of the superpixels [17]. Two parameters: i) k the desired number of superpixels, and ii) c the compactness factor have to be settled to run SLIC. Larger values of c results in more compact superpixels as the importance of spatial proximity is emphasized.

A detailed description of the SLIC algorithm working in CIELAB space can be found in [17], in this work also the authors demonstrate that the computational complexity of SLIC is linear on the number of pixels in the image.

Superpixels generated by SLIC reduce the influence of noise, preserve most of the edges of the images, and are approximated in size and shape. However, SLIC is not able to generate a hierarchical segmentation. Our approach fills this gap by proposing an adaptation of SLIC that makes the generation of a hierarchical multi-scale segmentation possible.

Based on the fact that in SLIC low values of k results in a coarser segmentation in comparison to finer segmentations obtained increasing the values of k, we propose to generate segmentations at different scales by varying the superpixels size and then combining resulting segmentations.

To establish the different segmentation scales, a vector $K = [k_1, k_2, \ldots, k_n]$ containing different values for the size of the superpixels in ascending order is defined. Each value of the vector corresponds to a different scale (i). Consequently, segments corresponding to a segmentation performed according to k_i will be larger than those of the segmentation according to k_{i+1}. A segment A^i in scale i is considered as the parent of a segment B^{i+1} in scale $i+1$, if the area of B^{i+1} can be entirely covered by segment A^i. The relationship parent-children among segments is shown in Figure 1. In order to ensure that the resulting segmentation process is hierarchical, segments in scale k_{i+1} are partitioned by segments in scale k_i, obtaining a new segmentation that generally contains a higher number of segments than in the segmentation achieved in scale k_{i+1}. To adjust the number of superpixels in scale $i+1$ to its original number of superpixels, smaller size segment are joined to the most similar neighboring segment belonging to the same parent. We use euclidean distance in the spectral space (color) to define the similarity between neighboring segments.

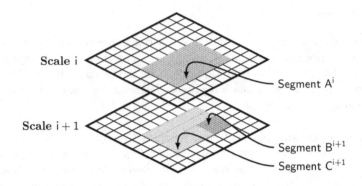

Fig. 1. Segment A^i is considered the parent of segments B^{i+1} and C^{i+1}, as both segments can be covered entirely by segment A^i. Each square of the grid represents an image pixel.

Evaluation. The evaluation process that we propose is based on the one proposed in [16] but adapting that one to deal with the hierarchical image segmentation process that has been followed. Thus, in our case we evaluate for each parent segment at the first scale those segments related to it instead of evaluating all the scene as in [16].

The main advantage of this scheme is that it provides different optimal scales for different image regions, against the results showed in [16], where an optimal scale for the whole image is provided. This fact is especially relevant in images where the variability inter-regions is high, as it is the case of natural land-covers in satellite images. A general overview of the evaluation scheme is shown in Figure 2.

In order to evaluate the segmentation, the following inter and intra-segments quality measures are used:

- the global Moran's index is used to measure the inter-heterogeneity between segments;
- the weighted variance is used for intra-segment homogeneity.

The global Moran's index measures the correlation among segments. It is used to evaluate the similarity of a segment with its neighborhood and is calculated as follows:

$$MI^p = \frac{n \sum_{i=1}^{n} \sum_{j=1}^{n} w_{ij}(y_i^p - \hat{y}^p)(y_j^p - \hat{y}^p)}{\sum_{i=1}^{n}(y_i^p - \hat{y}^p)^2 \left(\sum_{i \neq j} \sum w_{ij}\right)} \tag{1}$$

where:

- w is an adjacency matrix among neighbor regions that are relatives of the segment p at scale 1, therefore w_{ij} is equal to 1 if regions i and j are adjacent, otherwise, $w_{ij} = 0$.

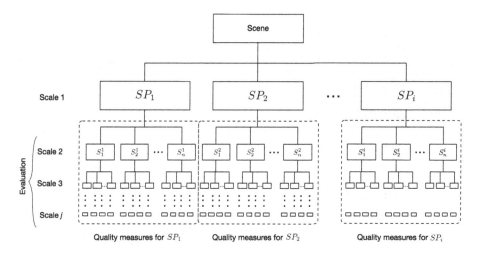

Fig. 2. General overview of the Evaluation process

- y_i is the spectral average of region i.
- n is the total number of regions at the analysis scale that are related to segment p.

In order to calculate the similarity at a particular scale level only those regions belonging to a certain parent segment p at scale 1 are considered.

The Global Moran's index ranges from the value -1, that corresponds to the perfect dispersion, to 1, that indicates a complete correlation. The 0 value corresponds to a random correlation.

The weighted variance is used for intra-segment quality measure, and is calculated as follows:

$$wVar^p = \frac{\sum_{i=1}^n a_i^p v_i^p}{\sum_{i=1}^n a_i^p} \qquad (2)$$

where v_i is the variance pixel that compound the segment i and a_i corresponds to the size of segment i. Values of variance closed to zero indicates homogeneity of a segment. Note that, when calculating this measure bigger size segments have a higher effect in the evaluation than those of a smaller size.

In this paper, the best scale is defined as one the value of the objective function F_i^p is closer to zero, being F_i^p calculated by adding the normalized (range 0-1) of intra and inter-segments measures.

The objective function is calculated as:

$$F_i^p = wVar_i^p + MI_i^p \qquad (3)$$

where p represents the region of the parent at the scale 1 to which the segments at the evaluation scale i are related.

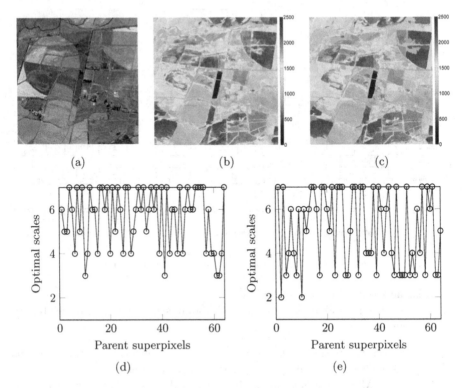

Fig. 3. (a) Color composition (Near Infrared-Green-Blue) of the analyzed scene. (b) and (c) Segmented images for the lineal and the dyadic succession of scales. (d) and (e) the corresponding optimal scales for each parent superpixel.

2.3   Experiments

Two experiments were carried out. In the first one, the number of superpixels associated with each scale was generated with a linear spacing. In the second one, the number of superpixels of each scale was generated with a dyadic spacing (2^n). The number of scales used in both experiments was 7, varying from 64 to 4096 superpixels.

3   Results and Discussion

A color composition of the analyzed multispectral scene is shown in Figure 3a. Figures 3b and 3c display the segmented images for the experiments where the scales were generated by a linear and dyadic spacing, respectively. Figures 3d and 3e depict the optimal scales for each parent superpixel (superpixel at the first scale). It can be observed that the number of segments generated by the linear spacing (Figures 3b and 3d) is higher than for the dyadic case (Figures 3c and 3e). This difference is due to the fact that linear spacing generates a higher number of superpixels in the intermediate scales, producing an abrupt change from coarsest to finest scale, and generating an over-segmentation in some homogeneous areas like the irrigation pivots. This is a problem when the results are used as information to improve the agricultural management.

4   Conclusions

In this work, a new methodology for optimal scale determination of a hierarchical segmentation method for satellite images has been proposed. An objective function, that combines weighted variance and global Moran's index, was used to select the selection of the optimal scale. The main outreach of this proposal is that it provides different optimal scales for different image regions, against the results showed in [12], where an optimal scale for the whole image is provided. This fact is especially relevant in images where the variability inter-regions is high, as it is the case of natural land covers in satellite images.

The proposed method allows users to benefit from detecting objects in a image at different scales. This allow a better comprehension of the land-cover, their objects and phenomena.

Finally, given the flexibility of the proposed methodology, is possible to apply in almost every domain where multi-scale images analysis are required (e.g. health, astronomy, biology).

Acknowledgements. A. García-Pedrero (grant 216146) and D. Fonseca-Luengo acknowledge the support for the realization of their doctoral thesis to the Mexican National Council of Science and Technology (CONACyT) and the National Commission for Scientific and Technological Research (CONICYT), respectively.

This work has been funded by the Centro de Recursos Hídricos para la Agricultura y la Minería (CONICYT/FONDAP/1513001).

References

1. Benz, U.C., Hofmann, P., Willhauck, G., Lingenfelder, I., Heynen, M.: Multiresolution, object-oriented fuzzy analysis of remote sensing data for gis-ready information. ISPRS-J. Photogramm. Remote Sens. 58(3), 239–258 (2004)
2. Homeyer, A., Schwier, M.: H.H.: A Generic Concept for Object-based Image Analysis. In: VISAPP 2010, pp. 530–533 (2010)
3. Blaschke, T.: Object based image analysis for remote sensing. ISPRS-J. Photogramm. Remote Sens. 65(1), 2–16 (2010)
4. Hay, G., Castilla, G.: Geographic object-based image analysis (GEOBIA): A new name for a new discipline. Object-Based Image Analysis, 75–89 (2008)
5. Chitiboi, T., Hennemuth, A., Tautz, L., Stolzmann, P., Donati, O.F., Linsen, L., Hahn, H.K.: Automatic detection of myocardial perfusion defects using object-based myocardium segmentation. In: Computing in Cardiology Conference (CinC), 2013, pp. 639–642 (2013)

6. Lucchi, A., Smith, K., Achanta, R., Knott, G., Fua, P.: Supervoxel-Based Segmentation of Mitochondria in EM Image Stacks With Learned Shape Features. IEEE Trans. Med. Imaging 31(2), 474–486 (2012)
7. Vieira, M., Formaggio, A., Rennó, C., Atzberger, C., Aguiar, D., Mello, M.: Object based image analysis and data mining applied to a remotely sensed landsat time-series to map sugarcane over large areas. Remote Sens. Environ. 123, 553–562 (2012)
8. Gao, Y., Mas, J.: A comparison of the performance of pixel-based and object-based classifications over images with various spatial resolutions. Online J. Earth Sci. 2(1), 27–35 (2008)
9. Yan, G., Mas, J.F., Maathuis, B., Xiangmin, Z., Van Dijk, P.: Comparison of pixel-based and object-oriented image classification approachesa case study in a coal fire area, Wuda, Inner Mongolia, China. Int. J. Remote Sens. 27(18), 4039–4055 (2006)
10. Levin, S.: The problem of pattern and scale in ecology: The Robert H, MacArthur award lecture. Ecology 73(6), 1943–1967 (1992)
11. Burnett, C., Blaschke, T.: A multi-scale segmentation/object relationship modelling methodology for landscape analysis. Ecol. Model. 168(3), 233–249 (2003)
12. Espindola, G., Camara, G., Reis, I., Bins, L., Monteiro, A.: Parameter selection for region-growing image segmentation algorithms using spatial autocorrelation. Int. J. Remote Sens. 27(14), 3035–3040 (2006)
13. Kim, M., Madden, M., Warner, T.A., et al.: Forest type mapping using object-specific texture measures from multispectral ikonos imagery: Segmentation quality and image classification issues. Photogramm. Eng. Remote Sens. 75(7), 819–829 (2009)
14. Zhou, W., Troy, A.: Development of an object-based framework for classifying and inventorying human-dominated forest ecosystems. Int. J. Remote Sens. 30(23), 6343–6360 (2009)
15. Trias-Sanz, R., Stamon, G., Louchet, J.: Using colour, texture, and hierarchial segmentation for high-resolution remote sensing. ISPRS-J. Photogramm. Remote Sens. 63(2), 156–168 (2008)
16. Johnson, B., Xie, Z.: Unsupervised image segmentation evaluation and refinement using a multi-scale approach. ISPRS-J. Photogramm. Remote Sens. 66(4), 473–483 (2011)
17. Achanta, R., Shaji, A., Smith, K., Lucchi, A., Fua, P., Susstrunk, S.: Slic superpixels compared to state-of-the-art superpixel methods. IEEE Trans. Pattern Anal. Mach. Intell. 34(11), 2274–2282 (2012)

3D Dendrite Spine Detection - A Supervoxel Based Approach

César Antonio Ortiz, Consuelo Gonzalo-Martín[1],
José Maria Peña[2], and Ernestina Menasalvas[1]

[1] Universidad Politécnica de Madrid, Centro de Tecnología Biomédica, Madrid
[2] Universidad Politécnica de Madrid, Departamento de Arquitectura y Tecnología
de Sistemas Informáticos, Madrid, Spain

Abstract. In neurobiology, the identification and reconstruction of dendritic spines from large microscopy image datasets is an important tool for the study of neuronal functions and biophysical properties. But the problem of how to automatically and accurately detect and analyse structural information from dendrites images in 3D confocal microscopy has not been completely solved. We propose an novel approach to detect and extract dendritic spines regardless their size o type, for images stacks result of 3D confocal microscopy. This method is based on supervoxel segmentation and their classification using a number of different, complementary algorithms.

Keywords: segmentation, supervoxels, dendrite, dendritic spine.

1 Introduction

Dendritic spines are small protrusions that emerge from dendrites. This structures define the inter-connectivity of the neural network. Many neuronal functions, such as learning and memory, are closely related with the appearance or disappearance of those neuronal connections. Hence, the dendritic spines study remains as an active, important area in neurobiology.

Usually, dendritic spines are manually labelled to analyse their morphological changes, which is very time-consuming and susceptible to operator bias, even with the assistance of computers. To deal with these issues, several methods have been proposed to automatically detect and measure the dendritic spines with no o little human interaction, but the accurate three-dimensional reconstruction of dedritic spines is already an open problem, specially over a wide volume where the compute resources needed becomes a challenger.

We propose an automated method to detect and extract dendritic spines regardless their size o type. It can be divided in four stages: volume segmentation, segment classification, classification results clustering and final cluster post-processing. This paper focus on the classification stage. Two different complementary classification algorithms based in supervoxel segmentation are proposed: first one uses local regularity in supervoxel space. Second one is based in the dendrite graph construction.

M. Kryszkiewicz et al. (Eds.): RSEISP 2014, LNAI 8537, pp. 359–366, 2014.

This paper is organized as follows: in the next section we review some relevant works in dendritic spine detection and reconstruction. Section 3 presents the bases of supervoxel dendrite segmentation and the segments classification and clustering algorithms. In section 4 the results of both algorithm are shown, followed by the conclusions in section 5.

2 Related Work

The development of automatic techniques for dendrite structures classification is associated to the improvement of microscopic imaging techniques that enable studies on increasingly larger amounts of data. Previous works on dendritic spine detection are usually divided into two groups, classification-based methods and dendrite centerline extraction-based methods.

Classification-oriented methods separate the elements in an image into different categories using trained classifiers based in the element and its neighbourhood characteristics. A simple voxel clustering classifier taking into account only the distance to the dendritic surface proposed by Rodriguez et al. in [1]. A more complex algorithm is utilized in [2] by Li et al. , based on a classifier using the surface curvature, the distance to the dendrite central area and the relation between surface and central area normals in the specific point.

Centerline extraction-based methods detect all the possible centerlines of the objects presented in the image stack. Dendritic spines will appears as small protrusions attached to the centerline areas. There are different approaches to the dendrite backbone extraction, usually based in more general skeletonization algorithms. W. Zhou, H. Li and X. Zhou in [3] propose a 3D level set based on local binary fitting model to extract the dendrite backbone. The work of Xiaoyin et al in [4] adapts the well known grassfire skeletonization method [5] to calculate the central region of the dendritic stem. In [6] is proposed the creation of a curvilineal structure using an approximation of the geodesic mean function defined by the surface of the dendrite. In [7], skeletonization is performed using a curvilinear structure detector, and the dendritic spines are detected through the fast marching algorithm, taking the central regions of spines as initial points. Also related with the dendrite centerline extraction-based approaches, some methods propose the use of model to fit the neuronal volume. For example, in Herzog et al. and Al-Kofahi et al. ([8] and [9]) a parametrical model based in cylindrical structures is employed as an approximation of the neuron topology.

3 Supervoxel Based Dendrite Detection

The concept of supervoxel is an extension of the concept of superpixel, first exposed in the works on binary classifiers by Ren and Malik in [10]. The idea behind the superpixels arises from the fact that the division of an image in pixels is not really a natural division, but simply an artefact of the camera that captures the images. A superpixel is commonly defined as a perceptually uniform region in the image. This definition is easily extended to a three dimensional environment.

The use of superpixel/supervoxel results not in an object segmentation of an image, but in an image oversegmentation composed by small, closely related areas. In any case, supervoxel segmentation can reduce the computational and memory costs of processing big images stack by several orders of magnitude.

There are many different techniques to generate supervoxels, but we will focus on the use of Simple Linear Iterative Clustering (SLIC, see [11]), an adaptation of the k-means clustering approach that take in account not only spatial proximity but also the local intensity similitude in the image. Despite its simplicity, SLIC adheres to boundaries as well as or better than other methods and the result has not big differences in supervoxels size, distributed evenly over the image.

A dendrite can be seen as a cylindrical structure presenting small protrusions, the dendritic spines. As previously mentioned, the SLIC generated supervoxels sizes are mostly uniform, which results in many of the supervoxels placed inside the dendritic stem, more or less regularly distributed. However, dendritic spines will generate fewer supervoxels, distributed between the dendritic spine head and the spine neck. An example of the segmentation of an slice can be seen in Fig.1

3.1 Supervoxel Based Classification

Using a supervoxel segmentation, we propose two different methods for dendritic spines detection based on the structure of neuronal dendrites. The ultimate goal is to combine different dendritic spines detection methods in order to reduce both the misses and false positives in supervoxel labelling.

As test dataset, stacks result of 3D confocal microscopy (from a Leica SP 5 with an 63x glycerol immersion objective) codified as 16 bits intensity images, each representing a slice of the volume, with a $0.13\mu m$x $0.13\mu m$x $0.4\mu m$resolution

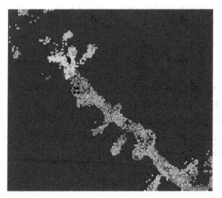
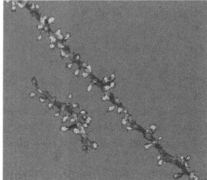

Fig. 1. Supervoxel segmentation of a dendrite image stack (slice, detail). Colours has not special meaning.

Fig. 2. Sum vector in the local neighbourhood regularity method (maximum projection). Bluish supervoxels indicate a shorter vector, reddish means longer vectors.

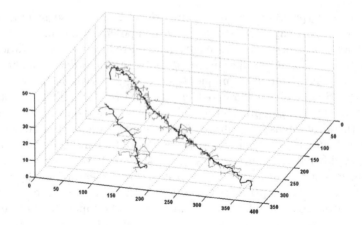

Fig. 3. Minimum spanning tree with the final graph representing the stem graph superimposed. Scale is in voxels.

are used. Although artefact as noise or photobleaching effects can be present, no further image post-processing has been applied.

As a first step, common to both proposed methods, we will separate dendrites from the background in the 3D confocal image stack, using a predefined threshold value to select supervoxels by its mean image intensity.

Detection by Local Neighbourhood Regularity. Inspired in the works of mitochondria segmentation in [12], we take advantage of the SLIC generated supervoxels distribution. Working on the space defined by the supervoxels, the neighbourhood of the supervoxels located inside a dendrite stem will be roughly regular. On the other hand, the neighbourhood of a supervoxel located inside a dendritic spine will be defined by the presence of the surrounding background and the dendritic spine neck. This results in:

$$\left\| \sum C_s - n\left(C_s\right) \right\| < \left\| \sum C_d - n\left(C_d\right) \right\| \tag{1}$$

where $n\left(x\right) = \{C_1, C_2, C_3, ..., C_n\}$ is the neighborhood of point x, C_s is a supervoxel part of the dendritic stem and C_d is a supervoxel part of a dendrite spine. This discrepancy is used to detect dendritic spines, as can be seen in Fig. 2.

Only local neighbourhood information is used to detect dendrites spine, so this method can be classified as a classification-oriented method.

Detection by Dendrite Stem Graph Distance. A different approach, a centerline extraction-based method, is founded on a graph describing the structure of the dendrite. We expect to simplify this graph to leave only a representation of the dendritic stem. The base graph is the result of connecting the centres of mass defined by supervoxels with its neighbours within a certain radius.

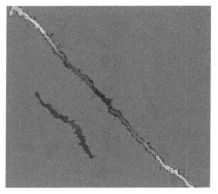

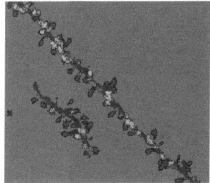

Fig. 4. Supervoxels in a dendrites labelled using the number of longest paths in a minimum spanning tree of which is part. A bluish supervoxels less paths, reddish more paths.

Fig. 5. Supervoxel distances to the dendritic stem in the dendrite stem graph distance method (maximum projection). A Bluish supervoxels indicate a closer element, reddish means farther.

In order to simplify this graph until only the representation of the dendritic stem remains, we again take advantage of the geometry of the dendrite. The minimum spanning tree of the base graph is found, the distance between its nodes assigned as the weight to each edge (see Fig. 3). As a dendrite stem is a bigger linear structure than the dendritic spines, the "trunk" of the minimum spanning tree is formed mainly by the nodes defined by supervoxels present in the dendritic stem. To isolate this "trunk", an algorithm loosely based in an ant colony optimization [13] is used. For each node of the graph, the longest possible path inside the minimum spanning tree is calculated. The tree structure of the graph ensures that the longest paths will traverse the minimum spanning tree "trunk". Nodes located inside the "trunk" of the tree will appear in many more paths than those inside branches, as shown in Fig. 4. This algorithm can find the dendritic stem despite dendrites bifurcations or the presence of multiples, isolated dendrites in the image stack. The distance of the supervoxels center of mass to the edges of the graph representing the stem can be used to classify supervoxels, as supervoxel in dendritic spines would be farther than those inside the dendritic stem. Result of this method is the Fig. 5.

3.2 Clustering

Since the use of supervoxels results in an image oversegmentation, the last step in our classifiers is to group supervoxels classified as dendritic spines. Clustering is done using maximal inscribed spheres, in a similar way as [14]. From the center of each candidate feature supervoxel, the largest spherical volume that does not include the center of a non candidate supervoxel is calculated. Supervoxels will be aggregated taking in account the overlaps between the calculated spheres. Each of these aggregates is labelled as a different feature, as show in Fig. 6.

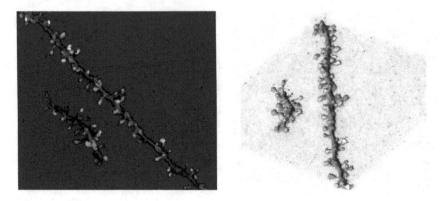

Fig. 6. Dendrite image with the clustered dendritic spine results of the local neighbourhood regularity method superimposed. Maximum projection (left) and volume reconstruction (right). Colours have not special meaning.

4 Results

Detection by local neighbourhood regularity is effective detecting small dentritic spines, included spines aligned with the z axis. However, larger dendritic spines could generate a supervoxels distribution similar to the distribution on the dendritic stem. In this case some supervoxels could be misclassified, as is seen in Fig. 7. There may also be false positives in areas whose geometry is not the usual dendritic geometry, as the ends, or, to a lesser extent, in bifurcations. . There

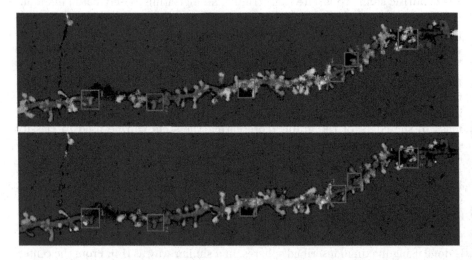

Fig. 7. Maximum projection of clustered dendritic spine detection (up) and detection by dendrite stem graph distance (down) results. Differences between method can be seen in results of poorly defined areas (1 and 2) and small dendritic spine close to the dendritic stem (3, 4, 5 and 6) are highlighter.

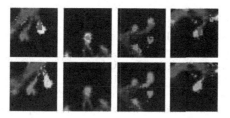

Fig. 8. Slice in different clustered dendritic spine detection results (Details). Partial dendritic spines detections by results of the local neighbourhood regularity method (upper row) are paired with the results of the detection by dendrite stem graph distance method (lower row).

may also be false positives in areas whose geometry is not the usual dendritic geometry, as the ends, or, to a lesser extent, in bifurcations.

On the other hand, using detection by dendrite stem graph distance allows to easily label large and medium-sized dendritic spines. However, as mentioned in [6] centerline extraction-based methods can be very sensitive to small changes in the object shape. The use of supervoxel minimizes this problem, as the skeletonizacion do not depends on voxels but a higher order object, as supervoxel are, but it is still possible to lose features close to the dendritic stem, especially in narrow or poorly defined areas within the dendritic volume. Some examples of those problems are show in Fig. 8.

5   Conclusions

Two different algorithms have been proposed in this paper, both part of a method to identify and reconstruct dendritic structures. The objective is the identification and separation of dendritic spines taking advantage of the results of different, complementary algorithms. The first method is able to detect small or ill defined dendritic spines, but can miss some parts on larger features, the second one is designed to reliably detect large and medium-sized dendritic spines.

This is a work in progress. Currently, an unoptimized Matlab implementation of the algorithms has an average computing time for detecting spines in a image stack of roughly 29 million of voxels (1365 x 590 x 36) is 22 minutes on an 2.40GHz Intel Core 2 Quad CPU with 4GB Memory. We expect further performance improvements in future reviews of the code.

This work will continue both improving the classifiers and defining the conditions for the integration of the classifiers result. Also, a previous image stack preprocessing stage has been proposed. Taking in account the flexibility proven by the segmentation and analysis of image stacks using supervoxels, the development and integration of new dendritic spine detectors is not discarded.

Acknowledgements. Authors thankfully acknowledge Dr. Ruth Benavides-Picciones and Dr. Isabel García, as well as the Cajal Blue Brain Project for the provision of microscopy images and their support in the evaluation of the results.

References

1. Rodriguez, A., Ehlenberger, D.B., Dickstein, D.L., Hof, P.R., Wearne, S.L.: Automated three-dimensional detection and shape classification of dendritic spines from fluorescence microscopy images. PLoS One 3(4), e1997 (2008)
2. Li, Q., Zhou, X., Deng, Z., Baron, M., Teylan, M.A., Kim, Y., Wong, S.T.C.: A novel surface-based geometric approach for 3d dendritic spine detection from multi-photon excitation microscopy images. In: ISBI, pp. 1255–1258. IEEE (2009)
3. Zhou, W., Li, H., Zhou, X.: 3d neuron dendritic spine detection and dendrite reconstruction. IJCAET 1(4), 516–531 (2009)
4. Xu, X., Cheng, J., Witt, R.M., Sabatini, B.L., Wong, S.T.C.: A shape analysis method to detect dendritic spine in 3d optical microscopy image. In: ISBI, pp. 554–557. IEEE (2006)
5. Leymarie, F.F., Levine, M.D.: Simulating the grassfire transform using an active contour model. IEEE Trans. Pattern Anal. Mach. Intell. 14(1), 56–75 (1992)
6. Janoos, F., Mosaliganti, K., Xu, X., Machiraju, R., Huang, K., Wong, S.T.C.: Robust 3d reconstruction and identification of dendritic spines from optical microscopy imaging. Medical Image Analysis 13(1), 167–179 (2009)
7. Zhang, Y., Zhou, X., Witt, R.M., Sabatini, B.L., Adjeroh, D.A., Wong, S.T.C.: Automated spine detection using curvilinear structure detector and lda classifier. In: ISBI, pp. 528–531 (2007)
8. Jaeger, S., Palaniappan, K., Casas-Delucchi, C.S., Cardoso, M.C.: Classification of cell cycle phases in 3d confocal microscopy using pcna and chromocenter features. In: Proceedings of the Seventh Indian Conference on Computer Vision, Graphics and Image Processing, ICVGIP 2010, pp. 412–418. ACM, New York (2010)
9. He, W., Hamilton, T.A., Cohen, A.R., Holmes, T.J., Pace, C., Szarowski, D.H., Turner, J.N., Roysam, B.: Automated three-dimensional tracing of neurons in confocal and brightfield images. Microsc. Microanal 9, 296–310 (2003)
10. Ren, X., Malik, J.: Learning a classification model for segmentation. In: Proceedings of the Ninth IEEE International Conference on Computer Vision, ICCV 2003, vol. 2, pp. 10–17. IEEE Computer Society, Washington, DC (2003)
11. Achanta, R., Shaji, A., Smith, K., Lucchi, A., Fua, P., Ssstrunk, S.: Slic superpixels compared to state-of-the-art superpixel methods. IEEE Transactions on Pattern Analysis and Machine Intelligence 34(11), 2274–2282 (2012)
12. Lucchi, A., Smith, K., Achanta, R., Knott, G., Fua, P.: Supervoxel-based segmentation of mitochondria in em image stacks with learned shape features. IEEE Trans. Med. Imaging 31(2), 474–486 (2012)
13. Dorigo, M., Maniezzo, V., Colorni, A.: The ant system: Optimization by a colony of cooperating agents. IEEE Trans. on Systens, Man and Cybernetics-Part B 26(1), 29–41 (1996)
14. Silin, D., Patzek, T.: Pore space morphology analysis using maximal inscribed spheres. Physica A: Statistical and Theoretical Physics 371(2), 336–360 (2006)

Histogram of Bunched Intensity Values Based Thermal Face Recognition

Ayan Seal[1], Debotosh Bhattacharjee[1], Mita Nasipuri[1],
Consuelo Gonzalo-Martín[2], and Ernestina Menasalvas[2]

[1] Computer Science and Engineering, Jadavpur University, India
[2] Center for Biomedical Technology, Universidad Politecnica de Madrid, Spain
`ayan.seal@gmail.com, debotosh@indiatimes.com,`
`mnasipuri@cse.jdvu.ac.in,`
`{consuelo.gonzalo,ernestina.menasalvas}@upm.es`

Abstract. A robust thermal face recognition method has been discussed in this work. A new feature extraction technique named as Histogram of Bunched Intensity Values (HBIVs) is proposed. A heterogeneous classifier ensemble is also presented here. This classifier consists of three different classifiers namely, a five layer feed-forward backpropagation neural network (ANN), Minimum Distance Classifier (MDC), and Linear Regression Classifier (LRC). A comparative study has been made based on other feature extraction techniques for image description. Such image description methods are Harris detector, Hessian matrix, Steer, Shape descriptor, and SIFT. In the classification stage ANN, MDC, and LRC are used separately to identify the class label of probe thermal face images. Another class label is also assigned by majority voting technique based on the three classifiers. The proposed method is validated on UGC-JU thermal face database. The matching using majority voting technique of HBIVs approach showed a recognition rate of 100% for frontal face images which, consists different facial expressions such as happy, angry, etc On the other hand, 96.05% recognition rate has been achieved for all other images like variations in pose, occlusion etc, including frontal face images. The highly accurate results obtained in the matching process clearly demonstrate the ability of the thermal infrared system to extend in application to other thermal-imaging based systems.

Keywords: Thermal face image, Histogram of Bunched Intensity Values, Minimum Distance Classifier, Linear Regression Classifier, Artificial Neural Network.

1 Introduction

In the last forty years, Automatic Face Recognition (AFR) has been extensively studied by the research community in the field of computer vision and applied pattern recognition. Visible spectrum image based AFR Systems have reached a significant level of maturity. However, some factors like change in illumination, pose, facial expression, and facial disguises, between other, provoke a serious

M. Kryszkiewicz et al. (Eds.): RSEISP 2014, LNAI 8537, pp. 367–374, 2014.

problem in visible spectrum image based AFR system. Pose factor can be resolved to a great extent by generating a model using an analysis-by-synthesis approach [1] or synthesized using a statistical method [2]. On the other hand, illumination problem can be solved by using image processing filtering [3]. However, this method is application specific. Sometimes, it works well and sometimes it doesn't. Multi-image face biometric has been used to solve the uncertain environmental illumination risks [4]. Illumination problem can also be solved by using infrared (IR) spectrum images. It is not an application specific approach. Thermal IR camera can capture IR spectrum. The technique by which an image is produced by an IR camera is known as thermal imaging or thermography. In this technique, an image is produced by invisible infrared light emitted from the objects surface itself. A thermal IR camera doesn't sense the temperature instead it captures the IR energy, which is transferred from an object surface to its environment and produces an image or thermography. Recently, researchers are using thermal IR camera for face recognition based on thermal face images. In this study, a recognition system for biometric security based on thermal face images has been developed since thermal face images have advantages over visible face images. The advantages of thermal face images have been discussed in [5],[6], [7], [8], [9], [10]. It is integrated approach that combines novel algorithms for extracting features, and matching these features has been developed. Matching algorithm depends on the quality of features. If the features have distinguishing ability, then the matching algorithm can easily distinguish an object from other objects. The detection of salient features, often referred to as keypoints or landmarks, is an important task in many computer vision applications, such as object tracking and object recognition. In this work, a new feature extraction technique named as Histogram of Bunched Intensity values (HBIVs) has been discussed in the context of thermal face recognition. A comparative studies have been made on the basis of different feature extraction techniques such as Harris detector [11], Hessian matrix [12], Steerable filters [13], SIFT [14]. After extraction of features from the face images using the above mentioned techniques, three well known classifiers named as five layer feed-forward backpropagation neural network (ANN) [15], Minimum Distance Classifier (MDC) [16], Linear Regression Classifier (LRC) [17] accept these features and try to measure the system's performance separately. In this work, an ensemble classifier [18], [19] has also been introduced for recognition of thermal face images. An ensemble of classifiers can integrate multiple component classifiers, such as ANN, MDC, LRC etc. All the experiments have been evaluated on UGC-JU thermal face database [20], which has been created at our laboratory to compare the reliability of the algorithms designed for feature extraction, and similarity measures. The rest of the paper is organized as follows. Section 2 explains the face data acquisition and preprocessing. Section 3 outlines a novel feature extraction technique i.e. HBIVs. An ensemble classifier is introduced in section 4. Section 5 contains experimental results and conclusion has been made in section 6.

2   Images Collection and Preprocessing

2.1   Thermal Face Image Acquisition

In the present work, thermal face images are acquired simultaneously under variable expressions, poses and with/without glasses by FLIR 7 at our own laboratory. Our FLIR 7 camera can capture thermal infrared spectrum (wavelength of $8\mu m - 14\mu m$).The name of the database is UGC-JU face database. Till now 84 individuals have volunteered for this photo shoots and each individual 39 different templates of RGB color images with happy, angry, sad, disgusted, neutral, fearful and surprised facial expressions are taken. Different pose changes about x-axis; y-axis and z-axis are also taken. Resolution of each image is 320×240, and the images are saved in JPEG format. A sample thermal IR image is shown in Fig. 1a.

Fig. 1. Thermal face image and its various preprocessing stages, a) A thermal face image. b) Corresponding grayscale image . c) Corresponding binary image, d) Largest component of the face skin region in the binary image, e) Extracted face skin region in the grayscale image and scaled to resolution 256×256 f) Restored image of Fig. 1e.

2.2   Image Preprocessing

The image preprocessing stage can be further divided into three subsections.

Image Binarization. The grayscale image of the sample image of Fig. 1a is shown in Fig. 1b. Then the grayscale images are converted into corresponding binary images. The binary image corresponding to the grayscale image of Fig. 1b, is shown in Fig. 1c.

Finding the Largest Component. The foreground of a binary image may contain more than one object. The largest foreground component has been extracted from binary image using Connected Component Labeling algorithm [21],[22]. This algorithm is based either on 4-conneted neighbours or 8-connected neighbours method [23]. The largest foreground component of Fig. 1c is shown in Fig. 1d. The largest foreground component as considered here face skin region using the pixel maps of the largest foreground component in the binary image, the face skin region in the corresponding grayscale image is identified which is then cropped and scaled to the size of 256×256 pixels. The cropped and resized grayscale face skin region of Fig. 1b is shown in Fig. 1e.

Restoration of Missing Data Using GappyPCA. After cropping the face region, it is important to locate the missing thermal information within facial area excluded during binarization process. False exclusion of thermal information is caused by the presence of inconsistent distribution of temperature statistics and makes the process of identifying persons using their thermal facial images difficult. In this work, a GappyPCA [24] based method is used to detect and then restored the missing thermal information within facial. The restored image after application of GappyPCA (GPCA) on face image of Fig. 1e is shown in Fig. 1f.

3   Features Extraction

Once the missing information is restored using GappyPCA, feature extraction comes into picture. Most facial feature extraction methods are sensitive to variations illumination, noise, orientation etc. A substantial body of work has been done in this area and several methods have been proposed for the detection, and description of landmarks. A new feature extraction method will be discussed here. Raw information i.e. pixel by pixel information sometimes may be helpful, which will be treated as features for further processing i.e. identify the class level of a particular object. It considers the local information only. The advantages of using pixel by pixel information is that spatial information could be maintained and no other processing need to be done in order to get the features. It is very simple but an effective approach. The problem of using such technique is that sometimes two similar objects may be treated as dissimilar objects due to pose, and shift changes. To overcome the limitation of using pixel by pixel information directly, histogram of an image could be considered as features. It is treated as global features. Spatial information might be lost of using histogram of an image. So, a new feature extraction technique named as Histogram of Bunched Intensity Values (HBIVs) has been introduced, which neither follows pixel by pixel information nor histogram of an image directly, rather it takes a middle man approach. A typical window $\frac{M}{\sqrt{M}} \times \frac{N}{\sqrt{N}}$ sized block has been traversed from the top left end to bottom right end of the face image in raster scan order with 50% overlap in horizontal and vertical directions because sometimes face may shift due to pose changes. Where, $M \times N$ is the size of the image. After filling up the missing information using GappyPCA, the size of the image is 256×256. and it has 256 graylevels. These are grouped in 16 bins. That allows to avoid surrounding temperatures changes. We propose to use as a feature the number of pixels in each window that having gray level of each bin. In this study, the first bin was not considered since first bin contains intensity values between 0 to 15 including 0 and 15 and lower intensity means background of an image. Here, the number of block means the number of features of a particular face image which could possibly be helped to find out its own identity. The total number of blocks could be found by the following equation:

$$nb = (\sqrt{M} - 1) \times \sqrt{N} + (\sqrt{N} - 1) \times \sqrt{M} + \sqrt{MN} \qquad (1)$$

where, nb is the total number of blocks of a face image. Number of features for this method could be found by the following equation:

$$nf = nb \times (bn - 1) \qquad (2)$$

where, nf is total number of features, nb is total number of blocks, and bn is number of bins. HBIVs is scale invariant in nature as the size of each face image is 256×256 before extracting the features of a face image. HBIVs is based on image histogram. The histogram of a rotated image is same as the histogram of an original image. So, HBIVs is rotation invariant. Pose change problem is basically known as translation problem. In HBIVs, a window has been traversed from the top left end to bottom right end of the corner image in raster scan order with 50% overlap in horizontal and vertical directions in order to get the final feature vector because sometimes face may shift due to pose changes. Thus, HBIVs is translation invariant. The thermal IR face images are illumination independent. Indirectly, HBIVs is illumination invariant in nature.

4   Ensemble Classifier

Classifier combination is now an active research topic in the domain of pattern recognition and machine learning. Many works have been published, which depicts the advantages of the combination of multiple classifiers over single classifier. In multiple classifier based system, feature set has to partition into several sets and each set feed into a different classifiers. Such multiple classifier systems are commonly known as classifier ensemble. Let, $x = [x_1, ..., x_n]$ be a feature vector of a particular face, which has n features. Let X is a matrix of size $P \times n$ where, P represents the number of number of training images. Each row having n features, which depicts a face. Let Y be vector with class labels for each face $Y = [y_1, ..., y_n]$. Let $C_1, ...C_L$ are the classifiers in the ensemble. In this work, the value of L is 3. It means three different classifiers namely, ANN, MDC, and LRC are used to construct classifier ensemble. Total training set X is divided into L disjoint subsets of equal size, where the first feature feeds into first classifier, second feature goes into second classifier and third enters into third classifier and this process is continued until all the features are exhausted. Then L disjoint subsets have been fed into C_i classifiers. Each subset has $\frac{n}{L} i.e. \frac{n}{3}$ features. So, three different matrices have been created for three different classifiers each of size P for training purpose. Depending on the three different classier final decisions has been made using majority voting technique.

5   Experimental Results

Experiments have been performed on UGC-JU thermal face database, created at our laboratory. The detail has been discussed in section 2.1. Some well known features detector techniques are used for extraction of features from the restored thermal face images for comparative studies. Such features descriptors are Harris detector, Hessian matrix, SIFT, Steer filter, Most of these feature extraction techniques are scale, translation, and rotation invariant. Here, total number of times the window traversed was 11040. So, the size of the feature vector is 1 × 11040 which corresponds to features of one face. In this fashion, features of each face are extracted and stored in a separated row of a matrix which, is known as feature matrix.In the classification stage total feature sets named as X is partitioned into two disjoint sets U and V respectively. The U set is used for training and V set is used for testing purpose. This process is known as 2-fold cross validation. Then the U set is fed into each of the classifiers separately for training purpose and V set is used for testing purpose. The obtained results are reported in Table 1. Then decision level fusion is being adopted on eleven sets of results, based on majority voting technique since majority voting technique will be more appropriate with the idea of using multiple results that each represents more or less stable information in the face area. Some of the results may give the incorrect, but still many others will give the correct decision. The outcome after using of majority voting technique is also plotted in Table 1. Then the total number of features i.e. 11040 is divided into three set each of size 11040/3=3680 and fed into three different classifiers which, worked parallel. The acquired output has been incorporated into Table 1. Two experiments have been performed

Table 1. Recognition performances (%) using different feature extraction techniques

Feature Extraction method	Classifier used	Frontal Face	All Face
Harris detector [25]	ANN	81.94	78.94
(Scale, rotation invariant)	Minimum distance	87.50	52.63
	LRC	93.06	71.49
	Majority voting(ANN,MD,LRC)	100.00	85.96
	Ensemble	95.83	67.11
Hessian matrix [26]	ANN	76.38	67.98
	Minimum distance	87.50	56.58
	LRC	93.06	73.25
	Majority voting(ANN,MD,LRC)	98.61	85.33
	Ensemble	93.06	68.42
SIFT	ANN	90.27	70.17
	Minimum distance	90.28	56.58
	LRC	95.83	74.12
	Majority voting(ANN,MD,LRC)	100.00	74.12
	Ensemble	98.61	72.81
Steer filter	ANN	69.44	61.40
	Minimum distance	86.11	53.51
	LRC	95.83	73.25
	Majority voting(ANN,MD,LRC)	100.00	67.71
	Ensemble	94.44	71.05
HBIVs (proposed method)	ANN	93.05	74.56
	Minimum distance	97.22	81.58
	LRC	100.00	93.42
	Majority voting(ANN,MD,LRC)	100.00	96.05
	Ensemble	98.61	87.28

namely frontal face images and all face images, where 39 different images have been considered.

6 Conclusions

Sometimes exact localization of local features such as points are very difficult due to noise present in the image using Harris detector, Hessian matrix, SIFT, Steer filter. To overcome the limitation of the problem, in this study a new approach named as Histogram of Bunched Intensity Values (HBIVs) is used in order to extract features from the thermal face images for their recognition. It neither considers local feature points nor global features like histogram. It is a middle man approach. The obtained results have been mentioned in section 5. The percentage of recognition rates have been reduced with varying block sizes other than 16×16. The proposed method is reasonable in terms of number of features compare to pixel by pixel information.

Acknowledgments. Authors are thankful to a project entitled Development of 3D Face Recognition Techniques Based on Range Images, funded by Deity, Govt. of India and DST-PURSE Programme at Department of Computer Science and Engineering, Jadavpur University, India for providing the necessary infrastructure to conduct experiments relating to this work. Ayan Seal is grateful to Department of Science & Technology (DST), Govt. of India for providing him Junior Research Fellowship-Professional (JRF-Professional) under DST-INSPIRE Fellowship programme [No: IF110591]. Ayan Seal is also thankful to Universidad Politecnica de Madrid, Spain for providing him scholarship under Erasmus Mundus Action 2 India4EU II.

References

1. Blanz, V., Vetter, T.: Face recognition based on tting a 3D morphable model. PAMI (2003)
2. Mohammed, U., Prince, S., Kautz, J.: Visio-lization: Generating novel facial images. ACM Trans. on Graphics (2009)
3. Nishiyama, M., Yamaguchi, O.: Face recognition using the classi ed appearance-based quotient image. AFG (2006)
4. Arandjelovic, O., Cipolla, R.: A pose-wise linear illumination manifold model for face recognition using video. CVIU (2009)
5. Wu, S., Fang, Z.-J., Xie, Z.-H., Liang, W.: Blood Perfusion Models for Infrared Face Recognition, pp. 183–207. School of information technology, Jiangxi University of Finance and Economics, China (2008)
6. Friedrich, G., Yeshurun, Y.: Seeing people in the dark: Face recognition in infrared images. In: BMVC (2003)
7. Pavlidis, I., Symosek, P.: The imaging issue in an automatic face/disguise detection system. In: CVBVS (2000)
8. Nicolo, F., Schmid, N.A.: A method for robust multispectral face recognition. In: ICIAR (2011)

9. Kong, S.G., Heo, J., Abidi, B.R., Paik, J., Abidi, M.A.: Recent advances in visual and in-frared face recognitiona review. Computer Vision Image Understanding 97, 103–135 (2005)
10. Manohar, C.: Extraction of Super cial Vasculature in Thermal Imaging, masters thesis. Dept. Electrical Eng., Univ. of Houston, Houston, Texas (December 2004)
11. Harris, C., Stephens, M.: A Combined Corner and Edge Detector. In: Proc. Fourth Alvey Vision Conf., pp. 147–151 (1988)
12. Gradshteyn, I.S., Ryzhik, I.M.: Hessian Determinants, 14, 6th edn. 14.314 in Tables of Integrals, Series, and Products, p. 1069. Academic Press, San Diego (2000)
13. Freeman, W., Adelso, E.: The Design and Use of Steerable Filters. IEEE Trans. Pattern Analysis and Machine Intelligence 13(9), 891–906 (1991)
14. Lowe, D.: Distinctive Image Features from Scale-Invariant Keypoints. International Journal of Computer Vision 60, 91–110 (2004)
15. Seal, A., Bhattacharjee, D., Nasipuri, M., Basu, D.K.: Minutiae based thermal face recognition using blood perfusion data. In: Processing, ICIIP (2011)
16. Gonzalez, R.C., Woods, R.E.: Digital Image Processing, 3rd edn. Prentice Hall (2002)
17. Naseem, R.T., Bennamoun, M.: Linear Regression for Face Recognition. IEEE Transactions on Pattern Analysis and Machine Intelligence (IEEE TPAMI) 32(11), 2106–2112 (2010)
18. Hansen, L., Salamon, P.: Neural netwok ensembles. IEEE Trans. Pattern Anal. Mach. Intell. 12(10), 993–1001 (1990)
19. Kuncheva, L.: Combining Pattern Classi ers: Methods and Algorithms. Wiley, Hoboken (2004)
20. Seal, A., Bhattacharjee, D., Nasipuri, M., Basu, D.K.: UGC-JU Face Database and its Benchmarking using Linear Regression Classier. Multimedia Tools and Applications (2013)
21. Bhattacharjee, D., Seal, A., Ganguly, S., Nasipuri, M., Basu, D.K.: A Comparative Study of Human thermal face recognition based on Haar wavelet transform (HWT) and Local Binary Pattern (LBP). Computational Intelligence and Neuroscience (2012)
22. Chen, Y.-T., Wang, M.-S.: Human Face Recognition Using Thermal Image. Journal of Medical and Biological Engineering 22(2), 97–102 (2002)
23. Morse, B.S.: Lecture 2: Image Processing Review, Neighbors. Connected Components, and Distance (1998-2004)
24. Everson, R., Karhunen, L.S.: Loeve procedure for gappy data. Journal of the Optical Society of America A 12(8), 1657–1664 (1995)
25. Ardizzone, E., Cascia, M., Morana, M.: Probabilistic Corner Detection for Facial Feature Extraction. In: Proceedings of the 15th International Conference on Image Analysis and Processing, pp. 461–470, ISBN: 978-3-642-04145-7
26. Alwakeel, M., Shaaban, Z.: Face Recognition Based on Haar Wavelet Transform and Principal Component Analysis via Levenberg-Marquardt Backpropagation Neural Network. European Journal of Scientic Research 42(1), 25–31 (2010) ISSN 1450-216X

Cost-Sensitive Sequential Three-Way Decision for Face Recognition

Libo Zhang[1], Huaxiong Li[1], Xianzhong Zhou[1], Bing Huang[2], and Lin Shang[3]

[1] School of Management and Engineering, Nanjing University, Nanjing,
Jiangsu, 210093, P.R. China
[2] School of Information Science, Nanjing Audit University,
Nanjing, Jiangsu, 211815, P.R. China
[3] Department of Computer Science and Technology, Nanjing University, Nanjing,
Jiangsu, 210093, P.R. China
mg1315013@smail.nju.edu.cn,
{huaxiongli,zhouxz,shanglin}@nju.edu.cn, hbhuangbing@126.com

Abstract. Recent years have witnessed an increasing interest in Three-Way Decision (TWD) model. In contrast to traditional two-way decision model, TWD incorporates a boundary decision, which presents a delay-decision choice when available information for a precise decision is insufficient. The boundary decision can be transformed into positive decision or negative decision with the increasing of available information, thus forming a sequential three-way decision process. In real-world decision problems, such sequential three-way decision strategies are frequently used in human decision process. In this paper, we propose a framework of cost-sensitive sequential three-way decision approach to simulate the human decision process in face recognition: a sequential decision process from rough granularity to precise granularity strategy. Both theoretic analysis and experimental verification are presented in this paper.

Keywords: sequential three-way decision, cost-sensitive learning, face recognition, decision-theoretic rough sets.

1 Introduction

In conventional two-way decision model [1], there are only two optional choices for a decision: positive decision or negative decision. The two-way model frequently lead to wrong decisions when available information is insufficient. To address this issue, Yao proposed Three-way decision model [2–4], which extends two-way decision theory by incorporating an additional choice: boundary decision. Three-way decision theory derives from decision-theoretic rough set model (DTRS) [5, 6], which presents a semantics explanation on how to determine a sample as positive, negative and boundary region. In recent years, three-way decision theory has been successfully applied in many fields [7–14].

Previous researchers mostly concentrate on static decision problems. In reality, however, the available information is always insufficient and new information is gradually acquired, so decisions are sequentially made accordingly. For example, in medical diagnosis decision problems, examinations may include X-rays,

M. Kryszkiewicz et al. (Eds.): RSEISP 2014, LNAI 8537, pp. 375–383, 2014.
© Springer International Publishing Switzerland 2014

Nuclear Magnetic Resonance Imaging(NMRI), Computed Tomography(CT) etc. Some examinations are effective but expensive, e.g. NMRI, so they may not be the first choices. Instead, the examination with a lower cost, e.g. CT, may have a priority, then the one with higher cost if necessary, thus forming a cost-sensitive sequential three-way decision strategy. Recently, Yao provided a framework of sequential three-way decisions based on probabilistic rough sets [15], and Li. et al. proposed a cost-sensitive sequential decision strategy [16].

Cost-sensitive sequential three-way decision can simulate the human decision procedure: from rough granularity (high cost) to precise granularity (low cost) in many fields, where face recognition is a typical one. For example, in criminal investigation face recognition, human can not immediately recognize a blurred face image, but they can be more determined if the blurred image can be sharpened with more available information. There are also many actual demands to study the application of sequential three-way strategy in vision recognition. For example, the cameras may be blurry to guarantee the search area while searching for a target, but if we find a suspected one, we will gradually adjust cameras for a higher accuracy to identify it. Motivated by this concern, in this paper, we propose a framework of cost-sensitive sequential three-way decision approach to simulate the human decision process in face recognition. The proposed model will present a granular-based new view on human vision mechanism as well as computer vision mechanism.

2   Cost-Sensitive Sequential Three-Way Decision Model

Two-way decision mode requires the deciders to make decisions promptly [1], thus a third choice denoted as delay decision may be appropriate when the available information is insufficient. Considering this, Yao proposed three-way decision model [2], whose decision set is $\mathcal{A} = \{a_P, a_N, a_B\}$, denoting $POS(X)$, $NEG(X)$ and $BND(X)$ decisions respectively. We assume that the data set subjects to decision consistency assumption, i.e., the accuracy of decision has a positive correlation with the amount of information, and the boundary region will disappear if all features (attributes) are used, namely all the samples are definitely classified into either positive region or negative region [16].

At the beginning of the sequential decision process, only a part of features are used and many samples will be classified into boundary region. With the increasing of the decision steps, some new features are added so that some samples which are previously divided into boundary region can be precisely classified into positive region or negative region. At last, the boundary region disappears and the loss decrease to a satisfying criterion. It is proved that the decision cost and the boundary set will reduce with the increasing of the decision steps from a global view [16]. According to the existing work, we propose a new general definition on cost-sensitive sequential three-way decision model, not based on the equivalence class, but on the data set, as presented in Definition 1.

Definition 1. *Let C be a feature set, which consists of all m available features, ranked in a fixed order. $\Omega = \{x_i\}(i = 1, 2 \cdots, n)$ denotes a sample set with n*

samples, and $Y = \{y_i\}$ *is the corresponding label set. Denote* $C^{(l)}$ *as a feature set including the first l-dimensional features (l ≤ m), and* $x^{(l)}$ *as the corresponding sample feature w.r.t.* $C^{(l)}$. *Then a cost-sensitive sequential three-way decision can be defined as a sequential decision series:*

$$SD = (SD_1, SD_2 \cdots SD_l \cdots SD_m) = (\phi^*(x^{(1)}), \phi^*(x^{(2)}) \cdots \phi^*(x^{(l)}) \cdots \phi^*(x^{(m)})),$$

where $\phi^*(x^{(l)})$ *is the optimal decision for* $x^{(l)}$ *in the l-th decision step.*

For a cost-sensitive sequential decision SD, the l-th optimal decision step $\phi^*(x^{(l)})$ is acquired based on the minimum risk Bayes decision rule. Only binary-class problem is concerned in this paper, i.e., $\Omega = \{X_P, X_N\}$, where all samples in positive set X_P are with label P, all in negative set X_N are with label N. Considering a cost-imbalanced problem, so let $\lambda_{PP}, \lambda_{BP}, \lambda_{NP}$ denote the respective costs of three decisions a_P, a_B, a_N for samples from X_P, and $\lambda_{PN}, \lambda_{BN}, \lambda_{NN}$ are the respective costs for samples from X_N, as presented in Table 1 [16].

Table 1. Decision Cost Matrix

Actual States	Decide $POS(X)$	Decide $BND(X)$	Decide $NEG(X)$
X (X_P)	λ_{PP}	λ_{BP}	λ_{NP}
$\neg X$ (X_N)	λ_{PN}	λ_{BN}	λ_{NN}

On the basis of Bayes and the decision-making conditional risk, the decision costs $R(a_i|x^{(l)})(i = P, B, N)$ can be computed as follows:

$$\begin{aligned} R(a_P|x^{(l)}) &= \lambda_{PP} P(X_P|x^{(l)}) + \lambda_{PN} P(X_N|x^{(l)}), \\ R(a_N|x^{(l)}) &= \lambda_{NN} P(X_N|x^{(l)}) + \lambda_{NP} P(X_P|x^{(l)}), \\ R(a_B|x^{(l)}) &= \lambda_{BP} P(X_P|x^{(l)}) + \lambda_{BN} P(X_N|x^{(l)}). \end{aligned} \tag{1}$$

where the conditional probabilities $P(X_P|x^{(l)})$ and $P(X_N|x^{(l)})$ are the respective possibilities of that under the information of the l-th step, a sample $x^{(l)}$ is divided into X_P or X_N. Then we find out the decision with minimum cost as the optimal strategy for l-th step, which can be presented as follows:

$$\phi^*(x^{(l)}) = \underset{\mathcal{D} \in \{a_P, a_N, a_B\}}{\operatorname{argmin}} R(\mathcal{D}|x^{(l)}). \tag{2}$$

3   Cost-Sensitive Sequential TWD for Face Recognition

Most traditional face recognition systems are two-way model and cost-blind. To simulate human decision process in face recognition, we incorporate the cost-sensitive sequential three-way decision model in Definition 1. Firstly, we introduce a Principal Component Analysis (PCA) based technique to describe the granularity of a face image. PCA is a well-known dimensionality reduction method, which seek to retain the maximum variance of the face images in the process of dimensionality reduction [17].

Assume the dataset $\Omega = \{x_i\}(i = 1, 2 \cdots, n)$ consists of n images and each image x_i has zero mean, so the covariance matrix of native data is: $S = \frac{1}{n} \sum\limits_{i=1}^{n} x_i x_i{}^T$. Suppose u_1 is an m-dimensional unit vector, then each image x_i can be projected to $u_1^T x_i$, and the variance of the projected sample data is:

$$var(u_1^T x_i) = \frac{1}{n} \sum_{i=1}^{n} (u_1^T x_i x_i^T u_1) = u_1^T \{\frac{1}{n} \sum_{i=1}^{n} (x_i x_i^T)\} u_1 = u_1^T S u_1, \qquad (3)$$

Under the normalization condition $u_1^T u_1 = 1$, we can maximize the projected variance $var(u_1^T x_i)$ with respect to u_1. Let d_1 be a Lagrange multiplier, it is transformed to an unconstrained maximization problem:

$$u_1 = \operatorname*{argmax}_{u_1^T u_1 = 1, u_1 \in R^m} \{u_1^T S u_1 + d_1(1 - u_1^T u_1)\}. \qquad (4)$$

By setting the derivative of u_1 equal to zero, we will get: $u_1^T S u_1 = d_1$. It means we can achieve the maximum variance of rotated data if u_1 is the eigenvector corresponding to the maximum eigenvalue of S. Similarly, We can find new directions(unit vectors) which is orthogonal to those already considered and also maximize the projected variance. Finally, we will get the optimal projection matrix $U = \{u_1, u_2 \cdots, u_l\}$, which are the eigenvectors corresponding to the first l largest eigenvalues $d_1, d_2 \cdots, d_l$ of S. The projected data with projection matrix U is the data in the lth step, and the percentage of retained variance is:

$$P_{var}^{(l)} = (\sum_{i=1}^{l} d_i)/(\sum_{i=1}^{n} d_i). \qquad (5)$$

Given a data set, we can reconstruct the image in each step with respective numbers of principal components, denoting the respective numbers of features (as in Definition 1). The eigenvalues reflect the utility of the respective eigenvectors and $P_{var}^{(l)}$ denotes the amount of information in the l-th step. The more components used in the reconstruction, the clearer the image will be. Therefore, it provides a practical technique to present images in different clearness and different granules. In reality, e.g. in the medical diagnosis problem, patient choose examinations may not strictly according to the utility from high to low. But in this paper, we only consider this decreasing sequence. A series of face images gradually ranging from rough granule to precise granule are shown in figure 1.

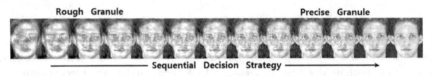

Fig. 1. Images from rough granularity to precise granularity

Then we should acquire the possibility of classification $P(X_P|x^{(l)})$ in formula (1). There are some classifiers that can be used for possibility estimation, such as logistic regression, softmax regression, naive Bayes and so on [18, 19]. Recently,

Zhang. et al. proposed a kNN-based method for possibility estimation, called cost-sensitive k-nearest neighbor classifier for multi-class classification($mckNN$) [20]. We adopt this method to obtain $P(X_P|x^{(l)})$ and substitute it into formula (1).

Considering a sample $x^{(l)}$ in the l-th sequential decision step, its k nearest neighbors can be found out and reordered as: $\{x_1', x_2'...x_k'\}$. Therefore, the labels of the k nearest neighbor form a label set $z = \{y_1', y_2'...y_k'\}$, which can be defined as a new feature for sample $x^{(l)}$. Then the posterior probability of $x^{(l)}$ with label y can be expressed as $P(y|z) = P(y|y_1', y_2'...y_k')$.

Under the general assumption that the k nearest neighbors are conditionally independent, the prior possibility $P(z|y)$ can be computed as follows:

$$P(z|y) = P(y_1', y_2'...y_k'|y) = P(y_1'|y)P(y_2'|y)...P(y_k'|y), \tag{6}$$

where $P(y_j'|y)$ represents the probability of that the sample with label y_j' is in feature z of the sample with label y [20]. Assume that there are s samples with label y in the training data set, and find out all of their k nearest neighbors, in which there are s_j samples with label y_j', then we can compute $P(y_j'|y) = (s_j)/(ks)$. The prior probability $P(y)$ can be presented as $P(y) = \frac{|D_y|}{|D|}$.

Based on Bayes theory, the posterior probability can be computed as:

$$P(y|z) = \frac{P(y)P(z|y)}{P(z)} = \frac{P(y)P(y_1'|y)P(y_2'|y)...P(y_k'|y)}{P(z)}. \tag{7}$$

Therefore, we obtain the posterior probabilities of sample $x^{(l)}$ with label $y = P$ and $y = N$: $P(X_P|x^{(l)}) = P(y = P|z)$, and $P(X_N|x^{(l)}) = P(y = N|z)$. To adopt a cost-sensitive three-way decision strategy, we substitute the posterior possibilities into formula (1), the decision costs can be computed as follows:

$$R(a_P|x^{(l)}) = \lambda_{PP}P(y = P|z) + \lambda_{PN}P(y = N|z),$$
$$R(a_N|x^{(l)}) = \lambda_{NP}P(y = P|z) + \lambda_{PN}P(y = N|z), \tag{8}$$
$$R(a_B|x^{(l)}) = \lambda_{BP}P(y = P|z) + \lambda_{BN}P(y = N|z),$$

and the optimal decision in the lth step can be rewritten as follows:

$$\phi^*(x^{(l)}) = \begin{cases} a_P & , \text{ if } R(a_P|x^{(l)}) \leq R(a_N|x^{(l)}), R(a_P|x^{(l)}) \leq R(a_B|x^{(l)}); \\ a_N & , \text{ if } R(a_N|x^{(l)}) \leq R(a_P|x^{(l)}), R(a_N|x^{(l)}) \leq R(a_B|x^{(l)}); \\ a_B & , \text{ if } R(a_B|x^{(l)}) \leq R(a_N|x^{(l)}), R(a_B|x^{(l)}) \leq R(a_P|x^{(l)}). \end{cases} \tag{9}$$

Remark: All the analysis above is based on the decision consistency assumption, which may hold in most cases for image analysis. In other words, the decision cost decrease w.r.t the decision step from a global view. However, from a local view, decision cost may increase even if the decision steps increase in some cases. Theoretical analysis and mathematical proof are presented in [16].

4   Experimental Analysis

This section presents an experimental analysis on the proposed cost-sensitive sequential three-way decision method. The performance is evaluated on facial

Fig. 2. All 14 different images of one subject from AR

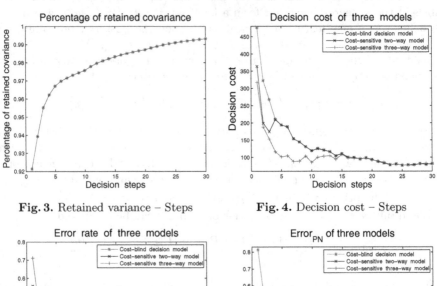

Fig. 3. Retained variance – Steps

Fig. 4. Decision cost – Steps

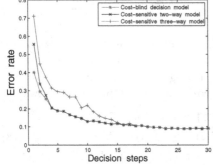

Fig. 5. Decision error rate – Steps

Fig. 6. Error rate in high-cost area – Steps

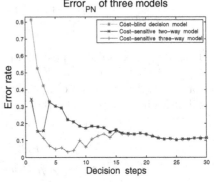

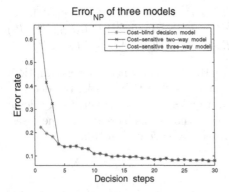

Fig. 7. Error rate in low-cost area – Steps

Fig. 8. Boundary Region – Steps

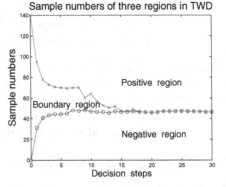

data set AR [21]. The AR database includes 126 subjects, each with 26 different face images. In this section, we select a subset of 100 subjects, each with 14 different grayed images, and Fig.2 shows all the 14 images of a selected subject.

Considering an imbalanced binary-class problem, we randomly select 70 subjects as set X_P(e.g. heathy people), and the remainder subjects as set X_N(e.g. patients). The percentages of retained variance for the 30 decision steps (points) are shown in figure 3. In kNN, we let $k = 3$. Assume that the cost of right classification is zero and the other costs are set as: $\lambda_{BP} = 1, \lambda_{NP} = 3, \lambda_{PN} = 12$, and $\lambda_{BN} = 2$. We adopted 10-fold cross-validation experiments on three methods: cost-blind two-way decision model, cost-sensitive two-way decision model and cost-sensitive three-way decision model. We repeated 20 times for all three methods and computed the average, as shown in Fig. 4 to Fig. 8.

Firstly, we investigate the tendency of decision costs and the variation of the boundary region with the increasing of decision steps based on three-way decision model. According to Fig.4, we can conclude that the decision cost decreases with the increasing of the decision steps from a global view, but the decision cost is non-monotonic with respect to the decision steps from a local view, i.e., the cost may abnormally increase along with increasing of the decision steps in some points. Then we measure the variation of sample numbers of positive region, boundary region and negative region with respect to decision steps in Fig. 8. We can find that the boundary region will globally decrease with the increasing of decision steps, and eventually it disappears, just as discussed above.

Secondly, we compare the performance of three methods. From Fig. 4 to Fig.7, we could draw the following conclusions: (1) Under the same conditions, the cost of TWD is less than that of cost-sensitive two-way model, and cost-sensitive two-way model behaves better than cost-blind two-way model, especially at the beginning of the sequence. (2)The error rate in high-cost area may explain the performance of TWD method, i.e., it decrease the error rate in high-cost area to sharply diminish the decision cost. (3) When the percentage of retained variance, i.e., the amount of information reach a certain value, both cost-sensitive methods perform no better than cost-blind two-way model.

5   Conclusion

In this paper, we propose a cost-sensitive sequential three-way decision strategy to simulate the decision process of human beings in face recognition. The proposed strategy present a new view on human vision mechanism as well as computer vision mechanism. In the future, we will further explore the cost-sensitive sequential three-way decision model for granule-based vision mechanism.

Acknowledgments. This research is supported by the National Natural Science Foundation of China under grant No. 71171107, 70971062, 71201076, 61170105, 71201133, the Natural Science Foundation of Jiangsu, China (BK2011564), and the Ph.D. Programs Foundation of Ministry of Education of China (20120091120004).

References

1. Duda, R.O., Stork, D.G., Hart, P.E.: Pattern classification. Wiley, New York (2000)
2. Yao, Y.Y.: The superiority of three-way decisions in probabilistic rough set models. Information Sciences 181, 1080–1096 (2011)
3. Yao, Y.Y.: Three-way decision: An interpretation of rules in rough set theory. In: Wen, P., Li, Y., Polkowski, L., Yao, Y.Y., Tsumoto, S., Wang, G. (eds.) RSKT 2009. LNCS, vol. 5589, pp. 642–649. Springer, Heidelberg (2009)
4. Yao, Y.Y.: Three-way decisions with probabilistic rough sets. Information Sciences 180, 341–353 (2010)
5. Yao, Y.Y.: Decision-theoretic rough set models. In: Yao, J.T., Lingras, P., Wu, W.-Z., Szczuka, M., Cercone, N.J., Ślęzak, D. (eds.) RSKT 2007. LNCS (LNAI), vol. 4481, pp. 1–12. Springer, Heidelberg (2007)
6. Yao, Y.Y., Wong, S.K.M., Lingras, P.: A decision-theoretic rough set model. In: Methodologies for Intelligent Systems, vol. 5, pp. 17–24 (1990)
7. Herbert, J.P., Yao, J.T.: Game-theoretic risk analysis in decision-theoretic rough sets. In: Wang, G., Li, T., Grzymala-Busse, J.W., Miao, D., Skowron, A., Yao, Y.Y. (eds.) RSKT 2008. LNCS (LNAI), vol. 5009, pp. 132–139. Springer, Heidelberg (2008)
8. Liu, D., Li, T., Liang, D.: Three-way decisions in dynamic decision-theoretic rough sets. In: Lingras, P., Wolski, M., Cornelis, C., Mitra, S., Wasilewski, P. (eds.) RSKT 2013. LNCS (LNAI), vol. 8171, pp. 291–301. Springer, Heidelberg (2013)
9. Li, H.X., Zhou, X.Z., Zhao, J.B., Huang, B.: Cost-sensitive classification based on decision-theoretic rough set model. In: Li, T., Nguyen, H.S., Wang, G., Grzymala-Busse, J., Janicki, R., Hassanien, A.E., Yu, H. (eds.) RSKT 2012. LNCS (LNAI), vol. 7414, pp. 379–388. Springer, Heidelberg (2012)
10. Jia, X.Y., Shang, L., Chen, J.J.: Attribute reduction based on minimum decision cost. Journal of Frontiers of Computer Science and Technology 5, 155–160 (2011) (in Chinese)
11. Li, H.X., Zhou, X.Z., Zhao, J.B., Liu, D.: Attribute reduction in decision-theoretic rough set model: A further investigation. In: Yao, J.T., Ramanna, S., Wang, G., Suraj, Z. (eds.) RSKT 2011. LNCS, vol. 6954, pp. 466–475. Springer, Heidelberg (2011)
12. Li, H.X., Zhou, X.Z.: Risk decision making based on decision-theoretic rough set: A three-way view decision model. International Journal of Computational Intelligence Systems 4, 1–11 (2011)
13. Liu, D., Li, H.X., Zhou, X.Z.: Two decades' research on decision-theoretic rough sets. In: Proc. 9th IEEE International Conference on Cognitive Informatics and Cognitive Computing (ICCI* CC), pp. 968–973. IEEE CS Press (2010)
14. Zhou, X.Z., Li, H.X.: A multi-view decision model based on decision-theoretic rough set. In: Wen, P., Li, Y., Polkowski, L., Yao, Y.Y., Tsumoto, S., Wang, G. (eds.) RSKT 2009. LNCS, vol. 5589, pp. 650–657. Springer, Heidelberg (2009)
15. Yao, Y.Y., Deng, X.F.: Sequential three-way decisions with probabilistic rough sets. In: Proc. 10th IEEE International Conference on Cognitive Informatics and Cognitive Computing (ICCI* CC), pp. 120–125. IEEE CS Press (2011)
16. Li, H.X., Zhou, X.Z., Huang, B., Liu, D.: Cost-sensitive three-way decision: A sequential strategy. In: Lingras, P., Wolski, M., Cornelis, C., Mitra, S., Wasilewski, P. (eds.) RSKT 2013. LNCS (LNAI), vol. 8171, pp. 325–337. Springer, Heidelberg (2013)

17. Turk, M., Pentland, A.: Eigenfaces for recognition. Journal of Cognitive Neuro-science 3, 71–86 (1991)
18. Chen, J., Huang, H., Tian, S., Qu, Y.: Feature selection for text classification with naïve bayes. Expert Systems with Applications 36, 5432–5435 (2009)
19. Kleinbaum, D.G., Klein, M.: Logistic regression: A self-learning text. Springer (2010)
20. Zhang, Y., Zhou, Z.H.: Cost-sensitive face recognition. IEEE Transactions on Pattern Analysis and Machine Intelligence 32, 1758–1769 (2010)
21. Martinez, A.M., Benavente, R.: The AR face database. CVC Technical Report 24 (1998)

Image Enhancement Based on Quotient Space

Tong Zhao[1], Guoyin Wang[1,2], and Bin Xiao[1]

[1] Chongqing Key Laboratory of Computational Intelligence,
Chongqing University of Posts and Telecommunications, Chongqing, 400065, China
[2] Institute of Electronic Information Technology, Chongqing Institute of Green
and Intelligent Technology, CAS Chongqing, 400714, China
hbjszt07@163.com, wanggy@ieee.org, xiaobin@cqupt.edu.cn

Abstract. Histogram equalization (HE) is a simple and widely used method in the field of image enhancement. Recently, various improved HE methods have been developed to improve the enhancement performance, such as BBHE, DSIHE and PC-CE. However, these methods fail to preserve the brightness of original image. To address the insufficient of these methods, an image enhancement method based on quotient space (IEQS) is proposed in this paper. Quotient space is an effective approach that can partitions the original problem in different granularity spaces. In this method, different quotient spaces are combined and the final granularity space is generated using granularity synthesis algorithm. The gray levels in each interval are mapped to the appropriate output gray-level interval. Experimental results show that IEQS can enhance the contrast of original image while preserving the brightness.

Keywords: histogram equalization, quotient space, image enhancement, granularity synthesis.

1 Introduction

Image enhancement is one of the most interesting and important issues in digital image processing field [1]. The main purpose of image enhancement is to bring out details that are hidden in an image, or to increase the contrast in a low contrast image. Histogram equalization (HE) is one of the most popular image enhancement methods in many image enhancement techniques. However, HE tends to introduce some annoying artifacts and unnatural enhancement. To overcome such a problem, many variants of histogram equalization have been proposed. Brightness preserving Bi-Histogram Equalization (BBHE) [2] has been proposed, which first segments the input histogram into two sub-histograms based on the mean of the input *images* brightness and then executes histogram equalization on each sub-histogram independently. Dualistic sub-image histogram equalization (DSIHE) [3] which is proposed by Wang et al also separates the input histogram into two subsections based on median value. Although these methods can preserve image brightness to a better extent than HE, the brightness improvement is still insufficient and the visualization of the result image is degraded.

M. Kryszkiewicz et al. (Eds.): RSEISP 2014, LNAI 8537, pp. 384–391, 2014.

For reducing overstretching of conventional HE, Power-constrained contrast enhancement (PC-CE) [4] has been proposed by Lee et al, which introduces a log-based histogram scheme and formulates an objective function that consist of the histogram-equalizing term. It can enhance contrast of original image but fail to preserve brightness of original image. Recently, the theory and model of granular computing based on quotient space have been proposed by Zhang et al [5]. The granularity synthesis algorithm provides a method to solve the problem at different granularity space. In this method, different quotient spaces are combined and the final granularity space is generated. Thus, the gray levels in each interval are mapped to the appropriate output gray-level interval.

2   Granular Theorem of Quotient Space

Granular computing is often regarded as an umbrella term to cover theories, methodologies techniques, and tools that make use of granules in complex problem solving [5]. Its essence is to analyze the problem from different sides (aspects) and granularity, and then to obtain a comprehensive or integrated solution (or approximate solution) [6]. The quotient space theory (QST), a new granular computing tool dealing with imprecise, incomplete and uncertain knowledge, has better representation power than other granular computing tools [7]. It combines different granular with the concept of mathematical quotient set and uses a triplet (X, f, T) to describe a problem space or simply a space, where X denotes the universe; f indicates the attribute function of X; T is the structure of X. For obtaining a new problem space, two general adopted principles are as following:

A. true-preserving

If the original problem space (X,f,T) has a solution then its corresponding coarse-grained space ([X],[f],[T]) should have a solution as well.

B. false-preserving

If a problem has no solution on its coarse-grain space $([X], [f], [T])$, then the original problem must have no solution.

Assume that $([X_1], [f_1], [T_1])$ and $([X_2], [f_2], [T_2])$ are two coarser problem spaces of $([X], [f], [T])$, where X_1 and X_2 are quotient spaces of X. Then the combination space $([X_3], [f_3], [T_3])$ of spaces $([X_1], [f_1], [T_1])$ and $([X_2], [f_2], [T_2])$ should satisfy:

(1)x_1 and x_2 are quotient spaces of x_3.

(2)f_1 and f_2 are projections of f_3 on x_1 and x_2, respectively.

(3)T_1 and T_2 are quotient topologies of x_3 with respect to X_1 and X_2, respectively.

(X_3, f_3, T_3) might need to satisfy some optimal criteria as well [8].

Equivalence relations are the basic concept of topology and are the key of the quotient space granular computing [9]. The methods of construction, including the combinations of domains and attributes respectively, are represented as follows:

A. The combination of Domains [10].

Suppose that (X_1, f_1, T_1) and (X_2, f_2, T_2) are quotient spaces of (X, f, T), R_1 and R_2 are two equivalence relations on X_1 and X_2, respectively. Then the combination of X_1 and X_2 can be represented by X_3, and the corresponding relationship is R_3, $xR_3y \Leftrightarrow xR_1y$ and xR_2y. So $X_3 = \{a_i \cap b_i | a_i \in X_1, b_i \in X_2\}$.

B. The combination of attributes [9].

Suppose that (X_1, f_1, T_1) and (X_2, f_2, T_2) are different quotient spaces of the problem space (X, f, T) then the combination space (X, f, T) satisfying the following conditions:

(1)$p_i f_i = f_i, i = 1, 2$where $p_i : (X, f, T) \rightarrow (X_i, f_i, T_i), i = 1, 2$
is called a natural projection.

(2)$D(f, f_1, f_2)$is a given judging criterion, shown as
$D(f_3, f_1, f_2) = \min D(f, f_1, f_2)$ or
$D(f_3, f_1, f_2) = \max D(f, f_1, f_2)$ (2)
Where f ranges over all attribute functions on X_3 that satisfy form (1)

3   Image Enhancement Based on Granularity Synthesis

The research of image enhancement based on granularity synthesis is a new issue in the field of image enhancement. The key question of image enhancement is how transforming the *pixels* gray levels in each input interval to the appropriate output gray-level interval. BBHE first divides the image histogram into two parts with the average gray level of the input-image pixels as the separation intensity. Two histograms are then independently equalized so as to solve the brightness preservation problem. DSIHE separates the histogram into two histograms which gains equal number of pixels based on the median value.These efforts improve histogram equalization and obtain the image enhancement at the end. Analyzing on the conceptions and various algorithms of image enhancement, we can conclude that the ideology of image enhancement and granularity theorem of quotient space are identical in essence. Therefore, we present the method of image enhancement based on quotient space. In this paper, we use the GMM to partition the distribution of the input image into a mixture of different gaussian components. The number of gaussian components can be determined by peaks in the fitting curve of original histogram which can be obtained by the method of Second Derivative. A gaussian component is set as a granular to indicate a granularity space $([X], [f], [T])$. It represents compact data with a dense distribution around the mean value of the component [11].

The gray-level distribution g(x), where $x \in X$, of an input image X (size H×W)can be modeled as a combination of N gaussian functions or granularity spaces (X_i, f_i, T_i), where X_i can donate a fine-grained region and f_i is attribute of a granular, i is the count of partitioned regions.

$$g(x) = \sum_{n=1}^{N} G(w_n)g(x|w_n) \qquad (1)$$

Where $g(x|w_n)$ is the n th component density and $G(w_n)$ is the prior proba-
bility of the data points generated from component of the mixture. And $G(w_n)$
satisfy the following constraints:

$$\sum_{n=1}^{N} G(w_n) = 1, 0 \le G(w_n) \le 1 \qquad (2)$$

Therefore, the component density functions are constrained to be gaussian
distribution functions, i.e.

$$g(x|w_n) = \frac{1}{\sqrt{2\pi\sigma_{w_n}^2}} e^{-\frac{(x-\mu_{w_n})^2}{2\sigma_{w_n}^2}} \qquad (3)$$

Where μ_{w_n} and σ_{w_n} are the mean and the variance of the n th component,
respectively. And they are regarded as attributes on universe of original problem.

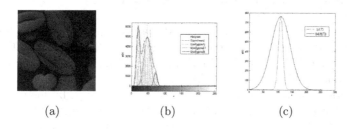

(a)   (b)   (c)

Fig. 1. (a) Pollen image, (b) corresponding histogram and granularity spaces and (c)
granularity synthesis

For specifying granularity synthesis technology, we select image *"pollen"* for
testing. Assume that $Fig.1(a)$ has a dynamic range of $[x_1, x_2]$ where $x(i,j) \in$
$[x_1, x_2]$. Fig.1(b) is the corresponding histogram of Fig.1(a) and the histogram of
original image is divided into several separate regions by three gaussian functions.
In order to solve the complicated problem, we treat every gaussian competent
as a granular and every granular is formed as a new problem space. Many granularity spaces $([X_i], [f_i], [T_i])$ compose a new coarse-grained granularity space.
Through clustering method, the problems can be transformed from a fine granularity space into a coarser granularity space. Suppose that $([X], [f], [T])$ denotes
a granular space of original problem, and $([X_1], [f_1], [T_1])$ indicates a new granular space. In the end, a coarse-size or fine-size granularity space will be generated
by increasing or decreasing the values of f. Therefore, it is essential to obtain appropriate partition points which can combine fine-size granularity spaces into a
final coarse-size granularity space as shown in Fig.1(b). In this paper, maximum-
likelihood-estimation techniques such as the expectation maximization algorithm

can be rightly used to find the appropriate intersection points so as to combine the granularity spaces into a coarse granularity space.

$$\zeta(X;\theta) = \prod_{\forall k} g(x_k;\theta) \tag{4}$$

$$\hat{\theta} = \underset{\theta}{argmax}\zeta(X;\theta) \tag{5}$$

According to (4) and (5), the maximum estimation of $\hat{\theta}$ can be achieved. Therefore, the significant intersection points are selected from all the possible intersections between the granularity components. Finally, we introduce a method called granularity synthesis technology to turn n fine granularity spaces into an appropriate coarser granularity space. And original pixels are mapped to appropriate output gray-level interval with Group Mapping Law(GML) widely used. Then an output image $Y = \{y(i,j)|1 < i < H, 1 < j < W\}$ is generated, which has a better visual quality than X.

4   Experimental Results and Discussions

In this section, we select two images named "$woman$" and "$Einstein$" for testing. It is shown that the results of experiments comparing the proposed IEQS method with HE, BBHE, DSIHE and PC-CE. Fig.2 and Fig.3 illustrate the image and its corresponding histograms, respectively.

Fig.2 shows the processed results of the original image by histogram equalization and proposed method. The result of HE [see Fig.2(b)] shows unpleasant artifacts in the background, decreasing contrast in the hair and also producing unnatural enhancement of the face. Comparing the results of BBHE, DSIHE, PC-CE with HE, experimental results obtained by four algorithms are very similarly performed. By observing Table 1, HE has a contrast value of 40.2418. BBHE and DSIHE have contrast values of 40.2119 and 40.4413 respectively. All of three methods tend to over enhance contrast of background. However, PC-CE with $\beta = 0$ has the reducing contrast value of 36.6115 and obtains smaller brightness value of 120.4541. Thus, it has not significantly improved the visualization of original image. By analyzing $Fig.2(h) - (k)$, the similarly results obtained by the different algorithms can be verified by the similarly histograms. In $Fig.2(f)$, the final processed image using IEQS with $f_{extended=2}$ (The variance of granular is increased by 2 times) is generated. It slightly darkens the $woman's$ hair and avoids the part of face is over enhanced. The merit of algorithm proposed keeps the details while improving the visualization of original image. In addition, IEQS provides brightness and contrast that are neither too high nor too low [see its values of woman in Table 1]. By analyzing the histograms in $Fig.2(h) - (j)$, the output histograms succeed in preserving the two main peaks of original histogram and achieve smooth distribution between high and low gray levels. But it loses a key peak on the right side of original histogram, which might lead to some details of original image missed. It can be seen in $Fig.2(b) - (d)$ that the results are visually unpleasing (a part of woman face is over enhanced). Comparing $PC - CE_{\beta=0}$ to HE, BBHE and DSIHE, PC-CE with $\beta = 0$ only preserves the part of feature on the right side peak in original histogram but it can not improve the visualization quality significantly [see Fig. 2(e)]. However, as shown in Fig.2(l),IEQS preserves the overall shape of the gray-level distribution and redistributions the gray levels. Thus retaining the natural look of the enhanced image shown in Fig.2(f).

Fig.3 shows the enhancement results for Einstein. In Fig.3(b), the result obtained by the conventional HE algorithm is shown, and its contrast is excessively stretched which can generate unnatural and non-existing objects in the processed image (the luminance in the part of face is over enhanced). Similarly HE as shown in Fig.3(c)-(e), the resulting images generated by BBHE, DSIHE and $PC - CE_{\beta=0}$ show some unnatural high intensity on the background and a part of face. The processed image with our method based on granularity synthesis algorithm (the variance of f increases 2 times) retains a natural looking output image. The output of $PC - CE_{\beta=0}$ and IEQS is shown in Fig.3(e) and (f). The difference in results is caused from the reason that $PC - CE_{\beta=0}$ can enhance contrast of original image while the proposed IEQS not only increases the contrast but also preserves the brightness. In addition, IEQS can effectively preserve the details globally(facial wrinkles and spots become more apparent). By analyzing the histograms of image processed by exist methods [see Fig.3(h)-(k)], the histograms of output image misses a key right side peak which can not achieve smooth distribution between high and low gray levels. Therefore, the enhancement results exhibit unnatural looking image (a part of Einstein' face is too bright). Since IEQS preserves the overall shape of original histogram and achieves a smoother distribution, IEQS performs better than exist methods and retains a natural visualization of original image. This can be verified by the visual results shown in Fig. 3(f).

The methods previous sections described which enhance the brightness of image. For further illustration, Table 1 shows the values of the brightness ($\mu = \sum_{l=1}^{L-1}(l \times P(l))$) and contrast ($\sigma = \sqrt{\sum_{l=1}^{L-1}(l - \mu) \times P(l)}$). By observing the absolute difference between the value of brightness in the original and the processed images. The method BBHE obtains the larger value of brightness (129.068 and 127.5775) and the smallest value of brightness (120.4514 and 110.5354) is obtained by our method. Therefore, We can conclude that: 1) the image produced by our proposed method are better in preserving the brightness of the original image. 2) The resulting brightness of our method is always larger than the brightness of original image. By observing the contrast values, we state that: the method $PC - CE_{\beta=0}$ produces the larger value of contrast while our method produces the smallest image contrast enhancement and enhances the contrast of original image effectively. Therefore, the value of our method denotes that we almost can not obtain image contrast enhancement, brightness preserving and natural looking image at the same time.

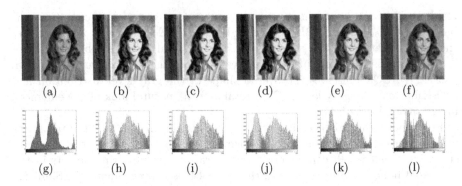

Fig. 2. Results for the woman image and histogram. Following from left to right: (a,g) original image, (b,h) HE, (c,i) BBHE, (d,j) DSIHE, (e,k) $PC-CE_{\beta=0}$, (f,l) IEQS with $f_{extended=2}$.

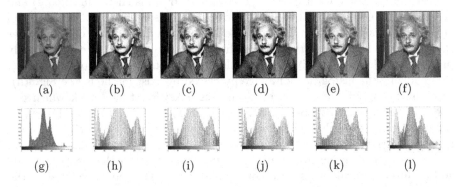

Fig. 3. Results for the Einstein image. Following from left to right: (a,g) original image; (b,h) HE; (c,i) BBHE; (d,j) DSIHE; (e,k) $PC-CE_{\beta=0}$; (f,l) IEQS with $f_{extended=2}$.

Table 1. Subjective quality test scores of different methods

Images	(methods)						indicators
	Original	HE	BBHE	DSIHE	PC-CE	IEQS	
Woman	26.5954	40.2418	40.2119	40.4413	36.6115	27.8689	contrast
	113.1303	128.517	129.068	124.428	123.0114	120.4514	brightness
Einstein	13.2552	26.9955	27.0844	27.0517	36.8936	26.9331	contrast
	108.4638	129.2900	127.5775	121.1472	125.5579	110.5354	brightness

5   Conclusion

In this paper, for improving the visuality of image, the algorithm of image enhancement based on quotient space (IEQS) is proposed. A triplet (X, f, T) is used to describe a problem space, which provides a good scheme for partitioning a complicated problem into several sub-granular spaces. Thus, the paper presents a model of image enhancement based on quotient space which constructs a fine granularity space for obtaining a natural-looking output image by using granularity synthesis algorithm. Experimental results demonstrate that the algorithm is effective. In future, we will make further research to improve contrast and to preserve brightness of image simultaneously based on granularity synthesis algorithm.

Acknowledgments. This work is supported by the Natural Science Foundation of China under Grant (Nos. 61272060, 61201383), the Key Natural Science Foundation of Chongqing of China under Grant (Nos. CSTC2013jjB40003, cstc2013jcyjA40048).

References

1. Kong, N.S.P., Ibrahim, H.: Color image enhancement using brightness preserving dynamic histogram equalization. IEEE Transactions on Consumer Electronics 54(4), 1962–1968 (2008)
2. Kim, Y.T.: Contrast enhancement using brightness preserving bi-histogram equalization. IEEE Transactions on Consumer Electronics 43(1), 1–8 (1997)
3. Wang, Y., Chen, Q., Zhang, B.: Image enhancement based on equal area dualistic sub-image histogram equalization method. IEEE Transactions on Consumer Electronics 45(1), 68–75 (1999)
4. Lee, C., Lee, C., Lee, Y.Y., et al.: Power-constrained contrast enhancement for emissive displays based on histogram equalization. IEEE Transactions on Image Processing 21(1), 80–93 (2012)
5. Zhang, L., Zhang, B.: The quotient space theory of problem solving. Fundamenta Informaticae 59(2), 287–298 (2004)
6. Wang, G., Zhang, Q.: Granular Computing based cognitive computing. In: 8th IEEE International Conference on Cognitive Informatics, ICCI 2009, pp. 155–161. IEEE, Hong Kong (2009)
7. Yao, J., Vasilakos, A.V., Pedrycz, W.: Granular computing: Perspectives and challenges, pp. 1–13 (2013)
8. Liang, Y., Mao, Z.: A Method of Segmenting Texture of Targets in Remote Sensing Images Based on Granular Computing. In: 2011 International Conference on Information Technology, Computer Engineering and Management Sciences (ICM), pp. 280–283. IEEE, Nanjing (2011)
9. Zou, B., Jia, Q., Zhang, L., et al.: Target detection based on granularity computing of quotient space theory using SAR image. In: 2010 17th IEEE International Conference on Image Processing (ICIP), pp. 4601–4604. IEEE, Hong Kong (2010)
10. Chen, X., Wu, Y., Cheng, H.: Quotient space granular computing for the Clickstream data warehouse in Web servers. In: 2010 International Conference on Computer and Communication Technologies in Agriculture Engineering (CCTAE), pp. 93–96. IEEE, Chengdu (2010)
11. Celik, T., Tjahjadi, T.: Automatic image equalization and contrast enhancement using Gaussian mixture modeling. IEEE Transactions on Image Processing 21(1), 145–156 (2012)

Author Index